Advances in Public Transport Platform for the Development of Sustainability Cities

Advances in Public Transport Platform for the Development of Sustainability Cities

Editors

Juan M. Corchado
Josep L. Larriba-Pey
Pablo Chamoso
Fernando De la Prieta

MDPI • Basel • Beijing • Wuhan • Barcelona • Belgrade • Manchester • Tokyo • Cluj • Tianjin

Editors

Juan M. Corchado
BISITE Research Group
University of Salamanca
Salamanca
Spain

Josep L. Larriba-Pey
Data Maagement Group
Polytechnic University of
Catalonia
Barcelona
Spain

Pablo Chamoso
BISITE Research Group
University of Salamanca
Salamanca
Spain

Fernando De la Prieta
BISITE Research Group
University of Salamanca
Salamanca
Spain

Editorial Office
MDPI
St. Alban-Anlage 66
4052 Basel, Switzerland

This is a reprint of articles from the Special Issue published online in the open access journal *Electronics* (ISSN 2079-9292) (available at: www.mdpi.com/journal/electronics/special_issues/ transport_platform_cities).

For citation purposes, cite each article independently as indicated on the article page online and as indicated below:

LastName, A.A.; LastName, B.B.; LastName, C.C. Article Title. *Journal Name* **Year**, *Volume Number*, Page Range.

ISBN 978-3-0365-3980-5 (Hbk)
ISBN 978-3-0365-3979-9 (PDF)

Contents

About the Editors

Juan M. Corchado

Juan Manuel Corchado, Full Professor with Chair at the University of Salamanca. He was Vice President for Research from 2013 to 2017 and the Director of the Science Park of the University of Salamanca. Chosen twice as the Dean of the Faculty of Science, he holds a PhD in Computer Sciences from the University of Salamanca and a PhD in Artificial Intelligence from the University of the West of Scotland. He is the director of a renowned research group called BISITE (Bioinformatics, Intelligent Systems and Educational Technology), which was created in the year 2000.

J.M Corchado is also the Director of the IOT Digital Innovation Hub and the President of the AIR Institute. He has been a Visiting Professor at the Osaka Institute of Technology since January 2015, and has been a Visiting Professor at the Universiti Malaysia Kelantan and a Member of the Advisory group on Online Terrorist Propaganda of the European Counter Terrorism Centre (EUROPOL).

He currently combines all his activities with the direction of Master's Programs in Master in Security, Digital Animation, Multiplatform Mobile Application Development, Digital Intelligence, Financial Technology, Knowledge Transfer and R&D Management, Digital Transformation, ICT, Internet of Things, Social Media, 3D Printing, Blockchain, Industry 4.0 y Smart Cities and Intelligent Buildings, at the University of Salamanca, and with his work as Editor-in-Chief of Journals such as *ADCAIJ* (Advances in Distributed Computing and Artificial Intelligence Journal), *OJCST* (Oriental Journal of Computer Science and Technology) or *Electronics* MDPI (Computer Science and Engineering section).

J.M. Corchado works on projects in the fields of Artificial Intelligence, Machine Learning, Blockchain, IoT, Fog Computing, Edge Computing, Smart Cities, Smart Grids, Sentiment Analysis, etc.

Josep L. Larriba-Pey

Founder of the DAMA-UPC research group within BarcelonaTech and director over the past 15 years. This group deals with the management of big data and has a strong relationship with industry, both at local and international levels, with strong players such as IBM or Oracle.

Founder of Sparsity Technologies, BarcelonaTech spin-out, and CEO for the past 10 years. The company has been recognized by the European Commission as the SME with most innovative potential in the European Community in 2015.

Sparsity's star product, Sparksee, is a Graph Database used in Mobile and Embedded systems as a unique, small software and data footprint, persistent, high-performance and reliable device for the storage and analysis of large amounts of linked data.

Sparsity has also created CIGO!, a mobility management platform to plan routes, interact with fleets through mobile devices and analyses geolocated data without the need for sensors, intensive server use or expensive hardware. What makes CIGO! different from other route planning and fleet management systems is:

- Offline GPS tracking technology.

- Data integration from potentially any source: Open Data, Mobile Apps, private city or company data, sensors, etc.

- Optimal route calculation according to customized criteria (not just shortest path).

Scenarios where CIGO! makes the difference: tow away services, tourism, taxi, home medical services, police management, cycling safe and healthy, public transport, electric vehicle, delivery areas management.

Pablo Chamoso

Pablo Chamoso Santos holds a PhD in Computer Engineering from the University of Salamanca. He obtained the titles of Technical Engineer in Computer Systems, Computer Engineer (i3 Award for the best final project of Castilla y León) and Official Master' s Degree in Intelligent Systems. In addition to these official degrees, he holds a Master's Degree in Systems Development for Electronic Commerce and a Master's Degree in Information and Communication Systems Management. Currently, he is an Associate Professor at the Department of Computer Science of the University of Salamanca. His PhD thesis focused on the Development of Platforms for the Deployment and Management of Smart Cities. This research is co-funded by the Junta de Castilla y León and the European Social Fund through the program "Excellent Science and Technological Leadership". He has been a research member of the renowned research group BISITE since March 2011, and is collaborator at IBSAL, member of the University Institute of Research in Art and Technology of Animation, member of the Institute of Electrical and Electronics Engineers (IEEE), member of the IEEE Smart Cities Community, member of the Professional Association of Computer Engineers of Castilla y León, member of the University Institute of Research and Art and Technology of Animation and secretary of the academic and quality committees of the PhD Program in Energy and Marine Propulsion at the University of Salamanca.

Fernando De la Prieta

Fernando de la Prieta Pintado is an Associate Professor at the University of Salamanca Department of Computer Science and Automation.

Dr. De la Prieta is equally well experienced in research and teaching. Over recent years, he has followed a clearly defined line of research, focusing on the integration of multi-agent organizations, machine learning and advanced architectures in different fields. He applied the results in both his doctoral thesis (for which he obtained an international PhD mention and an extraordinary PhD award) and in the projects he has been involved in. He has more than 50 publications in international journals, many of which have a JCR impact factor on the Web of Science database. His H index in Google Scholar is 27. Furthermore, he has published more than 100 articles in books and in the proceedings of prestigious international conferences, and around thirty of these publications have been published in conferences indexed according to the CORE ranking. He has worked on more than 90 research projects (16 of them were international and in several he has been the principal investigator). In addition, he has participated in more than 30 research contracts (Art. 83), in some of them as the principal investigator. As a result of his work, around 40 intellectual properties have been registered. He has had several stays abroad (pre- and post-doctoral) in Portugal, Japan and South Korea. He has also taken an active part in the organization of international conferences, some of them included in the CORE ranking: IEEE-GLOBECOM (core B), ICCBR (Core B), CEDI, PAAMS (core C), ACM-SAC (core B), IEEE-FUSION (core C), and others.

Preface to "Advances in Public Transport Platform for the Development of Sustainability Cities"

Modern societies demand high and varied mobility, which in turn requires a complex transport system adapted to social needs that guarantees the movement of people and goods in an economically efficient and safe way, but all are subject to a new environmental rationality and the new logic of the paradigm of sustainability. From this perspective, an efficient and flexible transport system that provides intelligent and sustainable mobility patterns is essential to our economy and our quality of life. The current transport system poses growing and significant challenges for the environment, human health, and sustainability, while current mobility schemes have focused much more on the private vehicle that has conditioned both the lifestyles of citizens and cities, as well as urban and territorial sustainability.

Transport has a very considerable weight in the framework of sustainable development due to environmental pressures, associated social and economic effects, and interrelations with other sectors. The continuous growth that this sector has experienced over the last few years and its foreseeable increase, even considering the change in trends due to the current situation of generalized crisis, make the challenge of sustainable transport a strategic priority at local, national, European, and global levels.

This Special Issue will pay attention to all those research approaches focused on the relationship between evolution in the area of transport with a high incidence in the environment from the perspective of efficiency, which has become one of the neuralgic centers of sustainability. This relates to producing, consuming, and moving people and goods better, with fewer resources and less environmental impact.

Juan M. Corchado, Josep L. Larriba-Pey, Pablo Chamoso, and Fernando De la Prieta
Editors

 *electronics*

Editorial

Advances in Public Transport Platform for the Development of Sustainability Cities

Juan M. Corchado [1], **Josep L. Larriba-Pey** [2], **Pablo Chamoso-Santos** [1] and **Fernando De la Prieta Pintado** [1,*]

[1]   BISITE Research Group, University of Salamanca, Edificio Multiusos I + D + I, 37007 Salamanca, Spain; corchado@usal.es (J.M.C.); chamoso@usal.es (P.C.-S.)

[2]   Data Management Group, Polytechnic University of Catalonia, 08034 Barcelona, Spain; larri@ac.upc.edu

*   Correspondence: fer@usal.es; Tel.: +34-677-522-678

Citation: Corchado, J.M.; Larriba-Pey, J.L.; Chamoso-Santos, P.; De la Prieta Pintado, F. Advances in Public Transport Platform for the Development of Sustainability Cities. *Electronics* **2021**, *10*, 2771. https://doi.org/10.3390/electronics10222771

Received: 4 November 2021
Accepted: 11 November 2021
Published: 12 November 2021

Publisher's Note: MDPI stays neutral with regard to jurisdictional claims in published maps and institutional affiliations.

1. Introduction

There is high and varied mobility in modern societies which requires a complex transport system that adapts to social needs and guarantees the movement of people and goods in an economically efficient and safe way. All this designed from the new perspective of environmental wellness and of the sustainability paradigm. From this viewpoint, an efficient and flexible transport system that provides intelligent and sustainable mobility patterns is essential to our economy and quality of life. The current transport system poses growing and significant challenges for the environment, human health, and sustainability. Existent mobility schemes focus excessively on the use of private vehicles which have conditioned the lifestyle of citizens in cities, as well as urban and territorial sustainability.

Transport is an important element of the sustainable development framework due to the growing environmental strain, the associated social and economic effects, and its interconnection with other sectors. The continuous growth that this sector has experienced over the last few years and its foreseeable future growth, even considering the change of trend caused by the current situation of generalized crisis, make the challenge of sustainable transport a strategic priority at local, national, European, and global levels.

2. The Present Issue

This special issue consists of sixteen papers covering important topics in the field of public transportation under the framework of smart cities.

The research community is now turning its attention to different areas such as optimization and prediction [1–5]. As evidenced in references [2,3,5], which have analyzed travel time data to evaluate the performance of a public transport system. Others have focused on the demand for different modes of transportation and interaction among them, including a proposal for minimizing the passengers' waiting times and maximizing the vehicles' occupancy ratios. The use of unmanned aerial vehicles for emergency situations is extensively described in [1] for search and rescue operations, surveillance, disaster monitoring, response to terrorist attacks. Finally, ref. [4] studied the influence of the economy on transportation systems.

Recommender Systems are also commonly used within the framework of transportation for sustainable cities. Hence, references [6,7] focused on offering improved usability and services based on multi-modal door-to-door passenger experiences to increase engagement. Other examples can be found in reference [8], where recommendation systems are designed to improve the passengers' experience and the drivers' profit. Finally, other approaches focused on educating the general public about this topic [9].

Other topics included in this special issue are energy consumption forecasting in sustainable cities [10] as well as the analysis of energy trading and the development of a trust model [11]. Security is also an important issue within public transpormation, in reference [12] the secure management of railway transportation systems has been analyzed.

Finally, analytical models using Machine Learning and Deep Learning have been explored as part of this special issue [13,14]. Also, two case studies, carried out in the city of Barcelona, Spain [15] and Taipei, Taiwan [15], have been described.

3. Conclusions

This special issue has paid attention to all the research approaches that focus on the relationship between the evolution of transportation and the new perspective of achieving environmental wellness and efficiency, which has become one of the cornerstones of sustainability. It revolves around producing, consuming, and transporting people and goods better, while using up fewer resources and having lower environmental impact.

Author Contributions: J.M.C., J.L.L.-P., P.C.-S. and F.D.l.P.P. worked together in the whole editorial process of the special issue, "Advances in Public Transport Platform for the Development of Sustainability Cities", published by journal Electronics. F.D.l.P.P. drafted this editorial summary. J.M.C., J.L.L.-P. and P.C.-S. reviewed, edited, and finalized the manuscript. All authors have read and agreed to the published version of the manuscript.

Funding: This research was funded by "Ministerio de Ciencia, Innovación y Universidades. Proyectos I + D + i «RETOS INVESTIGACIÓN» del Programa Estatal de I+D+i orientada a los retos de la sociedad", grant number RTI2018-095390-B-C32. This research was also funded by the Shift2Rail Joint Undertaking under the Europeans Union's Horizon 2020 Research and Innovation Programme, grant number 777640.

Acknowledgments: First of all, we would like to thank all researchers who submitted articles to this special issue for their excellent contributions. We are also grateful to all the reviewers who helped in the evaluation of the manuscripts and made very valuable suggestions to improve the quality of the contributions. We would like to acknowledge the editorial board of Electronics, who invited us to guest edit this special issue. We are also grateful to the Electronics Editorial Office staff who worked thoroughly to maintain the rigorous peer-review schedule and timely publication.

Conflicts of Interest: The authors declare no conflict of interest.

References

1. Saha, S.; Vasegaard, A.E.; Nielsen, I.; Hapka, A.; Budzisz, H. UAVs Path Planning under a Bi-Objective Optimization Framework for Smart Cities. *Electronics* **2021**, *10*, 1193. [CrossRef]
2. Hartmann Tolić, I.; Nyarko, E.K.; Ceder, A. Optimization of Public Transport Services to Minimize Passengers' Waiting Times and Maximize Vehicles' Occupancy Ratios. *Electronics* **2020**, *9*, 360. [CrossRef]
3. Martin, L.; Wittmann, M.; Li, X. The Influence of Public Transport Delays on Mobility on Demand Services. *Electronics* **2021**, *10*, 379. [CrossRef]
4. Sousa, R.; Lima, T.; Abelha, A.; Machado, J. Hierarchical Temporal Memory Theory Approach to Stock Market Time Series Forecasting. *Electronics* **2021**, *10*, 1630. [CrossRef]
5. Yuan, Y.; Shao, C.; Cao, Z.; He, Z.; Zhu, C.; Wang, Y.; Jang, V. Bus Dynamic Travel Time Prediction: Using a Deep Feature Extraction Framework Based on RNN and DNN. *Electronics* **2020**, *9*, 1876. [CrossRef]
6. García-Retuerta, D.; Rivas, A.; Guisado-Gámez, J.; Antoniou, E.; Chamoso, P. Reputation System for Increased Engagement in Public Transport Oriented-Applications. *Electronics* **2021**, *10*, 1070. [CrossRef]
7. Rivas, A.; González-Briones, A.; Cea-Morán, J.J.; Prat-Pérez, A.; Corchado, J.M. My-Trac: System for Recommendation of Points of Interest on the Basis of Twitter Profiles. *Electronics* **2021**, *10*, 1263. [CrossRef]
8. Wan, X.; Ghazzai, H.; Massoud, Y. A Generic Data-Driven Recommendation System for Large-Scale Regular and Ride-Hailing Taxi Services. *Electronics* **2020**, *9*, 648. [CrossRef]
9. Jordán, J.; Valero, S.; Turró, C.; Botti, V. Using a Hybrid Recommending System for Learning Videos in Flipped Classrooms and MOOCs. *Electronics* **2021**, *10*, 1226. [CrossRef]
10. Oliveira, P.; Fernandes, B.; Analide, C.; Novais, P. Forecasting Energy Consumption of Wastewater Treatment Plants with a Transfer Learning Approach for Sustainable Cities. *Electronics* **2021**, *10*, 1149. [CrossRef]
11. Andrade, R.; Wannous, S.; Pinto, T.; Praça, I. Extending a Trust model for Energy Trading with Cyber-Attack Detection. *Electronics* **2021**, *10*, 1975. [CrossRef]
12. Hatzivasilis, G.; Fysarakis, K.; Ioannidis, S.; Hatzakis, I.; Vardakis, G.; Papadakis, N.; Spanoudakis, G. SPD-Safe: Secure Administration of Railway Intelligent Transportation Systems. *Electronics* **2021**, *10*, 92. [CrossRef]
13. Taboada, G.L.; Han, L. Exploratory Data Analysis and Data Envelopment Analysis of Urban Rail Transit. *Electronics* **2020**, *9*, 1270. [CrossRef]

14. Hernández-Nieves, E.; Parra-Domínguez, J.; Chamoso, P.; Rodríguez-González, S.; Corchado, J.M. A Data Mining and Analysis Platform for Investment Recommendations. *Electronics* **2021**, *10*, 859. [CrossRef]
15. Mariñas-Collado, I.; Frutos Bernal, E.; Santos Martin, M.T.; Martín del Rey, A.; Casado Vara, R.; Gil-González, A.B. A Mathematical Study of Barcelona Metro Network. *Electronics* **2021**, *10*, 557. [CrossRef]

 *electronics*

Article

UAVs Path Planning under a Bi-Objective Optimization Framework for Smart Cities

Subrata Saha [1], **Alex Elkjær Vasegaard** [1], **Izabela Nielsen** [1,*], **Aneta Hapka** [2], **Henryk Budzisz** [2]

[1] Department of Materials and Production, Aalborg University, Fibigerstræde 16, DK 9220 Aalborg, Denmark; saha@m-tech.aau.dk (S.S.); aev@mp.aau.dk (A.E.V.)

[2] Faculty of Electronics and Computer Science, Koszalin University of Technology, 75-343 Koszalin, Poland; aneta.hapka@tu.koszalin.pl (A.H.); henryk.budzisz@tu.koszalin.pl (H.B.)

* Correspondence: izabela@mp.aau.dk

Abstract: Unmanned aerial vehicles (UAVs) have been used extensively for search and rescue operations, surveillance, disaster monitoring, attacking terrorists, etc. due to their growing advantages of low-cost, high maneuverability, and easy deployability. This study proposes a mixed-integer programming model under a multi-objective optimization framework to design trajectories that enable a set of UAVs to execute surveillance tasks. The first objective maximizes the cumulative probability of target detection to aim for mission planning success. The second objective ensures minimization of cumulative path length to provide a higher resource utilization goal. A two-step variable neighborhood search (VNS) algorithm is offered, which addresses the combinatorial optimization issue for determining the near-optimal sequence for cell visiting to reach the target. Numerical experiments and simulation results are evaluated in numerous benchmark instances. Results demonstrate that the proposed approach can favorably support practical deployability purposes.

Keywords: unmanned aerial vehicles (UAVs); multi-objective optimization; integer programming; GLPK; variable neighborhood search; search and rescue

Citation: Saha, S.; Vasegaard, A.I.; Nielsen, I.; Hapka, A.; Budzisz, H. UAVs Path Planning under a Bi-Objective Optimization Framework for Smart Cities. *Electronics* **2021**, *10*, 1193. https://doi.org/10.3390/electronics10101193

Academic Editors: Juan M. Corchado, Josep L. Larriba-Pey, Pablo Chamoso and Fernando De la Prieta

Received: 16 March 2021
Accepted: 13 May 2021
Published: 17 May 2021

Publisher's Note: MDPI stays neutral with regard to jurisdictional claims in published maps and institutional affiliations.

1. Introduction

The path planning problem for a set of Unmanned Aerial Vehicles (UAVs) has gained unprecedented interest from researchers and practitioners to develop intelligent systems and execute various tasks with minimum human intervention. With upgraded components such as cameras, sensors, or telemetry systems, UAV application is becoming an integral strategic part for emergency management; aerial photography; mountain rescue; smart farming; maritime search and rescue; information collection, post-disaster relief; homeland security, crowd management, etc. [1–3]. UAVs, in practice, has many significant advantages such as human workload reduction, high mobility, saving of valuable resources, etc. In the literature, the path planning problem is categorized in several ways according to problem characteristics. For example, according to the targets' reaction, one can classify the problem into two categories: one-sided vs. two-sided path planning problems. On the other hand, based on targets' motion, one can classify the situation as static vs. moving target search or open vs. closed-loop decision models based on the decision-making context [4–10].

In recent years, the utilization of UAVs has been becoming increasingly attractive in the context of Smart City Management solutions. Several key technologies are continuously integrated into smart cities operations, such as data collection and protection and intrusion detection technologies. In this regard, the application of UAVs to collect data or images is an economical and effective solution. UAVs operations can lead to a new paradigm for developing smart cities with a high-quality life and sustainable economic growth. For example, Felemban et al. [11] noted that UAVs could be used to detect the earlier signs of a stampede, congestion, and other crowd problems. The authors proposed a Priority-Based Routing Framework to increase the delivery speed of images during Hajj in Saudi

Arabia. Researchers found that UAVs can be helpful in policing systems to fight against crime [12]. It was reported that such UAV policing systems work well for extensive crime deterrence [13]. However, there are many challenges, and we highlight one of those where UAVs are deployed in search and rescue problems.

Due to the sequential decision-making nature, the fundamental search and rescue path planning problem is a non-deterministic polynomial-time problem (NP-hard) [14]. Therefore, researchers employ both exact algorithm and heuristic approaches alternatively to solve such complex decision-making problems. One can argue that the modern search theory originated from the pioneering works by the group of researchers, Stewart [8], Brown [15] and Benkoski et al. [16]. Researchers mainly focus on the allocation decision instead of the optimal sequential path generation. By assuming an exponential detection function, Stewart [8] formulated a network flow model to characterize a moving target detection problem and used the branch and bound method to find a near-optimal solution. Later, Eagle [4] formulated the model in a dynamic programming framework and utilized the Markov process to replicate target motion as state transitions. Washburn [17] made an effort to determine the best upper bound for a generalized path planning problem. After that, researchers progressively shifted their attention toward the evaluation of algorithm performance in more complex enshrinement [18]. However, the travel time in the earlier model was assumed as uniform. Lau et al. [19] relaxed this assumption and formulated a model where travel time among regions are non-uniform. Rogge and Aeyels [20] introduced the concept of a collaborative path planning problem where the search area consists of multiple moving targets with an arbitrary number of obstacles. Li et al. [21] studied energy-efficient rechargeable UAV deployment strategy to provide seamless coverage in urban areas and employed the two-stage particle swarm optimization (PSO) algorithm to solve the problem. Regarding other variants, Berger and Lo [22] introduced a mixed-integer programming model under a directed acyclic graph framework and used CPLEX software to find an optimal path. To overcome computational effort, Perez-Carabaza et al. [23] proposed a modified ant colony optimization (ACO) algorithm to investigate the nature of trajectories for a set of heterogeneous UAVs. Ye et al. [24] used an adaptive genetic algorithm (GA) to find the solution for a collaborative multiple task assignment problem with fixed-wing UAVs. The authors employed a robust encoding strategy to generate feasible chromosomes. Lu et al. [25] use the wolf pack algorithm (WPA) to solve the task assignment problem for UAVs. The authors found that WPA can outperform PSO and GA in terms of convergence speed and solution accuracy. Lou et al. [26] proposed a multi-swarm fruit fly optimization algorithm to find a solution for multi-UAV cooperative mission planning problem. However, Alhaqbani et al. [27] stated that a common problem in most of the metaheuristics is that those can perform poorly in regards to run time. More recently, Xiong et al. [28] introduced Voronoi-based Ant colony optimization algorithm combined with the Dijkstra's algorithm to investigate optimal trajectories. In recent years, various types of machine learning algorithms have been employed to obtain optimal deployment strategy, and we refer to the recent review works by [29] and [30] for detailed discussion in this aspect. In addition, we refer the following works for more discussion on path planning from various perspectives [31–37].

In this study, we use a modified Variable Neighborhood Search (VNS) meta-heuristic [38]. Since its inception, the algorithm has been employed in numerous fields such as network design problems in communication [39], facility location problem [40], data mining [41], timetabling and related manpower organization problems [42], single- and multi-objective job shop scheduling [43,44], vehicle routing problem [45] and bioinformatics [46] due to its user-friendliness, higher precision and robustness. The VNS systematically exploits the idea of neighborhood change iteratively to improve the initial solution inside the shaking and local search procedures [47,48]. Unlike other meta-heuristic approaches, parameter tuning is always an issue; the fundamental VNS algorithm and its extension version require few or, occasionally, no parameters. One significant advantage to the VNS-based approach for path planning is that it accommodates the path maneuverability through the path constructor (see Algorithm 1) operator. At the same time, the inherent shaking procedure seeks to overcome the possible local optima. The algorithm then attempts to improve the randomly changed path to catch a more rewarded path than the incumbent solution.

The cited literature's main disadvantage is that most authors only studied the problem as a single-objective optimization problem, e.g., maximizing the probability of finding targets, minimizing the path length, equal utilization of resources, etc. However, in a time-constrained decision-making context, only considering one objective may not lead to an acceptable outcome [49,50]. From a practical point of view, it is essential to handle several objectives simultaneously to obtain a pragmatic solution. Explicitly, the two most fundamental goals that need to be considered are maximization of finding the targets and minimizing the path length objective that can ensure minimum utilization of resources and implicitly ensure less operational time and energy consumption. It is challenging to find the ideal solution due to the conflicting nature of objective functions; therefore, researchers have proposed different approaches such as weighted sum [51], global criterion [52], goal programming [53], multi-choice goal programming [54], non-dominated sorting genetic algorithm II [55], fuzzy-two phase approach [56], etc., and the issue of a specific method largely depends on the decision-makers. Note that UAV path planning is itself an NP-hard problem [57]; thus, we use a simple weighted sum approach in this study. This study formulated the model as binary linear programming (BLP) formulation under a bi-objective optimization environment and proposed a modified VNS algorithm to find the solution. Numerical experiments were conducted to validate the overall framework. The key contributions of the study are as follows: First, a bi-objective optimization problem is proposed to obtain paths for multiple UAVs in a time-constrained environment. Second, a modified VNS algorithm is proposed, which is highly parallelizable and straightforward to understand. Moreover, the simulation study reveals that it can provide a solution within a reasonable time when the exact solver fails to provide a solution, and the performance for the algorithm is always higher compared to Dijkstra's algorithm, which is extensively used by several researchers [58,59]. Finally, a sensitivity analysis on the weight-space provide an overview regarding the importance of multi-objective formulation in the practical implementation of UAVs.

The paper is organized as follows. The mathematical model and corresponding assumption and notation are presented in Section 2. In Section 3, an overview is presented for the data generation. The solution procedure for the model is described in Section 4. A detailed overview of the VNS algorithm is also presented in this section. Extensive numerical experiments and validation of the proposed solution framework's effectiveness are presented in Section 5. Finally, Section 6 concludes by highlighting findings, limitations and future research directions.

2. Mathematical Model

Path planning and trajectory mapping for UAV is an important topic because of the incredible versatility and flexibility of UAVs that allow them to be employed in different operations. Although path planning goes before trajectory mapping, fundamentally, their characteristics are not entirely distinct. If point-to-point trajectories are measured, the two problem needs to be solved simultaneously if the initial and final positions are specified. One can define the path planning problem as finding a collision-free motion within a specified environment where initial and final locations are pre-defined. In this study, we use the cell decomposition method. In this method, the entire search space is subdivided into several regions (equal/unequal), called cells. The corresponding path will represent a connected graph and describe the adjacent relations between cells. Simultaneously, the trajectory planning problem is based on the input generated by the path planner. To plan a trajectory, commonly, a sequence of waypoints needs to be extracted. A kinematic inversion needs to be performed based on some decision-maker criteria such as minimizing total execution time, energy, distance, jerk, etc. In the present formulation, we ignore the effect of the kinematics of the UAV. We assume that a team of homogeneous UAVs is searching stationary targets in a pre-defined search region [60].

The search area is divided into an $N \times N$ grid describing possible target locations. The time duration for each cell visit, with equal size, is assumed as constant. The cell occupancy probabilities are generated initially, and, as we assume the targets to be stationary and non-moving, we omit the dynamics of a changing probability map. To maneuver its neighboring cells, any UAV can move in eight different directions{E, W, N, S, SE, SW, NE, NW}. However, at the cell where the UAVs start maneuvering is located, the UAVs are also allowed to hover. This mimics the possibility of early landing or later departure for some UAVs. A graph theory-based directed acyclic network representation is employed to streamline the setup. The entire graph is defined as $G_t = (V_t, E_t)$ for all t in a given time horizon T, V_t, the set of vertices, represent all possible locations $n \in N^* = \{1, \ldots, N^2 - 1, N^2\}$ at time $t \in T$. E_t, the set of edges, represents all the possible state transition related to each UAV between episodes t and $t + 1$. An adjacency matrix A defines the connectivity of G, $A_{tn'n} = 1$ if $v_{tn'} \in V_t$ and $v_{t'n} \in V_{t+1}$ are connected, else $A_{tn'n} = 0$. Consequently, a binary decision variable x_{ntr} is introduced to represent the cells n traversed at the respective time period t for the respective rth UAV.

The following notations are used to formulate the mathematical model:

N	the entire search region is divided into $N \times N$ number of cells with equal area in the grid, $n \in N^* = \{1, .., N^2 - 1, N^2\}$
T	set of time intervals with equal length defining the time horizon to explore a grid, $t \in \{0, 1, \ldots, |T| - 1\}$
R	number of UAVs, $r \in \{1, \ldots, R\}$
p_n	probability of actual target occupancy on cell n
x_{ntr}	state transition binary variable; $x_{ntr} = 1$, if the path of rth UAV investigates the nth cell in time period t, while $x_{ntr} = 0$, if that the corresponding cell is not visited
$F_{n'ntr}$	a binary matrix representation of the infeasible maneuvers. That is, $F_{n'ntr} = 1$ whenever $A_{tn'n} = 0$
Z_{ntr}	a binary binary matrix representation of all cells through the time horizon representing the same location
B_{ntr}	a binary matrix representation of the cells that can only be visited once
H_{ntr}	a binary matrix representation of all maneuvers performed in the time period t
S_{ntr}	a binary matrix representation of start and ending positions for rth UAV

Based on the above notation, the following mathematical model is proposed, where the first objective represents the cumulative probability of success for the total number of UAVs to be deployed and the second objective minimizes the total spent time performing the mission:

$$\max \quad f_1 = \sum_{r \in R} \sum_{t \in T} \sum_{n \in N^*} p_n x_{ntr} \tag{1}$$

$$\min \quad f_2 = \sum_{r \in R} \sum_{t \in T} \sum_{n \in N^*} \frac{x_{ntr}}{RT} \tag{2}$$

$$\text{s.t.}$$

$$\sum_{t \in T} \sum_{n \in N^*} F_{n'ntr} x_{ntr} \leq 1 \quad \forall r \in R \quad \forall n' \in N^* \tag{3}$$

Constraint (3) ensures that infeasible maneuvers cannot be performed between two consecutive time periods. The binary matrix F showcases each pair between consecutive cells n and n' that are infeasible for a given time period t. That is, if $F_{n'ntr} = 1$, then the two cells n and n' in time period t and $t + 1$, respectively, are not feasible in the same path for any r.

$$\sum_{r \in R} Z_{ntr} x_{ntr} \leq 1 \quad \forall n \in N^* \quad t \in T \tag{4}$$

Constraint (4) enforces a safety zone around each path, that is, a single agent r can only traverse a cell in a given time period. Note that the binary matrix Z showcases the decision variable's index that represents the same time period.

$$\sum_{r \in R} \sum_{t \in T} B_{ntr} x_{ntr} \leq 1 \quad \forall n \in N^* \tag{5}$$

Here, constraint (5) considers gathering images of a cell over multiple different time periods, where the binary B matrix showcase each index that represents the same cell. In this paper, we neglect the dynamics of changing probability, and we are not interested in obtaining a search path that acquires multiple images of the same cell. Note that we do not have to consider a conditional probability map that is dependent on the chosen paths because of this constraint, as the cumulative probability will be in the range of $[0,1]$.

$$\sum_{n \in N^2} H_{ntr} x_{ntr} = 1 \quad \forall n \in N^* \quad r \in R \tag{6}$$

In constraint (6), the binary H matrix ensures that the paths only allow a single maneuver to be performed per time period per UAV.

$$\sum_{r \in R} \sum_{t \in T} S_{ntr} x_{ntr} = 1 \quad \forall r \in R \tag{7}$$

Constraint (7) ensures that the complete path starts and ends in the designated time zones in the designated time periods.

$$x_{ntr} \in \{0,1\} \quad \forall n \in N^*, \forall t \in T, r \in \{1, \ldots, R\} \tag{8}$$

Finally, the above constraint (8) represent the decision and auxiliary variables.

3. Scenario Generation

The UAV-assisted SAR mission generally consists of multiple different phases, with the common goal of deploying as soon as possible when sufficient information about the mission is gathered. The UAV aspect is to either aid or collect information as fast as possible for the rescue team's job. In this research, the UAVs are only gathering information through images. Therefore, when generating the problem scenarios, we have to assume some information that later can be modified to accommodate real-world scenario. In general, the overall map is divided into an $N \times N$ grid where each cell is assumed to have the same area. Then, a probability map is generated where each cell is given a certain probability of containing the missing target. The probability map is generated randomly based on a given number of hotspots and corresponding spread (see Figure 1). To accommodate the problem scenario, the number of deployed UAVs also affects the size of the problem scenario. These are assumed to be taking off and landing in a specific grid cell. There is also denoted a time horizon with a given number of equidistant points in time, and the UAVs are then able to search an entire grid cell for each time period, and then go to one of their neighboring grid cells in the following time period. As mentioned in the Mathematical Modelling Section, the UAVs can move in all directions, but they can only hover (land) in the grid cell containing the UAV station. Note that this cell, therefore, should not have any gain or loss in terms of the objectives, e.g., probability of locating the target. Due to the problem complexity, we assume there to only be two hot spots with a spread of three and the UAV station to be located in grid cell $[0,0]$. The parameters assumed to affect the size of the problem scenario are the grid size, N, time horizon, T, and number of UAVs, R.

Note that the proposed division of the search area is analogous to the raster model, which is a data storage method used extensively in geographic information systems.

4. Solution Procedure

In this section, we explain the solution procedure and the selection of search parameters for the employed search method. The exact approach is often not applicable in large-scale scenarios, as it can even fail to deliver a feasible solution. In a time-restricted environment such as UAV-assisted search and rescue, this is not applicable. On the other side of the spectrum, a greedy approach does deliver a feasible solution, but it often lacks in performance. This is what we try to investigate with the deployed VNS approach. We evaluate the performance of the algorithm with Dijkstra's algorithm and exact solvers such as GNU Linear Programming Kit (GLPK) to establish its efficiency. However, before doing so, the following definitions should be presented.

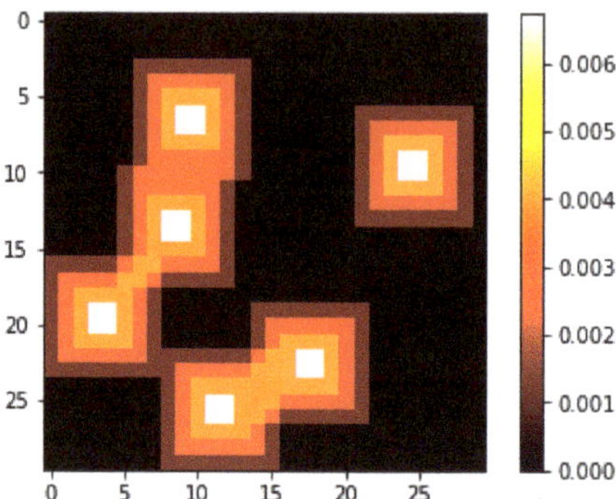

Figure 1. Probability map on a 30 × 30 grid given six hotspots with a spread of 3. The start and end cell (UAV station) is located in cell $[0, 0]$ in the upper left corner.

Definition 1. *Multiple objective optimization problems can be represented as follows:*

$$\begin{cases} max & (f_1(x), f_2(x), \ldots, f_k(x)) \\ min & (g_1(x), g_2(x), \ldots, g_r(x)) \\ s.t. & x \in X = \{x \mid h_t(x) \leq 0, t = 1, \ldots, m\} \end{cases}$$

where $x = (x_1, x_2, \ldots, x_n)$ are the decision variables; $f_i(x), (i = 1, \ldots, k)$ are maximization type objective functions; $g_j(x), (j = 1, \ldots, r)$ are minimization type objective function; $h_t(x), (t = 1, \ldots, m)$ are set of constraints [60].

Definition 2. *A decision plan $x^0 \in X$ is said to be a Pareto optimal solution to the multiple objective optimization problems if there does not exist another $y \in X$, such that $f_k(y) \leq f_k(x^0)$ for all k and $f_s(y) < f_s(x^0)$ for at least one s Wu et al. [61].*

From the perspective of the search and rescue problem, it is difficult to define the strict upper or lower bounds for the multi-objective setting problem. This is first because of the fuzzy nature of the multi-objective setting but also because of the complexity of obtaining a solution. Therefore, we incorporate both exact and inexact solution approachs to illustrate these issues.

4.1. Transforming Multi-Objective Framework into a Single-Objective One

When dealing with a multi-objective framework, several types of solution approaches can be applied, such as transforming the problem into a single-objective one, incorporating them through a lexicographic method, identifying the entire Pareto front to determine the trade-off among objective weightings, etc. Therefore, it generally comes down to whether the decision maker's preference is incorporated before, under or after exploring the solution space.

In a time-restricted environment such as search and rescue mission planning, it is of absolute necessity that a solution can be obtained in real-time. Therefore, we utilize the approach to transform the multi-objective framework into a single objective. For the bi-objective framework, the objectives do not have a fitting cost transform due to the respective units of the objectives. However, there is a range similarity in terms of the sum of them being between 0 and 1; a simple weighted average is, therefore, fitting to do this. Here, α represents the trade-off between the objectives [62].

$$f_{combined} = \alpha f_1 + (1 - \alpha) f_2 \tag{9}$$

Note that the naive weighted average can be controversial, and we therefore elaborate the use of this in Section 5 (for more information see, Wang [29]).

4.2. GLPK

We utilized the freely available GNU Linear Programming Kit (GLPK) package for the exact solution procedure. The GLPK package is used for large-scale mixed-integer linear programming problems [63]. It utilizes the branch-and-cut method for integer restriction of the decision variables, extending to the branch-and-bound and cutting plane method. The package is implemented in Python, where a maximum solution time is set to 12 min. In a general real-world setting, the ultimately allowed solution time in practice is likely to be lower, and this limit is therefore only set for illustrative purposes.

4.3. Dijkstra's Algorithm

A useful path can be established by implementing graph searching algorithms. In this direction, we utilize Dijkstra's Algorithm, which is extensively used in single-source shortest path problems with non-negative weights for each edge. In implementing the Dijkstra's algorithm for the path finding problem, it is imperative to introduce the constraint on revisiting nodes that represent the same location in different time periods. A way to incorporate this is when visiting the node (i.e., that node being the lowest distance in the queue), then not allowing it to go back after a defined safety period has passed. The set of nodes is then removed in the same way as the visiting node is removed from the queue. Here, the distance that is sought to be minimized is the cumulative score, while the graph traversed is the directed graph G, not allowing it to go backward in time. We refer to the works of Yuan et al. [58] and Sathyara et al. [59] for the detail overview of the algorithm.

4.4. Variable Neighborhood Search

The inexact solution procedure developed in this research is a two-step VNS method that incorporates the general approaches of the VNS but couples that with the known information of directed acyclic graph of feasible paths through a path construction algorithm. The general VNS is proposed by Mladenovic and Hansen [38] in 1997, and it represents a flexible framework for building heuristics to approximately solve combinatorial and non-linear optimization problems. The VNS search heuristic systematically changes its neighborhood structures to obtain a solution. It does so based on the following key observations [64]:

- A local optimum relative to one neighborhood structure is not necessarily a local optimum for another neighborhood structure.
- A global optimum is a local optimum concerning all neighborhood structures.
- Empirical evidence shows that all or a large majority of the local optima are relatively close to each other for many problems.

The ingredients of a variable neighborhood search heuristic include an improvement phase used to improve a given solution and a so-called shaking phase used to resolve local minima traps. The improvement phase, the shaking procedure and the neighborhood change step are executed alternately until a predefined stopping criterion. This research combined it with a path construct algorithm to obtain feasible solutions more quickly and ensure that it follows the stated constraints. The path construct algorithm can be found in the pseudo-code of Algorithm 1. This approach linearly goes through the available time horizon and selects the next maneuver through a weighted probability based on each alternative's respective score. It accompanies the constraint by removing feasible maneuvers and steers it back to the end position by narrowing the feasible maneuvers based on the Chebyshev and Manhattan distances to the end position. Note that this feature of steering the path back to the selected end position is necessary as the two-step VNS randomly selects new neighborhoods to investigate. The grid representation is, therefore, not enough to steer it back. The integrated VNS approach selects a random neighborhood to improve upon the path. It stops selecting new neighborhoods when a designated number of iteration have been investigated. The pseudocode of the algorithm is presented in Algorithms 1 and 2.

Algorithm 1: The path constructing algorithm: **path_constructor(x_original, t1, t2, rs, score)**

Result: feasible path in x
1 x_original := the path that should be updated
2 N, R, T := the dimensions of the problem (grid size, number of UAVs, size of time horizon)
3 t1, t2 := the start and end time where a new path between should be located
4 rs := the set of UAVs
5 x_new := x_original
6 x_new[:,:,[t1:t2]] := 0
7 **for** *r in rs* **do**
8   start = starting position
9   end = end position
10   **if** *start, end is empty* **then**
11   start := global start
12   end := global end
13   **end**
14      $t := t1 + 1$
15   infeasible_neighbor := []
16   infeasible_neighbor_T := []
17   **while** $t < t2$ **do**
18   prior := cell of t-1 maneuver
19   feasible_maneuver := all maneuvers inside grid
        `/* remove infeasible maneuvers                                */`
20   feasible_maneuvers := remove infeasible manuevers as indicated by infeasible_neighbor if t is in infeasible_neighbor_T
21   **if** *Manhatten distance from prior to end >= t2 - (t-1)* **then**
22   maneuver_distance = manhatten distance from possible neighbors to end + 1
23   **if** *maneuver_distance <= t2-t* **then**
24   remove maneuver from feasible_maneuvers
25   **end**
26   **end**
27   **if** *chebyshev distance from prior to end >= t2 - (t-1)* **then**
28   maneuver_distance = chebyshev distance from possible neighbors to end + 1
29   **if** *maneuver_distance <= t2-t* **then**
30   remove maneuver from feasible_maneuvers
31   **end**
32   **end**
33   **if** *maneuver in feasible_maneuvers has already been visited by other UAVs* **then**
34   remove maneuver from feasible_maneuvers
35   **end**
36   **if** *feasible_maneuvers is empty* **then**
37   **if** *t-1 is equal to t1* **then**
38   **RETURN**(x_original)
39   **end**
40   **else**
41   Add prior manuever to infeasible_neighbor
42   Add t-1 to infeasible_neighbor_T
43   remove prior from x_original
44                  $t := t-1$
45   **break**
46   **end**
47   **end**
        `/* select feasible maneuver based on weighted probability       */`
48   ranked_maneuvers := ARGSORT(score[feasible_manuevers])
49   weighted_maneuvers := ranked_maneuvers / SUM(ranked_maneuvers)
50   chosen := CHOOSE(feasible_manuevers, weighted_maneuvers, 1)
51   x_new[chosen] = 1
52   **end**
53 **end**
54 **RETURN**(x_new)

Algorithm 2: Pseudocode representing **VNS(score, N, R, T, neighborhood_size, nmax, kmax, tmax)**

```
1   h! Result: best path P
2   score := score for each cell
3   N, R, T := the dimensions of the problem (grid size, number of UAVs, size of time horizon)
4   neighborhood_size := size of searched neighborhood
5   nmax := maximum number of neighborhood changes
6   kmax := maximum searches per neighborhood
7   tmax := total maximum runtime in seconds
8   n := 0
9   x_best := path_constructor(ZEROS(N,R,T), t1=0, t2=end, rs=[0,1], score)
10  score_best := SUM(x_best * score)
11  while n<nmax do
12      k := 0
13      while k<kmax do
14          n1 := RANDOM(0,T)
15          n2 := min(T,n1+neighborhood_size)
16          nr := RANDOM(0,R)
17          x_temp := path_constructor(x_best, t1=n1, t2=n2, rs=nr, score)
18          score_temp := SUM(x_best * score)
19          if score_temp > score_best then
20              x_best := x_temp
21              score_best := score_temp
22              k := 0
23          end
24          k := k+1
25      end
26      n := n+1
27  end
28  RETURN(x_best)
```

5. Experiments

All numerical experiments were executed with Intel Core i5-8250 CPU with 1.60 GHz processors and 8.00 GB RAM for performance evaluation. For numerical verification, we model the probability map through two hotspots with a spread of two cells.

5.1. Sensitivity of VNS Parameters

The VNS algorithm has three different parameters indicating the search depth, i.e., neighborhood, $nmax$ and $kmax$, defining the size of the neighborhood each search considers; the maximum number of searched neighborhoods; and the number of searches per neighborhood. The results are shown in Table 1.

Table 1. Average performance, standard deviation and average runtime for 100 different runs with different neighborhood parameter.

Neighborhood Parameter	Relative Performance ($f_{combined}$)	Standard Deviation ($f_{combined}$)	Runtime (s)
0.250	0.630	0.086	13.239
0.333	0.820	0.033	20.094
0.417	0.837	0.036	27.870
0.500	0.860	0.025	35.917
0.583	0.908	0.027	39.621
0.667	0.921	0.027	42.898
0.750	0.893	0.043	47.320
0.833	0.862	0.035	49.739

The performance in Table 2 illustrates the change in deviation and runtime when modifying the *nmax* and *kmax* parameter, but it should be noted that the computation of these could easily be parallelized. In the parameter indicating the neighborhood's size, we can see that there is not a unified result showing which size of a neighborhood to chose. Therefore, we choose to further extend the algorithm by randomly selecting a length within the range of 0.3 to 0.9 for each neighborhood change. This furthers the shake and improvement steps of the VNS, as both local and global solutions will be investigated.

Table 2. Average relative performance of the Variable Neighborhood Search (VNS) method compared to the exact GNU Linear Programming Kit (GLPK) approach for 100 different runs with different nmax and kmax settings.

	Avg. Performance (Relative to GLPK)						Avg. Runtime (In Seconds)					
nmax\kmax	5	15	25	35	45	55	5	15	25	35	45	55
50	0.778	0.844	0.847	0.865	0.907	0.889	1.279	3.634	6.006	8.895	10.885	13.773
100	0.777	0.865	0.889	0.885	0.885	0.847	2.335	7.114	11.527	16.675	22.014	26.318
150	0.931	0.890	0.870	0.931	0.950	0.933	3.662	11.177	16.975	24.241	32.656	37.795
200	0.926	0.931	0.843	0.823	0.864	0.912	5.356	14.136	23.159	32.099	42.245	54.182
250	0.867	0.779	0.911	0.933	0.891	0.865	6.185	16.359	28.851	39.850	55.093	62.888
500	0.932	0.867	0.913	0.911	0.869	0.975	12.262	33.660	59.775	87.177	102.260	126.429
1000	0.934	0.912	0.910	0.868	0.871	0.932	23.976	70.143	111.092	156.020	211.473	269.177
1500	0.886	0.846	0.928	0.928	0.867	0.913	32.441	103.338	171.001	248.853	323.948	360.003
2000	0.927	0.849	0.911	0.912	0.956	0.912	45.564	133.198	229.218	338.928	360.003	360.007
2500	0.869	0.976	0.911	0.912	0.974	0.912	61.237	175.410	298.774	360.004	360.002	360.002

5.2. Performance and Runtime for VNS, Dijkstra, and GLPK

GLPK is an exact approach and is therefore significantly slower, but it also yields the optimal solution. However, the GLPK is not able to solve any of the larger problem scenarios. The performance and runtime for the three approaches on different scenario sizes relative to grid size N, time horizon T and the number of UAVs R can be seen in Figure 2. Note that, when a solution approach reaches the time limit, the time is noted, while its performance is not.

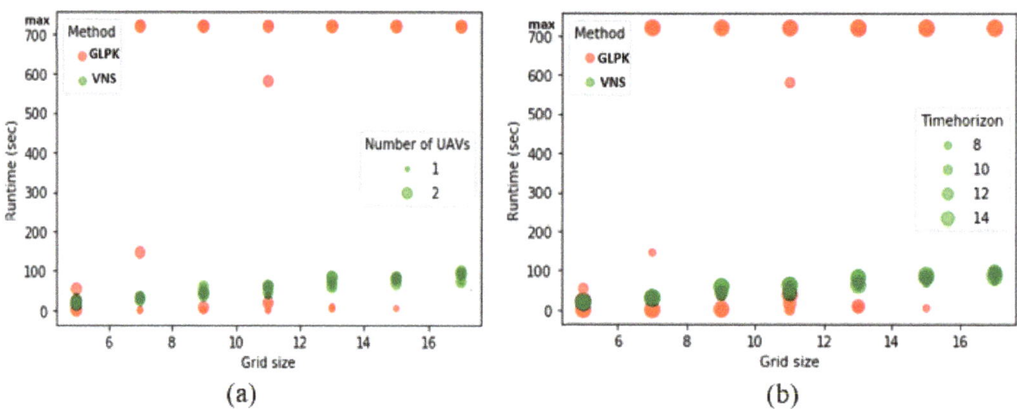

Figure 2. (**a**) The relative performance of the Variable Neighborhood Search (VNS) and Dijkstra algorithm compared to the optimal solution found by the GNU Linear Programming Kit (GLPK) approach is shown. Note that many experiments do not yield a relative performance as GLPK could not obtain a solution. (**b**) The runtime for the three approaches is presented, demonstrating relation to different grid sizes and time horizons. The exact value of performance measures is presented in Table 3.

Table 3. The performance of the respective solution approaches on different scenarios. Note that GLPK could not obtain a solution on some of the scenarios. This is illustrated by (-), while its runtime reached the limit of 720 s.

Grid Size	Time Horizon	No. of UAVs	Performance ($f_{combined}$)			Relative Performance	
N	T	R	VNS	Dijkstra	GLPK	$\frac{f_{|VNS}}{f_{|GLPK}}$	$\frac{f_{|Dijkstra}}{f_{|GLPK}}$
5	10	1	0.225	0.018	0.242	0.928	0.074
5	10	2	0.137	0.324	0.416	0.329	0.778
5	14	1	0.310	0.206	0.357	0.868	0.576
5	14	2	0.454	0.124	0.454	1.000	0.273
5	18	1	0.332	0.199	0.484	0.686	0.412
5	18	2	0.371	0.172	0.428	0.865	0.406
5	22	1	0.460	0.128	0.460	1.000	0.280
5	22	2	0.400	0.185	0.481	0.830	0.384
7	10	1	0.075	0.046	0.102	0.730	0.450
7	10	2	0.082	0.019	0.135	0.612	0.143
7	14	1	0.163	0.037	0.178	0.912	0.208
7	14	2	0.119	0.045	0.143	0.831	0.316
7	18	1	0.366	0.140	0.385	0.951	0.364
7	18	2	0.214	0.061	-	-	-
7	22	1	0.554	0.005	0.554	1.000	0.009
7	22	2	0.344	0.016	-	-	-
9	10	1	0.184	0.057	0.222	0.832	0.258
9	10	2	0.319	0.188	0.344	0.927	0.546
9	14	1	0.096	0.035	0.101	0.947	0.351
9	14	2	0.026	0.011	-	-	-
9	18	1	0.346	0.244	0.371	0.932	0.656
9	18	2	0.431	0.003	-	-	-
9	22	1	0.446	0.237	0.496	0.899	0.479
9	22	2	0.458	0.300	-	-	-
11	10	1	0.185	0.007	0.185	1.000	0.042
11	10	2	0.283	0.099	0.296	0.958	0.336
11	14	1	0.273	0.042	0.297	0.916	0.141
11	14	2	0.408	0.162	0.427	0.955	0.380
11	18	1	0.242	0.029	0.297	0.812	0.098
11	18	2	0.272	0.003	-	-	-
11	22	1	0.447	0.191	0.447	0.999	0.428
11	22	2	0.545	0.002	-	-	-
13	10	1	0.097	0.056	0.111	0.876	0.504
13	10	2	0.110	0.027	-	-	-
13	14	1	0.223	0.068	0.248	0.899	0.273
13	14	2	0.347	0.274	-	-	-
13	18	1	0.348	0.002	0.348	1.000	0.008
13	18	2	0.522	0.005	-	-	-
13	22	1	0.373	0.089	-	-	-
13	22	2	0.547	0.047	-	-	-
15	10	1	0.111	0.042	0.111	0.999	0.384
15	10	2	0.085	0.028	-	-	-
15	14	1	0.095	0.009			
15	14	2	0.039	0.028	-	-	-
15	18	1	0.072	0.043	-	-	-
15	18	2	0.112	0.039	-	-	-
15	22	1	0.092	0.004	-	-	-
15	22	2	0.107	0.031	-	-	-
17	10	1	0.010	0.008			
17	10	2	0.024	0.081	-	-	-
17	14	1	0.492	0.068	-	-	-
17	14	2	0.573	0.355	-	-	-
17	18	1	0.372	0.163	-	-	- -
17	18	2	0.448	0.003	-	-	-
17	22	1	0.249	0.001	-	-	-
17	22	2	0.141	0.033	-	-	-

The performance clearly indicates that the GLPK is generally faster for small problem scenarios with a single UAV. However, it cannot even obtain a solution whenever there are two UAVs to consider or the grid size or time horizon is larger. The relative performance of VNS indicates that, for larger problem scenarios, it will perform within 20% of the optimal solution, while, for smaller problem scenarios, it performs within 50% of the optimal. The latter is perhaps because VNS searches with a neighborhood size that is too small relative to the grid size, so it will never get out of the local optima. However, it does not seem to be an issue for the larger problem scenarios. Similarly, the Dijkstra approach seems to decrease in performance relative to the exact approach when the scenario size increases. This is probably due to the greedy nature of the method, as it does not want to investigate areas that require it to cross a section of cells without any probability of success. The results also showcase the complexity of the large-scale problem scenarios in UAV-assisted search and rescue missions. Overall, Figure 2 demonstrates that the VNS outperforms GLPK and Dijkstra's algorithm in the perspective of relative performance measures for most of the instances.

5.3. Sensitivity of Objective Weighting for the GLPK

Figure 3 illustrates the sensitivity to changes in the trade-off between objectives represented by modifying α. The sensitivity analysis sheds light on the change in the optimal path for different trade-offs. Figure 3 shows that the UAVs for alpha equal to 0 and 0.1 clearly stay in take-off and landing zone for the entire time horizon for both UAVs or just for one UAV. This is because the score for each grid cell outside the take-off zone is too high to consider. Finally, Figure 4 shows that the optimal path changes for almost all different alpha settings. However, the pattern of each path seems to follow the same structure because the path is sensitive to the parameter α, which also justifies the multi-objective formulation of the problem.

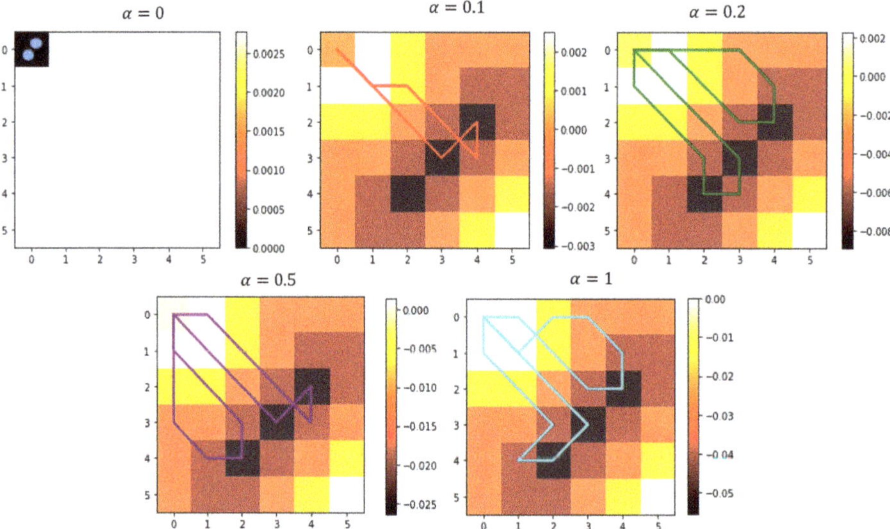

Figure 3. The corresponding route in 2D generated by GLPK for different weightings of alpha on the corresponding scoring map. Note that the illustrated paths is for two UAVs on a 6×6 grid with a time horizon of 10 and start and end in grid cell $[0, 0]$. In addition, for alpha = 0.1, the second UAV stays in the start cell.

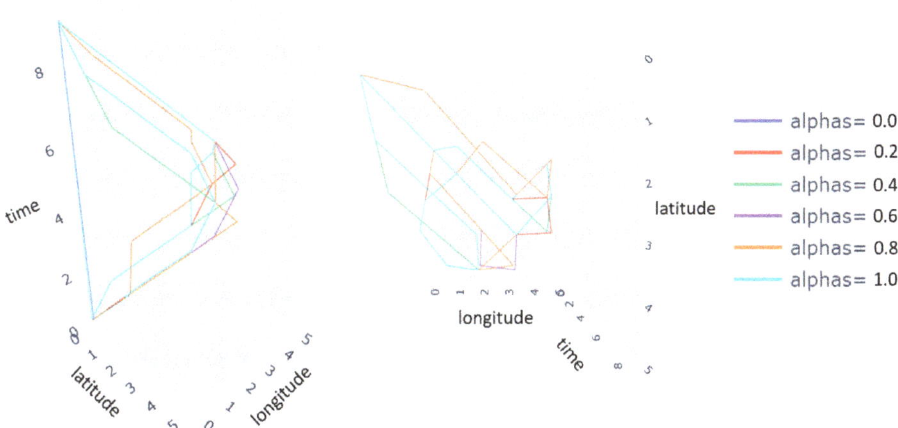

Figure 4. The corresponding route in three dimension generated by GLPK for different weights of alpha. The longitude and latitude axes represent the possible maneuvers on the grid, while time illustrates the time dimension.

5.4. Benefits and Adverse Circumstances Associated with Multi-Objective Framework

The results on the sensitivity clearly showcase some of the dangers when incorporating the bi-objective framework on the UAV pathfinding. It is very difficult to see which alpha enforces that all equipment will be employed and not spending too much time in the landing zone. Clearly, the solution procedure should allow UAVs to return before time, but it is very difficult to identify when it is too early to specify through the alpha parameter.

There is similarly a robustness issue when introducing the multi-objective framework as objectives can be conflicting, and a solution can satisfy an objective that is not of our interest. In the case of this paper, we are clearly interested in searching as many high probability cells as possible in as little time as possible. However, indicating how little time is too much is very difficult in the presented setting. The last thing one wants to introduce is nervousness in the scheduling, so some rules about searching different areas could be of advantage.

Nevertheless, introducing these additional objectives clearly brings us closer to the optimal goal. For these search and rescue Missions, we are interested in accumulating the highest probability of locating the missing target. We are, however, also interested in doing it as quickly as possible by obtaining the best quality images possible. Similarly, there could be a chance that the missing target has a higher probability of survival in some regions than others, which is why we also are interested in locating the target alive. Therefore, additional objectives other than the ones considered in this research could be introduced.

6. Conclusions

The smart city concept is almost around last couple of decades, and one of the critical concepts is to integrate cutting-edge technology without raising costs in improving environmental sustainability and life expectancy. In this direction, we proposed a multi-objective path planning and trajectory mapping problem under the mixed integer programming problem framework for a set of homogeneous UAVs deployed to search for static targets. A graph theory-based directed acyclic network representation is employed to reduce complexities and track the inward and outward movement of each UAV from its respective present cell location by ensuring flow conservation. A modification of the basic VNS algorithm is proposed and implemented in two phases to find the solution. In the first phase, a path is generated and in the second phase, trajectory mapping is done sequentially by considering constraints associated with the problem environment. Numerical simulation on synthetic experimental settings demonstrates that the proposed approach can reduce computational complexity and provide a solution within reasonable amount of time compared to the exact solver. Moreover, it is found that the exact solver is unable to provide a solution within a time threshold. When we compare the relative performance of VNS with GLPK or Dijkstra's algorithm, it was found that Dijkstra's algorithm's performance is relatively lower as the grid size increases, which justifies the efficiency of the proposed algorithm. To our best knowledge, this is the first work to explore the path for multiple UAVs by using a bi-objective VNS algorithm. Considering the numerical evaluation, one can conclude that the approach presented in this study is a better alternative than the exact solver, and methodology can contribute to intelligent systems.

For future work, we intend to extend the proposed approach to calculate paths for finding moving targets. We assumed altitude differentiation from the perspective of collision avoidance. We ignored constraints such as fuel, sensor capacity, search pattern, etc., those need to be integrated to formulate a robust path planning model. We compared the outcome of proposed solution approach with exact solver, therefore one can employ other algorithms such as particle swarm optimization [65], bat algorithm [66], A^* algorithm [59], machine learning (ML) algorithms [29] etc. to compare the performance of the proposed VNS algorithm. Finally, one can use a multi-criterion decision-making algorithm [67] to incorporate customizable preferences of decision-makers robustly to take advantage of the inherent flexibility while setting weights.

Author Contributions: Conceptualization, S.S. and A.E.V.; methodology, S.S. and A.E.V.; software, A.E.V.; validation, I.N., A.H. and H.B.; formal analysis, I.N.; investigation, A.E.V.; resources, A.H. and H.B.; writing—review and editing, S.S., A.E.V. and I.N.; and project administration, S.S. All authors have read and agreed to the published version of the manuscript.

Funding: This research received no external funding.

Data Availability Statement: The data used in this study are available on request from the corresponding author.

Conflicts of Interest: The authors declare no conflict of interest.

References

1. Nielsen, I.E.; Dang, V.Q.; Nielsen, P.; Pawlewski, P. Scheduling of Mobile Robots with Preemptive Tasks, Advances in Intelligent Systems and Computing. In *Distributed Computing and Artificial Intelligence, 11th International Conference*; Springer: Berlin, Germany, 2014; pp. 19–29.
2. Sung, I., Nielsen, P. Zoning a service area of unmanned aerial vehicles for package delivery services. *J. Intell. Robot. Syst.* **2020**, *97*, 719–731. [CrossRef]
3. Thibbotuwawa, A.; Bocewicz, G.; Radzki, G.; Nielsen, P.; Banaszak, Z. UAV Mission planning resistant to weather uncertainty. *Sensors* **2020**, *20*, 515. [CrossRef]
4. Eagle, J.N. 1984. The optimal search for a moving target when the search path is constrained. *Oper. Res.* **1984**, *32*, 1107–1115. [CrossRef]
5. Puglicoe, L.D.P.; Guerriero, F.; Zorbas, D.; Razafindralambo, T. Modelling the mobile target covering problem using flying drones. *Optim. Lett.* **2016**, *10*, 1021–1052. [CrossRef]
6. Samaniego, F.; Sanchis, J.; García-Nieto, S.; Simarro, R. Recursive Rewarding Modified Adaptive Cell Decomposition (RR-MACD): A Dynamic Path Planning Algorithm for UAVs. *Electronics* **2019**, *8*, 306. [CrossRef]
7. San, Juan, V.; Santos, M.; Andújar, J.M. Intelligent UAV map generation and discrete path planning for search and rescue operations. *Complexity* **2018**. [CrossRef]
8. Stewart, T.J. Search for a moving target when searcher motion is restricted. *Comput. Oper. Res.* **1979**, *6*, 129–140. [CrossRef]
9. Wang, J.; Jiang, C.; Han, Z.; Ren, Y.; Maunder, R.G.; Hanzo, L. Taking drones to the next level: Cooperative distributed unmanned-aerial-vehicular networks for small and mini drones. *IEEE Veh. Technol. Mag.* **2017**, *12*, 73–82. [CrossRef]
10. Yuan, H.; Xiao, C.; Zhan, W.; Wang, Y.; Shi, C.; Ye, H.; Jiang, K.; Ye, Z.; Zhou, C.; Wen, Y. Target detection, positioning and tracking using new UAV gas sensor systems: Simulation and analysis. *J. Intell. Robot. Syst.* **2019**, *94*, 871–882. [CrossRef]
11. Felemban, E.; Sheikh, A.A.; Naseer, A. Improving response time for crowd management in Hajj. *Computers* **2021**, *4*, 46. [CrossRef]
12. Miyano, K.; Shinkuma, R.; Shiode, N.; Shiode, S.; Sato, T.; Oki, E. Multi-UAV allocation framework for predictive crime deterrence and data acquisition. *Internet Things* **2020**, *11*, 100205. [CrossRef]
13. Huang, S.; Gui, J.; Wang, T.; Li, X. Joint Mobile Vehicle–UAV Scheme for Secure Data Collection in a Smart City. *Ann. Telecommun.* **2020**, 1–22. Available online: https://www.researchgate.net/publication/344005230_Joint_mobile_vehicle-UAV_scheme_for_secure_data_collection_in_a_smart_city (accessed on 10 March 2021). [CrossRef]

14. Trummel, K.E.; Weisinger, J.R. The complexity of the optimal searcher path problem. *Oper. Res.* **1986**, *34*, 324–327. [CrossRef]
15. Brown, S.S. Optimal search for a moving target in discrete time and space. *Oper. Res.* **1980**, *28*, 1275–1289. [CrossRef]
16. Benkoski, S.J.; Monticino, M.G.; Weisinger, J.R. A survey of the search theory literature. *Nav. Res. Logist.* **1991**, *38*, 469–494. [CrossRef]
17. Washburn, A.R. Branch and bound methods for a search problem. *Nav. Res. Logist.* **1998**, *45*, 243–257. [CrossRef]
18. Lau, H.; Huang, S.; Dissanayake, G. Optimal search for multiple targets in a built environment. In Proceedings of the 2005 IEEE/RSJ International Conference on Intelligent Robots and Systems, Edmonton, AB, Canada, 2–6 August 2005; pp. 3740–3745.
19. Lau, H.; Huang, S.; Dissanayake, G. Discounted mean bound for the optimal searcher path problem with non-uniform travel times. *Eur. J. Oper. Res.* **2008**, *190*, 383–397. [CrossRef]
20. Rogge, J.A.; Aeyels, D. Multi-robot coverage to locate fixed and moving targets. In Proceedings of the 2009 IEEE Control Applications (CCA) & Intelligent Control (ISIC), St. Petersburg, Russia, 8–10 July 2009; pp. 902–907.
21. Li, X.; Yao, H.; Wang, J.; Xu, X.; Jiang, C.; Hanzo, L. A near-optimal UAV-aided radio coverage strategy for dense urban areas. *IEEE Trans. Veh. Technol.* **2019**, *68*, 9098–9109. [CrossRef]
22. Berger, J.; Lo N. An innovative multi-agent search-and-rescue path planning approach. *Comput. Oper. Res.* **2015**, *53*, 4–31. [CrossRef]
23. Perez-Carabaza, S.; Besada-Portas, E.; Lopez-Orozco, J.A.; Jesus, M. Ant colony optimization for multi-UAV minimum time search in uncertain domains. *Appl. Soft Comput.* **2018**, *62*, 789–806. [CrossRef]
24. Ye, F.; Chen, J.; Tian, Y.; Jiang, T. Cooperative task assignment of a heterogeneous multi-UAV system using an adaptive genetic algorithm. *Electronics* **2020**, *4*, 687. [CrossRef]
25. Lu, Y.; Ma, Y.; Wang, J.; Han, L. Task assignment of UAV swarm based on Wolf Pack algorithm. *Appl. Sci.* **2020**, *23*, 8335. [CrossRef]
26. Luo, R.; Zheng, H.; Guo, J. Solving the multi-functional heterogeneous UAV cooperative mission planning problem using multi-swarm fruit fly optimization algorithm. *Sensors* **2020**, *18*, 5026. [CrossRef]
27. Alhaqbani, A.; Kurdi, H.; Youcef-Toumi, K. Fish-inspired task allocation algorithm for multiple unmanned aerial vehicles in search and rescue missions. *Remote Sens.* **2021**, *1*, 27.
28. Xiong, C.; Chen, D.; Lu, D.; Zeng, Z.; Lian, L. Path planning of multiple autonomous marine vehicles for adaptive sampling using Voronoi-based ant colony optimization. *Robot. Auton. Syst.* **2019**, *115*, 90–103. [CrossRef]
29. Wang, J.; Jiang, C.; Zhang, H.; Ren, Y.; Chen, K.C.; Hanzo, L. Thirty years of machine learning: The road to Pareto-optimal wireless networks. *IEEE Commun. Surv. Tutor.* **2020**, *22*, 1472–1514. [CrossRef]
30. Kundid Vasić, M.; Papić, V. Multimodel Deep Learning for Person Detection in Aerial Images. *Electronics* **2020**, *9*, 1459. [CrossRef]
31. Li, Y.; Yuan, X.; Zhu, J.; Huang, H.; Wu, M. Multiobjective Scheduling of Logistics UAVs Based on Variable Neighborhood Search. *Appl. Sci.* **2020**, *10*, 3575. [CrossRef]
32. Lo N.; Berger, J.; Noel, M. Toward optimizing static target search path planning. In Proceedings of the 2012 IEEE Symposium on Computational Intelligence for Security and Defence Applications, Ottawa, ON, Canada, 11–13 July 2012; pp. 1–7.
33. Nielsen, L.D.; Sung, I.; Nielsen, P. Convex decomposition for a coverage path planning for autonomous vehicles: Interior extension of edges. *Sensors* **2019**, *19*, 4165. [CrossRef]
34. Perez-Carabaza, S.; Scherer, J.; Rinner, B.; López-Orozco, J.A.; Besada-Portas, E. UAV trajectory optimization for Minimum Time Search with communication constraints and collision avoidance. *Eng. Appl. Artif. Intell.* **2019**, *85*, 357–371. [CrossRef]
35. Raap, M.; Meyer-Nieberg, S.; Pickl, S.; Zsifkovits, M. Aerial vehicle search-path optimization: A novel method for emergency operations. *J. Optim. Theory Appl.* **2017**, *172*, 965–983. [CrossRef]
36. Sitek, P.; Wikarek, J.; Rutczyńska-Wdowiak, K.; Bocewicz, G.; Banaszak, Z. Optimization of capacitated vehicle routing problem with alternative delivery, pick-up and time windows: A modified hybrid approach. *Neurocomputing* **2021**, *423*, 670–678. [CrossRef]
37. Thibbotuwawa, A.; Bocewicz, G.; Zbigniew, B.; Nielsen, P. A solution approach for UAV fleet mission planning in changing weather conditions. *Appl. Sci.* **2019**, *9*, 3972. [CrossRef]
38. Mladenovic, N.; Hansen, P. Variable neighborhood search. *Comput. Oper. Res.* **1997**, *24*, 1097–1100. [CrossRef]
39. Loudni, S.; Boizumault, P.; David, P. On-line resources allocation for ATM networks with rerouting. *Comput. Oper. Res.* **2006**, *33*, 2891–2917. [CrossRef]
40. Geiger, M.J.; Wenger, W. On the assignment of students to topics: A Variable Neighborhood Search approach. *Socio-Econ. Plan. Sci.* **2010**, *44*, 25–34. [CrossRef]
41. Brusco, M.J.; Singh, R.; Steinley, D. Variable neighborhood search heuristics for selecting a subset of variables in principal component analysis. *Psychometrics* **2009**, *74*, 705. [CrossRef]
42. Schilde, M.; Doerner, K.F.; Hartl, R.F.; Kiechle, G. Metaheuristics for the bi-objective orientation problem. *Swarm Intell.* **2009**, *3*, 179–201. [CrossRef]
43. Anghinolfi, D.; Paolucci, M. Parallel machine total tardiness scheduling with a new hybrid metaheuristic approach. *Comput. Oper. Res.* **2007**, *34*, 3471–3490. [CrossRef]
44. Qian, B.; Wang, L.; Huang, D.X.; Wang, X. Multi-objective flow shop scheduling using differential evolution. In *Intelligent Computing in Signal Processing and Pattern Recognition*; Springer: Berlin/Heidelberg, Germany, 2006; pp. 1125–1136.
45. Fleszar, K.; Osman, I.H.; Hindi, K.S. A variable neighborhood search algorithm for the open vehicle routing problem. *Eur. J. Oper. Res.* **2009**, *195*, 803–809. [CrossRef]

46. Montemanni, R.; Smith, D.H. Construction of constant GC-content DNA codes via a variable neighborhood search algorithm. *J. Math. Model. Algorithms* **2008**, *7*, 311. [CrossRef]
47. Hansen, P.; Mladenović, N.; Pérez, J.A.M. Variable neighborhood search: Methods and applications. *Ann. Oper. Res.* **2010**, *175*, 367–407. [CrossRef]
48. Ribeiro, C.C.; Aloise, D.; Noronha, T.F.; Rocha, C.; Urrutia, S. An efficient implementation of a VNS/ILS heuristic for a real-life car sequencing problem. *Eur. J. Oper. Res.* **2008**, *191*, 596–611. [CrossRef]
49. Gosiewski, Z.; Kwaśniewski, K. Time Minimization of Rescue Action Realized by an Autonomous Vehicle. *Electronics* **2020**, *9*, 2099. [CrossRef]
50. Lanillos, P.; Yañez-Zuluaga, J.; Ruz, J.J.; Besada-Portas, E. A bayesian approach for constrained multi-agent minimum time search in uncertain dynamic domains. In Proceedings of the 15th Annual Conference on Genetic and Evolutionary Computation, Amsterdam, The Netherlands, 6–10 July 2013; pp. 391–398.
51. Peng, Z.H.; Wu, J.P.; Chen, J. Three-dimensional multi-constraint route planning of unmanned aerial vehicle low-altitude penetration based on coevolutionary multi-agent genetic algorithm. *J. Cent. South Univ. Technol.* **2011**, *18*, 1502. [CrossRef]
52. Sakawa, M.; Yano, H.; Nishizaki, I.; Nishizaki, I. *Linear and Multiobjective Programming with Fuzzy Stochastic Extensions*; Springer US: New York, NY, USA, 2013.
53. Chen, L.; Peng, J.; Zhang, B. Uncertain goal programming models for bicriteria solid transportation problem. *Appl. Soft Comput.* **2017**, *51*, 49–59. [CrossRef]
54. Chung, C.K.; Chen, H.M.; Chang, C.T.; Huang, H.L. On fuzzy multiple objective linear programming problems. *Expert Syst. Appl.* **2018**, *114*, 552–562. [CrossRef]
55. Deb, K.; Pratap, A.; Agarwal, S.; Meyarivan, T.A.M.T. A fast and elitist multiobjective genetic algorithm: NSGA-II. *IEEE Trans. Evol. Comput.* **2002**, *6*, 182–197. [CrossRef]
56. Moon, I.; Jeong, Y.J.; Saha, S. Fuzzy bi-objective production-distribution planning problem under the carbon emission constraint. *Sustainability* **2016**, *8*, 798. [CrossRef]
57. Davoodi, M.; Panahi, F.; Mohades, A.; Hashemi, S.N. Multi-objective path planning in discrete space. *Appl. Soft Comput.* **2013**, *13*, 709–720. [CrossRef]
58. Yuan, Z.; Yang, Z.; Lv, L.; Shi, Y. A Bi-Level Path Planning Algorithm for Multi-AGV Routing Problem. *Electronics* **2020**, *9*, 1351. [CrossRef]
59. Sathyaraj, B.M.; Jain, L.C.; Finn, A.; Drake, S. Multiple UAVs path planning algorithms: A comparative study. *Fuzzy Optim. Decis. Mak.* **2008**, *7*, 257. [CrossRef]
60. Nielsen, I.E.; Bocewicz, G.; Saha, S. Multi-Agent Path Planning Problem Under a Multi-objective Optimization Framework. In *DCAI 2020: Distributed Computing and Artificial Intelligence, Special Sessions, 17th International Conference, Proceedings of the International Symposium on Distributed Computing and Artificial Intelligence, L'Aquila, Italy, 17–19 June 2020*; Springer: Cham, Switzerland, 2020; pp. 5–14.
61. Wu, Y.K.; Liu, C.C.; Lur, Y.Y. Pareto-optimal solution for multiple objective linear programming problems with fuzzy goals. *Fuzzy Optim. Decis. Mak.* **2015**, *14*, 43–55. [CrossRef]
62. Marler, R.T.; Arora, J.S. Survey of multi-objective optimization methods for engineering. *Struct. Multidiscip. Optim.* **2004**, *26*, 369–395. [CrossRef]
63. Makhorin, A. GLPK (GNU Linear Programming Kit). 2008. Available online: http://www.gnu.org/s/glpk/glpk.html (accessed on 12 March 2019).
64. Hidalgo-Paniagua, A.; Vega-Rodríguez, M.A.; Ferruz, J. Applying the MOVNS (multi-objective variable neighborhood search) algorithm to solve the path planning problem in mobile robotics. *Expert Syst. Appl.* **2016**, *58*, 20–35. [CrossRef]
65. Jeong, Y.; Saha, S.; Chatterjee, D.; Moon, I. Direct shipping service routes with an empty container management strategy. *Transp. Res. Part E Logist. Transp. Rev.* **2018**, *118*, 123–142. [CrossRef]
66. Saha, S.; Chatterjee, D.; Sarkar, B. The ramification of dynamic investment on the promotion and preservation technology for inventory management through a modified flower pollination algorithm. *J. Retail. Consum. Serv.* **2021**, *58*, 102326. [CrossRef]
67. Vasegaard, A.E.; Picard, M.; Hennart, F.; Nielsen, P.; Saha, S. Multi criteria decision making for the multi-satellite image acquisition scheduling problem. *Sensors* **2020**, *20*, 1242. [CrossRef]

 *electronics*

Article

Optimization of Public Transport Services to Minimize Passengers' Waiting Times and Maximize Vehicles' Occupancy Ratios

Ivana Hartmann Tolić [1,*], Emmanuel Karlo Nyarko [2] and Avishai (Avi) Ceder [3,4,5]

[1] Department of Software Engineering, Faculty of Electrical Engineering, Computer Science and Information Technology Osijek, J.J. Strossmayer University of Osijek, 31000 Osijek, Croatia
[2] Department of Computer Engineering and Automation, Faculty of Electrical Engineering, Computer Science and Information Technology Osijek, J.J. Strossmayer University of Osijek, 31000 Osijek, Croatia; karlo.nyarko@ferit.hr
[3] Transportation Research Institute, Technion—Israel Institute of Technology, Faculty of Civil and Environmental Engineering, Haifa 32000, Israel; a.ceder@auckland.ac.nz
[4] Department of Civil and Environmental Engineering, University of Auckland, Auckland 1142, New Zealand
[5] School of Traffic and Transportation, Beijing Jiaotong University, Beijing 100044, China
* Correspondence: ivana.hartmann@ferit.hr

Received: 21 January 2020; Accepted: 14 February 2020; Published: 20 February 2020

Abstract: Determining the best timetable for vehicles in a public transportation (PT) network is a complex problem, especially because it is just necessary to consider the requirements and satisfaction of passengers as the requirements of transportation companies. In this paper, a model of the PT timetabling problem which takes into consideration the passenger waiting time (PWT) at a station and the vehicle occupancy ratio (VOR) is proposed. The solution aims to minimize PWT and maximize VOR. Due to the large search space of the problem, we use a multiobjective particle swarm optimization (MOPSO) algorithm to arrive at the solution of the problem. The results of the proposed method are compared with similar results from the existing literature.

Keywords: optimization models; timetable; passenger waiting time; vehicle occupancy ratio

1. Introduction

In modern transportation systems, the greatest challenge is to minimize, in general terms, energy consumption, and maximize economic, technological and social goals. The problem of optimizing and finding the best timetable for public transportation (PT) vehicles has been known for years [1–3]. Recent research has provided more efficient algorithms which have achieved better results by modifying known mathematical models and modifying or combining various known algorithms [4–6]. Planning of PT is a highly complex task which is usually analyzed via two different aspects: minimization of the passenger waiting time (PWT) at the station and optimization of the number and/or sizes of vehicles. The train timetabling problem (TTP) has recently been studied, and the main problem in this field is to determine a periodic or non-periodic timetable, which satisfies the capacities of vehicles and limits of operations [7–10]. Some of the greatest challenges in waiting time (WT) minimization models are to find the optimal number of vehicles and to find the optimal route and minimize travel time [11]. The goal of every PT service should be to attract more people to use it by reducing the use of private cars, which is directly related to reducing traffic congestion, decreasing the number of car accidents and reducing pollution. The use of PT services by passengers depends on three elements; namely, travel time (walking, waiting and riding times), fare (ticket and other related services' costs) and convenience (a comfortable walk, waiting under a shelter, having a seat on the vehicle, air-conditioning on the vehicle, etc.). However, for the PT operator to maximize profits, operational costs need to be minimized. The PT operator's requirements can be achieved by designing an efficient network (the more transfers the network has the more efficient it is), adopting a quality timetable, efficiently schedule the vehicles and maximize the vehicle occupancy ratio (VOR). This paper proposes a model of the PT timetabling problem whose solution improves PT operations planning in terms of timetabling and vehicle scheduling; that is, changes to the departure times and assignment of the PT vehicles, so as to reduce total PWT and increase VOR (thereby minimizing operational costs of the vehicles for the PT operator). Due to the complexity of the problem, a multiobjective particle swarm optimization (MOPSO) algorithm is used in order to minimize PWT while maximizing VOR. The paper is structured as follows: Section 2 extensively describes the state-of-the-art in the timetabling problem. Section 3 elaborates on the problem statement. Section 4 presents the proposed method, and Section 5 contains two numerical examples. Section 6 presents the analysis and discussion of the results. Section 7 contains the concluding remarks.

2. State-of-the-Art

In order to increase the productivity and efficiency of transport services on the one hand, and customer satisfaction on the other , four main analytical methods for determining the number of vehicles needed during the relevant period are presented in [1,3,12]. The first two methods in these papers ensure the average maximum daily occupancy of the vehicles during a given period. The other two methods elaborate on the capacities of the vehicles, which will never be exceeded with an additional constraint on the part of the route which is loaded more than the required availability. Numerical calculation of the average PWT at a station with a limited capacity of the intercity transport vehicles is described in [13]. It is concluded that for a more accurate representation of PWT at the station, the reliability of supply services, passenger behavior and characteristics of the transportation system should be considered. The passenger transport problem is always observed from the viewpoint of costs and improvement of business transport companies. On the contrary, passenger satisfaction is rarely taken into account, especially when creating the timetable [14]. Passenger satisfaction is directly related to the total travel time; the shorter the travel time the more satisfied the passenger is and vice-versa. Excluding the travel speed of the vehicle, the total travel time can be reduced by reducing PWT and the number of passenger transfers. Hence, in this paper, we use the term passenger satisfaction to refer primarily to PWT. A multicriteria approach to timetabling problems, with an emphasis on minimizing PWT at the station, is proposed in [15,16]. The authors in these papers analyze two criteria; namely, empty seat penalty (empty seat kilometers or empty seat hours) and approximate PWT at the station. The problem is solved using a multiobjective label-correcting algorithm that results in 43% saving of PWT with an acceptable load of the vehicle. In order to ensure fast and energy efficient urban rail transport, a nonlinear problem of minimizing the total travel time and energy consumption is presented in [17]. The model reduces the cost (number of trains and energy consumption) and improves passenger satisfaction (reduced PWT and the number of transfers, thereby reducing the total travel time). Optimization of energy consumption and PWT can also be found in [18]. The authors in this paper propose a bi-objective timetable optimization model to minimize PWT and energy consumption. The genetic algorithm (GA) is used for generating a solution with reduced total energy consumption and total passenger waiting time in comparison to the real timetable. The timetable synchronization problem (TSP) refers to the problem of waiting time during a transfer, which is usually solved using a branch and bound (B&B) method. In order to speed up the execution time, the optimization based heuristic method (OHM) is developed and compared with B&B in [19]. It is concluded that OHM is much faster and optimizes the problem more efficiently. The scheduling problem is usually based on maximization of the number of synchronized vehicles arriving at the transfer station or minimization of the total waiting time at the station. For solving the latter problem, a genetic algorithm with local search is used in [20]. The model is applied to a small bus network and the cost of the waiting time is reduced by 9.5%. An optimization model for synchronizing a timetable is proposed in [21]; i.e., for minimizing the maximum PWT during the transfer and reducing the worst time of the transfer. Mathematical models with PWT as the objective function at the transfer station include a set of mixed integer programming (MIP) models [22]. The model can be solved using a conventional MIP solver such as CPLEX solver (B&B) if there are fewer than 50 lines and by using genetic algorithms for a greater network. Additionally, a solution of the timetable synchronization problem is proposed in [23]. The authors develop a multicriteria optimization model which takes the vehicle scheduling and passenger demand assignment into account. In order to solve the problem to obtain a set of Pareto-efficient solutions, a novel deficit function (DF)-based sequential search method is proposed. Minimization of the average transfer time in the periodic scheduling of trains (PRTS—periodic railway timetable scheduling problem) is solved using an improved differential evolution (DE) algorithm with dual population [24]. A comparison of the presented model against the B&B method and greedy-based heuristic algorithm for using the PRTS simulation algorithm to solve the problem of schedules shows that the given model provides a better indicator for PRTS problem and numerical indicators of optimization functions. The problem of minimizing PWT at the station and the cost of vehicle occupancy is solved using the genetic algorithm in [25]. The models developed for solving the problem of minimizing PWT at the station are defined with preserving the flow of passengers waiting at the station, as shown in [26,27]. The problem of PWT at the station occurs when there is a delay in the timetable. One possible solution is to include additional time delays according to the probability theory; i.e., the objective function includes an exponential distribution of delay, $Exp()$ with the expected delay, as shown in [28,29]. In [30], the problem of minimizing train delays while maximizing the total satisfaction during a traffic jam (for example, a vehicular crash or similar) is solved using a heuristic algorithm. Timetable optimization is based on the optimal departure time of vehicles from the station for each line of vehicles in order to reduce PWT. A model which minimizes PWT and distinguishes a direct vehicle transfer from walking from one station to another is solved using a heuristic algorithm, the same as in [31]. If the headway is reduced, the average PWT can be reduced as well. Minimization of the sum of headways results in minimization of the average PWT. The average PWT is equal to the ratio of the sum of headway squares and the sum of headways during which a traveler arrives. A numerical method for solving this issue is described in [32].

In order to rationalize departure intervals as much as possible, make the bus journey quicker and minimize PWT, an optimization bus schedule model is proposed in [33]. The authors in this paper consider vehicle overtaking, the limit of the vehicle's capacity and the uncertainty of passenger choice for a bus type (traditional bus or rapid bus). They propose two methods for the solution: a hybrid method of traditional PSO (HPSO) and a combination of GA and PSO named GAPSO. Table 1 summarizes a literature review and shows the details of the studies presented in this section.

Table 1. Literature review.

Paper	Year	Objective	Constraints	Solution Method	Case
[28]	2006	Minimize the waiting time	Running time, departure time, buffer time, spacing time	Linear programming	real
[19]	2008	Minimize the interchange waiting times	Running time, dwell times, trip times, headways, turnaround	B&B, heuristic	real
[22]	2010	Minimize the waiting time	Headway bounds, extra dwell times	Genetic algorithm	real
[15]	2012	Minimize the expected passenger waiting time and Minimize the discrepancy from a desired occupancy level on the vehicles	Headway bounds, vehicle capacity	A multi-objective label-correcting algorithm	real

Table 1. *Cont.*

Paper	Year	Objective	Constraints	Solution Method	Case
[25]	2012	Minimize the waiting time	Headway, in-train passengers, passenger demands, fleet-size, number of boarded passengers	Genetic algorithm	real
[24]	2013	Minimize the waiting time	Traveling time	Evolutionary algorithm	real
[31]	2014	Minimize the waiting time	Headway, departure time	Tabu Search algorithm	test
[17]	2015	Minimize the total travel time of all passengers and the energy consumption of the trains using a weighted sum strategy	Headway, train capacity	Evolutionary algorithms	test
[21]	2015	Minimize the maximal passenger waiting time	Headway, departure time, running time	Genetic algorithm	real
[32]	2015	Minimize the waiting time	Headway	Analytical	test
[33]	2017	Minimize the waiting time	Headway	Improvements of the Genetic Algorithm and Particle Swarm Optimization	test
[23]	2018	Vehicle scheduling problem with the transit assignment	Headway, the number of vehicle departures, fleet size	Heuristic	test
[18]	2019	Minimize the PWT and energy consumption	Headway	Genetic algorithm	real
This paper		Minimize the waiting time and maximize vehicles' occupancy	Headway, passenger demands, number of boarded passengers	Particle swarm optimization	test

3. Problem Statement

In order to further elaborate on the effects of departure times and the assignments of PT vehicles on PWT and VOR, a representative network with stations A, B, C, D and E is considered [23] (see Figure 1). The PT network has two terminals (a and b), two routes ($r_{a \rightarrow b}$, $r_{b \rightarrow a}$) and one transfer stop (node C). An estimated origin-destination (OD) demand matrix is shown in Table 2.

Table 2. Simple example origin destination matrix [23].

From \ To	A	B	C	D	E
A	0	100	50	70	80
B		0	0	0	50
C	50	20	0	80	60
D	60	0	0	0	0
E	80	100	20	60	0

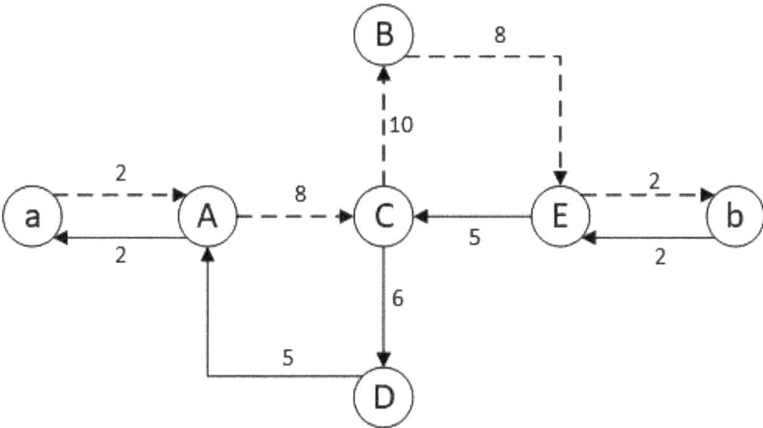

Figure 1. Example of a small network with running times (based on [23]).

The maximum load on route $r_{a \to b}$ is the load on route segment C-B (360 passengers), and the maximum load on route $r_{b \to a}$ is the load on route segment C-D (340 passengers). Assuming that a vehicle with a suitable capacity is used for service ACBE1, all 360 passengers on route segment C-B will be satisfied; i.e., transferred from their respective origins to destinations. For a case in which the vehicle does not have a suitable capacity, some passengers will be left at one of the stations, depending on the vehicle capacity. These passengers may choose to wait for the next service or use another type of transport (e.g., taxi). This decision is based upon the time needed to wait for the next service. In order to perform some example calculations, let us assume that the desired occupancy (d_o) for both routes is set to 70 passengers and that we have a given set of the number of departures per hour (four for route $r_{a \to b}$ and four for route $r_{b \to a}$). After service ACBE1 leaves Station A, 230 passengers are left behind and have to wait for the next service. Assuming that all 50 passengers for Station C get onboard at Station A, after service ACBE1 leaves Station C, 60 passengers are left behind. Considering the number of passengers left in Station A by the previous service (ACBE1), after service ACBE2 leaves Station A, 460 passengers are left behind. Assuming that all 50 passengers for Station C get onboard at Station A, after service ACBE2 leaves Station C, 120 passengers are then left behind (this number includes those passengers left behind from the previous service). The vehicle occupancy ratio for services ACBE1 and ACBE2 is the same: [1 1 1]. Each element of the occupancy ratio array is given by the ratio of the number of passengers in the vehicle to the vehicle capacity and is defined for each OD pair of the service—in this case AC, CB, BE. The amount of PWT for a given station is defined as the product of the number of passengers left behind by the previous service and the time needed for the current service or vehicle to arrive (plus an additional constant term which takes into consideration the number of passengers arriving at the station in the meantime—see Equation (11)). Hence, if the time between consecutive services is 15 min, i.e., service ACBE2 leaves 15 min after ACBE1, the amount of PWT at Station A for service ACBE2 is 3450 *passengers · min*, while that of Station C is 900 *passengers · min*. Assuming that the headway between services is too long, it can be assumed that the remaining passengers will find other means of transportation; this is suitable neither for the service operator nor for the passengers. A cost effective solution for the service operator would be if route ACBE were to be such that the same vehicle and crew can return and perform both services within an hour. Otherwise, if the operator sends more vehicles per time period, additional costs are incurred (more vehicles and crew members are needed). Supposing the operator has passenger cars or coaches that can be connected (assuming this is a tram or rail line), then, in this example, if the capacity is 350 ($5 \cdot 70$), no passengers are left behind at Station A, which is good. However, the vehicle occupancy ratio for both ACBE1 and ACBE2 is now [0.8571 1 0.54286]. The numbers of passengers left at the stations are $[0 \quad 10 \quad 0]$ and $[0 \quad 20 \quad 0]$ respectively for the two consecutive services. For these reasons, it is necessary to make optimal decisions in order to satisfy the operators business interests and passenger needs at the same time.

4. Proposed Model

In order to describe the proposed model, an overview of the notation and variables used is provided in Table 3.

Table 3. Notation.

GENERAL	
OD	Origin-destination
OD_s	Set of OD pairs for a given service s
S_{OD}	Set of services for a given OD pair
$M = \{1, 2, \ldots, m\}$	Set of vehicles
$N = \{1, 2, \ldots, n\}$	Set of stations
L	Set of lines
$v_{s,c}$	Vehicle v with capacity c, for service s
INDEXES	
$s \in S_{OD}$	Service s from the set of services
$i \in N$	Station i from the set of stations
$v \in M$	Vehicle v from the set of vehicles
$l \in L$	Line l from the set of lines
VARIABLES	
$t^d_{s,i}$	Departure time of the vehicle at station i for service s
$t^a_{s,i}$	Arrival time of the vehicle at station i for service s
d_o	Desired occupancy
$dwell_{s,i}$	Dwelling time at station i for service s
$run_{s,i}$	Running time – traveling time between adjacent stations i and $i+1$ for service s
run_s	Sum of running time between all adjacent stations for service s
$H_{i,s}$	Headway – difference between departure times at station i for consecutive services s and $s+1$
$T_{s,i}$	Difference between departure times of vehicle v between adjacent stations i and $i+1$ for a given service s
T_i	Time horizon for i-th station
$P(v_{s,c}, i)$	Total number of passengers in vehicle v in station i
P_{ij}	Number of passengers entering the vehicle in station i heading for destination j (ij OD pair)
$P(v_{s,c}, kj)$	Number of passengers who entered the vehicle in station k with traveling to j
$p^{out}(v_{s,c}, i)$	Number of passengers exiting vehicle v of service s at station i
$p^{in}(v_{s,c}, i)$	Number of passengers entering vehicle v of service s at station i
$p^{curr}(v_{s,c}, i)$	Number of passengers in vehicle v of service s arriving at station i
$p^{stay}(v_{s,c}, i)$	Number of passengers remaining at station i after vehicle v of service s leaves the station
w_i	Average waiting time at station i
k_i	Average number of passengers per time
$Z_{i,s}$	Amount of PWT at station i for service s
η	Total vehicles' occupancy ratio for all services for a given line
$\tau_{v,s}$	Average vehicle occupancy ratio for service s

The OD demand formulation of the number of passengers is given as input data in order to calculate PWT for a given station. According to Figure 1, the OD pairs are AB, AC, AD, AE, CD, CA, CB, CE, BE, EC, EB, ED, EA and DA; while the possible lines are AC, ACB, ACBE, ACD, etc. Each line (l) has multiple services per day; for example, line ACBE has services ACBE1, ACBE2, ACBE3, etc.

Hence, a set of services is defined for each OD pair; i.e., $S_{ACBE} = ACBE1, ACBE2, ACBE3, \ldots$ Each service s depends on the:

- Departure time of the vehicle from station i $(t^d_{s,i})$;
- Vehicle assignment—$v_{s,c}$ = the vehicle v of service s with given capacity c.

The set of OD pairs for each service s is denoted as OD_s, and the set of services for each OD pair is denoted as S_{OD}.

Let us suppose that the parameters are as follows. There are six OD pairs for one line ACBE: AC, AB, AE, CB, CE and BE (i.e., $OD_{\{ACBE\}} = \{AC, AB, AE, CB, CE, BE\}$). The number of passengers entering service s at station i $(P^{in}(v_{s,c}, i))$ is lower than or equal to the sum of the number of passengers for all OD pairs in station i at departure time t in service s:

$$P^{in}(v_{s,c}, i) \leq \sum_{j=i+1}^{n} P_{ij} \tag{1}$$

where P_{ij} is the number of passengers who arrive at station i and are traveling to station j (input data from OD matrix). An inequality sign in Equation (1) signifies that is not necessary that all passengers at station i board vehicle v.

4.1. Assumptions, Variables, Parameters Which Are Time Dependent

First, a set of vehicles with different capacities is considered. The set of vehicles is marked with $M = 1, 2, \ldots, m$ (v-th vehicle is marked with index v). This is important in order to track the number of used vehicles in the network at a certain time. The time between two consecutive services in station i with the same OD pair is defined as:

$$H_{i,s} = t^d_{s,i} - t^d_{(s-1),i}, \quad \forall s \in S_{OD} \tag{2}$$

which is the objective of timetable generation and is a dependent variable of the model.

When the vehicle arrives at the final station of the network for one service, this vehicle is free to be used for another service. The traveling time between adjacent stations is defined as running time $(run_{s,i})$ for each vehicle v:

$$run_{s,i} = t^a_{s,(i+1)} - t^d_{s,i}. \tag{3}$$

For each vehicle v at each station i of the service s, the dwelling time $(dwell_{s,i})$ is defined:

$$dwell_{s,i} \leq t^d_{s,i} - t^a_{s,i}. \tag{4}$$

The time for boarding or alighting is considered as a constant for now. The difference between departure times of vehicle v between consecutive stations (running time + dwelling time) for a given service s (see Figure 2) is defined by:

$$T_{s,i} = t^d_{s,i+1} - t^d_{s,i}, \quad \forall s \in S_{OD}. \tag{5}$$

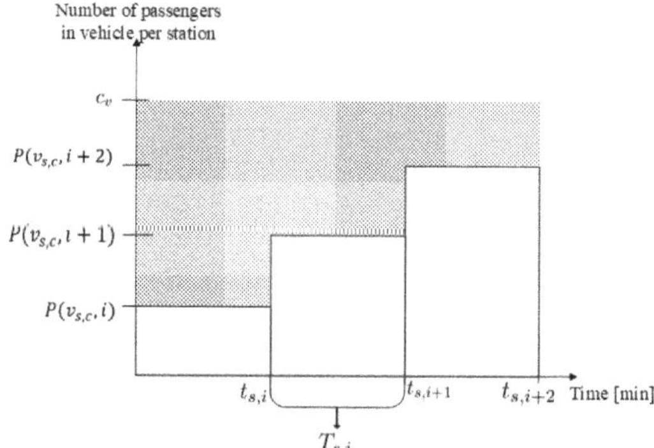

Figure 2. Vehicle occupancy depending on time, between consecutive stations for a given service *s*.

4.2. Assumptions, Variables, Parameters and Sets—Passengers, PWT and VOR

For each OD pair, we have passenger variables that depend on departure times for stations $(t^d_{s,i})$. The number of passengers in *i*-th station for *j*-th OD-pair (e.g., AB) for all stations for one line (at departure time $t^d_{s,i}$), for a given vehicle $v_{s,c}$ is shown in Figure 2. The total number of passengers in vehicle *v* in station *i* is defined by:

$$P(v_{s,c}, i) = P^{curr}(v_{s,c}, i) + P^{in}(v_{s,c}, i) - P^{out}(v_{s,c}, i) \tag{6}$$

where $P^{out}(v_{s,c}, i)$ is the number of passengers leaving service *s* at station *i* and $P^{in}(v_{s,c}, i)$ is the number of passengers entering service *s* at station *i*. The following assumptions are made for the number of passengers entering service *s* at station *i*: (a) The maximum number of passengers that can board the service depends on the currently available capacity of the service; i.e., the current number of free seats. (b) Passengers board the vehicle at station *i* according to the direct proportional distribution determined by the current number of passengers for each OD pair of station *i* (station *i* being the origin).

Those assumptions are necessary in order to simplify the calculation of Equation (6); i.e., P^{out} for a given station will always be the maximum possible for that station.

The number of passengers in vehicle *v* before station *i* (entered in station *k*) and with destination *j* (curr = current) is:

$$P^{curr}(v_{s,c}, i) = \sum_{k=1}^{i-1} \sum_{j=i+1}^{n} P(v_{s,c}, kj) \tag{7}$$

$$P(v_{s,c}, i) = \sum_{k=1}^{i-1} \sum_{j=i+1}^{n} P(v_{s,c}, kj) + P^{in}(v_{s,c}, i) - P^{out}(v_{s,c}, i); \tag{8}$$

i.e., passengers entering before station *i* and exiting after station *i* (station *i* excluded).

The average number of passengers at a station between consecutive services.

$$k_i = \frac{\sum_{s \in S_{OD}} P(v_{s,c}, i)}{T_i} \tag{9}$$

where T_i is the time horizon for station *i*. The average waiting time at station *i* is given by [23]

$$w_i = \frac{E[H_i]}{2} \left[1 + \frac{Var[H_i]}{E^2[H_i]} \right] \tag{10}$$

where $E[H_i]$ and $Var[H_i]$ are, respectively, the mean and variance of headway time between vehicles at station i. The amount of PWT in station i for every service (s) is given by

$$Z_{i,s} = P^{stay}(v_{s,c}, i) \cdot H_{i,s} + k_i H_{i,s} w_i, \tag{11}$$

where $P^{stay}(v_{s,c}, i)$ is the number of passengers left at station i because they could not enter service $s - 1$. The amount of PWT (Equation (11)) at station i is the sum of the amount of waiting time at station i until the next service arrives at the station ($P^{stay}(v_{s,c}, i) \cdot H_{i,s}$)) and the average amount of waiting time at station i given by $k_i H_{i,s} w_i$ (for all OD pairs). It should be noted that this constant term is a measure of passengers arriving at the station between services with the average waiting time (w_i) for these passengers. This is based on the assumptions that (a) passengers randomly arrive at the station and (b) the arrivals of vehicles are uneven but occur in predetermined time intervals.

In the multiobjective optimization algorithm implemented in this paper, average normalized values of PWT and VOR are used. The average normalized amount of PWT is defined by:

$$Z_{av,norm} = \frac{\sum_{i,s} Z_{i,s}^{norm}}{n \cdot m} \tag{12}$$

where the normalized amount of PWT is defined by $Z_{i,s}^{norm} = \dfrac{Z_{i,s} - Z_{min}}{Z_{max} - Z_{min}}$ for a given station i and service s, with Z_{max} being the amount of PWT for the maximal difference between departure times of consecutive services for a given station using vehicles of minimal capacity, and Z_{min} is the amount of PWT for the minimal differences between departure times of consecutive services for a given station using vehicles of maximal capacity. n is the number of stations and m is the total number of vehicles used (i.e., number of services). Equation (12) represents the objective function for minimizing the amount of PWT.

The vehicle occupancy ratio for service s is defined by:

$$\tau_{v,s} = \frac{\sum_{i=1}^{n} P^{curr}(v_{s,c}, i) \cdot run_{s,i}}{n \cdot c_v \cdot run_s}. \tag{13}$$

where $run_s = \sum_{i=1}^{n} r_{s,i}$ is the sum of running time between all adjacent stations for service s.

The total vehicle occupancy ratio (VOR) for all services across all stations is given by:

$$\eta = \sum_{s \in S} \tau_{v,s}. \tag{14}$$

The average normalized value of η, $\eta_{av,norm}$ is defined by:

$$\eta_{av,norm} = \frac{\eta}{n \cdot m} \tag{15}$$

Equation (15) represents the objective function for maximizing VOR. The profit margin of a transport company increases as the vehicles' occupancy ratio increases.

4.3. Objective Function

The optimization problem aims at minimizing the amount of PWT while maximizing VOR. The variable PWT is used in the optimization model in order to satisfy users of PT. On the another hand, VOR (Equation (15)), used in the optimization model, is based on vehicles' occupancy for all services in time horizon. The variables included in the model are discrete so the model can be solved using combinatorial optimization methods.

The objective function of the model is defined as:

$$\min_{x} F(x) = (f_1(x), f_2(x)) \tag{16}$$

where x is decision variable in the solution space of dimension $2|S_{OD}|$. $|S_{OD}|$ represents the number of services in S_{OD}. The first $|S_{OD}|$ elements in x represent the departure times of the services from the first station on the line, while the second $|S_{OD}|$ elements in x represent the corresponding vehicle capacities of the services. $f_1(x)$ corresponds to the inverse average normalized value of η, i.e., $(1 - \eta_{av,norm})$ given by (15), and $f_2(x)$ corresponds to the average normalized amount of PWT, $Z_{av,norm}$, given by (12). The objective function (16) minimizes two goals and it is solved using multiobjective optimization. It minimizes $f_1(x)$ and $f_2(x)$, which correspond to maximizing VOR and minimizing the amount of PWT. The objective function of the model has the following constraints:

$$P(v_{s,c}, i) \leq c_v, \qquad \forall v \in \mathcal{M}, \forall i \in N \tag{17a}$$

$$t^d_{s-1,i} < t^d_{s,i}, \qquad \forall s \in S_{OD} \tag{17b}$$

$$T_{s,i} = t^d_{s+1,i} - t^d_{s,i}, \qquad \forall i \in N, \forall s \in S_{OD} \tag{17c}$$

$$H_{i,s} = t^d_{s,i} - t^d_{s-1,i}, \qquad \forall i \in N, \forall s \in S_{OD} \tag{17d}$$

$$k_i \leq \frac{c_v}{\sum_{s \in S} H_{i,s}}, \qquad \forall i \in N, \forall s \in S_{OD} \tag{17e}$$

In order to simplify the model and future calculations, it is assumed that the number of passengers in j-th vehicle cannot be more than c_v, $\forall v \in M$, as shown in constraint (17a). Constraint (17b) implies that the departure time of vehicle v in service s has to be after the previous departure time of the same OD pair served. Equation (17c) expresses the difference between departure times of vehicle v between consecutive stations in one service s. Equation (17d) expresses the difference of the departure time between two consecutive services in station i with the same OD pair. Constraint (17e) expresses that the average arrival rate for the OD pair at station i is less than or equal to the average maximum capacity rate.

5. Results

Due to the complexity of the objective function (16) and the large search space, we propose to use a heuristic optimization algorithm to determine a suitable solution. Such algorithms have shown to be suitable in such optimization problems, especially those involving large search space [34–37]. The efficiency and suitability of the various heuristic optimization algorithms in solving a whole range of complex problems have been receiving a lot of attention from academia for many years. Most of the available heuristic optimization algorithms mainly fall into two categories—swarm intelligence algorithms and evolutionary algorithms. The main representatives of these categories are particle swarm optimization (PSO) and the genetic algorithm (GA), respectively. In this paper, the multiobjective particle swarm optimization (MOPSO) algorithm, proposed in [38], was used. The algorithm in [38] extends the standard PSO algorithm to solve multiobjective optimization problems by utilizing an additional repository of particles to help the main set of particles in their search. The exploratory capabilities of the algorithm are also enhanced by a specific mutation operator that is incorporated. A preliminary comparison by the authors of this paper, of an implementation of the aforementioned algorithm, was made with a brute force search for the Pareto-optimal set solution of a model of a much simpler PT timetabling problem, and the same results were obtained. As a result, the implemented version was deemed suitable enough. A comparison of the MOPSO algorithm implemented in this paper with other PSO algorithms for multiobjective optimization, such as that proposed in [39], will be considered in future research, especially with respect to the accuracy and convergence rate in solving the proposed objective function. Additionally, a comparison of other heuristic optimization algorithms regarding their suitability in solving the proposed objective function will also be considered.

The proposed optimal solution from the Pareto-optimal set, obtained using the implemented MOPSO algorithm, was determined based on the multicriteria decision making method. The technique for order of preference by similarity to ideal solution (TOPSIS), as proposed in [40], was used for solving traffic problems. In this section, two experimental problems based on input data and assumptions given in [23] are provided and analyzed. In both experiments, the results obtained are compared using the proposed method and assumptions and the results from [23].

5.1. Experiment 1

The first experiment consists of a simple passenger transportation line (Figure 1) involving three sets of a number of departures (q = 4, q = 5 and q = 6) for route $r_{a \rightarrow b}$ with four stations and two sets of a number departures (q = 4 and q = 5) for route $r_{b \rightarrow a}$. The input data during the given time period (7 a.m.–8 a.m.) consist of the average travel times, and the constant time needed for alighting and boarding is set to 0.5 minutes. An estimated OD demand matrix is presented in Table 2. The running times between each station for each service are presented in Figure 1. Other details of the experimental setup, which define the search space, are as follows:

- Desired occupancy of each vehicle: 70;
- Possible departure time (in minutes) at the first terminal is defined by the time intervals for each service respectively:

  - $[0, 15], [16, 30], [31, 45], [46, 60]$
  - $[0, 12], [13, 24], [25, 36], [37, 48], [49, 60]$
  - $[0, 10], [11, 20], [21, 30], [31, 40], [41, 50], [51, 60]$.

Table 4 presents the combined results of the proposed method using MOPSO algorithm and the results from the literature. The results are displayed using the following parameters: the number of passengers left at a station, waiting time and amount of PWT. Each of these are displayed as a matrix with the number of columns representing the number of stops and the number of rows representing the number of services. Based on the input data and results in [23], the departure times from the terminals a and b, needed for comparison, are presented in Table 4, in three sets for the number of departures for route $r_{a \rightarrow b}$ and two sets for the number of departures for route $r_{b \rightarrow a}$.

With respect to the MOPSO algorithm, all experiments were performed using 200 particles (population size) and 500 generations. The detailed results are presented in Table 4.

In order to compare the results, the waiting time and amount of PWT for each solution are presented. As shown, the waiting time for the next service is shorter with the MOPSO algorithm although it is an uneven timetable. Hence, the amount of PWT based on (11) is better when using MOPSO. Although VOR is maximized (it is not displayed in Table 4, but it can be deduced by looking at the number of passengers left at the station and taking into consideration the fixed desired occupancy), the amount of PWT is not acceptable because of the high number of passengers left at the station and the long waiting time for consecutive service. From the obtained results, it can be concluded that passengers will choose another type of PT. In order to provide an example with a more user-oriented PT service, Experiment 1 is expanded as presented in Experiment 2.

5.2. Experiment 2

The second experiment is a bit more complex. All the parameters and conditions are the same as in Experiment 1 with two changes: the PT line is a train or tram line and the vehicles are passenger cars or coaches each having a capacity of 70. It is also assumed that a maximum of seven passenger cars can be connected. The aim of this experiment is to reduce the amount of PWT and maximize VOR at the same time. MOPSO was used to determine the optimal parameters. With respect to the MOPSO algorithm, all experiments were performed using 200 particles (population size) and 500 generations. The proposed optimal solution was found by the TOPSIS method. The obtained Pareto-optimal set and the proposed optimal solution is displayed in Figure 3, The detailed results are presented in Table 5.

Comparing the results in Experiment 2 with those in Experiment 1, the waiting time for an uneven headway is acceptable when more than one passenger car or coach is used per service. The number of passengers left at the station is reduced, and it is assumed that passengers will wait for the next service.

Table 4. Comparison of the results obtained in Experiment 1 using the proposed method and those obtained in the literature [23].

$d_o = 70$			Dep. Time	$Z_{av,norm}$	Number of Passengers Left at the Station			Waiting Time			Amount of PWT (Equation (11))		
$r_{a \to b}$	$q = 4$	Proposed method	[7:15,	0.142	230	152	8	13	12	12	4386.45	2813.87	336.74
			7:28,		460	304	16	10	9	9	5674.19	3684.52	339.03
			7:38,		690	456	24	8	7	7	6379.35	4163.61	335.23
			7:46]		920	608	32	90	89	89	92,467.74	60,520.65	4491.29
		Results according to [23]	[7:15,	0.161	230	152	8	15	14	14	5700	3630	495
			7:30,		460	304	16	15	14	14	9150	5910	615
			7:45,		690	456	24	15	14	14	12,600	8190	735
			8:00]		920	608	32	90	89	89	96,300	62,820	5130
	$q = 5$	Proposed method	[7:12,	0.126	230	152	8	12	11	11	4239.73	2711.84	342.62
			7:24,		460	304	16	11	10	10	6416.42	4157.85	402.07
			7:35,		690	456	24	8	7	7	6506.49	4239.89	356.41
			7:43,		920	608	32	6	5	5	6259.86	4091.92	315.31
			7:49]		1150	760	40	72	71	71	91,678.38	60,047.03	4359.73
		Results according to [23]	[7:12,	0.141	230	152	8	12	11	11	4560	2904	396
			7:24,		460	304	16	12	11	11	7320	4728	492
			7:36,		690	456	24	12	11	11	10,080	6552	588
			7:48,		920	608	32	12	11	11	12,840	8376	684
			8:00]		1150	760	40	72	71	71	93,600	61,200	4680
	$q = 6$	Proposed method	[7:10,	0.114	230	152	8	10	9	9	3635.37	2321.22	302.56
			7:20,		460	304	16	10	9	9	5935.37	3841.22	382.56
			7:30,		690	456	24	10	9	9	8235.37	5361.22	462.56
			7:40,		920	608	32	7	6	6	7374.76	4816.85	379.79
			7:47,		1150	760	40	4	3	3	5134.15	3360.49	249.02
			7:51]		1380	912	48	60	59	59	90,812.20	59,527.32	4215.37
		Results according to [23]	[7:10,	0.127	230	152	8	10	9	9	3800	2420	330
			7:20,		460	304	16	10	9	9	6100	3940	410
			7:30,		690	456	24	10	9	9	8400	5460	490
			7:40,		920	608	32	10	9	9	10,700	6980	570
			7:50,		1150	760	40	10	9	9	13,000	8500	650
			8:00]		1380	912	48	60	59	59	91,800	60,120	4380
$r_{b \to a}$	$q - 4$	Proposed method	[7:15,	0.169	190	178	21	13	12	12	3680.26	3291.52	552.29
			7:28,		380	355	41	10	9	9	4730.97	4301.94	624.84
			7:38,		570	533	62	8	7	7	5304.77	4865.55	667.87
			7:46]		760	711	83	90	89	89	76,778.71	70,757.42	9403.55
		Results according to [23]	[7:15,	0.192	190	178	21	15	14	14	4800	4245	765
			7:30,		380	355	41	15	14	14	7650	6900	1065
			7:45,		570	533	62	15	14	14	10,500	9570	1380
			8:00]		760	711	83	90	89	89	80,100	73,440	10,170
	$q = 5$	Proposed method	[7:12,	0.150	190	178	21	12	11	11	3562.43	3171.81	547.95
			7:24,		380	355	41	11	10	10	5355.56	4854.49	722.28
			7:35,		570	533	62	8	7	7	5414.95	4954.54	693.30
			7:43,		760	711	83	6	5	5	5201.22	4783.91	645.97
			7:49]		950	888	103	72	71	71	76,094.59	70,150.86	9191.68
		Results according to [23]	[7:12,	0.168	190	178	21	12	11	11	3840	3396	612
			7:24,		380	355	41	12	11	11	6120	5520	852
			7:36,		570	533	62	12	11	11	8400	7656	1104
			7:48,		760	711	83	12	11	11	10,680	9792	1356
			8:00]		950	888	103	72	71	71	77,760	71,496	9576

Table 5. Detailed results obtained in Experiment 2 using the proposed method.

				$1-\eta_{av,norm}$	Vehicle Occupancy			Number of Free Seats			Number of Passengers Left at the Station			$Z_{av,norm}$	Waiting Time			Amount of PWT (Equation (11))		
$r_{a\to b}$	q = 4	Dep. time	[7:15, 7:25, 7:37, 7:46]	0.080	1	1	1	0	0	0	230	152	8	0.498	10	9	9	3348.39	2149.03	254.73
					1	1	0.52	0	0	236	40	136	0		12	11	0	1738.06	2386.84	209.68
		d_o	[70, 490, 70, 490]		1	1	1	0	0	0	270	288	8		9	8	8	3373.55	3158.13	229.26
					1	1	0.52	0	0	236	80	272	0		90	89	0	16635.48	30141.29	1572.58
	q = 5	Dep. time	[7:12, 7:21, 7:32, 7:40, 7:49]	0.064	1	1	1	0	0	0	230	152	8	0.057	9	8	8	3125.07	2001.04	247.84
					1	1	0.52	0	0	236	40	136	0		11	10	0	1729.53	2269.72	214.92
		d_o	[70, 490, 70, 490, 350]		1	1	1	0	0	0	270	288	8		8	7	7	3097.84	2866.7	220.31
					1	1	0.52	0	0	236	80	272	0		9	8	0	1775.07	3081.04	175.84
					1	1	1	0	0	0	310	424	8		72	71	71	30760.54	35592.32	1982.76
	q = 6	Dep. time	[7:10, 7:19, 7:29, 7:36, 7:45, 7:51]	0.080	1	1	1	0	0	0	230	152	8	0.040	9	8	8	3245.49	2073.29	267.91
					1	1	0.52	0	0	236	40	136	0		10	9	0	1669.51	2121.71	211.59
					1	1	1	0	0	0	270	288	8		7	6	6	2778.66	2549.20	204.11
		d_o	[70, 490, 70, 490, 70, 490]		1	1	0.52	0	0	236	80	272	0		9	8	0	1862.56	3133.54	190.43
					1	1	1	0	0	0	310	424	8		6	5	5	2621.71	3001.02	174.95
					1	1	0.52	0	0	236	120	408	0		60	59	0	14817.07	29050.24	1269.51
$r_{b\to a}$	q = 4	Dep. time	[7:15, 7:25, 7:37, 7:46]	0.066	1	1	1	0	0	0	190	178	21	0.060	10	9	9	2808.60	2513.87	419.68
					1	1	0.63	0	0	154	30	194	0		12	11	0	1450.32	3208.65	251.61
		d_o	[70, 420, 70, 490]		1	1	1	0	0	0	220	371	20		9	8	8	2797.74	3999.48	368.71
					0.98	1	0.6	10	0	196	0	350	0		0	89	0	8177.42	38104.84	1887.10
	q = 5	Dep. time	[7:12, 7:22, 7:31, 7:41, 7:49]	0.055	1	1	1	0	0	0	190	178	21	0.061	10	9	9	2910.14	2595.88	443.11
					1	1	1	0	0	0	310	323	3		9	8	8	3699.12	3641.29	236.80
		d_o	[70, 140, 490, 70, 490]		1	1	0.57	0	0	212	80	306	0		10	9	0	1810.14	3875.88	233.11
					1	1	1	0	0	0	270	484	21		8	7	7	2968.11	4524.70	354.49
					1	1	0.60	0	0	194	40	469	0		72	71	0	10152.97	39642.32	1678.38

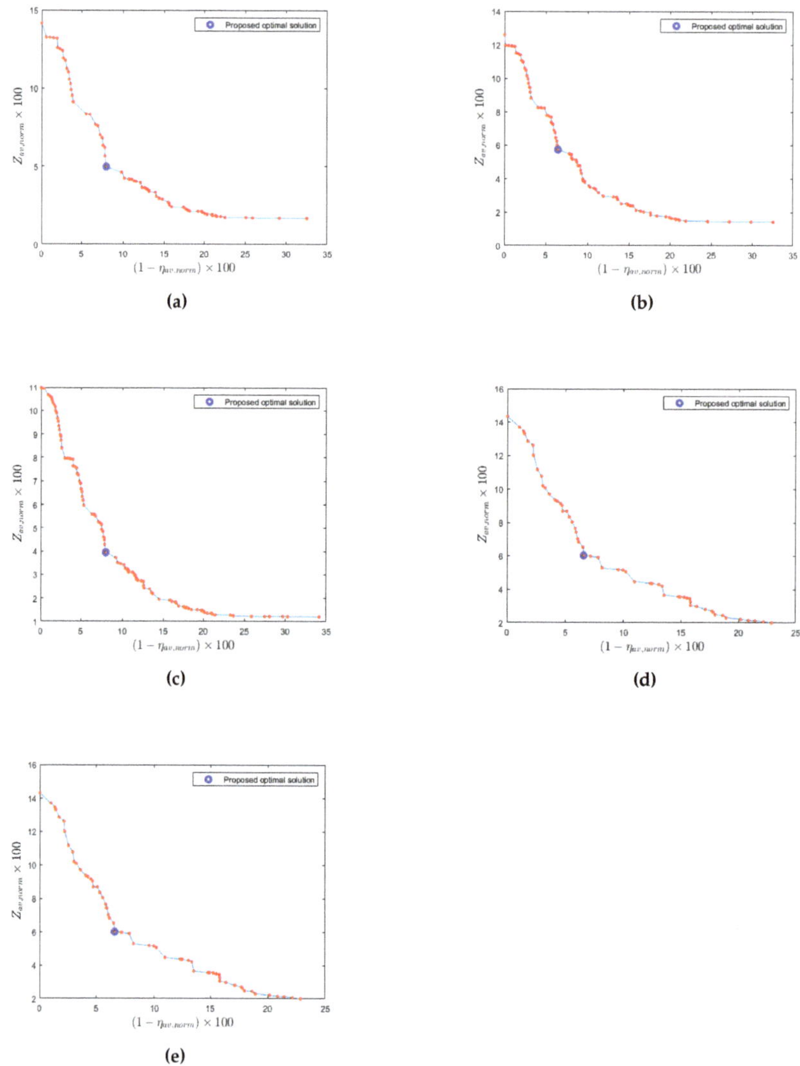

Figure 3. Obtained Pareto-optimal set and the proposed optimal solutions. (**a**–**c**) The solutions for $r_{a \rightarrow b}$ and $q = 4$, $q = 5$, $q = 6$, respectively. (**d**,**e**) Solutions for $r_{b \rightarrow a}$ and $q = 4$, $q = 5$, respectively.

6. Analysis and Discussion

In Experiment 1 (Table 4), the results indicate that by optimizing the timetable using the proposed method, an uneven timetable is obtained compared to the solution obtained by [23]. However, the amount of PWT is lower . In Experiment 2, it is assumed that the PT line is a train or tram line and that a maximum of seven coaches, each of capacity 70, can be used. The results obtained using the proposed method show a drastic improvement of the amount of PWT (i.e., a decrease in the value of $Z_{av,norm}$).

Table 6 shows the results that are obtained when the desired vehicle occupancy obtained in Experiment 2 is combined with the departure time obtained in Experiment 1. As was expected, both $(1 - \eta_{av,norm})$ and $Z_{av,norm}$ values are decreased (implying an increase in VOR and decrease in the amount of PWT) compared to the solutions obtained by [23] in Experiment 1.

Table 6. Results obtained when vehicle occupancies of Experiment 2 are combined with corresponding departure times of Experiment 1.

		$r_{a\to b}$					$r_{b\to a}$			
		Dep.Time	d_o	$1-\eta_{av,norm}$	$Z_{av,norm}$		**Dep.Time**	d_o	$1-\eta_{av,norm}$	$Z_{av,norm}$
$q=4$	Proposed method	[7:15, 7:28, 7:38, 7:46]	70 490 70 490	0.080	0.050	Proposed method	[7:15, 7:28, 7:38, 7:46]	70 420 70 490	0.066	0.061
	Results according to [23]	[7:15, 7:30, 7:45, 8:00]	70 490 70 490	0.080	0.064	Results according to [23]	[7:15, 7:30, 7:45, 8:00]	70 420 70 490	0.066	0.078
$q=5$	Proposed method	[7:12, 7:24, 7:35, 7:43, 7:49]	70 490 70 490 70	0.095	0.040	Proposed method	[7:12, 7:24, 7:35 , 7:43 , 7:49]	70 140 490 70 490	0.055	0.061
	Results according to [23]	[7:12, 7:24, 7:36, 7:48, 8:00]	70 490 70 490 70	0.095	0.048	Results according to [23]	[7:12, 7:24, 7:36, 7:48, 8:00]	70 140 490 70 490	0.055	0.073
$q=6$	Proposed method	[7:10, 7:20, 7:30, 7:40, 7:47, 7:51]	70 490 70 490 70 490	0.080	0.040					
	Results according to [23]	[7:10, 7:20, 7:30, 7:40, 7:50, 8:00]	70 490 70 490 70 490	0.080	0.046					

For example, for route $r_{a\to b}$, when four departures and the possibility of using more passenger cars or coaches are considered, the value indicating the amount of PWT (0.050) is lower than that obtained by [23] (0.064), while the value representing the vehicles' occupancy ratio (0.080) is the same for both algorithms (Table 6). If the desired vehicle occupancy is fixed, i.e., $d_o = 70$, the amount of PWT when using the proposed method is 0.142, while it is 0.161 when using the input given in [23] (Table 4). From these results, it can be concluded that it is more appropriate to use more passenger cars or coaches and an uneven timetable for the input data during the particular time period.

A qualitative analysis of the proposed timetable, with respect to the headway, is performed using the following indicators—the average headway (AH) and the expected waiting time (EWT), as recommended in [12]. AH and EWT of randomly arriving passengers, for all sets of departures and routes, are presented in Table 7. For example, for route $r_{a\to b}$ and six departures during the given time period, the expected PWTs, when using the proposed method, are 4.45 and 4.23 for Experiments 1 and 2 respectively, while the expected PWT is 5 when using the procedure in [23]. The presented qualitative analysis confirms that the timetable obtained using the proposed model and optimized using MOPSO is more appropriate compared to the timetable obtained in literature [23].

Table 7. Average headway and expected waiting time.

			Average Headway	Expected Waiting Time
$r_{a \to b}$	q = 4	Proposed method – exp 1 / exp 2	10.33	5.37/5.24
		[23] – exp 1	15	7.5
	q = 5	Proposed method – exp. 1 / exp. 2	9.25	4.93/4.68
		[23] – exp 1	12	6
	q = 6	Proposed method – exp 1 / exp 2	8.2	4.45/4.23
		[23] – exp 1	10	5
$r_{b \to a}$	q = 4	Proposed method – exp 1 / exp 2	10.33	5.37/5.24
		[23] – exp 1	15	7.5
	q = 5	Proposed method – exp 1 / exp 2	9.25	4.93/4.66
		[23] – exp 1	12	6

7. Conclusions

This paper presents a new model for determination of PT timetable. The model is formulated using a multiobjective optimization model to optimize VOR and PWT. Thus, this model takes into account the satisfaction of transport companies and passengers. The MOPSO algorithm is used in optimizing the model. The best solution in the Pareto-optimal set was found by the TOPSIS method. Practical implementation of the proposed model is presented using two numerical examples. PWT and VOR indices are used to present the performances of the proposed model in comparison to the similar results in the existing literature. Experiment 1 uses a simple passenger transportation line involving three sets of a number of departures (q = 4, q = 5 and q = 6) for route $r_{a \to b}$ with four stations and two sets of number departures (q = 4 and q = 5) for route $r_{b \to a}$. Experiment 2 has the same parameters and conditions as Experiment 1, and the additional assumptions that the PT line is a train or tram line and the vehicles are passenger cars or coaches. In both experiments, the proposed model using MOPSO algorithm shows better performances; i.e., shorter PWT and greater VOR. The case study based on the operation data from the existing literature shows that the proposed approach can reduce the average PWT by 10.54% for all sets with differing numbers of departures for two routes during the given time period. Based on the presented results, it can be concluded that the presented model, which uses the MOPSO algorithm to determine the optimal timetable, has an advantage in comparison to the existing models in scientific literature, which makes it suitable for scientists and practitioners in the field of PT. Further research will initially involve verification of the model using data from a real network scenario. We plan to modify the proposed model by including an extra constraint that ensures that the operation duration remains unchanged, especially if the number of vehicles or services is kept constant. We also plan to take into consideration boarding and alighting times, transfer stations and scheduling several lines.

Author Contributions: Conceptualization, I.H.T., E.K.N. and A.C.; methodology, I.H.T., E.K.N. and A.C.; software, I.H.T. and E.K.N.; formal analysis, I.H.T., E.K.N. and A.C.; investigation, I.H.T.; data curation, I.H.T. and E.K.N.; visualization, Ivana Hartamnn Tolić; writing—original draft preparation, I.H.T.; writing—review and editing, E.K.N. and A.C.; supervision, E.K.N. All authors have read and agreed to the published version of the manuscript.

Funding: This research received no external funding.

Conflicts of Interest: The authors declare no conflict of interest.

Abbreviations

The following abbreviations are used in this manuscript:

PT	Public transportation
PSO	Particle swarm optimization
PWT	Passenger waiting time
VOR	Vehicles' occupancy ratio
TTP	Train timetabling problem
WT	Waiting time
MOPSO	Multiobjective particle swarm optimization
GA	Genetic algorithm
TSP	Timetable synchronization problem
B&B	Branch and bound
OHM	Optimization based heuristic method
MIP	Mixed integer programming
PRTS	Periodic railway timetable scheduling problem
DE	Differential evolution
HPSO	Hybrid method of traditional PSO
OD	Origin-destination
TOPSIS	Technique for order of preference by similarity to ideal solution
AH	Average headway
EWT	Expected waiting time

References

1. Ceder, A. Bus frequency determination using passenger count data. *Transp. Res. Part A General* **1984**, *18*, 439–453. [CrossRef]
2. Ceder, A.; Wilson, N.H. Bus network design. *Transp. Res. Part B Methodol.* **1986**, *20*, 331–344. [CrossRef]
3. Ceder, A. Methods for creating bus timetables. *Transp. Res. Part A General* **1987**, *21*, 59–83. [CrossRef]
4. Isaai, M.T.; Singh, M.G. Hybrid applications of constraint satisfaction and meta-heuristics to railway timetabling: A comparative study. *IEEE Trans. Syst. Man Cybern. Part C (Appl. Rev.)* **2001**, *31*, 87–95. [CrossRef]
5. ShangGuan, W.; Yan, X.H.; Cai, B.G.; Wang, J. Multiobjective optimization for train speed trajectory in CTCS high-speed railway with hybrid evolutionary algorithm. *IEEE Trans. Intell. Transp. Syst.* **2015**, *16*, 2215–2225. [CrossRef]
6. Park, Y.B. A hybrid genetic algorithm for the vehicle scheduling problem with due times and time deadlines. *Int. J. Prod. Econ.* **2001**, *73*, 175–188. [CrossRef]
7. Caprara, A.; Fischetti, M.; Toth, P. Modeling and solving the train timetabling problem. *Oper. Res.* **2002**, *50*, 851–861. [CrossRef]
8. Siebert, M.; Goerigk, M. An experimental comparison of periodic timetabling models. *Comput. Oper. Res.* **2013**, *40*, 2251–2259. [CrossRef]
9. Shafia, M.A.; Sadjadi, S.J.; Jamili, A.; Tavakkoli-Moghaddam, R.; Pourseyed-Aghaee, M. The periodicity and robustness in a single-track train scheduling problem. *Appl. Soft Comput.* **2012**, *12*, 440–452. [CrossRef]
10. Bussieck, M.R.; Kreuzer, P.; Zimmermann, U.T. Optimal lines for railway systems. *Eur. J. Oper. Res.* **1997**, *96*, 54–63. [CrossRef]
11. Yalçınkaya, Ö.; Bayhan, G.M. Modelling and optimization of average travel time for a metro line by simulation and response surface methodology. *Eur. J. Oper. Res.* **2009**, *196*, 225–233. [CrossRef]
12. Ceder, A. *Public Transit Planning and Operation: Modeling, Practice And Behavior*; CRC Press, Boca Raton, FL, USA, 2016.
13. Chang, S.J.; Hsu, S.C. Modeling of passenger waiting time in intermodal station with constrained capacity on intercity transit. *Int. J. Urban Sci.* **2004**, *8*, 51–60. [CrossRef]
14. Baita, F.; Pesenti, R.; Ukovich, W.; Favaretto, D. A comparison of different solution approaches to the vehicle scheduling problem in a practical case. *Comput. Oper. Res.* **2000**, *27*, 1249–1269. [CrossRef]

15. Hassold, S.; Ceder, A. Multiobjective approach to creating bus timetables with multiple vehicle types. *Transp. Res. Rec.* **2012**, *2276*, 56–62. [CrossRef]

16. Hassold, S.; Ceder, A. Public transport vehicle scheduling featuring multiple vehicle types. *Transp. Res. Part B Methodol.* **2014**, *67*, 129–143. [CrossRef]

17. Wang, Y.; Tang, T.; Ning, B.; van den Boom, T.J.; De Schutter, B. Passenger-demands-oriented train scheduling for an urban rail transit network. *Transp. Res. Part C Emerg. Technol.* **2015**, *60*, 1–23. [CrossRef]

18. Sun, H.; Wu, J.; Ma, H.; Yang, X.; Gao, Z. A Bi-Objective Timetable Optimization Model for Urban Rail Transit Based on the Time-Dependent Passenger Volume. *IEEE Trans. Intell. Transp. Syst.* **2019**, *20*, 604–615. [CrossRef]

19. Wong, R.C.W.; Yuen, T.W.Y.; Fung, K.W.; Leung, J.M.Y. Optimizing Timetable Synchronization for Rail Mass Transit. *Transp. Sci.* **2008**, *42*, 57–69. [CrossRef]

20. Wu, Y.; Tang, J.; Yu, Y.; Pan, Z. A stochastic optimization model for transit network timetable design to mitigate the randomness of traveling time by adding slack time. *Transp. Res. Part C Emerg. Technol.* **2015**, *52*, 15–31. [CrossRef]

21. Wu, J.; Liu, M.; Sun, H.; Li, T.; Gao, Z.; Wang, D.Z. Equity-based timetable synchronization optimization in urban subway network. *Transp. Res. Part C Emerg. Technol.* **2015**, *51*, 1–18. [CrossRef]

22. Shafahi, Y.; Khani, A. A practical model for transfer optimization in a transit network: Model formulations and solutions. *Transp. Res. Part A Policy Pract.* **2010**, *44*, 377–389. [CrossRef]

23. Liu, T.; Ceder, A.A. Integrated public transport timetable synchronization and vehicle scheduling with demand assignment: A bi-objective bi-level model using deficit function approach. *Transp. Res. Part B Methodol.* **2018**, *117*, 935–955. [CrossRef]

24. Zhong, J.H.; Shen, M.; Zhang, J.; Chung, H.S.H.; Shi, Y.H.; Li, Y. A differential evolution algorithm with dual populations for solving periodic railway timetable scheduling problem. *IEEE Trans. Evolut. Comput.* **2012**, *17*, 512–527. [CrossRef]

25. Niu, H.; Zhang, M. An optimization to schedule train operations with phase-regular framework for intercity rail lines. *Discret. Dyn. Nat. Soc.* **2012**, *2012*. [CrossRef]

26. Barrena, E.; Canca, D.; Coelho, L.C.; Laporte, G. Exact formulations and algorithm for the train timetabling problem with dynamic demand. *Comput. Oper. Res.* **2014**, *44*, 66–74. [CrossRef]

27. Barrena, E.; Canca, D.; Coelho, L.C.; Laporte, G. Single-line rail rapid transit timetabling under dynamic passenger demand. *Transp. Res. Part B Methodol.* **2014**, *70*, 134–150. [CrossRef]

28. Vansteenwegen, P.; Van Oudheusden, D. Developing railway timetables which guarantee a better service. *Eur. J. Oper. Res.* **2006**, *173*, 337–350. [CrossRef]

29. Vansteenwegen, P.; Van Oudheusden, D. Decreasing the passenger waiting time for an intercity rail network. *Transp. Res. Part B Methodol.* **2007**, *41*, 478–492. [CrossRef]

30. Corman, F.; D'Ariano, A.; Pacciarelli, D.; Pranzo, M. Bi-objective conflict detection and resolution in railway traffic management. *Transp. Res. Part C Emerg. Technol.* **2012**, *20*, 79–94. [CrossRef]

31. Parbo, J.; Nielsen, O.A.; Prato, C.G. User perspectives in public transport timetable optimisation. *Transp. Res. Part C Emerg. Technol.* **2014**, *48*, 269–284. [CrossRef]

32. Berrebi, S.J.; Watkins, K.E.; Laval, J.A. A real-time bus dispatching policy to minimize passenger wait on a high frequency route. *Transp. Res. Part B Methodol.* **2015**, *81*, 377–389. [CrossRef]

33. Li, J.; Hu, J.; Zhang, Y. Optimal combinations and variable departure intervals for micro bus system. *Tsinghua Sci. Technol.* **2017**, *22*, 282–292. [CrossRef]

34. Hussain, B.; Khan, A.; Javaid, N.; Hasan, Q.U.; A Malik, S.; Ahmad, O.; Dar, A.H.; Kazmi, A. A Weighted-Sum PSO Algorithm for HEMS: A New Approach for the Design and Diversified Performance Analysis. *Electronics* **2019**, *8*, 180. [CrossRef]

35. Zuo, X.; Li, B.; Huang, X.; Zhou, M.; Cheng, C.; Zhao, X.; Liu, Z. Optimizing hospital emergency department layout via multiobjective tabu search. *IEEE Trans. Autom. Sci. Eng.* **2019**, *16*, 1137–1147. [CrossRef]

36. Wu, Q.; Zhou, M.; Zhu, Q.; Xia, Y.; Wen, J. MOELS: Multiobjective Evolutionary List Scheduling for Cloud Workflows. *IEEE Trans. Autom. Sci. Eng.* **2019**, *17*, 166–176. [CrossRef]

37. Gao, K.; Cao, Z.; Zhang, L.; Chen, Z.; Han, Y.; Pan, Q. A review on swarm intelligence and evolutionary algorithms for solving flexible job shop scheduling problems. *IEEE/CAA J. Autom. Sin.* **2019**, *6*, 904–916. [CrossRef]

38. Coello, C.A.C.; Pulido, G.T.; Lechuga, M.S. Handling multiple objectives with particle swarm optimization. *IEEE Trans. Evol. Comput.* **2004**, *8*, 256–279. [CrossRef]
39. Lv, Z.; Wang, L.; Han, Z.; Zhao, J.; Wang, W. Surrogate-assisted particle swarm optimization algorithm with Pareto active learning for expensive multi-objective optimization. *IEEE/CAA J. Autom. Sin.* **2019**, *6*, 838–849. [CrossRef]
40. Wang, Z.; Rangaiah, G.P. Application and analysis of methods for selecting an optimal solution from the Pareto-optimal front obtained by multiobjective optimization. *Ind. Eng. Chem. Res.* **2017**, *56*, 560–574. [CrossRef]

Article

The Influence of Public Transport Delays on Mobility on Demand Services

Layla Martin [1,*] , Michael Wittmann [2,*] and Xinyu Li [3]

1   Operations, Planning, Accounting and Control, School of Industrial Engineering, Eindhoven University of Technology, 5615 AP Eindhoven, The Netherlands
2   Chair of Automotive Technology, TUM School of Engineering and Design, Technical University of Munich, Boltzmannstr. 15, 85748 Garching, Germany
3   Faculty of Engineering, McGill University, Montreal, QC H3A 0G4, Canada; xinyu.li@mail.mcgill.ca
*   Correspondence: l.martin@tue.nl (L.M.); michael.wittmann@tum.de (M.W.)

Abstract: Demand for different modes of transportation clearly interacts. If public transit is delayed or out of service, customers might use mobility on demand (MoD), including taxi and carsharing for their trip, or discard the trip altogether, including a first and last mile that might otherwise be covered by MoD. For operators of taxi and carsharing services, as well as dispatching agencies, understanding increasing demand, and changing demand patterns due to outages and delays is important, as a more precise demand prediction allows for them to more profitably operate. For public authorities, it is paramount to understand this interaction when regulating transportation services. We investigate the interaction between public transit delays and demand for carsharing and taxi, as measured by the fraction of demand variance that can be explained by delays and the changing OD-patterns. A descriptive analysis of the public transit data set yields that delays and MoD demand both highly depend on the weekday and time of day, as well as the location within the city, and that delays in the city and in consecutive time intervals are correlated. Thus, demand variations must by corrected for these external influences. We find that demand for taxi and carsharing increases if the delay of public transit increases and this effect is stronger for taxi. Delays can explain at least 4.1% (carsharing) and 18.8% (taxi) of the demand variance, which is a good result when considering that other influencing factors, such as time of day or weather exert stronger influences. Further, planned public transit outages significantly change OD-patterns of taxi and carsharing.

Keywords: carsharing; data analysis; delays; demand; public transit; taxi

Citation: Martin, L.; Wittmann, M.; Li, X. The Influence of Public Transport Delays on Mobility on Demand Services. *Electronics* **2021**, *10*, 379. https://doi.org/10.3390/electronics 10040379

Academic Editor: Juan M. Corchado
Received: 15 January 2021
Accepted: 28 January 2021
Published: 4 February 2021

Publisher's Note: MDPI stays neutral with regard to jurisdictional claims in published maps and institutional affiliations.

1. Introduction

Metropolitan areas suffer as a consequence of a car-centric city layout. Roads are frequently congested [1], air quality decreases [2], and valuable space for active mobility is restricted [3]. Thus, city planners aim to incentivize as many travelers as possible to use rail and public transit. As travelers minimize their own transportation cost—comprised of travel time and fare—[4], it is paramount to understand the impact of the different factors of the cost function. The travel time depends on the scheduled transit time and delays. While the scheduled transit time is usually comparatively low, delays can severely impact both the actual and the perceived travel time [5]. Nowadays, public transit competes with rising MoD services [6], such as ride-hailing and carsharing, since vehicle ownership is lower in metropolitan than in rural areas ([7], p. 35). While delays also occur in MoD systems, they do not propagate as severely as in public transit, due to missed connecting trains, and they are perceived less severe by travelers [8]. Thus, surveys indicate that users switch to road-based individual transportation if delays increase or public transit is unavailable [9], resulting in additional demand. This additional demand (i) increases road congestion and (ii) results in additional planning effort for MoD operators. They must react by moving vehicles to locations with increased demand (rebalancing, dispatching), and they may have to increase their fleet size to accommodate those peak demands. Otherwise, the unserved demand increases, which results in lost sales on a short horizon [10], and they may also influence customer satisfaction and retention on a longer horizon [11]. The changing demand patterns during the COVID-19 pandemic pose an additional challenge, but they also permit us to study the system under a different demand profile.

This paper studies the influence of public transit delays on the demand of MoD services. As such, it helps MoD operators to increase the quality of service by improving their demand predictions and, consequentially, their operational strategies (e.g., [12]), and gives public transit authorities first insights into how their delays impact road traffic, and eventually congestion. A better understanding of the influence of public transit delays—and, thus, the quality of service of public transit—eventually helps to carry over a demand prediction tool that was developed in one location to another (as [13] attempt for bikesharing services).

This work is the first data-driven approach for establishing the correlation between public transit delays and demand for carsharing and taxi services. From the data, we establish a lower bound of how much taxi and carsharing substitute public transit by measuring the number of additional taxi and carsharing trips for increasing delays, both over a period of 10 months and exemplarily for closures of the main tracks, and both for individual stations and the entire city center.

This suggests that

- a lower delay and/or higher coverage of the public transit system can result in a smaller necessary fleet (by up to 2.2% and up to 0.5% of the total fleet size, respectively), and most likely result in fewer traffic; and,
- carsharing and taxi operators can improve their demand prediction performance by including information on planned public transit closures.

In the following, we first review related work in Section 2, and describe the data collection process and general statistics in the data presented in Section 3. Section 4 analyzes the data to establish how demand MoD-systems and public transit delays are connected. Section 5 discusses the results and concludes the paper.

2. Related Work

This work is related to two different streams of literature: demand prediction for carsharing and taxi services, as well as substitution and complementarity between different modes of shared and public transit.

[14] review different works focusing on demand prediction for carsharing. They conclude that research still lacks in-depth knowledge about the intricacies of demand processes. [15] investigate how taxi, Uber, and Lyft demand increases during severe rain in New York City, as well as the price elasticity of Uber and Lyft during those periods. They find that the total number of rides increases, but the number of taxi rides is only weakly correlated with rain, which suggests that the additional demand is due to Uber and Lyft's pricing strategy. The impact of weather on demand is clearer than the influence of public transit delays, since rain can be modeled as a Boolean variable, while the demand increase may depend on the extent of the delay. [16] study how events influence taxi demand in New York City using online information (web mining demand hot spots from social media). They find that the frequency that events were mentioned has a significant impact on the taxi demand. However, the influence of events differs from the influence of public transit delays, as events only affect the origin of a trip, not origin-destination pairs, and since events are known earlier than delays. [17] show that this event information can improve the performance of demand predictors. Thus, we cannot directly adapt models for measuring the influence of weather or events to the influence of delays. [18] show that socio-demographic features have an impact on carsharing demand in Munich and Berlin (Germany), [19] study the impact of these factors on New York City taxi demand, and [20] investigate the impact of socio-demographic factors on ride-hailing demand in California. All three studies find that socio-demographics can explain parts of the demand variances. However, socio-demographics usually do not change within the observation period, unlike

public transit delays. Therefore, we must develop a new methodology to measure the impact of a cardinal variable that varies during the observation period, but it has strong correlation with some of the previously studied variables.

Ref. [19] also explore the influence of demographic and socioeconomic factors on the taxi passenger demand in New York. The results clearly indicate that a relationship between public transit accessibility and taxi demand exists. Taxi trips in this study occur more often if public transit is more accessible. The authors note that a finding of whether this relationship is competitive or complementary could not be determined from their results. Additionally, they do not consider if and how taxi demand varies, depending on the temporal availability of public transit. Ref. [21] study the demand patterns for taxi and Uber on a coarse-grained level. They find that similar external factors impact the demand for taxi and Uber. Ref. [22] state that users of free-floating carsharing are more likely to have a public transit subscription than the control group which suggests complementarity. Ref. [23] consider the spatio-temporal availability of public transit, but do not extend their analysis beyond examples during an outage at a central location. Naturally, such a dependency implies that if the required time for one of the transit modes increases, the demand for this mode should decrease. Ref. [23] gives some examples that support that public transit outages increase the demand for alternative transportation modes in the city of Vancouver, but do not extend this to a city-wide experiment over a longer period of time. Ref. [24] observe that, for the most common carsharing trips (origin-destination pairs) in Madrid, traveling by car is only slightly faster than public transit, but significantly more expensive, which suggests a customer preference for traveling by car, but they do not investigate whether the number of trips increases even further if public transit becomes less available (due to outages or delays). For early carsharing adopters, ref. [25] find that, on average, over seven European and North American cities, 40% of carsharing trips could have been performed more quickly by public transit, which suggests that carsharing, in fact, substitutes public transit. However, they do not include delays or outages, as well as the waiting time for the next train. Several studies indicate that, on the customer side, there is a notable difference in the perceived and real waiting time—especially in the case of unestimated delays. Ref. [26] found that passengers who did not have knowledge about the actual schedule perceived waiting times significantly longer than passengers with knowledge. Ref. [27] confirm that users already more frequently decide against using public transit for an entire trip if only a single trip segment is sub-optimal (e.g., long waiting times, slow transit). Ref. [28] study the modal choice of travelers in Taiwan using a general estimation equation, and find that intermodal transportation accessibility has a positive influence of public transit ridership at those stations where available. However, ref. [28] do not study the influence of the public transit operator's availability and punctuality on ridership (and alternative choices).

This paper is the first to analyze the influence of public transit delays and outages on MoD demand in a data-driven fashion (as compared to simulation- and survey-based research), both on a city level and more granular per public transit station (when compared to single occasions). This provides substantial additional insights, but it also requires a new methodology.

3. Reference Datasets

This study uses three different datasets for the city of Munich in the period from May 2019 to March 2020: downtime and delays for public transit, vehicle movement data of a major carsharing provider (only until December 2019), and taxi customer trip data of a local taxi agency. The data are discretized and filtered.

3.1. Public Transit Data

We query current departures (scheduled and actual departure time) of all suburban railway (S-Bahn) and underground (U-Bahn) lines at all stops in 5 min. intervals to obtain an understanding about delays and outages in the public transit system. The delay δ_{it} can then be calculated as the difference between the scheduled and actual departure time. Both of the times are only given in minutes, thus delays are only reported if they are at least 60 s, and delays are rounded down to the next integer. For our analysis, longer delays are more relevant, reducing the impact if rounding down. Throughout this paper, we indicate how we handle integrality. For each station i and time frame t, we report both the average delay δ_{it} and maximum delay $\max(\delta_{it})$ of all departures. Outages are derived from comparing the number of departures in a given interval to comparable intervals on other days (maximum over all weeks). Additionally, we collected information regarding track closures during the analysis period from public authorities.

Table 1 lists the basic statistics on the data set and the delays. Even in the short observation period of ≈ 10 months, the number of departures is in the order of 12 M (This number includes all departures of the same train and, thus, may seem to be excessively high at first). Thus, the impact of those instances with a very high delay is negligible. The mean delays (in minutes) are low (slightly above or below 1 min.), and the majority of departures is not delayed (contrary to what customers perceive). Figure 1 depicts the average of delays (suburban railway and underground) during an example week (week 42/2019, 14–20 October) at station Marienplatz. During rush hour, most of the lines are slightly delayed, and the morning rush hour incurs more severe delays than the evening rush hour. Major delays (>6 min.) occur infrequently, and on a more random pattern.

Table 1. Data Description Public Transit.

	Suburban Railway	Underground
# stations	125	98
# lines	7	8
# records	4,566,500	7,683,766
mean delay in min	1.168	0.837
max delay in min	354	1431
# records $1 >$ delay	2,131,051	3,890,972
# records $3 >$ delay ≥ 1	1,601,061	2,894,220
# records $6 >$ delay ≥ 3	618,356	759,421
# records delay ≥ 6	216,032	139,153

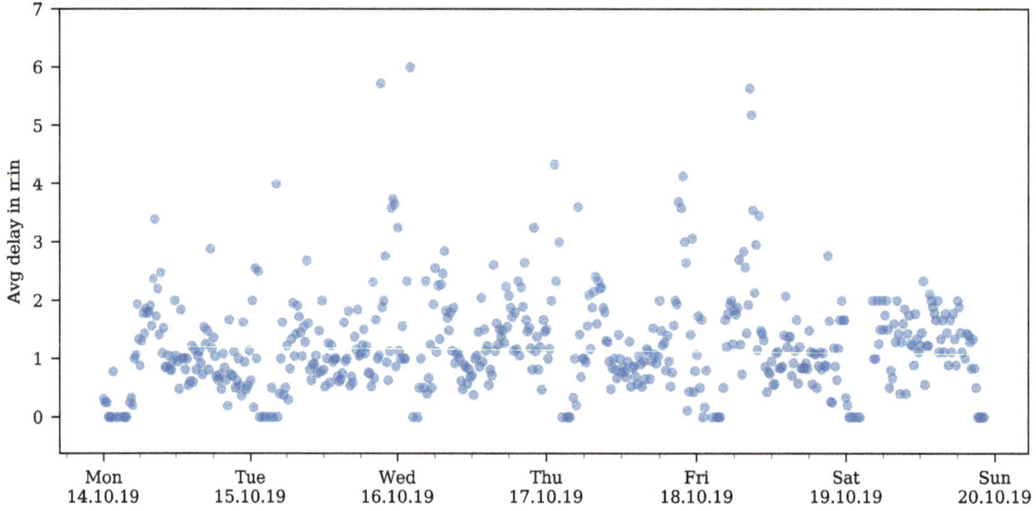

Figure 1. Average of Delays at Marienplatz during Example Week.

If an above average delay occurs at some point in time t, there is a high probability (>55%) that there will also be an above average delay at time $t + 15$ min.. If the delay at time t is in the 80th percentile ($\%^{\tau(t)}(\delta_{ti}) \geq 0.8$, where $\%^{\tau(\cdot)}(\cdot)$ is the percentile function that assigns the delay percentile among comparable time frames ($\tau(t)$) at station i), above average delays occur even more frequently, and delays persist longer. Figure 2 depicts the delay persistency, i.e., the probability that a delay occurs at the same station a given time interval after another delay.

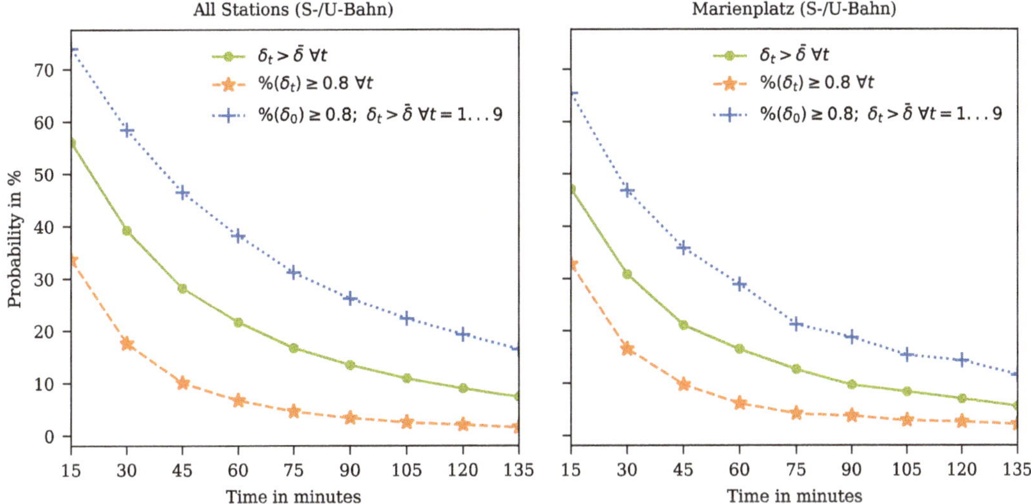

Figure 2. Delay Persistency in Munich and at Marienplatz.

When comparing the entire Munich city center to Marienplatz, it, at first, seems surprising that the central location with a very dense time table and high utilization has a lower delay persistency, but Marienplatz also has four tracks while most stations only have two tracks. Delays propagate almost independently on different tracks. This delay propagation is visible from Figure 3.

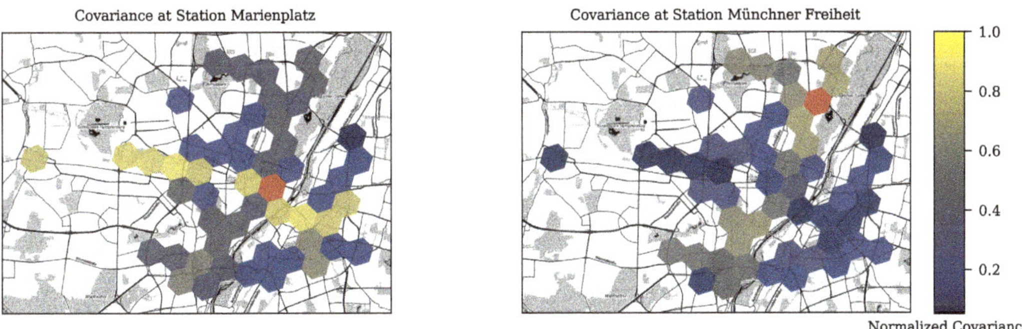

Figure 3. Delay Propagation in the Public Transit Network.

It shows the probability that an above average delay at Marienplatz (left) or Münchner Freiheit (right) correlates with an above average delays at other stations. Yellow shading refers to a high covariance, blue shading to a low covariance. Clearly, stations along the same line have a higher probability of delays and, at Marienplatz, this mainly affects the East-West connection (S-Bahn), not so much the North-South connection (U-Bahn). The covariance between the selected station and stations along other lines is low, and the remaining covariance can be due to external influences, customers transferring lines, or intersecting lines. From the strong spatial and temporal differences in delays, we conclude that our spatial and temporal resolution is reasonable.

3.2. Carsharing Data

Carsharing data for one of the largest Munich free-floating carsharing operators has been collected while using webscraping techniques since April 2018. The scraping ended mid-January 2020 when the API was discontinued. Every 5 min., the current location of all available vehicles (not rented or reserved by customers, or blocked by the operator) was recorded. Because of the data collection method, outages appear, and the data are cleared accordingly.

Movements of vehicles are created by recording the last location of the vehicle before becoming "invisible" and the first location after re-appearing in the data stream. The data collection method does not allow for us to differentiate between customer trips and rebalancing operations, but there should be significantly less rebalancing operations than customer trips, and the literature reports that rebalancing mainly occurs during the night [29]. We remove time windows during the night, as described later.

The data set contains >1.5 M trips. Roughly 20% of all data points are missing due to outages in the data collection. The reasons for outages include power outages, network connection loss, or, otherwise, discontinued service on the collection server, unavailability of the API, and service downtime of the carsharing service provider. On average, 131 trips occurred per hour, with a maximum of 653 trips during an one-hour interval, and the number of trips highly depends on the time of day and weekday.

Figure 4 shows an example weekly pattern for week 42/2019 (14–20 October).

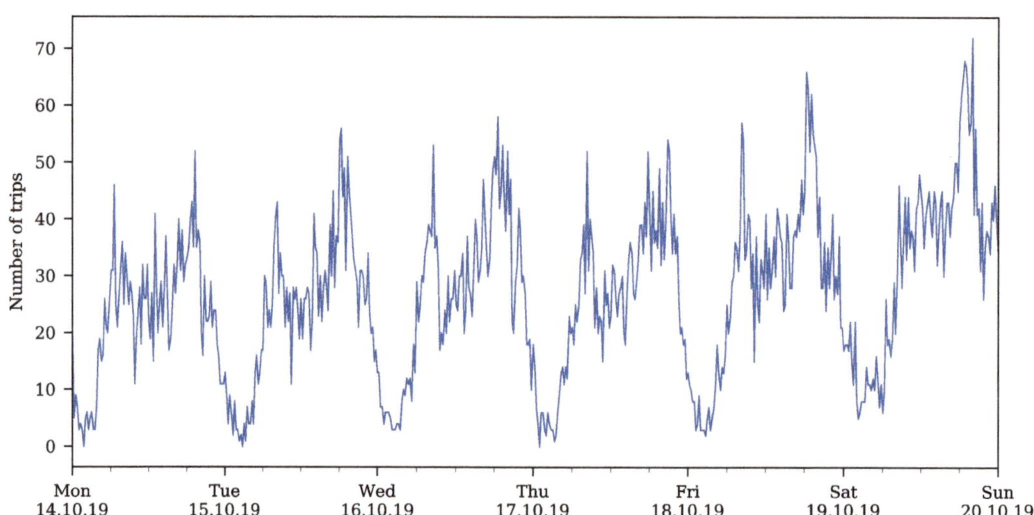

Figure 4. Number of Carsharing Trips during Example Week.

Demand is aggregated in one-hour intervals and it is reported at the beginning of the interval. The demand follows a daily pattern with more demand during the evening rush hour than the morning rush hour, and slightly decreasing demand during the course of the week.

3.3. Taxi Data

This study makes use of floating car data from a local taxi agency to derive the passenger demand for taxi services in Munich. A fleet of 550 taxis served ≈10 M customer trips between 2015 and 2020. The data are being continuously retrieved from the fleet management interface, which is usually used for dispatching by the local taxi agency [30]. The data are directly provided by the dispatching agency with full information about trip start, trip end, and driven route.

The data set contains >3.8 M trips in the observation period between April 2018 and March 2020. On average, 252 trips are recorded per hour (at most, 740 trips per hour).

Figure 5 shows an example weekly pattern for week 42/2019 (14–20 October).

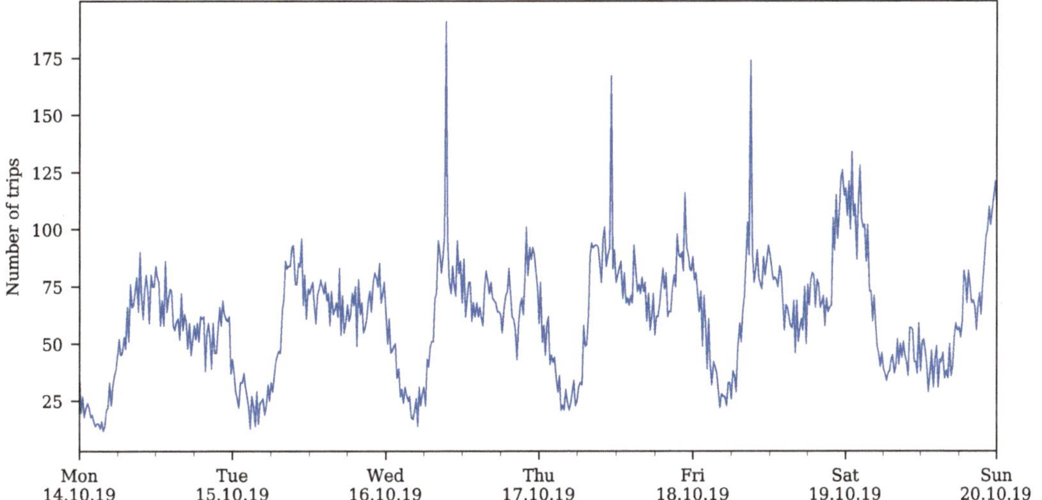

Figure 5. Number of Taxi Trips during Example Week.

The demand pattern follows the same high level trend as carsharing demand, but the highest demand peaks occur on the weekend rather than at the beginning of the week. The morning peak is more pronounced, and the daily afternoon demand peaks occur slightly later in the day than in the carsharing system.

3.4. Data Discretization and Filtering

For comparability, we discretize the area in hexagons with an edge length and radius of 461 m using Uber's Hierarchical Spatial Index H3 [31]. Imposing a maximum walking distance of 461 m is in alignment with literature [32]. Temporally, we discretize the carsharing and taxi data into one-hour intervals on a rolling scheme, creating data points every 15 min. Every data point then contains the total number of trips—the demand d_{it}— occurring in the 60 min. after a delay. One hour is a reasonable time frame for potential impacts and in line with delay persistency (Figure 2). This (i) increases the amount of available data points and, therefore, reduces random variances of our results compared to one-hour intervals without rolling time windows and (ii) smoothes the demand pattern as compared to 15-min. intervals, as one can see in Figure 6 for the carsharing and taxi demand.

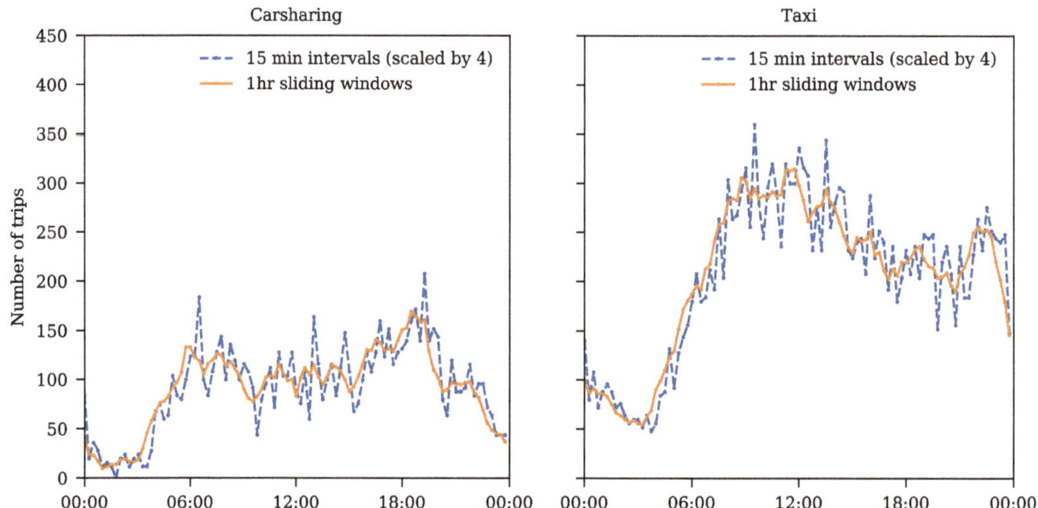

Figure 6. Number of Carsharing and Taxi Trips during Example Week with and without Smoothing.

Because public transit delays persist for some time and, as these delays take time to manifest in the taxi demand, using sliding time windows is also advantageous to be able to record longer-time impacts on the demand.

Obviously, carsharing and taxi demand can only be evaluated against public transit delays if the public transit service is scheduled to run. Further, an influence can only be measured to a statistically significant level if the average number of carsharing and taxi trips is sufficiently high. This does not exclude temporary outages, but it does exclude nights, as public transit is not operating between 1:30–4:30 AM, and the number of departures decreases substantially during the late evening. We exclude the time frame 10:00 PM–5:00 AM, to be safe against startup and end-of-horizon effects, and the low demand during the night. Omitting this longer period of time also makes it more probable that vehicle movements in carsharing are customer trips, rather than vehicle rebalancing (since vehicle rebalancing in carsharing systems mainly occurs during the night [29]). Subsequently, 13.0% of all public transit departures, 14.6% of all carsharing trips, and 31.6% of all taxi trips are omitted.

Additionally, to be able to measure the effect of public transit on carsharing and taxi demand, we exclude those hexagons without a public transit stop (suburban railway or underground) and those outside the Munich city highway "Mittlerer Ring" (except for Pasing station which is the west-most end of the suburban railway main tracks). Thus, we consider demand in the 53 hexagons that are depicted in Figure 7 with blue squares for U-Bahn stations, green dots for S-Bahn stations, and orange rhombuses for stations with both U-Bahn and S-Bahn connections.

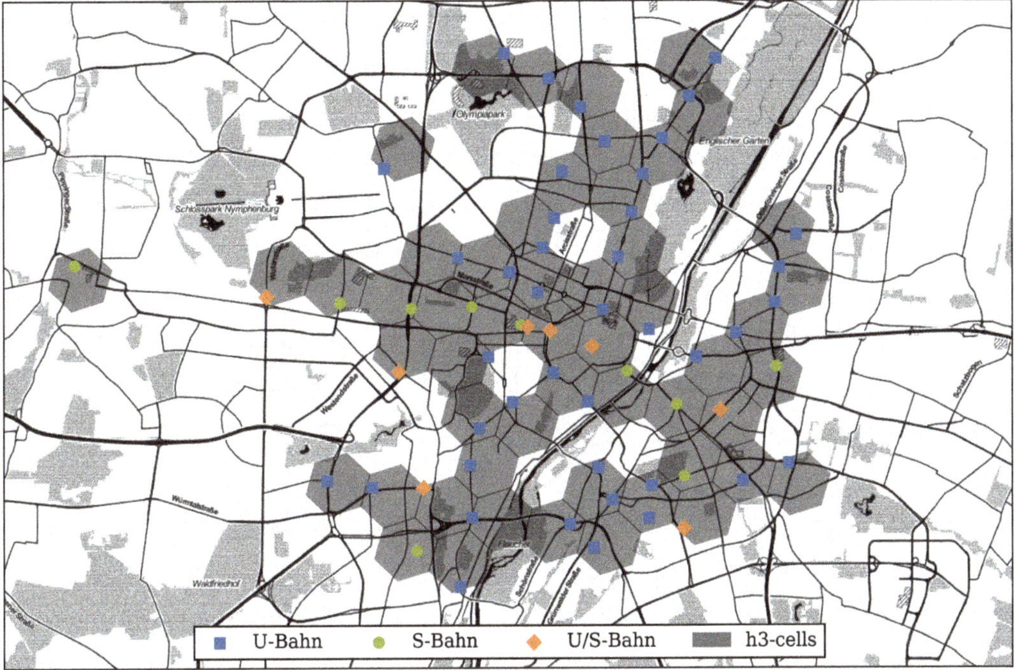

Figure 7. Considered Hexagons (in grey shading) with Suburban Railway (green dots), Underground (blue squares) stations, or a Combination of both (orange rhombuses).

3.5. Censored Demand

Demand for MoD is subject to censoring [33,34]: if no vehicle is available, one cannot record demand, and a straight-forward model tends towards underestimating demand. Outages of the carsharing service might correlate with public transit disruptions. Thus, none of the approaches that have been suggested in literature can be applied, since we measure increased delay which also impacts the demand censoring. Instead, we split the data set in those data points with and without censored demand (assuming that censoring can apply only if supply is 0). This occurs more frequently in remote locations with low demand. For the high demand location Marienplatz, censoring might have occurred in up to 24% of all data points. Such censored demand only occurs for the carsharing service, but not the taxi service, as taxi street-hailing is less common in Munich [35], and as Munich has a significant oversupply in taxis [36].

4. Analysis

Using the previously described data sets, we analyze how public transit delays influence demand for MoD services. In particular, we give a high level relation, calculate the fraction of demand variation that can be explained by public transit delays, analyze the varying demand patterns during outages, the probability of having no vehicles available depending on delays and outages, and the demand changes during the COVID-19 pandemic. We use this analysis to estimate the additional demand, traffic, and necessary increase in the fleet size due to delays and outages.

All of the numerical analyses are implemented in Python 3 with (among others) Numpy, Scikit Learn, and Gurobi. The experiments are performed on an Ubuntu server.

4.1. High Level Relation

Station-timeframe tuples with high delays (mean delay $\delta_{it} \geq 3$ min.) more frequently result in high taxi and carsharing demand than tuples with lower maximum delays. Tuples are clustered by the observed demand relative to the mean for this station-timeframe tuple (in 2% bins), and the observed maximum delay (no delay, up to 3 min. delay, and higher delays).

Figure 8 reports the relative frequency for each tuple by means of a cumulative distribution function (CDF). It is obvious that the higher the delay, the more frequent high demand instances appear. While this indicates some dependency, it does not yet show how delays and demand correlate.

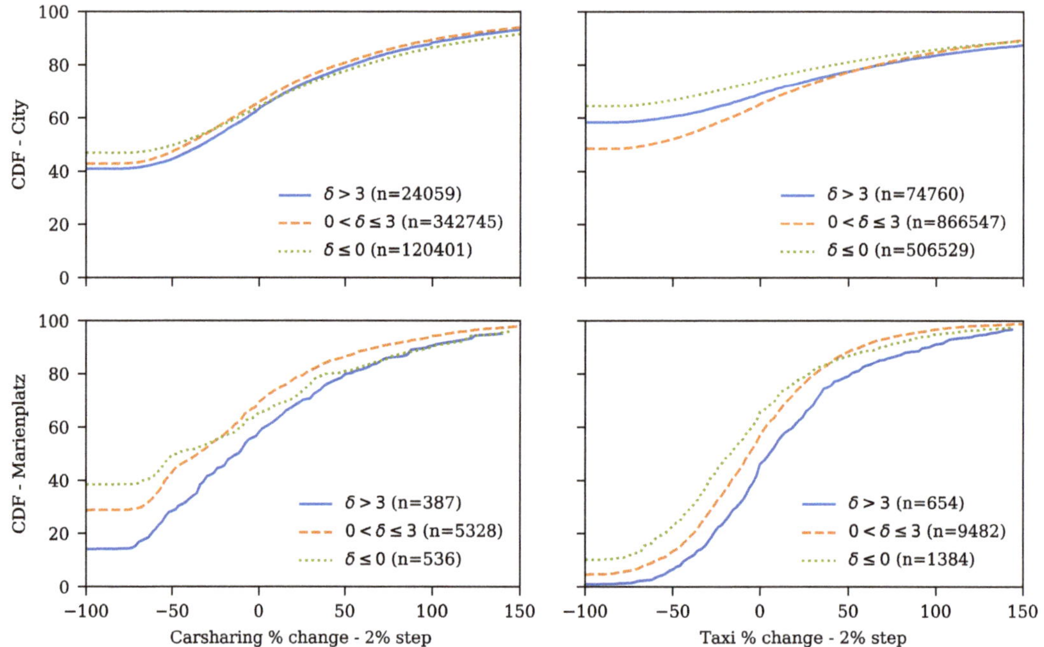

Figure 8. Maximum Delay in the Entire City and at Marienplatz.

To this end, we further observe that any increase in public transit delays entails an increase of the demand for taxi and carsharing services. We group the data points by delay (in intervals: $[0, 1)$, $[1, 3)$, $[3, 6)$, $[6, 10)$, $[10, 20)$, $[20, 60)$, $[60, \infty)$, with the last two being aggregated for carsharing due to a low number of data points). The lower number of high delay data points is a consequence of filtering potentially censored demand points in the carsharing data. Figure 9 depicts the boxplots of trip deviations from mean for carsharing and taxi. For ease of exposition, the boxplots do not contain outliers. The trip deviation from mean increases from 0.0% to 13.5% for carsharing, and −2.1% to 50.0% for taxi.

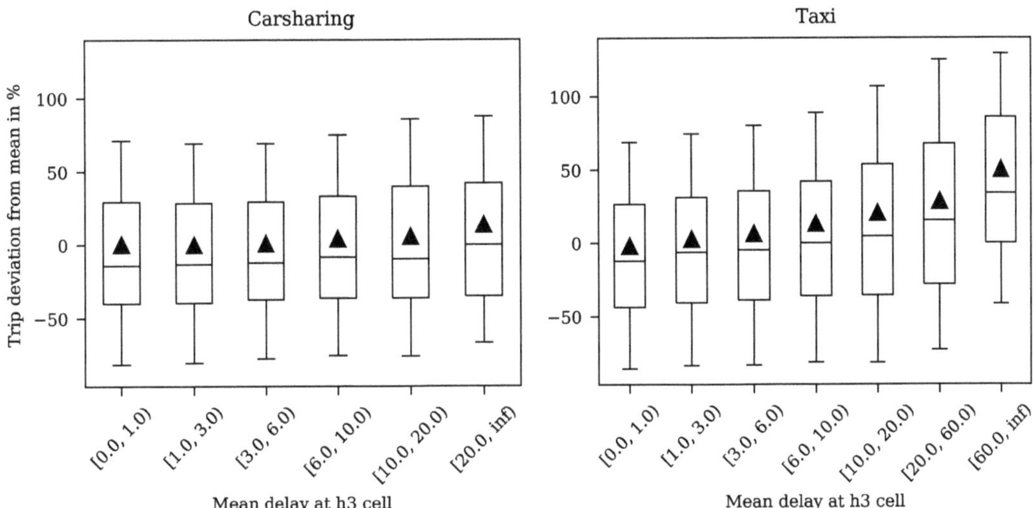

Figure 9. Increased Demand during Delays.

Thus, it is clear that MoD demand and public transit delay are correlated, even though other external factors (time of day, station, weather, events) also exert a strong influence.

4.2. Explained Demand Variance Due to Delays

We showed that the demand for MoD increases if delays occur. However, delays are not the only factor that can explain variances in the MoD demand, and some randomness is intrinsic to the system. In order to measure the explanatory power of delays on MoD demand, we assume that demand at a station during a given timeframe can be predicted using the mean value as a baseline, and measure how much the variance decreases when correcting the trip counts for the delay. Therefore, we filter the dataset for observations in which a mean delay ≥ 3 is observed. Thus, we define a lookup function $f(\delta_{it})$, which returns the mean percentage trip deviation for each delay bin.

The data points are transformed into the deviation dev_{it} from mean $\mu_{i\tau(t)}$ for location i and timeframe t.

$$\mathrm{dev}_{it} = \frac{d_{it}}{\mu_{i\tau(t)}} - 1$$

resulting in the "basic" data set S, and potentially corrected by

$$\hat{\mathrm{dev}}_{it} = \frac{d_{it} \cdot (1 - f(\delta_{it})) - \mu_{i\tau(t)}}{\mu_{i\tau(t)}}$$

resulting in the "corrected" data set $\hat{S}$.

Each set of data points S and $\hat{S}$ can be represented as a density function. The histograms for the density functions can be found in Figure 10.

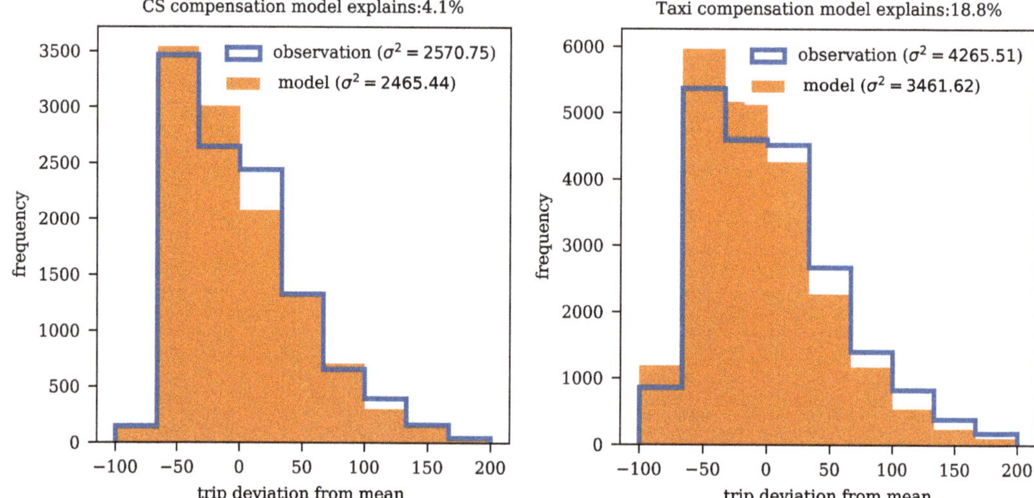

Figure 10. Explained Demand Variance due to Delays.

We individually compute the variance in each data set as

$$\sigma^2 = \sum_{it} \left(\mathrm{dev}_{it} - \mu_{\mathrm{dev}}\right)^2$$

$$\hat{\sigma}^2 = \sum_{it} \left(\hat{\mathrm{dev}}_{it} - \mu_{\hat{\mathrm{dev}}}\right)^2,$$

where μ_{dev} and $\mu_{\hat{\mathrm{dev}}}$ refer to the average over all dev_{it} and $\hat{\mathrm{dev}}_{it}$, respectively. The relative decrease in variance from the basic to corrected data set (S to $\hat{S}$) corresponds to the fraction of demand variation that can be explained by delays in the public transit network.

$$1 - \frac{\hat{\sigma}^2}{\sigma^2}$$

Thus, we can explain 4.1% of the variance by delays for carsharing and 18.8% for taxi. Consequentially, demand prediction accuracy can be improved by considering the delays during periods of high delay. At first, explaining 4.1% of carsharing demand may not seem much, but, when considering the abundant number of influencing factors (weather, events, ...), uncertainties of human behavior, the fact that delays do not vary too much, and spatial differences in the explanatory power, explaining 4.1% is already significantly improving the situation of MoD operators in a business with low profit margins. The explanatory power of delays on taxi demand variation is significantly higher. Technically, this is a consequence of larger maximum correction factors returned by the function $f(\delta_{it})$, as readily visible from Figure 9. It appears as if users rather switch to taxi than to carsharing if delays occur. This suggests that carsharing customers, on average, have a lower valuation of time than taxi customers, which is backed by general intuition. It does not necessarily mean that the user groups of taxi and public transit are overlapping more than the user groups of carsharing and public transit, but might rather point towards a lower willingness to wait among taxi users.

4.3. Changing Demand Patterns in Presence of Known Outages

While one might assume that MoD demand increases during an outage (and such a correlation has been reported by [23]), we cannot support this assumption based on our data. In Munich, the total demand does not significantly change on days with outages.

This might be due to the fact that outages are commonly known before, and travelers, therefore, either forgo trips, or use private vehicles or bikes. This indicates that, while taxi and carsharing are an alternative to public transit if the delay was unexpected, customers tend to use a different mode of transportation or omit trips if the delay was known before.

Instead, demand patterns (given by trip origins) change. For every day in the observation period, we compare the relative taxi demand per origin-destination pair to the previous year by means of a Wasserstein distance. The Wasserstein distance (also known as the Earth mover's distance) refers to the "work necessary" to transform one distribution into another. We compare the difference between two vectors v_t and v'_t, where v_t refers to the relative demand distribution in the previous year for some timeframe t

$$v_{it} = \frac{d_{it}}{\sum_j d_{jt}},$$

and analogous for v' (because outages occur for a longer period of time, and since random demand imbalances can occur within the day with Integer demands, we aggregate timeframes t to a daily level). To increase comparability, we compare any day in the current year to the closest day in the previous year, which is the same weekday (e.g., Monday 4 November 2019 to Monday 5 November 2018), and omit days that were a public holiday in either year. The Wasserstein distance per timeframe t is then calculated by solving the following linear program of a transportation problem

$$\min \sum_{i,j} \Delta_{ij} x_{ij}$$
$$v_{it} + \sum_j x_{ij} = v'_{it}$$
$$x_{ij} \geq 0 \qquad\qquad \forall i,j$$

where the decision variables x_{ij} refer to the amount of demand that is "shifted", and Δ_{ij} refers to the Euclidean distance between locations i and j (representing the "difference" between demand patterns, i.e., the transportation cost).

There are 21 days on which the main line of the S-Bahn was closed in one direction at one station, and 10 instances where the main line was closed in both directions. While taxi data are available for all days in the observation period, carsharing data are only available on 20 and six of these days, respectively.

We test whether the Wasserstein distances increase (or decrease) during outages when comparing an instance with an outage 2019 and no outage 2018 using a two-tailed Welch's t-test. Table 2 lists the results for carsharing and taxi. Carsharing demand patterns significantly ($\alpha = 5\%$) differ in presence of delays, given by a significantly increasing Wasserstein distance, for one-directional and two-directional closures, both independently and jointly. For taxi demand patterns, the null hypothesis (mean Wasserstein distances do not differ, travel patterns are similar) cannot be rejected for uni-directional closures (In this case, the mean Wasserstein distance even decreased insignificantly.) and all closures. Wasserstein distances for taxi demand patterns significantly increase (at a significance level of 10%) when the main track was closed in both directions. Thus, we conclude that demand patterns change in the presence of outages, and MoD operators should include this information in demand prediction at the local level.

Table 2. Results of the Welch's t-Test for Changing Demand Patterns during Outages.

	No Outage	Outage	No Out.	1-Way	No Out.	2-Way
Carsharing						
mean	25,498	30,878	25,498	29,496	25,498	35,486
variance	4.92E7	3.95E7	4.92E7	3.60E7	4.92E7	2.78E7
# observations	131	26	131	20	131	6
t-statistic	−3.91		-2.71		−4.47	
$p(T \geq t)$	3.71E-4		1.13E-2		4.25E-3	
crit. value	1.69		2.45		2.05	
Taxi						
mean	16,751	17,338	16,751	16,234	16,751	19,656
variance	2.83E7	2.42E7	2.83E7	2.20E7	2.83E7	2.30E7
# observations	304	31	304	21	304	10
t-statistic	−0.82		0.32		−1.99	
$p(T \geq t)$	0.42		0.75		7.41E-2	
crit. value	2.02		2.06		2.23	

4.4. Decrease of Demand along Lines during COVID-19

During the recent COVID-19 pandemic, demand decreased significantly (average number of taxi trips from 9 March to 3 May 2020 decreased by a factor of 3.5 as compared to the previous year; this decrease is significant at $\alpha = 0.05$ using a Welch's t-test). Reduced mobility is one of the key levers to reduce the spread of a pandemic, as [37] show for Italy. Surprisingly, the relative demand along public transportation lines also decreased (15.9% of all trips vs. 20.8%). This result is significant according to a Welch's t-test at all commonly used significance levels (see the results in Table 3).

Table 3. Results of the Welch's t-Test for Fraction of Trips Along Lines.

	Before COVID	During COVID
mean	0.159	0.208
variance	4.12E-4	2.25E-4
# observations	56	708
t-statistic	−17.7	
$p(T \geq t)$	1.54E-25	
crit. value	2.00	

When combined with the decreasing delays during this period (see Table 4), this provides anecdotal evidence that the increased punctuality made it unnecessary to choose alternative modes of transportation.

Table 4. Changing Public Transit Delay before and during COVID.

	Before COVID		During COVID	
	S-Bahn	U-Bahn	S-Bahn	U-Bahn
mean delay in min	1.168	0.837	0.979	0.612
% 1 > delay	46.7	50.6	47.3	59.9
% 3 > delay ≥ 1	35.1	37.7	37.5	33.5
% 6 > delay ≥ 3	13.5	9.9	12.0	6.1
% delay ≥ 6	4.7	1.8	3.2	0.5

4.5. Potential for Fleet Size Reduction

Public authorities can reduce congestion in the road network as well as the necessary fleet size of MoD operators by improving the punctuality of the public transit system, since a higher delay of the public transit service entails higher demand for MoD.

If all the delays were 0, the average demand for taxi would decrease by 2.2%, and the average demand for carsharing would decrease by 0.5%, as given by the shift to the leftmost bin in in Figure 9. If all delays were reduced by 50%, the average demand for taxi would decrease by 1.6%, and the average demand for carsharing would decrease by 0.5%. These values are computed by artificially reducing the average delay for each time interval at each location, and correcting the observed demand according to the corresponding carsharing/taxi trip deviation.

These demand reductions can serve as an upper bound for potential fleet size reductions. The actual fleet size reduction may be less due to risk pooling in the presence of stochastic demand. It stands to reason that the number of trips in privately owned cars also reduces, even though the exact values might differ. This indicates that road-congestion might be alleviated by reducing delays of the public transit operator, and making it more reliable.

5. Discussion and Conclusions

In this paper, we study the effect of public transit delays on MoD operators, i.e., carsharing and taxi. If customers judge trips based on some combination of travel time and travel cost, demand for the "outside option" (e.g., MoD) will increase if the travel time for public transit increases. We conduct large-scale experiments using carsharing and taxi trip data and public transit departure data for 10 months in Munich, Germany.

Demand for MoD increases if public transit delays increase. The mean demand for carsharing varies by up to 13.5% and the mean taxi demand varies by up to 52.1%, depending on the extent of the delay.

4.1% of carsharing and 18.8% of taxi demand variance can be explained if public transit is delayed. Thus, it seems as if carsharing customers valued travel time less than taxi customers. Because public transit delay is only one of many influencing factors (besides weather, events, and others), the explanatory power is high.

If the public transit operator were delayed less, the necessary carsharing and taxi fleet could be reduced by up to 0.5% and 2.2%, respectively. Even if delays did not vanish entirely, the number of taxi and carsharing trips could decrease substantially. Thus, public authorities might improve the public transit with the goal of reducing the congestion in their road network. Improvements of the public transit operator to alleviate congestion exceeds all approaches (e.g., [38]) for regulations of MoD discussed in existing literature, and poses an interesting line for future research.

Customers adapt their travel patterns if the public transit service is not operating. Unlike existing literature, we do not find that demand increases during outages. Most likely, a substantial number of travelers decide to delay their trip until the end of the outage, as outages are known upfront. Among the remaining travelers, origin-destination pairs change significantly, given by an increasing Wasserstein distance when comparing the origin–destination distribution to the previous year.

A few comments are in order: we only measure correlation, but no cause–effect relationship. It could also be possible that an increase in carsharing and taxi usage increases the delay for public transit. However, from an application point of view, this is unrealistic for rail traffic, and even the maximum number of carsharing vehicles and taxis should not incur significant delays for road-based public transit (bus, rail replacement services). Additionally, both public transit delays and carsharing/taxi demand might be dependent on an external influence that we did not correct for. While we cannot prove that no external source caused the correlation, a causal relation is the most likely explanation. Further, we must mention that the measured effect is minimal. This is because the S-Bahn has very similar and rather low delays on most instances. It is possible that some passengers choose

MoD, rather than public transit already due to the current mean delay. Our method cannot capture this and, therefore, only returns a lower bound on the influence of public transit delays on MoD demand. Because the data set only permits integer delays and since delays are subject to external influences, the effects remain minimal. More precise data would permit a more extensive analysis. However, this approach is important, since it allows third parties, such as policy makers or new market entrants, in order to measure the effect with data they have available, or can easily collect. The evidence that carsharing and taxi can help in increasing accessibility is rather anecdotal. Insights can be strengthened in future research if data are available prior and posterior to opening new lines in the public transit system. In future work, our results can be used to approximate a customer choice function in a data-driven fashion.

Author Contributions: Conceptualization, L.M. and M.W.; methodology, L.M. and M.W.; software, L.M., M.W. and X.L.; validation, L.M., M.W. and X.L.; investigation, L.M., M.W. and X.L.; resources, M.W.; data curation, L.M., M.W. and X.L.; writing—original draft preparation, L.M. and M.W.; writing—review and editing, L.M. and M.W.; visualization, M.W. and X.L.; supervision, L.M.; project administration, L.M. and M.W. All authors have read and agreed to the published version of the manuscript.

Funding: The work of Layla Martin and Xinyu Li was supported by Deutsche Forschungsgemeinschaft as part of the Research Training Group 2201 (Advanced Optimization in a Networked Economy). The work of Xinyu Li was partially supported by the German Academic Exchange Service. The work of Michael Wittmann was independently funded by the Chair of Automotive Technology.

Acknowledgments: The authors would like to thank IsarFunk Taxizentrale GmbH & Co. KG for providing real world taxi data.

Conflicts of Interest: The authors declare no conflict of interest.

References

1. INRIX. *INRIX Verkehrsstudie: Stau Verursacht Kosten in Milliardenhöhe*; INRIX: Kirkland, WA, USA, 2020. Available online: https://inrix.com/press-releases/2019-traffic-scorecard-german/ (accessed on 5 November 2020).
2. Banister, D. Cities, mobility and climate change. *J. Transp. Geogr.* **2011**, *19*, 1538–1546. [CrossRef]
3. Strong Towns. We Should Be Building Cities for People, Not Cars. 2018. Available online: https://www.strongtowns.org/journal/2018/7/2/we-should-be-building-cities-for-people-not-cars (accessed on 28 February 2020).
4. Liu, Y.; Bansal, P.; Daziano, R.; Samaranayake, S. A framework to integrate mode choice in the design of mobility-on-demand systems. *Transp. Res. Part C Emerg. Technol.* **2019**, *105*, 648–665. [CrossRef]
5. Hess, D.; Brown, J.; Shoup, D. Waiting for the Bus. *J. Public Transp.* **2004**, *7*, 67–84. [CrossRef]
6. Le Vine, S.; Lee-Gosselin, M.; Sivakumar, A.; Polak, J. A new approach to predict the market and impacts of round-trip and point-to-point carsharing systems: Case study of London. *Transp. Res. Part D Transp. Environ.* **2014**, *32*, 218–229. [CrossRef]
7. Nobis, C.; Kuhnimhof, T. *Mobilität in Deutschland—MiD Ergebnisbericht*; Technical Report; Infas, DLR, IVT and Infas 360: Bonn, Berlin, 2018.
8. Van Exel, N.J.A.; Rietveld, P. Perceptions of public transport travel time and their effect on choice-sets among car drivers. *J. Transp. Land Use* **2010**, *2*, 75–86. [CrossRef]
9. Rayle, L.; Dai, D.; Chan, N.; Cervero, R.; Shaheen, S. Just a better taxi? A survey-based comparison of taxis, transit, and ridesourcing services in San Francisco. *Transp. Policy* **2016**, *45*, 168–178. [CrossRef]
10. Albiński, S.; Fontaine, P.; Minner, S. Performance analysis of a hybrid bike sharing system: A service-level-based approach under censored demand observations. *Transp. Res. Part E Logist. Transp. Rev.* **2018**, *116*, 59–69. [CrossRef]
11. Basciftci, B.; Ahmed, S.; Shen, S. Distributionally robust facility location problem under decision-dependent stochastic demand. *Eur. J. Oper. Res.* **2020**, doi:10.1016/j.ejor.2020.11.002. [CrossRef]
12. Salazar, M.; Lanzetti, N.; Rossi, F.; Schiffer, M.; Pavone, M. Intermodal autonomous mobility-on-demand. *IEEE Trans. Intell. Transp. Syst.* **2020**, *21*, 3946–3960. [CrossRef]
13. Guidon, S.; Reck, D.J.; Axhausen, K. Expanding a(n) (electric) bicycle-sharing system to a new city: Prediction of demand with spatial regression and random forests. *J. Transp. Geogr.* **2020**, *84*, 102692. [CrossRef]
14. Vosooghi, R.; Puchinger, J.; Jankovic, M.; Sirin, G. A critical analysis of travel demand estimation for new one-way carsharing systems. In Proceedings of the 2017 IEEE 20th International Conference on Intelligent Transportation Systems (ITSC), Yokohama, Japan, 16–19 October 2017; pp. 199–205.
15. Brodeur, A.; Nield, K. An empirical analysis of taxi, Lyft and Uber rides: Evidence from weather shocks in NYC. *J. Econ. Behav. Organ.* **2018**, *152*, 1–16. [CrossRef]

16. Markou, I.; Kaiser, K.; Pereira, F.C. Predicting taxi demand hotspots using automated Internet Search Queries. *Transp. Res. Part C Emerg. Technol.* **2019**, *102*, 73–86. [CrossRef]
17. Markou, I.; Rodrigues, F.; Pereira, F.C. Real-Time Taxi Demand Prediction using data from the web. In Proceedings of the IEEE 2018 21st International Conference on Intelligent Transportation Systems (ITSC), Maui, HI, USA, 4–7 November 2018; pp. 1664–1671.
18. Schmöller, S.; Weikl, S.; Müller, J.; Bogenberger, K. Empirical analysis of free-floating carsharing usage: The Munich and Berlin case. *Transp. Res. Part C Emerg. Technol.* **2015**, *56*, 34–51. [CrossRef]
19. Yang, C.; Gonzales, E.J. Modeling taxi trip demand by time of day in New York City. *Transp. Res. Rec.* **2014**, *2429*, 110–120. [CrossRef]
20. Alemi, F.; Circella, G.; Mokhtarian, P.; Handy, S. What drives the use of ridehailing in California? Ordered probit models of the usage frequency of Uber and Lyft. *Transp. Res. Part C Emerg. Technol.* **2019**, *102*, 233–248. [CrossRef]
21. Correa, D.; Xie, K.; Ozbay, K. Exploring the taxi and Uber demand in New York City: An empirical analysis and spatial modeling. In Proceedings of the 96th Annual Meeting of the Transportation Research Board, Washington, DC, USA, 8–12 January 2017.
22. Kopp, J.; Gerike, R.; Axhausen, K.W. Do sharing people behave differently? An empirical evaluation of the distinctive mobility patterns of free-floating car-sharing members. *Transportation* **2015**, *42*, 449–469. [CrossRef]
23. Tyndall, J. Free-floating carsharing and extemporaneous public transit substitution. *Res. Transp. Econ.* **2019**, *74*, 21–27. [CrossRef]
24. Ampudia-Renuncio, M.; Guirao, B.; Molina-Sánchez, R.; de Álvarez, C.E. Understanding the spatial distribution of free-floating carsharing in cities: Analysis of the new Madrid experience through a web-based platform. *Cities* **2020**, *98*, 102593. [CrossRef]
25. Sprei, F.; Habibi, S.; Englund, C.; Pettersson, S.; Voronov, A.; Wedlin, J. Free-floating car-sharing electrification and mode displacement: Travel time and usage patterns from 12 cities in Europe and the United States. *Transp. Res. Part D Transp. Environ.* **2019**, *71*, 127–140. [CrossRef]
26. Hall, R.W. Passenger waiting time and information acquisition using automatic vehicle location for verification. *Transp. Plan. Technol.* **2001**, *24*, 249–269. [CrossRef]
27. Vande Walle, S.; Steenberghen, T. Space and time related determinants of public transport use in trip chains. *Transp. Res. Part A Policy Pract.* **2006**, *40*, 151–162. [CrossRef]
28. He, Y.; Zhao, Y.; Tsui, K.L. Modeling and Analyzing Impact Factors of Metro Station Ridership: An Approach Based on a General Estimating Equation. *IEEE Intell. Transp. Syst. Mag.* **2020**, *12*, 195–207. [CrossRef]
29. Weikl, S.; Bogenberger, K. A practice-ready relocation model for free-floating carsharing systems with electric vehicles—Mesoscopic approach and field trial results. *Transp. Res. Part C Emerg. Technol.* **2015**, *57*, 206–223. [CrossRef]
30. Jäger, B.; Wittmann, M.; Lienkamp, M. Analyzing and modeling a City's spatiotemporal taxi supply and demand: A case study for Munich. *J. Traffic Logist. Eng.* **2016**, *4*, doi:10.18178/jtle.4.2.147-153. [CrossRef]
31. Uber Engineering. H3—A Hexagonal Hierarchical Geospatial Indexing System. 2020. Available online: https://h3geo.org/ (accessed on 28 February 2020).
32. Jorge, D.; Correia, G. Carsharing systems demand estimation and defined operations: A literature review. *Eur. J. Transp. Infrastruct. Res.* **2013**, *13*, doi:10.18757/ejtir.2013.13.3.2999. [CrossRef]
33. Afian, A.; Odoni, A.; Rus, D. Inferring unmet demand from taxi probe data. In Proceedings of the 2015 IEEE 18th International Conference on Intelligent Transportation Systems, Las Palmas de Gran Canaria, Spain, 15–18 September 2015; pp. 861–868.
34. Negahban, A. Simulation-based estimation of the real demand in bike-sharing systems in the presence of censoring. *Eur. J. Oper. Res.* **2019**, *277*, 317–332. [CrossRef]
35. Deutscher Taxi- und Mietwagenverband e.V. *BZP Geschäftsbericht 2017/2018*; Technical Report; Deutscher Taxi- und Mietwagenverband e.V.: Frankfurt am Main, Germany, 2020.
36. Wittmann, M.; Neuner, L.; Lienkamp, M. A Predictive Fleet Management Strategy for On-Demand Mobility Services: A Case Study in Munich. *Electronics* **2020**, *9*, 1021. [CrossRef]
37. La Gatta, V.; Moscato, V.; Postiglione, M.; Sperli, G. An Epidemiological Neural network exploiting Dynamic Graph Structured Data applied to the COVID-19 outbreak. *IEEE Trans. Big Data* **2020**, doi:10.1109/TBDATA.2020.3032755. [CrossRef]
38. Yan, X.; Levine, J.; Zhao, X. Integrating ridesourcing services with public transit: An evaluation of traveler responses combining revealed and stated preference data. *Transp. Res. Part C Emerg. Technol.* **2019**, *105*, 683–696. [CrossRef]

 *electronics*

Article

Hierarchical Temporal Memory Theory Approach to Stock Market Time Series Forecasting

Regina Sousa, Tiago Lima, António Abelha and José Machado *

ALGORITMI Research Center, School of Engineering, Gualtar Campus, University of Minho, 4710-057 Braga, Portugal; regina.sousa@algoritmi.uminho.pt (R.S.); a77788@alunos.uminho.pt (T.L.); abelha@di.uminho.pt (A.A.)
* Correspondence: jmac@di.uminho.pt

Abstract: Over the years, and with the emergence of various technological innovations, the relevance of automatic learning methods has increased exponentially, and they now play a key role in society. More specifically, Deep Learning (DL), with the ability to recognize audio, image, and time series predictions, has helped to solve various types of problems. This paper aims to introduce a new theory, Hierarchical Temporal Memory (HTM), that applies to stock market prediction. HTM is based on the biological functions of the brain as well as its learning mechanism. The results are of significant relevance and show a low percentage of errors in the predictions made over time. It can be noted that the learning curve of the algorithm is fast, identifying trends in the stock market for all seven data universes using the same network. Although the algorithm suffered at the time a pandemic was declared, it was able to adapt and return to good predictions. HTM proved to be a good continuous learning method for predicting time series datasets.

Keywords: time series forecasting; HTM; regression; machine intelligence; deep learning

Citation: Sousa, R.; Lima, T.; Abelha, A.; Machado, J. Hierarchical Temporal Memory Theory Approach to Stock Market Time Series Forecasting. *Electronics* **2021**, *10*, 1630. https://doi.org/10.3390/electronics10141630

Academic Editors: Juan M. Corchado, Josep L. Larriba-Pey, Pablo Chamoso and Fernando De la Prieta

Received: 24 March 2021
Accepted: 6 July 2021
Published: 8 July 2021

Publisher's Note: MDPI stays neutral with regard to jurisdictional claims in published maps and institutional affiliations.

1. Introduction

1.1. Contextualization

HTM can be described as the theory that attempts to describe the functioning of the neocortex, as well as the methodology that intends to provide machines with the capacity to learn in a human way [1].

The neocortex is defined as the portion of the human cerebral cortex from which comes the highest cognitive functioning, occupying approximately half the volume of the human brain. The neocortex is understood by four main lobes with specific functions of attention, though, perception, and memory. These four regions of the cortex are the frontal, parietal, occipital, and temporal lobes. The frontal lobe's responsibilities are the selection and coordination of behavior. The parietal lobe is qualified to make decisions in numerical cognition as well as in the processing of sensory information. The occipital lobe, in turn, has a visual function. Finally, the temporal lobe has the functions of sensory as well as emotional processing and dealing with all significant memory. Thus, the algorithm that is presented intends to create a transposition of this portion of the brain, creating a machine with "true intelligence" [2].

The HTM is built based on three of the main characteristics of the neocortex. Thus, it is a system of memory, with temporal patterns and the construction of regions according to a hierarchical structure.

Starting with the first region, the encoder deals with all of the sensory component. This will receive the data in their raw form, converting them into a set of bits, that will later be transformed into a Sparse Distributed Representation (SDR). Transposing into the human organism, the SDRs correspond to the active neurons of the neocortex. Thus, a 1 bit represents an active neuron while a 0 bit represents an inactive neuron. This transformation is achieved by transforming the data into a set of bits while maintaining the semantic characteristics essential to the learning process. One of the characteristics that proved to be quite interesting is that similar data entries, when submitted to the encoding process, create overlapping SDRs; that is, with the active bits placed in the same positions. Another important characteristic is that all SDRs must have a similar dimensionality and sparsity (the ratio between the number of bits at 1 and the total number of bits) [3]. A certain percentage of sparsity will result in a system's ability to handle noise and under sampling.

The second region, Spatial Pooler (SP), is responsible for assigning the columns according to a fixed number, where each column corresponds to a dendritic segment of the neuron that connects to the input space created by the region described above, the encoder. Each segment has a set of synapses, that can be initialized at random, with a permanence value. Some of these synapses will be active (when connected to a bit with value 1) and consequently will be driven in such a way as to inhibit other columns in the vicinity. Therefore, the SP is responsible for creating an SDR of active columns. This transformation follows the Hebbian learning rule that for each input, the active synapses are driven by inhibiting the inactive synapses. The thresholds dictate whether a synapse is active or not.

The third region, Temporal Memory (TM), starts from the result of the previous two, finding patterns in the sequence of SDRs in order to determine a prediction for the next SDR. At the beginning of the process, all the cells of the active column are also active; however, the region TM is responsible for activating a subset of cells of those same columns when a context is predicted. In case there is no forecast, all the cells remain active. The activation of the previously mentioned subsets of cells is carried out because only in this way can the same entry be represented according to different contexts.

Finally, the classifier is the region in which a decoder calculates the overlap of the predicted cells of the SDR obtained, selecting the one with more overlaps and comparing it with the actual value (if known) [4,5].

Figure 1 describes the typical process of an HTM network.

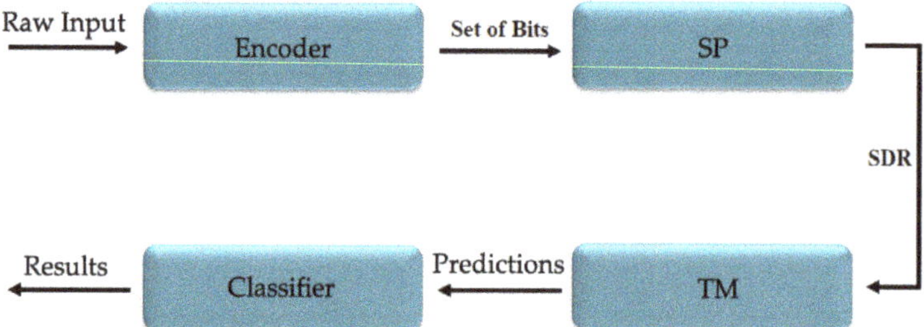

Figure 1. HTM network topology.

1.2. Motivation

HTM is built in three main features of the neocortex: it is a memory system with temporal patterns and its regions are organized in a hierarchical structure. There are many biological details that the theory ignores in case they have no relevance for learning. In short, this approach includes Sparse Distributed Representation (SDR)s, its semantical and mathematical operations, and neurons along the neocortex capable of learning sequences and enabling predictions; these systems learn in a continuous way, with new inputs through time and with flows of information top-down and bottom-up between its hierarchical layers, making them efficient in detecting temporal anomalies. The theory relies on the fact that by mimicking the neocortex, through the encoding of data in a way that gives it a semantic meaning, activating neurons sparsely in an SDR through time will give these systems a power to generalize and learn, not achieved to date with other classic approaches of AI. It is expected to achieve better results and conclusions, while being an intelligence with a higher flexibility when put up against adverse contexts.

1.3. Objectives

The idea of this paper was born from the scope previously mentioned, with the objective to study applications of the HTM theory that are still largely unknown to the pattern learning and recognition community; the applications being studied range from audio recognition, image classification, and time series forecasting with public datasets, that may someday help in anomaly detections in medicine, hospital management, or to act in case of urgency matters. In order to have the confidence to use these systems daily, there is the need for the introduction of new technologies, supported by an AI system with a higher generalization capacity to the ones already in place. With this in mind, the objectives are the following:

- Test and analyze the applications of the HTM theory;
- Compare the HTM theory results against traditional ML technics in terms of:
    - Accuracy and other classification or regression metrics;
    - Computing power/time required;
    - Amount and type of data required;
    - Noise robustness of the algorithms;
    - Possibility to justify the obtained results.

2. State of the Art

Predicting stock market performance is a very challenging task. Even people with an excellent understanding of statistics and probability have difficulty in doing so. Numerous factors combine to make stock prices so volatile that forecasting is at first sight impossible. Adding to all this complexity are all of the political and social factors. Therefore, this article intends to elaborate on a theory and its algorithm on stock market forecasting, determining the future value of a given company's shares. Nevertheless, several studies aim to accept the challenge, and while some statistical and Machine Learning algorithms achieve significant results, the search for closer to ideal results is underway [1,6,7].

There are numerous application fields where HTM can be applied and can produce excellent results. For example, smart cities and their use of sensors, actuators, and mobile devices produce huge streams of data daily, that should be exploited towards innovative solutions and applications [8]. These streams of data are essential for an HTM network that is continuously learning; thus, a problem such as stock market prediction is a good indicator of if HTM can be used in such a scenario, such as in smart cities.

The paper "Forecasting S&P 500 Stock Index Using Statistical Learning Models" [9] defines the primary objective as the forecast of the S&P 500 index movement, using statistical learning models such as logistic regression and naïve Bayes. In this work, an accuracy of 62.51% was obtained. Regarding the dataset, the data were collected between 2004 and 2014, and a transformation of daily prices into daily returns was performed. Similarly, the model described in [10] collects the stock price every 5 min by calculating its return using data for the years 2010 to 2014 from the South Korean stock market. However, in this study, a three-level Deep Neural Network (DNN) model was chosen, using four different representation methods: raw data, Principal Component Analysis (PCA), autoencoder, and restricted Boltzmann machine.

In 2018, ref. [11] proposed a two-stream gated Gated Recurrent Unit (GRU) model and a sentiment word embedding trained on a financial news dataset in order to predict the directions of stock prices by using not only daily S&P 500 stock prices but also a financial news dataset and sentiment dictionary, obtaining an accuracy of 66.32%. More recently, as presented in the article [12], a long short-term memory (LSTM) network was used to predict the future trend of stock prices based on the price history of the Brazilian stock market. However, the accuracy was only 55.9%.

In the same year, in [13], a LSTM network was also used, using an S&P 500 data set for the period from 17 December 2010 to 17 January 2013. In the published document, the objective was well clarified, and it was intended to predict the value of the following day, based on the last 30 days; the mean absolute percentage error (MAPE) obtained was 0.0410%.

In [14], three different models were proposed to forecast stock prices using data from January 2009 to October 2019: autoregressive integrated moving average (ARIMA), simple moving average (SMA), and Holt–Winters method. The SMA model had the best forecasting performance, with a MAPE of 11.456808% in the test data (January to October 2019).

Another DL approach, by [15], made use of Wavelet Transform (WT), Stacked AutoEncoder (SAE) and LSTM in order to create a network for the stock price forecasting of six different markets at different development stages (although it was not clear which companies' data were used); similarly to [16], 12 technical indicators were taken from the data. The WT component had the objective of eliminating noise, the SAE of generating "deep high-level features", and the LSTM would take these features and forecast the next day closing price. With 5000 epochs and the dataset divided into 80% for training, 10% for validation, and 10% for testing, the average MAPE obtained in six years was of 0.011% for the S&P 500 index.

With the increase in the availability of streaming time series data came the opportunity to model each stream in an unsupervised way in order to detect anomalous behaviors in real-time. Early anomaly detection requires that the system must process data in real-time, favoring algorithms that learn continuously. The applications of HTM have been focused on the matter of anomaly detection. In [17], a comparison between an HTM algorithm against others such as Relative Entropy, K-Nearest Neighbor (KNN), Contextual Anomaly Detector (CAD), CAD Open Source Edition (OSE), Skyline, in the anomaly detection of various datasets of the Numenta Anomaly Benchmark107(NAB) was made. HTM demonstrated that it is capable of detecting spatial and temporal anomalies, both in predictable and noisy domains.

In addition, in [18], an HTM network was compared against ARIMA, Skyline, and a network based on the AnomalyDetection R package developed by Twitter, using real and synthetic data sets. Not only were good precision results obtained using the HTM, but there was also a significant reduction in processing time. In [19], it is claimed that most anomaly detection technics perform poorly with unsupervised data; with this in mind, 25 datasets from the NYSE stock exchange, with historical data of 23 years, were analyzed by an HTM network in order to detect anomaly points. However, no explanation of the parameters used was made and no ground truth is known, making it hard to make conclusions. A synthetic dataset was also used, with known anomaly points—the network failed to detect when the values were too low, only detecting when the data were multiplied by 100—possibly by a faulty encoding process.

Leaving the anomaly detection domain, in 2016, [20] used a HTM model to predict the New York City taxi passenger count 2.5 h in advance, with aggregated data at 30-min intervals, obtaining a MAPE of 7.8%, after observing 10,000 data records, lower than other LSTM models used in the study. By including this reference, it is intended to demonstrate that HTM can be used in various contexts and with quite significant results in most cases. In 2020, ref. [21] used recurrent neural networks, such as LSTM and GRU, to solve the same problem of taxi passenger counting. On this approach, through hyper-parametric tuning and careful data formatting, it is stated that both the GRU model and the LSTM model exceeded the HTM model by 30% in lower runtime.

Kang et al. [22], compared the efficiency in memory and time consumption of an HTM network with a modified version of the network for a continuous multi-interval prediction (CMIP) in order to predict stock price trends based on various intervals of historical data without interruption; the conclusions were that the modified version was more efficient in memory and time consumption for this problem, although no conclusions were taken in terms of accuracy of the predictions.

In 2013, Gabrielsson et al. [16], used a genetic algorithm in order to optimize the parameters of two networks: HTM and Artificial Neural Network (ANN); with two months of the S&P 500 index data (open, close, high, low, and volume) aggregated by the minute, 12 technical indicators were extracted and fed to the networks. The problem was converted into a classification one, with training, validation, and test datasets, where the classifier was binary—price will or will not rise—following a buy-and-hold trading mechanism. The Profit and Loss (PnL) was used as a performance measure, where the HTM model achieved more than three times the profit obtained by the ANN network.

The arrival of the Covid-19 pandemic brought uncertainty to the financial markets around the globe. According to [23], an increase of 1% in cumulative daily Covid-19 cases in the US results in approximately 0.01% of an accumulative reduction in the S&P 500 index after one day and 0.03% after one month. In [24], a variety of economic uncertainty measures were examined, showing this same uncertainty; also, it was observed that there is a lack of historical parallelism of this phenomenon, due to the suddenness and enormity of the massive job losses. Both studies suggest that the peak of the negative effects in the stock market was observed during March 2020.

3. Why Hierarchical Temporal Memory?

The HTM starts from the assumption that everything the neocortex decides to do is based on both memories as well as the sequence of patterns; this algorithm is based on the theory of a thousand brains. Among many other things, this theory tries to suggest mechanisms to explain how the cortex represents objects as well as their behavior. HTM is the algorithmic implementation of this theory. The great goal is then to understand how the neocortex works and build systems on that same principle. In particular, this method focuses on three main properties:

- Sequence learning;
- Continuous learning;
- Sparse distributed representations.

This method is relatively recent when compared, for example, to neuronal network techniques. Therefore, it is important to highlight the advantages of HTM and why it was chosen. It should be noted that all the statements presented here were based on authors presented in the state of the art.

In short, the reasons why HTM was chosen are:

1. HTM is the most proven model for the construction of intelligence such as brain intelligence;
2. Although it presents some complexity, it is a scalable and comprehensive model for all the tasks of the neocortex;
3. The neuronal networks are based on mathematics while HTM is inspired fundamentally in the biology of the brain;
4. HTM is more noise-tolerant than any other technique presented until today, due to the sparse distribution representations of raw input;
5. It is a fault-tolerant model;
6. It is variable in time, since it is dependent of state as well as of the context it is presented;
7. It is an unsupervised model;
8. Only a small quantity of data are required;
9. No training/testing datasets are required;
10. Few hyper-parameters tuning—most of the parameters from the algorithms are general to the theory and fall into a specific range of values.

However, as in all methods ever presented, there are already trade-offs:

1. The optimization of HTM for GPU can be difficult;
2. HTM is not a mathematically sound solution as the neural network;
3. This theory is recent and therefore still under construction;
4. There are relatively few applications made so far, and although the community is growing, it is not as vast as the neural networks' community.

4. Data and Methods

Since it was not possible to find a representative dataset of the intended case studies, such as ozone values and traffic in cities, among others, the work was applied to time series forecasting of the close values in the stock market, for seven of the S&P 500 index companies: Amazon, Google, HCA Healthcare, Disney, McDonald's, Johnson & Johnson, and Visa.

4.1. Dataset

The selection of a dataset as well as the features to be used may be determinant for the success of the research work. Therefore, these were well thought out, and a script to obtain stock fluctuations for various companies was made, pulling data from Yahoo Finance, ranging from 3 January 2006 until 18 September 2020. Seven datasets were created, each related to an S&P 500 company: Amazon, Google, HCA Healthcare, Disney, McDonald's, Johnson & Johnson, and Visa; the HCA Healthcare dataset only had data from 10 March 2011, and the Visa dataset from 19 March 2008.

To choose from the S&P 500 list of companies, two parameters were considered: first the market capitalization and then the weight index. Companies are typically divided according to market capitalization: large-cap ($10 billion or more), mid-cap ($2 billion to $10 billion), and small-cap ($300 million to $2 billion). Market capitalization refers to the total dollar value of a company's outstanding shares. The market capitalization represents the product between stock price and outstanding shares:

$$Market - Cap = Stock\ Price \times Outstanding\ Shares \tag{1}$$

The S&P 500 uses a market capitalization weighting method, giving a higher percentage allocation to the companies with the highest market capitalization. Therefore, we chose the companies that represented several S&P 500 list levels with the following market capitalization and indexes [24]. The companies chosen are displayed in Table 1.

Table 1. Companies Market Capitalization and Indexes.

Company	Market Capitalization (Billion $)	S&P500 Index
Amazon	1233.4	4.4
Google	1752.64	1.7
Johnson & Johnson	395.3	1.3
Visa	383.9	1.2
Disney	195.3	1.0
McDonalds	139.5	0.5
HCA	164.46	0.14

With this in mind, the seven companies were chosen due to their familiar popularity and because they represent a wide range of business areas—although they did not represent the entire S&P 500 index, these seven datasets were a good sample for the present study, which pretended to investigate how well the HTM theory adjusts to the stock market forecasting, using the same network for different datasets. Another particularity considered was the inclusion of data after the declaration of the Covid-19 pandemic by the World Health Organization (WHO) on 11 March 2020.

The seven datasets had the same fields: date, open, high, low, close, volume and name. Two points were considered: the units of the Open, High, Low, and Close are in USD and the name corresponds to the name of the stock, not of use for forecasting.

Table 2 describes all columns present in the dataset. On the Table 3, it is shown the maximum values of each parameter per company and on the Table 4, the minimum values of the same parameters. A first comparative analysis can be made where it is verified that although all Amazon columns start with significantly lower values than Google, the company's growth was so positive that it ended up surpassing Google with higher values.

Table 2. Description of Dataset columns.

Column	Description
date	Day of the values taken from the stock market
open	Price of the stock at market open
high	Highest price reached in the day
low	Lowest price reached in the day
close	Price of the stock at market close
volume	Number of shares traded
name	The stock's ticker name

Table 3. Maximum values of each parameter.

Company	High	Low	Open	Close	Volume
Amazon	3495	3467	3547	3400	104,329,200
Google	1800	1540	1609	1442	82,151,100
Johnson & Johnson	157	154	153	148	98,440,200
Visa	207	205	212	205	337,533,600
Disney	145	144	148	135	87,048,500
McDonalds	220	211	229	218	86,981,300
HCA	148	147	147	148	81,150,000

Table 4. Minimum values of each parameter.

Company	High	Low	Open	Close	Volume
Amazon	29	26	34	36	881,300
Google	270	235	135	518	520,600
Johnson & Johnson	52	51	53	49	2,323,800
Visa	13	12	13	13	2,188,800
Disney	120	19	20	19	2,165,700
McDonalds	34	35	32	40	963,299
HCA	19	23	20	26	258,800

When plotting the close values for both companies, corresponding to the stock price at the close of the market, it can be observed that there has been a significant increase over the years. By looking at the Figure 2 it can be concluded that, although Amazon presented lower close values at the beginning of 2006, it recovered the difference, obtaining higher values than Google at the end of 2017. The datasets present different patterns and growths, hence the importance of using different companies for this study.

4.1.1. Hierarchical Temporal Memory Network

All data present in the dataset were uploaded to a HTM network which was developed using a python library called Numenta Platform for Intelligence Computing (NUPIC). NUPIC is a machine intelligence platform that allows the implementation of machine intelligence algorithms.

No pre-processing was carried out to the data because they were already very concise and consistent, without any missing or out of range values; also, the network should be able to interpret anomalies on the data and be resistant to noise.

The various regions of the network present the parameters in Tables 5–8.

The parameters presented in the previous tables were one of the most important processes of choice throughout the investigation. While, for example, inputWidth is a value required to guarantee the encoding of data, columnCount, numActiveColumns, boost, and others were carefully tested in order to choose the best one. Therefore, specifically for data encoding, importance was given to the days of the week and the season. The remaining values are numeric and adapted to the value scales.

As for the SP, the default values were maintained for the following parameters: globalInhibition, localAreaDensity, potentialPct, synPermConnected, synPermActiveInc, and synPermInactiveDec. The remaining parameters: numActiveColumnsPerInhArea, column-Count, and boostStrength were tested and adapted in order to obtain the least possible error.

For the TM region, the parameters tested and adapted according to the results were: cellsPerColumn, maxSynapsesPerSegment, and maxSynapsesPerCell. The remaining parameters were left at the default values: newSynapseCount, initialPerm, permanenceInc, permanenceDec, maxAge, globalDecay, minThreshold, activationThreshold, outputType, and pamLength.

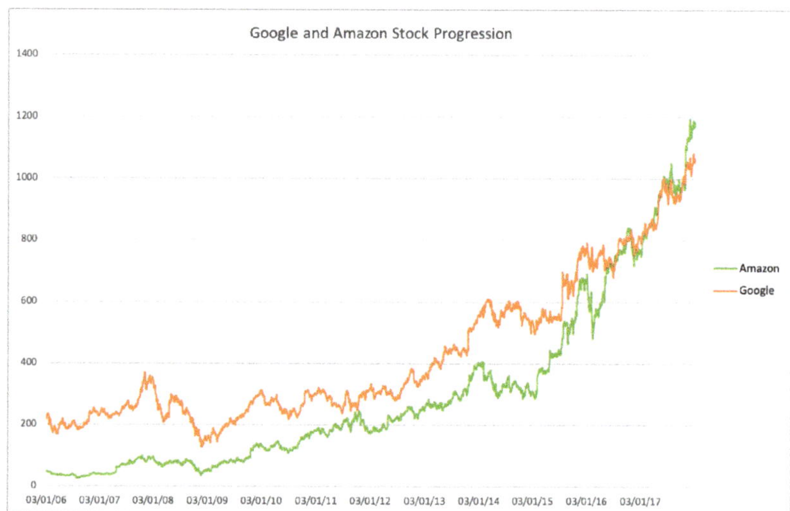

Figure 2. Google and Amazon close value progression in the dataset.

Table 5. Maximum values of each parameter.

Input	Type of Encoder	Parameters
date	DataEncoder	season = dayOfWeek = 3
open	RandomDistributedScalarEncoder	Resolution = 0.5
high	RandomDistributedScalarEncoder	Resolution = 0.5
low	RandomDistributedScalarEncoder	Resolution = 0.5
close	RandomDistributedScalarEncoder	Resolution = 0.5
volume	RandomDistributedScalarEncoder	Resolution = 200
date	DataEncoder	season = dayOfWeek = 3

Table 6. SP Region Parameters.

Parameter	Value
inputWidth	2033
columnCount	4096
GlobalInhibition	1
localAreaDensity	−1
numActiveColumnsPerInhArea	160
potencialPCT	0.85
synPermConnected	0.1
synPermInactiveDec	0.04
synPermInactiveDec	0.005
boostStrength	3

Table 7. TM Region Parameters.

Parameter	Value
inputWidth	2033
columnCount	4096
cellsPerColumn	64
newSynapseCount	20
initialPerm	0.21
permanenceInc	0.1
permanenceDec	0.1
maxAge	0
globalDecay	0
maxSynapsesPerSegment	64
maxSegmentsPerCell	256
minThreshold	12
activationThreshold	16
outputType	Normal
pamLength	1

Table 8. Classifier Region Parameters.

Parameter	Value
Type	SDRClassifier
Alpha	0.25
Steps	1.5

Many of these parameters were left as default, such as the ones related to the synaptic permanence and decay, since they represent the biological link between the known theory of how the neocortex works and its applicability to the network.

4.1.2. Metrics and Evaluation

This study aims to predict the next day's close value of the market for a given company. Three metrics were used to compute the results: root mean square error (RMSQ), MAPE, and absolute average error (AAE) [25].

$$RMSE = \sqrt{\frac{1}{N} \sum_{i=1}^{n} \left(\frac{\hat{x}_i - x_i}{x_i} \right)^2} \tag{2}$$

$$MAPE = \frac{1}{n} \sum_{i=1}^{n} \left| \frac{(\hat{x}_i - x_i)}{x_i} \right| \times 100 \tag{3}$$

$$AAE = \frac{\frac{1}{n} \left(\sum_{i=1}^{n} |\hat{x}_i - x_i| \right)}{\left(\frac{1}{n} \sum_{i=1}^{n} x_i \right)} \tag{4}$$

Since the HTM is supposed to be a continuous learning theory, there are no training/validation/test sets; the data are learned and predicted continuously. To access the learning, the metrics were taken on three moments: to the entire dataset, 365 days before the declaration of the Covid-19 pandemic, and after the declaration. With these three moments, it is possible to gain a better understanding of how quick (in terms of input data needed) the algorithm is to achieve good previsions, while inferring how it adapts to dramatic changes in the input data (in this case, as a consequence of the pandemic).

5. Results

The results were obtained by forecasting the value 'close', concerning the next day, of the stock market for seven different data sets, using the same parameters in the algorithm.

Table 9 shows the values MAPE, RMSE, and AAE obtained for the three different moments, explained in the previous section:

Table 9. Minimum values of each parameter.

Company	Total			365 Days Before Pandemic			After Pandemic		
	MAPE	RMSE	AAE	MAPE	RMSE	AAE	MAPE	RMSE	AAE
Amazon	1.6067	18.8210	8.3973	1.4366	36.4290	25.1878	2.0034	66.3107	51.6145
Google	1.2537	11.9884	6.7297	1.2393	21.8242	14.6977	1.9062	35.3589	25.3776
Johnson & Johnson	0.7320	1.1232	0.6765	0.8593	1.8820	1.1664	1.4166	3.0066	1.9545
Visa	1.2811	1.5859	0.8056	1.1627	2.8037	1.9013	2.1031	5.3197	3.7043
Disney	1.1345	1.2230	0.7204	1.1339	2.1880	1.4229	2.3098	3.5803	2.5156
McDonalds	0.8637	1.5649	0.8731	0.8791	2.6392	1.7090	1.8123	5.3295	3.2158
HCA	1.4504	1.7493	1.0308	1.3894	2.6339	1.8133	3.0940	4.5305	3.1896

In the following graphics (Figures 3–9), the predicted vs. actual values are displayed along the time axis. The algorithm kept a good performance, following the trends of market 'close' value through time, for all datasets. As expected, the algorithm suffered in its previsions around the time of the declared pandemic; however, it was able to achieve some stability afterwards, in line with the possible stability that the stock market can offer in such an unstable time.

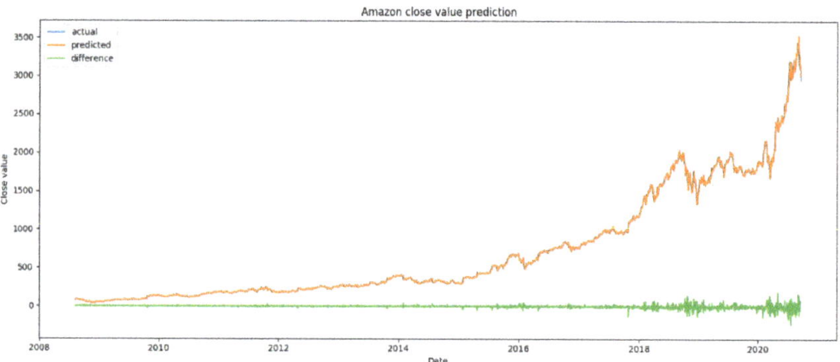

Figure 3. Amazon 'close' value prediction through time.

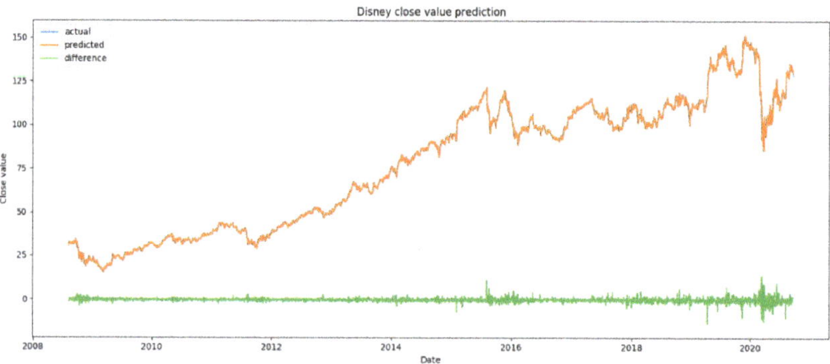

Figure 4. Disney 'close' value prediction through time.

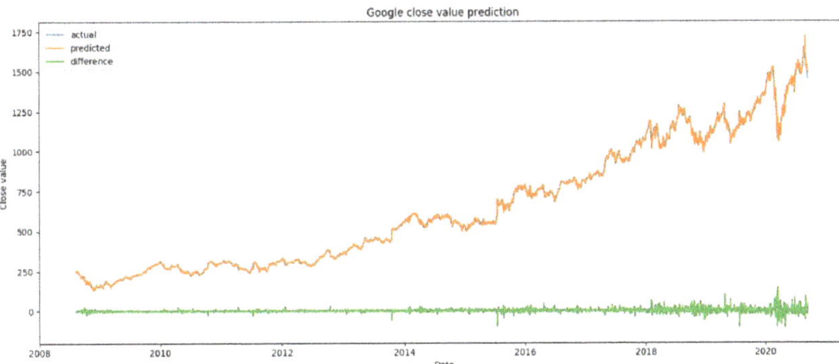

Figure 5. Google 'close' value prediction through time.

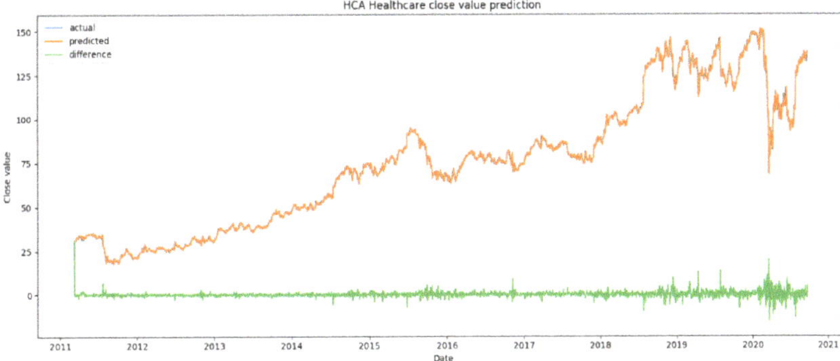

Figure 6. HCA 'close' value prediction through time.

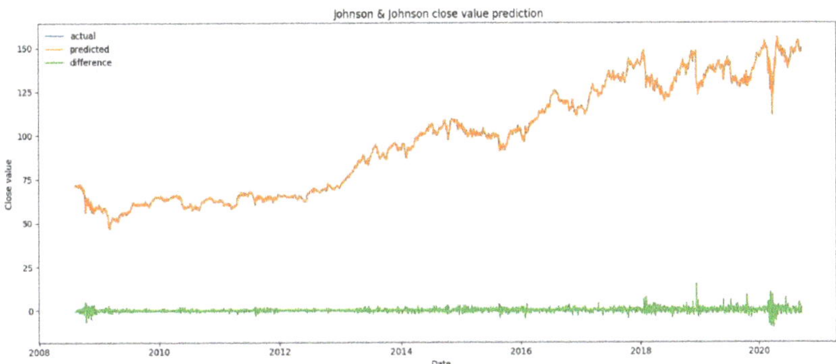

Figure 7. Johnson & Johnson 'close' value prediction through time.

It is also visible by the analysis of the graphics presented that although the value dropped significantly at the beginning of 2020, there is a trend of a continuous rise of the stock.

It is possible to infer that the algorithm learned the patterns quickly, making predictions that were very close to the actual ones with few data. The MAPE values were lower for every dataset in the more stable period before the pandemic, except for the McDonald's and Visa datasets, which received better results in the total period. All MAPE values increased for the post-pandemic period, although not as much for the Amazon dataset—this can be explained by the more stable stock pricing in this company. In general, the RMSE and AAE values increased through time; since these are not percentage metrics, and the data are not normalized, this increase can be explained by the higher 'close' values in the stock market in the last few years across all datasets.

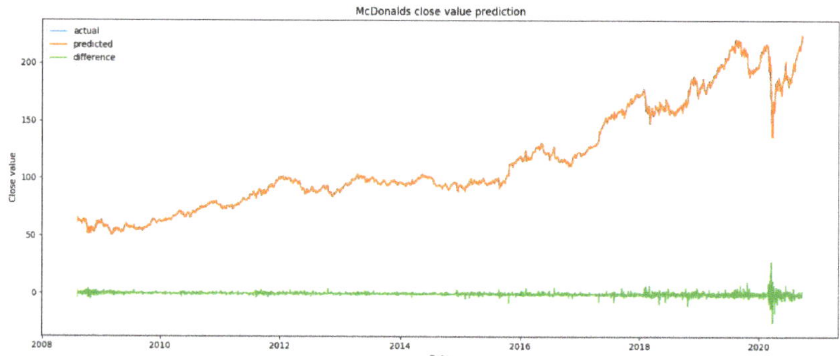

Figure 8. McDonald's 'close' value prediction through time.

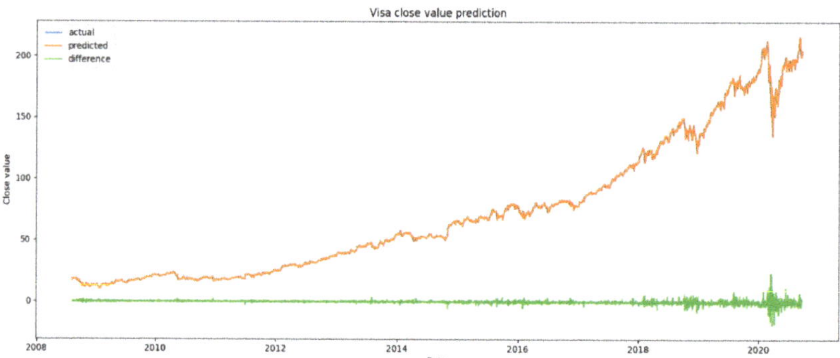

Figure 9. Visa 'close' value prediction through time.

The results obtained in this experiment were very promising, showing that the HTM theory provides a solid framework for time series forecasting, achieving good predictions with few data. Furthermore, the algorithm maintained a good performance across the various datasets: through time, being robust to temporal noise, a bigger complexity of data, and a disruption in the input data caused by the pandemic.

Because of the way HTM works, it is hard to make a rigorous comparison with other methods, which normally divide datasets into training and testing batches.

Besides, in this study, the data used are specific to some S&P 500 companies, ranging from 3 January 2006 until 18 September 2020, contrary to what is observed in the literature, where the time range is typically smaller and no designation of the companies is made—although, some comparisons and findings can be discerned. In [13], the SMA network obtained a MAPE of 11.45% for only a short period of a year, a value worse than what was obtained in the present study for any company for the whole time period available on the datasets. The other two studies presented previously on Section 2, [12,14], related to the forecasting of the next day 'close' value using different LSTM networks, obtained better MAPE values. However, it cannot be stated that these networks perform better, since only a small percentage of the datasets are used for testing and rely on massive training sessions. These methods do not rely on an online continuous learning mechanism such as HTM.

6. Discussion and Conclusions

The advancements of how our brains work biologically may lead to new and revolutionary ways of achieving a true machine intelligence, the aim of the HTM theory. This theory should evolve through the years and help the science community to solve problems typically solved by Machine Learning; specifically Deep Learning in the last few years.

The proposed HTM network obtained good results in the time series forecasting of close values of the stock market, for seven different datasets, through time, proving it can be a great methodology to make predictions while being robust to noise in the data, both in a temporal and spatial axis. It is shown that the network can adapt to different datasets in the same range of problems, with no different hyper-parameter tuning, unlike LSTM and other Deep Learning models; this attribute of HTM models is linked to the known properties of the human cortical neurons and the representation of SDR. Another key difference from other Deep Learning models is that HTM learns continuously, without the need for a specific training dataset; the model learns and predicts continuously. The known experiments where the 'close' value of the stock market is predicted use a classic approach, where training/validation/test dataset tuning is applied to the comparison between models, which is difficult in terms of prediction accuracy; moreover, classically, the data are normalized and suffer a lot of data pre-processing, contrary to the HTM network, where the raw input is only transformed into an SDR, keeping its semantic characteristics.

7. Future Work

As the HTM theory develops, bringing new perspectives of the human intelligence and learning process, such as grid cells [26], it should grab more attention from the data science community, as it will provide a great framework for intelligence and learning.

With regards to future work, there are several possibilities that stand out:

- The combination of this theory with other methods of machine learning. In this way, high dimensional temporal learning problems requiring pre-processing and feature extraction can be solved before creating a sparse representation of the raw input;
- The application of this theory, or even the combination mentioned above, to the in-depth study of the impact of the pandemic on stock prediction;
- The extension of the application of this theory or combination to all S&P 500 companies and to other markets.

We believe that this approach has the most value, since not only does it prove that it is possible to obtain good results with HTM, but it also encourages future research and applications in this same field.

Author Contributions: Each of the authors made substantial contributions to the conception of the article, pleasantly approving the submitted version. Conceptualization, R.S., T.L., A.A. and J.M.; methodology, R.S., T.L., A.A. and J.M.; software, T.L.; validation, A.A. and J.M.; formal analysis, R.S.; investigation, R.S., T.L., A.A. and J.M.; resources, A.A. and J.M.; data curation, R.S., T.L., A.A. and J.M.; writing–original draft preparation, R.S., T.L.; writing–review and editing, R.S.; visualization, R.S., T.L.; supervision, A.A. and J.M.; project administration, J.M.; funding acquisition, J.M. All authors have read and agreed to the published version of the manuscript.

Funding: This work is funded by "FCT—Fundação para a Ciência e Tecnologia" within the R&D Units Project Scope: UIDB/00319/2020. The grant of R.S. is supported by the European Structural and Investment Funds in the FEDER component, through the Operational Competitiveness and Internalization Programme (COMPETE 2020). [Project n. 039479. Funding Reference: POCI-01-0247-FEDER-039479].

Acknowledgments: We thank the administrative staff of the University of Minho for their availability.

Conflicts of Interest: The authors declare no conflict of interest.

Abbreviations

AAE	Absolute Average Error
ANN	Artificial Neural Network
ARIMA	Autoregressive integrated moving average
CAD	Contextual Anomaly Detector
CMIP	Continuous Multi-Interval Prediction
DNN	Deep Neural Network
GRU	Gated Recurrent Unit
HTM	Hierarchical Temporal Memory
KNN	K-Nearest Neighbor
LSTM	Long short-term memory
MAPE	Mean Average Percentage Error
NAB	Numenta Anomaly Benchmark
NUPIC	Numenta Platform for Intelligent Computing
NYSE	New York Stock Exchange
PCA	Principal Component Analysis
PNL	Profit and Loss
RMSE	Root Mean Square Error
SAE	Stacked Autoencoder
SDR	Sparse Distributed Representation
SMA	Simple Moving Average
SP	Spatial Pooler
SVM	Support vector machine
TM	Temporal Memory
WHO	World Health Organization
WT	Wavelet Transform

References

1. Neto, C.; Brito, M.; Peixoto, H.; Lopes, V.; Abelha, A.; Machado, J. Prediction of Length of Stay for Stroke Patients Using Artificial Neural Networks. In *World Conference on Information Systems and Technologies*; Springer: Berlin/Heidelberg, Germany, 2020; pp. 212–221.
2. Ghazanfar, A.A.; Schroeder, C.E. Is neocortex essentially multisensory? *Trends Cogn. Sci.* **2006**, *10*, 278–285. [CrossRef] [PubMed]
3. Purdy, S. Encoding data for HTM systems. *arXiv* **2016**, arXiv:1602.05925.
4. Cui, Y.; Ahmad, S.; Hawkins, J. Continuous Online Sequence Learning with an Unsupervised Neural Network Model. *Neural Comput.* **2016**, *28*, 2474–2504. [CrossRef] [PubMed]
5. Maltoni, D. Pattern Recognition by Hierarchical Temporal Memory. 2011. Available online: http://dx.doi.org/10.2139/ssrn.3076121 (accessed on 12 January 2021).
6. Neves, J.; Martins, M.R.; Vilhena, J.; Neves, J.; Gomes, S.; Abelha, A.; Machado, J.; Vicente, H. A Soft Computing Approach to Kidney Diseases Evaluation. *J. Med. Syst.* **2015**, *39*, 1–9. [CrossRef] [PubMed]
7. Neves, J.; Vicente, H.; Esteves, M.; Ferraz, F.; Abelha, A.; Machado, J.; Machado, J.; Neves, J.; Ribeiro, J.; Sampaio, L. A Deep-Big Data Approach to Health Care in the AI Age. *Mob. Netw. Appl.* **2018**, *23*, 1123–1128. [CrossRef]
8. Liu, C.; Wang, J.; Xiao, D.; Liang, Q. Forecasting S&P 500 Stock Index Using Statistical Learning Models. *Open J. Stat.* **2016**, *6*, 1067–1075. [CrossRef]
9. Chong, E.; Han, C.; Park, F.C. Deep learning networks for stock market analysis and prediction: Methodology, data representations, and case studies. *Expert Syst. Appl.* **2017**, *83*, 187–205. [CrossRef]
10. Minh, D.L.; Sadeghi-Niaraki, A.; Huy, H.D.; Min, K.; Moon, H. Deep Learning Approach for Short-Term Stock Trends Prediction Based on Two-Stream Gated Recurrent Unit Network. *IEEE Access.* **2018**, *6*, 55392–55404. [CrossRef]
11. Nelson, D.M.Q.; Pereira, A.C.M.; de Oliveira, R. Stock market's price movement prediction with LSTM neural networks. In Proceedings of the 2017 International Joint Conference on Neural Networks (IJCNN), Anchorage, AK, USA, 14–19 May 2017; pp. 1419–1426.
12. Liu, H.; Long, Z. An improved deep learning model for predicting stock market price time series. *Digit. Signal Process.* **2020**, *102*, 102741. [CrossRef]
13. Kulkarni, D.; Jadha, D.; Dhingra, D.D. Time Series and Data Analysis and for Stock and Market Prediction. In Proceedings of the 3rd International Conference on Innovative Computing and Communication, Ho Chi Minh City, Vietnam, 14–17 January 2020.
14. Bao, W.; Yue, J.; Rao, Y. A deep learning framework for financial time series using stacked autoencoders and long-short term memory. *PLoS ONE.* **2017**, *12*, e0180944. [CrossRef] [PubMed]
15. Gabrielsson, P.; König, R.; Johansson, U. Evolving Hierarchical Temporal Memory-Based Trading Models. In *European Conference on the Applications of Evolutionary Computation*; Springer: Berlin/Heidelberg, Germany, 2013; pp. 213–222.

16. Ahmad, S.; Lavin, A.; Purdy, S.; Agha, Z. Unsupervised real-time anomaly detection for streaming data. *Neurocomputing* **2017**, *262*, 134–147. [CrossRef]
17. Wu, J.; Zeng, W.; Yan, F. Hierarchical Temporal Memory method for time-series-based anomaly detection. *Neurocomputing* **2018**, *273*, 535–546. [CrossRef]
18. Anandharaj, A.; Sivakumar, P.B. Anomaly Detection in Time Series data using Hierarchical Temporal Memory Model. In Proceedings of the 2019 3rd International conference on Electronics, Communication and Aerospace Technology (ICECA), Coimbatore, India, 12–14 June 2019; pp. 1287–1292.
19. Cui, Y.; Surpur, C.; Ahmad, S.; Hawkins, J. A comparative study of HTM and other neural network models for online sequence learning with streaming data. In Proceedings of the 2016 International Joint Conference on Neural Networks (IJCNN), Vancouver, BC, Canada, 24–29 July 2016; pp. 1530–1538.
20. Struye, J.; Latré, S. Hierarchical temporal memory and recurrent neural networks for time series prediction: An empirical validation and reduction to multilayer perceptrons. *Neurocomputing* **2020**, *396*, 291–301. [CrossRef]
21. Kang, H.-S.; Diao, J. An Integrated Hierarchical Temporal Memory Network for Continuous Multi-Interval Prediction of Stock Price Trends. In *Software and Network Engineering*; Springer: Berlin/Heidelberg, Germany, 2012; Volume 413, pp. 15–27.
22. Yilmazkuday, H. COVID-19 Effects on the S&P 500 Index. Available online: https://papers.ssrn.com/sol3/papers.cfm?abstract_id=3555433 (accessed on 2 January 2021).
23. Altig, D.; Baker, S.; Barrero, J.M.; Bloom, N.; Bunn, P.; Chen, S.; Davis, S.; Leather, J.; Meyer, B.; Mihaylov, E.; et al. Economic Uncertainty Before and During the COVID-19 Pandemic. *J. Public Econ.* **2020**, *191*, 104274. [CrossRef] [PubMed]
24. Stock Market News. 2021. Available online: https://www.marketwatch.com (accessed on 2 January 2021).
25. Hong, W.C.; Li, M.W.; Fan, G.F. *Short-Term Load Forecasting by Artificial Intelligent Technologies*; MDPI-Multidisciplinary Digital Publishing Institute: Basel, Switzerland, 2019.
26. Klukas, M.; Lewis, M.; Fiete, I. Efficient and Flexible Representation of Higher-Dimensional Cognitive Variables with Grid Cells. 2020. Available online: https://journals.plos.org/ploscompbiol/article?id=10.1371/journal.pcbi.1007796 (accessed on 2 January 2021).

 *electronics*

Article

Bus Dynamic Travel Time Prediction: Using a Deep Feature Extraction Framework Based on RNN and DNN

Yuan Yuan [1,2], Chunfu Shao [1,*], Zhichao Cao [3], Zhaocheng He [4], Changsheng Zhu [5], Yimin Wang [4] and Vlon Jang [6]

[1] Key Laboratory of Transport Industry of Big Data Application Technologies for Comprehensive Transport, Beijing Jiaotong University, Beijing 100044, China; 13114240@bjtu.edu.cn
[2] School of Automotive and Transportation, Shenzhen Polytechnic College, Shenzhen 518055, China
[3] School of Transportation and Civil Engineering, Nantong University, Nantong 226000, China; caozhichao@bjtu.edu.cn
[4] Guangdong Provincial Key Laboratory of Intelligent Transportation System, Sun Yat-sen University, Guangzhou 510006, China; hezhch@mail.sysu.edu.cn (Z.H.); wangyim6@mail2.sysu.edu.cn (Y.W.)
[5] School of Intelligent Equipment, Shandong University of Science and Technology, Huangdao District, Qingdao 266590, Shandong Province, China; zhuchangsheng@sdust.edu.cn
[6] Singularity Cloud, Beijing 100000, China; wangqingbaidu@gmail.com
* Correspondence: cfshao@bjtu.edu.cn

Received: 24 September 2020; Accepted: 29 October 2020; Published: 8 November 2020

Abstract: Travel time data is an important factor for evaluating the performance of a public transport system. In terms of time and space within the nature of uncertainty, bus travel time is dynamic and flexible. Since the change of traffic status is periodic, contagious or even sudden, the changing mechanism of that is a hidden mode. Therefore, bus travel time prediction is a challenging problem in intelligent transportation system (ITS). Allowing for a large amount of traffic data can be collected at present but lack of precisely-conducting, it is still worth exploring how to extract feature sets that can accurately predict bus travel time from these data. Hence, a feature extraction framework based on the deep learning models were developed to reflect the state of bus travel time. First, the study introduced different historical stages of bus signaling time, taxi speed, the stop identity (ID) of spatial characteristics, and real-time possible arrival time, signified by fourteen spatiotemporal characteristic values. Then, an embedding network is proposed to leverage a wide and deep structure to mate the spatial and temporal data. In order to meet the temporal dependence requirements, an attention mechanism for a Recurrent Neural Network (RNN) was designed in this research in order to capture the temporal information. Finally, a Deep Neural Networks (DNN) was implemented in this research in order to achieve the dynamic bus travel time prediction. Two case studies of Guangzhou and Shenzhen were tested. The results showed that the performance of the algorithm was more efficient than that of the traditional machine-learning model and promoted by 4.82% compared to the deep neural network applied to the initial feature space. Moreover, the study visualized the weighted cost of attention on the bus's travel time features during a certain running state. Therefore, the study demonstrated the proposed model enabled to understand the characteristic data of transit travel time with visualization.

Keywords: dynamic bus travel time prediction; wide and deep; data fusion; attention; recurrent neural network; deep neural networks

---

1. Introduction

Bus travel time prediction is an important component of an intelligent transport system (ITS). The precise capturing of real-time travel information facilitates the choice of an optimal route by a traveler. Additionally, with unforeseen events occurring, traffic managers adjust departure schedules in real time to ensure the service quality of a system [1,2]. Nevertheless, the travel time of the same bus route in the same city is dynamic due to the nature of bus operation because of frequent traffic congestion, traffic accidents, and road construction. Therefore, it is necessary to focus on a real-time and dynamic bus travel time prediction model in depth in order to further improve traffic efficiency.

Bus travel time prediction has three dependencies. (1) Time dependence [3]: Due to the strong periodicity of passenger demand, bus scheduling also has a certain periodicity. Moreover, bus travel time also depends on the tendency of recent historical travel times. (2) Spatial dependence: The travel time of a particular line is influenced not only by the current traffic state variables of the running line but also by the traffic state variables of the entire bus line [4]. (3) Exogenous dependence: Some exogenous variables, weather conditions, and emergencies may have a great impact on traffic timing prediction [5]. However, driven by big traffic data, a challenge arises: can one gain broad utilization of the latent knowledge hidden in big traffic data in order to predict bus travel time?

Currently, the original statistical-based parameter models (such as K-Nearest Neighbor (KNN) or ARIMA) or machine learning models (such as Support Vector Machine (SVM)) are experiencing more and more difficulty in meeting the requirements of big data in some areas, while the research field of neural networks is active [6,7]. Recently, the neural network shallow prediction model has been used in most scenarios [8]. However, these models have limitations when dealing with large historical data sets and complex nonlinear functions [9].

Deep learning integrates multi-layer architecture and regression to extract inherent features in an end-to-end way. Based on the analysis of a large amount of real-time and historical traffic data, a deep neural network model can deal with the nonlinear characteristics of traffic data and obtain more precise prediction results [10]. However, real-time dynamic bus travel time prediction is very complex, and it involves complex space-time features [11,12]. Moreover, the potential traffic status and traffic events are in a hidden mode. Therefore, the development of a deep learning model is not well suited to capturing the deeper characteristics of bus travel time effectively [13].

For the critical issue of interpreting the space-time features of bus travel time, data-driven methods and neural network methods have been doubted to have this ability [5]. However, there have been a few research literature references that have focused on the diverse traffic features affecting the final prediction of bus travel time. Therefore, this research aimed to explore a new methodology for handling a large number of spatio-temporal features by using deep learning models for the prediction of bus travel time.

In order to solve the problem of focusing on big data feature extraction for bus travel time prediction, in this study, a dynamic real-time bus travel time prediction method was proposed based on a deep learning feature extraction framework and data fusion. In this research, bus travel times were divided into running times and dwelling times, and Global Positioning System (GPS) speeds were added for taxis and buses, as well as travel times based on real-time speeds in order to predict dynamic bus travel times, as indicated in Figure 1. In summary, the main contributions of the proposed approach are those reported below.

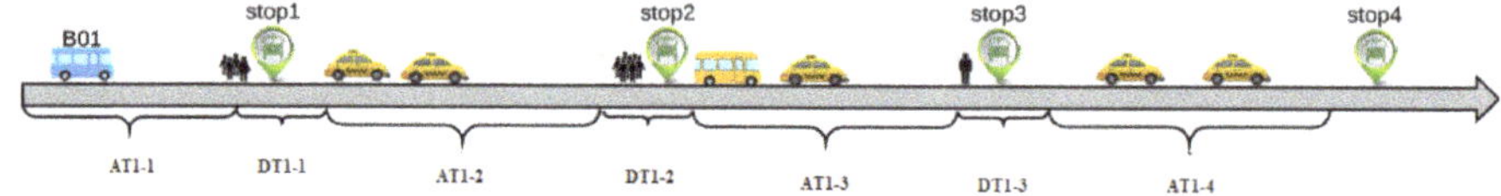

Figure 1. Examples of bus and taxi traveling process.

Based on the prediction of bus travel time, in this research, a new heterogeneous feature extraction framework was proposed based on the recurrent neural network (RNN) model of embedding wide and deep (WD) and an attention mechanism. The framework was proposed in order to gain a deep understanding of the spatio-temporal features and intrinsic connections of the characteristics related to bus travel time and to visualize the connections.

Fourteen spatial and temporal features were introduced, including stop Identities (IDs) as special characteristics, bus dwelling times at different historical levels, real-time GPS bus speeds with real-time possible transit times obtained based on real-time bus speeds as temporal features. These features have not been analyzed together in previous surveys. Lastly, multiple super positions of the RNN and Deep Neural Networks (DNNs) were employed to reduce the residual heterogeneous data fusion and real-time dynamic bus travel time prediction. A novel system for real-time dynamic bus travel time prediction was offered.

To verify the model's stability and generalization ability, the model was tested on the datasets of the Guangzhou No. 226 bus and the Shenzhen No. 113 bus. These buses ran along the main roads in large urban centers. Both of the experiments achieved good results. Other studies never tested their models in different cities.

2. Literature Review

Ever since the rapid development of deep learning methods occurred, the potential for processing large-scale high-dimensional data has been maturing [3,10,14–18].

Recurrent Neural Network (RNN), which is a distinctive construction of deep learning models, is widely used to solve sequence problems [19]. This type of network extends a DNN by repeatedly connecting hidden layers in different timestamps. In this network structure, memory units can dynamically model sequence data. Lately, some studies in the field of transportation has begun to seek RNN to solve the problem of time series predictions, such as traffic flow [20], traffic speed [10], and travel time prediction [21]. Petersen et al. (2019) and He et al. (2020) developed an RNN architecture for the prediction of bus travel times. They demonstrated that the network could capture long-term time dependencies in traffic data, as shown in Table 1 [6,22].

Table 1. A comparison of travel time prediction approaches.

Paper	Model	Feature Extraction	Classification	Factors			Number of Cities	Size of Cities
				AVL	Speed	Dwelling Time		
[23]	SVM, ANN	Cluster	Temporal	Yes	Bus	Historical mean	1 city	megacity
[4]	SVM ANN KNN	No	Temporal	Yes	Taxi	Predicted	1 city	megacity
[22]	CNN RNN	No	Spatio-temporal	Yes	No	No	1 city	metropolis
[6]	RNN	Cluster	Temporal	Yes	No	Historical mean	1 city	megacity
[5]	DNNS	PCA, Cluster	Spatio-temporal	Yes	No	No	1city	metropolis
Ours	RNN DNNs	Embedding Wide and- Deep Attention	Spatio-temporal	Yes	Bus and taxi	Multiple stages	2 cities	Megacities

Note: AVL means automatic vehicle location.

Deep Neural Networks (DNN) has deep fully connected neural layers. An individual DNN does not require the manual extraction of features, and it learns in a supervised way. For our specific problem, the factors that caused congestion, queue delays, and traffic flow came from the fuzzy interaction with complex features. DNN is a multi-layer deep structure that can extract features from data and reveal important potential or hidden structures. Furthermore, DNN provides a powerful and new way to learn how these features interact. Abdollahi et al. (2020) trained a deep, multi-layer perceptron to predict bus travel time [5].

Although the exploration of deep learning models with applications to bus travel time prediction has achieved delightful results, there are still some limitations in these fields. A comparison of the latest bus travel time prediction studies is shown in Table 1.

There are few existing studies on bus travel time prediction using deep learning methods. It has been even rarer to study real-time dynamic bus travel time prediction. In the only studies, although the deep learning methods had a powerful ability to handle large amounts of data and high-dimensional data, the gap between large-scale data and its shallow structure, the gap between full connectivity and rich features [5,13], and the hidden patterns of potential traffic states and traffic events made it difficult for the above models to derive representative features from the rich feature data set. In other words, there has been a lack of systematic, perfect, and in-depth feature learning. Therefore, it is necessary to develop a deep-seated deep learning architecture that fully reflects the features of bus travel time prediction.

The existing studies of the prediction of bus travel time with feature learning still belong to the category of shallower feature learning. Examples include geospatial feature analysis, principal component analysis (PCA), and unsupervised learning algorithms (K-Means) to extract spatial features, and a deep-stacked auto-encoder (SAE) to represent low-dimensional features [5,23]. Using the deep structure of a Recurrent Neural Network (RNN) in time, the historical sequence information was automatically remembered in the model structure [6,22]. The spatial features of the data were extracted from the Convolutional Neural Network (CNN) for use by the Long Short-Term Memory (LSTM) network [22]. DNNs were also used to predict bus travel times after feature extraction [5]. However, most of the research on bus travel time has been shallow in terms of the feature learning structures [5,6,23], lack of feature learning [4,22], or lack of feature learning depth and related breadth. Therefore, it is of great significance to develop a deep feature extraction structure that fully reflects the characteristics of travel time.

The study proposed a neural network that integrated embedded, wide and deep algorithm, and attention mechanisms, and introduced them into a dynamic bus travel time prediction model for design. The extraction framework made use of the non-static space-time correlation existing in urban public transport networks and discovered complex models that traditional methods could not capture. Our study also visualized the RNN model to interpret the impact of various spatial-temporal features on the prediction of dynamic bus travel times, which challenged the traditional neural network approach in the public transport field.

3. Prediction Model

3.1. Feature Extraction Framework

The underlying feature extraction framework was proposed. The framework was composed of Embedding, Wide and Deep, and Attention models.

3.1.1. Embedding

One-hot encoding is one of the most common methods used in dealing with discrete data. Taking Wednesday as an example, it is the third day of a week, and (0, 0, 1, 0, 0, 0, 0) is used to represent three out of seven. One-hot encoding treats each dimension independently, but these representations might not be capable of catching the similarity of each variable; for example, Saturday and Sunday during peak periods might be similar. Additionally, one-hot encoding is too sparse, which is difficult for a deep learning model to deal with [24]

Embedding is a particularly effective method to solve the problems mentioned above, which can be formalized into the following expression:

$$embedding = map\left(X \in R^{N \times 1} \rightarrow X_E \in R^{N \times d}\right),$$

where N denotes the words, d is the embedding size, X is the feature, X_E is the recoded features, and R is the data feature set.

Similar to the data structure mentioned in Section 3.2, the features hours (time of data), day (day of week), and distance, which was used as the station ID instead of bus travel distance, were discrete data features. In order to capture more similarity for each feature, the study implement an embedding model for each feature, as shown in Figure 2.

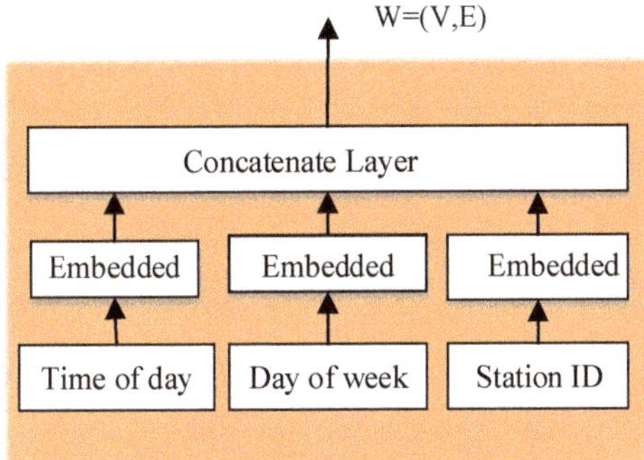

Figure 2. Embedding extraction of discrete features.

3.1.2. Wide and Deep

The bus travel time prediction task included both discrete features and continuous features. The dimensions of the discrete features were much smaller than those of continuous features, and the model would be more susceptible to the impact of continuous data if these features were directly input into the deep model for training. To solve this problem, our study were inspired by the designation of Wide and Deep, shown in Figure 3, for which the core idea was to combine the memory ability of the linear model with the generalization ability of the deep model. In this study, discrete features were applied, such as hours, day, and distance to the wide side, and continuous features were applied to the deep side.

a.   The Wide Component

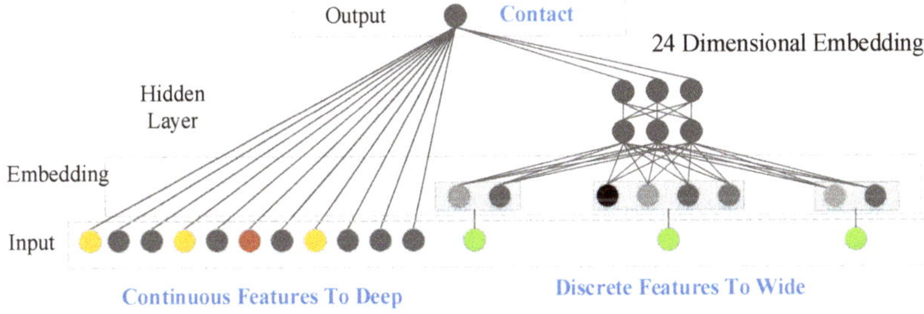

Figure 3. Illustration of the Wide and Deep model improved based on Cheng et al. (2016) [25].

Since the wide side had a high memory ability, it could be used to map the interrelationships after the embedding of discrete features turned into continuous features. Therefore, the discrete features were input into the wide side.

The wide side emphasized features that had often co-occurred in the past, also known as "frequent co-existence features." For example, "Monday", "7:30–9:30," and "station 2–3" often appeared together. The relationship between these three terms allowed us to explain why they occurred so often together. In fact, the memory could be effectively captured by adding interaction items to a broad learning model. The wide side is a generalized linear model for which the form is

$$embedding_{fd} = W_{V \times N}^{T} \times one\_hot(f_d) + b \tag{1}$$

where, $embedding_{fd}$ denotes the predicted discrete outputs, which were treated as traffic state features, W is a $V \times N$ matrix, and V is the set size of the corresponding discrete features. $one\_hot(f_d)$ is the one-hot encoding corresponding to the discrete features.

b. The Deep Component

Generalization is the use of new feature interactions that have occurred rarely or never in historical data, such as "V3 = 40.1" rarely co-occurring with "DT1 = 3" at the same time. Therefore, the wide side could not be used to predict situations that had occurred rarely or never in historical data. However, deep neural networks could find correlations between invisible features.

The deep side had strong feature generation ability, so continuous features were input into the deep side. This allowed the model to capture correlations between different continuous features. The learning model for the expression of continuous features can be expressed as follows:

$$feature_{fc} = W^T_{M \times N} \times f_c + b \tag{2}$$

where, $feature_{fc}$ represents the Continuous features in the bus operation data, W is the vector $M \times N$, M is the size of the continuous features, N is the size of the embedding, f_c is the hidden layer of the neural network, and b is the offset.

c. Joint Training of the Wide and Deep Model

Finally, the features calculated from the two branches were spliced together to obtain the features extracted from the original data. These features can be expressed as

$$feature_f = embedding_{fd} \oplus feature_{fc} \tag{3}$$

where, $feature_f$ is formed by combining discrete features and continuous features, and $\oplus$ is the split joint.

3.1.3. Attention Mechanism

In this study, the attention mechanism was introduced into the task, and our attention-based RNN model that used spatial-temporal features to predict dynamic bus travel times and capture the importance of spatial-temporal features at different locations was proposed.

The attention model performed element-wise multiplication with each feature matrix to obtain a weighted feature matrix, as shown in Figure 4:

$$attn\_feature_t = attn \otimes feature_t. \tag{4}$$

The goal of the attention model was to learn an attention weight matrix $attn\_featureT_t$. In this study, an RNN model was proposed in which h_t was used to learn weights at different states. Each element could be interpreted as the relative importance of $T^f(feature_f atT)$. The activation function *sigmoid* between the output and the hidden layer could limit the output to between 0 and 1:

$$attn\_featureT_t = sigmoid \times \left(W^T h_t(T^f) + b \right) \tag{5}$$

In the formula, W is a T^f matrix, and h_t is a mapping between the input and hidden neurons. In this study, Long Short-Term Memory (LSTM) or Gated Recurrent Unit (GRU) was used in a fully-connected RNN network. Then the spatial-temporal matrix T^f of the historical bus journey point by point was multiplied with the attention matrix $attn\_featureT_t$ (as shown in Formula (2) to obtain a weighted bus journey time matrix for further learning. Therefore, the final attended feature was

$$attn\_T_{rnn} = attn\_featureT_t \otimes T^f \tag{6}$$

In the formula, $attn\_T_{rnn}$ is the weighted eigenvector, and $\otimes$ represents the multiplication of the corresponding elements one by one.

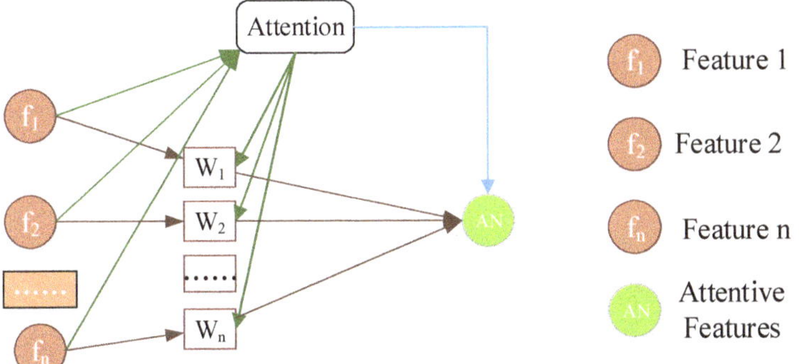

Figure 4. Framework of attention model.

The RNN was used to model the series data, and the RNN hidden features were used to weigh the features. The formation process of the weighted travel time matrix is shown in Figure 5.

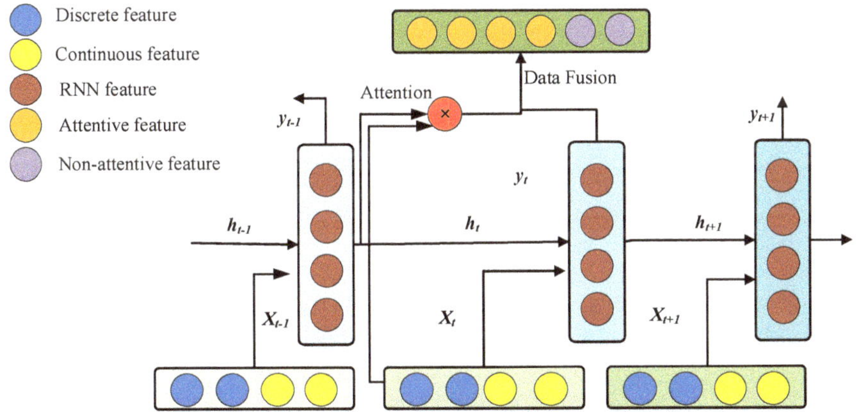

Figure 5. Illustration of the RNN and the attention model.

3.2. Dynamic Bus Travel Time Prediction

The bus travel time prediction procedures underlying are formed by embedding module, Wide and Deep module, RNN and DNN.

Step1: embedding model compresses and encodes discrete data, extracting correlations between discrete features.

Step2: since the wide side had a high memory ability, it was used to map the interrelation ships after the embedding of discrete features turned into continuous features. The deep side captures correlations between different continuous features. The features calculated from the two branches were spliced together to obtain the features extracted from the original data. The wide and deep module that fused discrete and continuous features are shown in Figure 6.

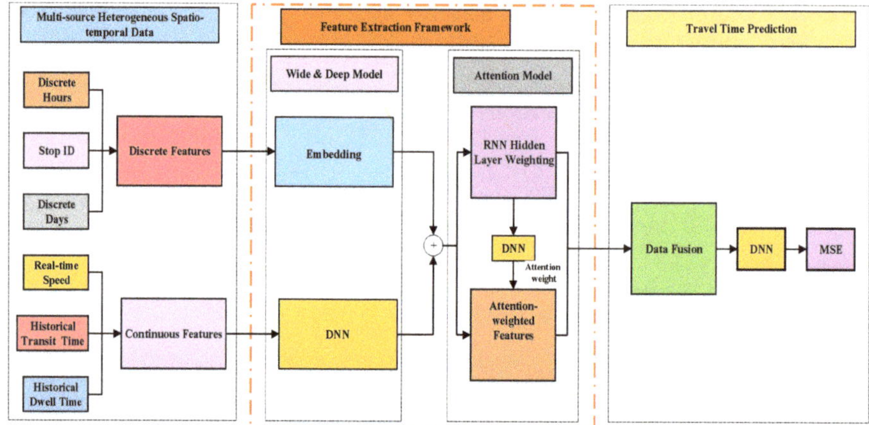

Figure 6. Overview of the model developed for the bus dynamic travel time prediction.

Step3: all of the features were weighted by the attention module. The weighted features were fed into the RNN and DNN models.

DNN is a fully connected deep learning model, which has better ability to obtain the optimal solution. However, there is an insoluble problem with fully connected DNN: it is impossible to model changes in time series. In a normal fully connected network, the hidden layer of DNN can only receive the input at the upper layer at the current moment, while in RNN, the output of neurons can act directly on itself in the next period. In other words, the hidden layer of a recursive neural network can not only receive the input of the previous layer, but also get the input of the current hidden layer at the previous moment. The significance of this change is that it makes the neural network capable of historical memory [26]. In principle, an infinite amount of historical information is well suited for tasks with long-term relevance, such as speech and language. The memory function of RNN is particularly suitable for memorizing and mining sequence data. The multiple combinations and superposition of DNN and RNN can capture the characteristics of the permissible sequence in bus travel time prediction and obtain the optimal solution. Meanwhile, the residual errors can be eliminated by multiple combinations and superposition.

Then Mean Squared Error (MSE) was used in this study to train the model to predict the bus dynamic travel time, as shown in Formula (7):

$$loss_t = \frac{1}{2}\left|target_t - \widetilde{target_t}\right|^2,\tag{7}$$

In the formula, $loss_t$ is a loss function, $target_t$ is the real travel time at time t, and is the predicted value at time t.

The model was built to be end to end, and all of the parameters in the model were trained together. Our general training process is listed as Algorithm 1.

Algorithm 1: Process for Model Training

1	Input:
2	Discrete data f_d in the training dataset
3	Continuous data f_c in the training dataset
4	Travel time $target_t$ in the training dataset
5	Output:
6	Step 1: Initialize all parameters
7	Step 2: Feature extraction
8	Step 3: Feed f_d to the Embedding model to obtain $embedding_{fd}$
9	Step 4: Feed f_c to the DNN model to obtain $feature_{fc}$
10	Step 5: Concatenate all $\{embedding_{fd}, feature_{fc}\}$ as $feature_f$
11	Step 6: For all states in a series of traffic data, do
12	$SAMPLES = [[f_1, f_2, f_3, \ldots, f_T]]$
13	End for stage I feature extraction
14	Training Algorithm
15	Repeat
16	Step 7: Randomly choose a batch of samples in $SAMPLES$
17	For each state of the above
18	Get RNN output $output_t$
19	Get Attention features $attn\_feature_t$
20	End of stage II feature extraction
21	Concatenate both $\{output_t, att\_feature_t\}$
22	Get forecasted travel time $target_t$
23	Compute loss on and $target_t$ using the mean standard error
24	Back propagation
25	Until convergence
26	End training

---

4. Data Collection and Feature Definition

4.1. Data Collection

We evaluated our approach using a large number of buses and taxi GPS data, as well as the bus Automatic Vehicle Monitoring (AVL) data collected by the Transport Department of Guangzhou and Shenzhen in the south of China, which are metropolises with populations of over 14.9 million people and 13.2 million people, respectively.

The bus travel time prediction could be divided into a main road with a signal and a road without a signal. Our experiment in Guangzhou and Shenzhen included different signal periods for multiple intersections connected to each other, which was more challenging for the accuracy of urban main road prediction [27].

To test the No. 226 Bus line in Guangzhou City (23.2 km, 28 stations), the dates for 27 sections and the corresponding areas were collected from 5 October 2014, to 9 November 2014. The No. 226 bus ran through the artery roads (such as Huangpu Road and Dongfeng Road). The running time of the vehicles was 6:00–22:00, and the departure time was 10 min.

The Shenzhen data set used data for the No. 113 bus (19.5 km, 23 stations) with 23 sections and the corresponding areas collected from 20 March 2018, to 5 August 2018. The buses ran through the main road, ShenNan Avenue. The running time of the vehicles is 6:10–23:00, and the departure time was about 4–8 min.

4.2. Features and Definition

Firstly, the main reason why existing estimation approaches could not achieve excellent accuracy is the fact that the travel times are impacted by various factors, such as different weather conditions [28,29], temporal variation of peak and off-peak hours [4,30,31], boarding passenger information [32–34], and real-time traffic conditions [35,36]. Some work focus on analyzing the impacts of different factors. In the study of He [37], the traffic state reports from Twitter information is added as additional data support to predict travel times. The results show that knowing real-time traffic condition helps to increase the estimation accuracy. From the analysis results of the above studies, we can observe that the traffic conditions are uncertain and important for travel time prediction [38]. However, bus GPS data are usually infrequent. Especially, the penetration rate of buses in the traffic network is low at low speed. It is less insensitive to irregular traffic conditions than taxis. It can be observed that only limited studies exist that analyses the influence of real-time traffic flow conditions on bus travel times and the correlation between them [4].

Secondly, the data of Shenzhen city is of 2016. In 2016, the working hours of bus lanes in Shenzhen were from 7:30 am to 9:30 am and from 17:30 pm to 19:30 pm on weekdays (except statutory holidays).Taxes are usually allowed to travel on bus lanes during non-bus lane working hours. In addition, in the field observation of taxi operation, it is found that sometimes passengers will park in the bus lane when getting on or off the taxi. As a result, taxis sometimes run on bus lanes.

Moreover, the data of Guangzhou comes from the time when bus lanes have not been implemented. Therefore, at that time, buses and taxis were traveling together. Therefore, bus GPS and taxi GPS were taken into account when considering the traffic status. Additionally, Different studies have different definitions of real-time. Nikolas Julio [39] defined the dynamic travel time prediction as 10 min when studying the use of traffic shock waves and machine learning algorithms to predict bus speed in real time. Qichongb [40] Predicts bus real-time travel time basing on both GPS and RFID data based on the assumption that the traffic flow keeps the same level in an interval of 30 min although he collects GPS data every 30 s. Hans [41] forecasts Real-time bus route state using particle filter and mesoscopic modeling with four loop detectors installed along the same corridor. Archived data provides access to volume and occupancy information collected approximately every minute. In order to predict the dynamic bus travel time, this paper adds the real-time GPS speed data of the bus every 20 s to the feature for dynamic bus travel time prediction, which effectively improves the prediction accuracy.

Allowing for the data of Shenzhen city based on 2016-year when the exclusive hours of bus lanes in Shenzhen were from 7:30 am to 9:30 am and from 17:30 pm to 19:30 pm on weekdays (except statutory holidays). Besides, taxes are indeed allowed to travel on bus lanes during non-exclusive-bus operating hours. In addition, in the field observation of taxi operation, it is found that sometimes passengers will park in the bus lane when getting on or off the taxi. As a result, taxis always run on bus lanes. Moreover, the fundamental data derived from bus GPS and taxi GPS were taken into account in the paper, which are assumed to represent the traffic status of the PT and road transit, respectively.

We selected fourteen characteristic data sets related to bus travel time prediction, including discrete data, continuous data, spatial data, and time data, as shown in Table 2. In this paper, the features of exited studies is refined into multiple stages, and expanded to the speed of bus and taxi rather than that of one kind vehicle, thus making it more comprehensive to reflect the traffic state.

The existed studies on dynamic travel time using deep learning model, especially the dynamic bus travel time, has not yet been considered. Therefore, based on the above eigenvalues, we added the real-time speed collected within 20s of the bus prediction time into the deep learning model proposed in this paper to predict the dynamic bus travel time.

Table 2. Features and definition.

Features	Data type	Temporal/Spatial	Definition
dt1	Continuous	Temporal	Average bus dwell time within 30 min.
dt2	Continuous	Temporal	Average bus dwell time in 30 min at this point in the last week.
dt3	Continuous	Temporal	Average bus dwell time within 30 min on the same day of the last week.
at1	Continuous	Temporal	Average bus travel time within 30 min.
at2	Continuous	Temporal	Average bus travel time within 30 min of the last week.
at3	Continuous	Temporal	Average bus travel time within 30 min on the same day of the last week.
V1	Continuous	Temporal	Average velocity of the probe vehicles within 5 min.
V2	Continuous	Temporal	Average velocity of the probe vehicles in 5 min at this point in the last week.
V3	Continuous	Temporal	Average velocity of the probe vehicles within 5 min on the same day of the last week.
Real-time speed	Continuous	Temporal	The real-time bus speed was used to reflect the real-time traffic status in the bus lane.
Possible time	Continuous	Temporal	The distance between stops divided by the real-time speed of the bus
Day	Discrete	Temporal	Day of the week, day = {1,2,3,4,5,6,7}.
Hours	Discrete	Temporal	Hours of the day, hours = {8,9 … ,20}.
Stop-ID	Discrete	Spatial	This reflected the spatial relationship between stops and the impact of different stops on the bus travel time prediction.

5. Evaluation

5.1. Establishment of the Experiment

5.1.1. Platform Configuration

The experimental platform hardware components used in this study were an Intel Core i7 8700 @3.2 GHz and 32G DDR4 Memory. The platform software was Centos7.5. Our experiment was operated using Python 3.6.8 and TensorFlow 1.10.0.

The experimental data can be found in Section 3, and the data features are shown in Table 2.

5.1.2. Missing Data

In the process of collecting data, missing data could not be avoided. In our experiments for this study, the records with missing values were discarded because of the use of an RNN, but all the states for a whole line were not discarded for the study. Instead, our study put the site information in as a discrete feature on the wide side for feature learning, which has been talked about previously. Therefore, for the entire bus travel time sequence, there may have been a sequence, such as 1-2-5-8-10, for which a station with missing data was dropped.

5.1.3. Hyper Parameters

In the experiment, the size of embedding for the research was set to four at the wide side, and the number of hidden DNN nodes at the deep side was set to 16. Finally, the features were concatenated. Each state was converted to 28-dimensional features.

5.2. Evaluation Criteria

Three metrics are often used to evaluate the performance of traffic prediction models. They are the mean absolute percentage error (MAPE) [10], mean absolute error (MAE) [4,10,13], and root mean square error (RMS) [10,13]:

$$MAE = \frac{1}{n} \sum_{i=1}^{n} \left| x_i - \widetilde{x_i} \right|, \tag{8}$$

$$MAPE = \frac{100}{n} \sum_{i=1}^{n} \left| \frac{x_i - \widetilde{x_i}}{x_i} \right|, \tag{9}$$

$$RMSE = \sqrt{\frac{1}{n}\sum_{i=1}^{n}(x_i - \widetilde{x}_i)^2},\tag{10}$$

where, N is the number of samples, i is the number of stations, x_i is the real bus travel time, and $\widetilde{x}_i$ is the predicted bus travel time.

The MAE and MAPE are indicators of regression tasks. Compared with the MAE and MAPE, the RMSE is more sensitive to outliers, and it can amplify larger prediction deviations. It is often used to compare the stability of different prediction models. The MAPE provides prediction errors based on the percentage difference between observed and predicted bus travel times as a measure of the prediction accuracy of the statistical prediction methods. These performance indicators provide a deep understanding of the nature of prediction errors [10].

5.3. Experiment Results

This section describes the evaluation of the accuracy of our approach for this study based on six types of experiments with our proposed model compared to the existing models.

5.3.1. Different Method

To study different methods of accuracy, the following results were compared for the test set of MAPE indicators in this research: The historical average estimate HAV, using only historical information and not joining the floating vehicle average speed information SVR1, historical information with the floating vehicle average speed SVR2 and a prior probability distribution of the bus travel time, the use of Bayesian theory to modify the SVR2-Bayes theorem of the SVR2 experiment results, the linear model, a neural network based on depth within the DNN [5] and RNN [6], and our proposed methods. See Table 3 and Figure 7 [5,6].

Table 3. Results for GuangZhou261.

Error Index	HAV	SVR	SVR1	SVR 2	ANN	SVR 2-Bayes	SVR 2-Bayes2	Our Study
MAPE (%)	18.5	17.98	17.98	16.4	17.23	14.93	14.29	**8.43**

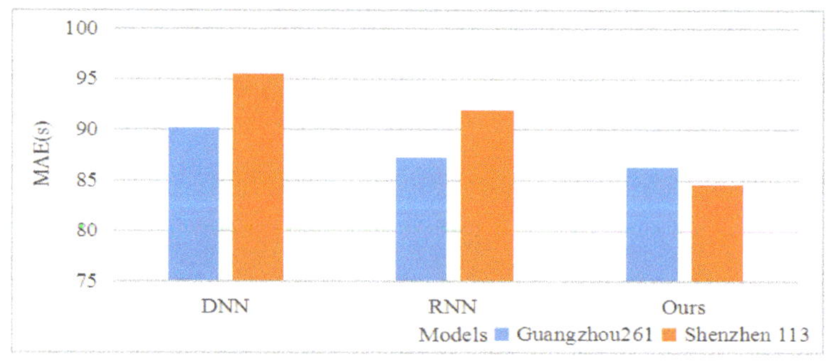

Figure 7. Comparison of different datasets

Table 3 indicates that the DNN [5] and RNN [6] represented relatively advanced artificial intelligence algorithms. The DNN had nearly 5% more absolute promotion than the SVR2-Bayes2 model did. After replacing the model with the RNN, the MAPE could be further reduced by capturing the interdependence between different sites, indicating that the spatial-temporal relationship between sites had a certain impact on the prediction accuracy. However, the RNN itself did not pay attention to the importance of different features in different states. Therefore, our study combined the RNN with Wide and Deep and attention mechanisms to form a feature extraction framework. The accuracy of the proposed RNN network was further reduced by 0.5% and relatively improved by 5% compared with the RNN network alone.

Based on Guangzhou PT center dataset, in order to study the changes in the travel time at different times for each station, the predicted and true values of the bus travel time model for 8:00 AM were randomly selected. It can be seen from Table 4 that our model could reduce the MAPE by 4–7% compared with svr2-bayes2, indicating that our algorithm had a good performance during the peak or flat peak times.

Table 4. MAPE values for different methods in different periods.

Time \ MAPE	SVR2 (%)	SVR2-Bayes (%)	SVR2-Bayes2 (%)	Our Study (%)
Morning (08:00–09:00)	15.99	14.21	13.83	9.51
Evening (17:00–19:00)	16.24	13.69	13.62	6.43
Flat peak	14.37	14.38	12.25	5.21

5.3.2. Different Dataset

The MAE and MAPE are indicators of regression tasks. For different scenarios, we use MAE and MAPE two standards to evaluate the error. First, for a complete bus line, compare the data sets of Guangzhou and Shenzhen with RNN, DNN and our own algorithm, and use MAE to evaluate the error time to compare the operation of the entire line. This is important for traffic managers or bus scheduling and dispatching personnel. The MAPE provides prediction errors based on the percentage difference between observed and predicted bus travel times as a measure of the prediction accuracy of the statistical prediction methods. Then, for the morning peak, evening peak and peace peak, we use MAPE for comparison and evaluation of the stability of different prediction models. These performance indicators provide a deep understanding of the nature of prediction errors. The results as below.

(1) With using the data for the No. 261 bus in Guangzhou and No. 113 bus in Shenzhen, we verify the generalization performance of our model. The results showed that the proposed algorithm in this research had better performance for the whole routes than the DNN or the RNN for both the Guangzhou and Shenzhen datasets, as shown in Figure 7.

(2) We chose the peak period of the working day with a more complex traffic state and the flat peak period with a simple traffic state as the research period, and we compared the algorithms. It can be seen from Table 5 that for the maximum morning rush hour of the MAPE in Guangzhou, the error of the one-hour bus journey time was about 6.5 min. For the maximum evening peak of the MAPE in Shenzhen, the error of the one-hour bus journey time was about 5.6 min, which indicated that this model had good generalization ability, and it solved the problem that the proposed deep learning algorithms were suitable for the traffic states of different cities.

Table 5. Comparison of the results of the model for Guangzhou and Shenzhen.

Time \ MAPE	Guangzhou (%)	Shenzhen (%)
Morning (08:00–09:00)	10.79	8.74
Evening (17:00–19:00)	9.11	9.33
Flat peak	8.17	7.97

5.3.3. Real-Time Bus Speed Information for Prediction

In order to get closer to the application scenario, our study divided their model into two scenarios in the prediction process. The two scenarios comprised a model based entirely on historical data and a model based on bus real-time speed parameter correction. Our study use Model-Hist and Model-Real for the two scenarios.

The hyper parameters of these two different models were the same. The only difference was that in the model based on real-time sensor data correction, the speed of a real-time bus sensor was added to the model as a feature on the deep side. The results used only historical data and real-time bus speed data with historical data, as shown in Table 6.

Table 6. Comparative experiment for the historical data model and the real-time data model.

Error Indexes	Model-Hist	Model-Real
MAPE (%)	8.14	3.32
MAE	84.61	38.85
RMSE	108.9	51.49

After the addition of real-time bus speed data, the MAPE of the bus travel time forecast decreased by 4.82% compared with the historical data alone. This indicated that the MAPE value obviously decreased after considering the real-time speed of the bus, which in turn indicated that the real-time speed information of the bus had a great influence on the bus travel time prediction.

This was consistent with the view that the speed data of a taxi could reflect the traffic status [4], but it was also valuable to add the real-time speed of a bus to reflect the traffic status of bus lines. Because the bus often ran in the bus lane, the combination of the real-time speed of the bus, the historical speed of the taxi, and the historical speed of the bus could reflect the traffic status of the bus route more comprehensively and accurately. It also confirmed the work of Ma et al. (2019), who said that in their future work they would focus on using an existing taxi or another type of traffic data to estimate the newly designed or sparsely recorded bus travel time [4].

Figure 8 shows the data for about one week from 9:00 to 10:00 for the morning bus travel time of the site actual arrival time and the predicted travel time and the cumulative error figure. With the bus real-time speed in the model, the bus travel time gap between the predicted values and the real value was relatively small.

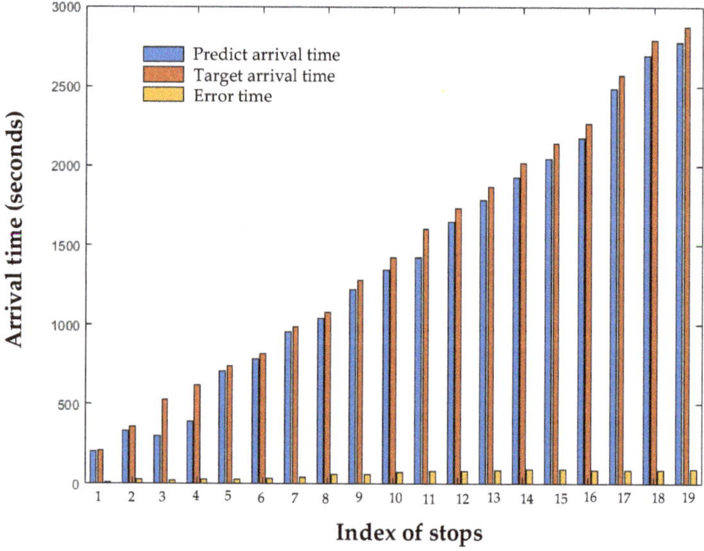

Figure 8. Actual/predicted arrival time and cumulative error diagram for each site.

5.3.4. Wide and Deep

In order to enable the model to capture as many differences between discrete and continuous features as possible, our study introduced the WD (Wide and Deep) model into the RNN model. With compared the influence of this module to the results for two different data sets in Guangzhou and Shenzhen, the results of the comparison are shown in Table 7.

Table 7. The impact of Wide and Deep.

Method	Guangzhou		Shenzhen with Real-Time Speed	
	Without W&D	With W&D	Without W&D	With W&D
MAPE (%)	8.81	8.43	3.42	3.32
MAE	87.27	86.31	40.16	38.85
RMSE	119.95	120.11	54.09	51.49

The model with the WD module was improved for the Guangzhou and Shenzhen data to different degrees, which proved that the discrete data and the continuous data played different roles in the model. The wide side could effectively memorize discrete features, while the deep side could effectively generalize continuous features.

5.3.5. Attention

The reason for using attention-based temporal and spatial architecture was that there was spatial-temporal correlation among the traffic variables that predicted the bus travel times. For the task of bus travel time prediction, our study thought that the spatial-temporal relationships of the data might have different influences on the prediction results. Therefore, in this study, the attention module was used to weight the spatial-temporal features. Under the condition of fixed hyper parameters, the effects of adding attention mechanism or removing the attention mechanism on the prediction results of the model were compared in this research.

As shown in Table 8, the results showed that the attention (Attn) model was helpful for improving the accuracy of the bus travel time prediction whether the data sets of Guangzhou or Shenzhen were used and whether the historical data was used alone or combined with the real-time bus speed.

Table 8. Influence of the attention mechanism on the prediction results for Guangzhou and Shenzhen.

Method	Guangzhou with Hist		Shenzhen with Real-Time Speed	
	Without Attn	With Attn	Without Attn	With Attn
MAPE (%)	8.53	8.43	3.42	3.32
MAE	87.67	86.31	39.33	38.85
RMSE	124.32	120.11	52.29	51.50

To verify the mechanism of attention, in this study, the weighted coefficient of attention for the bus's travel time features was visualized during a certain running state. Figure 9 shows the heat map of the spatial feature temporal features. In the visual feature map, the red areas represent higher response values, and the blue areas represent lower response values. By analyzing the attention scores learned by the attention model described in Section 4.1, our study were able to learn the view of the proposed method for the propagation mechanism of bus travel time prediction.

To further understand the propagation mechanism learned by the attention model used in the proposed method, the evolution of the attention scores was analyzed with respect to the impact on different bus stop IDs and the influence in the whole bus travel time prediction. Generally, it can be seen from Figure 9 that whether discrete (day, hour, stop ID) or continuous (v, dt, at, real-time speed, possible time) features were used, and whether temporal (v, dt, at, real-time speed, possible time, day, hour) or spatial (stop ID) features were used; all of the features had different impacts on the prediction of bus travel times and bus stops.

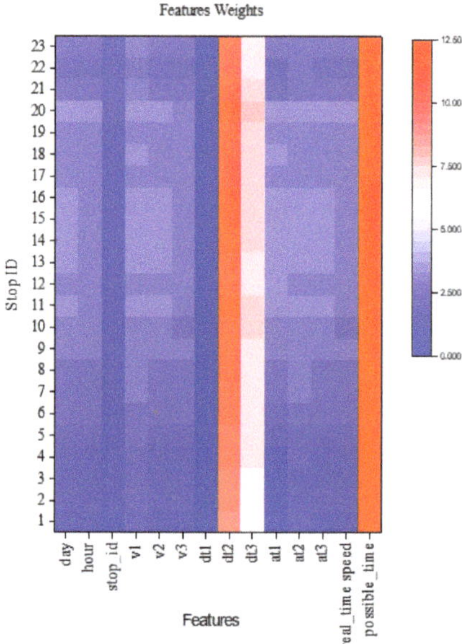

Figure 9. Average features weighting matrix.

As shown in Figure 9, our study could observe that temporal feature dt2 (average bus dwell time in 30 min at this point in the last week.) performed a rather important function in the model. This reflected the fact that there was an influence from the complex boarding mode, the bus dwell time was very unstable [4], and there were different basic modes, which had a significant impact on the total travel time. Additionally, dt2 had a different impact on different bus stops. In comparison, dt3 (average bus dwell time within 30 min on the same day of the last week) also had a moderate impact and dt1 (average bus dwell time within 30 min) had almost no impact on different bus stops.

Compared with previous studies, Ma et al. (2019) [4] did not forecast the dwelling time of each bus stop as a part of the total travel time of a bus [4], and Xu (2017) [23] and He et al. (2020) [6] did not use the full historical average bus dwelling times [6,24]. According to our heat map, shown in Figure 9, DT2 and DT3 have a greater weight on the prediction of bus travel time. DT2 is Average bus dwell time in 30 min at this point in the last week. DT3 is average bus dwell time within 30 min on the same day of the last week. It indicates that the bus travel time is influenced by the periodicity of dwelling time. Hence, it was a better decision to choose DT2 and DT3 simultaneously because the two features of bus dwelling time worked well in the prediction of bus travel time. Based on real-time information, it was important for the accuracy of real-time bus travel time prediction, especially the possible real-time transit times converted from real-time bus speeds. However, previous studies only considered the real-time bus speeds [23], rather than the possible real-time transit times. Furthermore, many research works of traffic prediction have emphasized the importance of spatial information [5,22]. The spatial feature of a bus stop ID had a certain impact on bus travel time prediction, and it had different influences on different bus stops. However, the impact is less prominent yet.

5.3.6. Hyper Parameters

In the experiment, the influences of different operation units of GRU and LSTM on the prediction results were compared for the study, as shown in Table 9.

Table 9. The influence of different hyper parameters on the model.

Error Indexes	Model-LSTM	Peepholes	Model-GRU
MAPE (%)	3.33	3.32	3.34
MAE	39.15	38.85	39.20
RMSE	52.05	51.50	52.27

Although the LSTM model was better than the GRU model, different computing units had little influence on the final prediction results of the model, which may have been because none of the computing units could capture the characteristics of migration between different states.

6. Conclusions

From the perspectives of time and space, the bus travel times of public transportation are dynamic/uncertain. The gap between a massive amount traffic data and its shallow features and the gap between full connection and rich features make it difficult to obtain representative features from datasets with rich features. The potential traffic state and traffic events belong to a hidden mode, so travel time prediction is a challenging problem of ITS. Therefore, it is of particular importance to develop a deep-seated architecture that fully reflects the characteristics of transit travel time.

We proposed an embedded network lever WD structure to solve the spatial data and designed an attention mechanism for the RNN to capture the temporal information. Finally, the system used the deep neural network model composed of the RNN and the DNN. The model could capture the non-static spatiotemporal correlation of the urban bus travel time. This enabled the model to generalize the learning model in the cross-temporal and spatial prediction. The model could be used to predict the dynamic travel times of buses. Its effect was better than those of the historical average method, traditional SVR model, SVR-Bayes optimization model, single DNN [5], and RNN [6,21], as shown in Figure 7. The main contributions of this study were as follows.

- Based on the prediction of bus travel time, the study proposed a new heterogeneous feature extraction framework based on the RNN model of embedding, WD, and an attention mechanism in order to gain a deep understanding of the space-time features and intrinsic connections of the characteristics related to bus travel time, and to visualize these features and connections, as shown in Tables 7 and 8, Figure 9.
- Fourteen historical spatial and temporal features were introduced, including the stop IDs as spatial features, bus dwelling times at different historical stages, and real-time GPS bus speeds with real-time possible transit times obtained based on real-time bus speeds as temporal features as shown in Table 2. Especially, the real-time bus speeds is important to improve the dynamic bus travel time prediction, which can be seen in Table 6. These features have not been studied together in previous studies
- The multiple super composition of the RNN and DNNs were carried out to reduce the residual heterogeneous data fusion and real-time dynamic bus travel time prediction. A new scheme for real-time dynamic bus travel time prediction was provided as shown in Figures 7 and 8.
- To verify the model's stability and generalization ability, the model was tested on the data sets of the Guangzhou 226 bus and the Shenzhen 113 bus. The buses ran on the main roads in big cities. Both of the experiments achieved good results. Few of the existing studies tested their models in different cities as shown in Figure 7 and Table 5.

For future work, our study will keep exploring the presented systems in the following directions. In addition to further improving the accuracy of the model, we will extend from one bus line to the bus lines of the entire road network. The existing models tested individual bus routes. The comparison can prove the validity of the model, but our study hold the point that more factors need to be considered. Therefore, it is a feasible choice to try to input the entire road network as a model. With the development of in-deep learning technology, this effect can be achieved through a deep image convolution network of reference image processing [42], which is an important direction of our future research. The effect of missing data on the prediction is obvious. When the missing rate was more than 5%, the performance of the model decreased significantly when only speed was used as the input. When using the multi-attribute fusion, the model had good performance, not only when the error value was low but also when the error growth rate was low, particularly when compared with the model of Liu et al. (2018) [10]. In this research, the missing data were not considered thoroughly enough. In the future, more kinds of sensors (such as ground loops, videos, and geomagnetism) can be considered in order to repair the missing data and to further improve the accuracy.

Author Contributions: Wrote the manuscript, Y.Y. and C.S.; provided relevant information, discussed the data, and corrected the manuscript, Z.C., Z.H. and C.Z.; revised the manuscript, Y.Y., C.S., Z.C., Y.W. and V.J. All authors have read and agreed to the published version of the manuscript.

Funding: This research was supported by National Natural Science Foundation of China (Grant No. 51678044), National Natural Science Foundation of China (Grant No. 52072025), Joint Funds of the National Natural Science Foundation of China (U1811463), National Natural Science Foundation Youth Fund (Y820631001). The study also had the support of the Guangdong Key Laboratory of Intelligent Transportation of Sun Yat-Sen University and the Shenzhen Transportation Committee.

Conflicts of Interest: The authors declare no conflict of interest.

References

1. Cao, Z.; Ceder, A. Autonomous shuttle bus service timetabling and vehicle scheduling using skip-stop tactic. *Transp. Res. Part C Emerg. Technol.* **2019**, *102*, 370–395. [CrossRef]
2. Cao, Z.; Ceder, A.; Zhang, S. Real-time schedule adjustments for autonomous public transport vehicles. *Transp. Res. Part C Emerg. Technol.* **2019**, *109*, 60–78. [CrossRef]
3. Yu, H.; Wu, Z.; Wang, S.; Wang, Y.; Ma, X. Spatiotemporal Recurrent Convolutional Networks for Traffic Prediction in Transportation Networks. *Sensors* **2017**, *17*, 1501. [CrossRef]
4. Ma, J.; Chan, J.; Ristanoski, G.; Rajasegarar, S.; Leckie, C. Bus travel time prediction with real-time traffic information. *Transp. Res. Part C Emerg. Technol.* **2019**, *105*, 536–549. [CrossRef]
5. Abdollahi, M.; Khaleghi, T.; Yang, K. An integrated feature learning approach using deep learning for travel time prediction. *Expert Syst. Appl.* **2020**, *139*, 112864. [CrossRef]
6. He, P.; Jiang, G.; Lam, S.-K.; Sun, Y. Learning heterogeneous traffic patterns for travel time prediction of bus journeys. *Inf. Sci.* **2020**, *512*, 1394–1406. [CrossRef]
7. Laña, I.; Del Ser, J.; Velez, M.; Vlahogianni, E. Road Traffic Forecasting: Recent Advances and New Challenges. *IEEE Intell. Transp. Syst. Mag.* **2018**, *10*, 93–109. [CrossRef]
8. Lv, Y.; Duan, Y.; Kang, W.; Li, Z.; Wang, F.-Y. Traffic Flow Prediction With Big Data: A Deep Learning Approach. *IEEE Trans. Intell. Transp. Syst.* **2014**, *16*, 1–9. [CrossRef]
9. Bengio, Y. *Learning Deep Architectures for AI*; Now Publishers Inc.: Boston, MA, USA, 2009.
10. Liu, Q.; Wang, B.; Zhu, Y. Short-Term Traffic Speed Forecasting Based on Attention Convolutional Neural Network for Arterials. *Comput. Civ. Infrastruct. Eng.* **2018**, *33*, 999–1016. [CrossRef]
11. Li, L.; Li, Y.; Li, Z. Efficient missing data imputing for traffic flow by considering temporal and spatial dependence. *Transp. Res. Part C: Emerg. Technol.* **2013**, *34*, 108–120. [CrossRef]
12. Tan, H.; Feng, J.; Feng, G.; Wang, W.; Zhang, Y.-J. Traffic Volume Data Outlier Recovery via Tensor Model. *Math. Probl. Eng.* **2013**, *2013*, 1–8. [CrossRef]
13. Wu, Y.; Tan, H.; Qin, L.; Ran, B.; Jiang, Z. A hybrid deep learning based traffic flow prediction method and its understanding. *Transp. Res. Part C Emerg. Technol.* **2018**, *90*, 166–180. [CrossRef]
14. Casas, N. Deep deterministic policy gradient for urban traffic light control. *arXiv* **2017**, arXiv:1703.09035.
15. El Hatri, C.; Boumhidi, J. Fuzzy deep learning based urban traffic incident detection. *Cogn. Syst. Res.* **2018**, *50*, 206–213. [CrossRef]
16. Gu, Y.; Lu, W.; Qin, L.; Li, M.; Shao, Z. Short-term prediction of lane-level traffic speeds: A fusion deep learning model. *Transp. Res. Part C Emerg. Technol.* **2019**, *106*, 1–16. [CrossRef]
17. Wang, Y.; Zhang, D.; Liu, Y.; Dai, B.; Lee, L.H. Enhancing transportation systems via deep learning: A survey. *Transp. Res. Part C Emerg. Technol.* **2019**, *99*, 144–163. [CrossRef]

18. Xu, C.; Ji, J.; Liu, P. The station-free sharing bike demand forecasting with a deep learning approach and large-scale datasets. *Transp. Res. Part C Emerg. Technol.* **2018**, *95*, 47–60. [CrossRef]
19. Tsoi, A.C.; Back, A. Discrete time recurrent neural network architectures: A unifying review. *Neurocomputing* **1997**, *15*, 183–223. [CrossRef]
20. Li, C.; Wang, J.; Ye, X. Using a Recurrent Neural Network and Restricted Boltzmann Machines for Malicious Traffic Detection. *NeuroQuantology* **2018**, *16*, 823–831. [CrossRef]
21. Duan, Y.; Lv, Y.; Wang, F.Y. Travel Time Prediction with LSTM Neural Network. In Proceedings of the IEEE 19th International Conference on Intelligent Transportation Systems (ITSC), Rio de Janeiro, Brazil, 1–4 November 2016.
22. Petersen, N.C.; Rodrigues, F.; Pereira, F.C. Multi-output bus travel time prediction with convolutional LSTM neural network. *Expert Syst. Appl.* **2019**, *120*, 426–435. [CrossRef]
23. Xu, H.; Ying, J. Bus arrival time prediction with real-time and historic data. *Clust. Comput.* **2017**, *20*, 3099–3106. [CrossRef]
24. Liu, Y.; Liu, Z.; Jia, R. DeepPF: A deep learning based architecture for metro passenger flow prediction. *Transp. Res. Part C Emerg. Technol.* **2019**, *101*, 18–34. [CrossRef]
25. Cheng, H.T.; Koc, L.; Harmsen, J.; Shaked, T.; Shah, H. Wide & Deep Learning for Recommender Systems. In Proceedings of the 1st Workshop on Deep Learning for Recommender Systems, Boston, MA, USA, 15 September 2016.
26. Liu, Y.; Wang, Y.; Yang, X.; Zhang, L. Short-term travel time prediction by deep learning: A comparison of different LSTM-DNN models. In Proceedings of the IEEE 20th International Conference on Intelligent Transportation Systems (ITSC), Yokohama, Japan, 16–19 October 2017.
27. Oh, S.; Byon, Y.J.; Jang, K.; Yeo, H. Short-term travel-time prediction on highway: A review on model-based approach. *KSCE J. Civ. Eng.* **2017**, *22*, 298–310. [CrossRef]
28. Cheng, Z.; Chow, M.Y.; Jung, D.; Jeon, J. A big data based deep learning approach for vehicle speed prediction. In Proceedings of the IEEE 26th International Symposium on Industrial Electronics (ISIE), Edinburgh, UK, 19–21 June 2017.
29. Ma, Z.; Koutsopoulos, H.N.; Ferreira, L.; Mesbah, M. Estimation of trip travel time distribution using a generalized Markov chain approach. *Transp. Res. Part C Emerg. Technol.* **2017**, *74*, 1–21. [CrossRef]
30. Kumar, B.A.; Vanajakshi, L.; Subramanian, S. Pattern-based bus travel time prediction under heterogeneous traffc conditions. In Proceedings of the 93rd Annual Meeting—Transportation Research Record, Washington, DC, USA, 12–16 January 2014.
31. Watkins, K.E.; Ferris, B.; Borning, A.; Rutherford, G.S.; Layton, D. Where Is My Bus? Impact of mobile real-time information on the perceived and actual wait time of transit riders. *Transp. Res. Part A Policy Pr.* **2011**, *45*, 839–848. [CrossRef]
32. Chien, S.I.; Ding, Y.; Wei, C. Dynamic Bus Arrival Time Prediction with Artificial Neural Networks. *J. Transp. Eng.* **2002**, *128*, 429–438. [CrossRef]
33. Rahman, M.; Wirasinghe, S.; Kattan, L. Analysis of bus travel time distributions for varying horizons and real-time applications. *Transp. Res. Part C Emerg. Technol.* **2018**, *86*, 453–466. [CrossRef]
34. Yang, M.; Chen, C.; Wang, L.; Yan, X.; Zhou, L. Bus arrival time prediction using support vector machine with genetic algorithm. *Neural Netw. World* **2016**, *26*, 205–217. [CrossRef]
35. Brakewood, C.; Macfarlane, G.S.; Watkins, K. The impact of real-time information on bus ridership in New York City. *Transp. Res. Part C Emerg. Technol.* **2015**, *53*, 59–75. [CrossRef]
36. Xinghao, S.; Jing, T.; Guojun, C.; QiChong, S. Predicting bus real-time travel time basing on both GPS and RFID data. In Proceedings of the 13th COTA International Conference of Transportation Professionals (CICTP), Shenzhen, China, 13–16 August 2013.
37. He, J.; Shen, W.; Divakaruni, P.; Wynter, L.; Lawrence, R. Improving traffic prediction with tweet semantics. In Proceedings of the 23rd International Joint Conference on Artificial Intelligence, Beijing, China, 3–9 August 2013.
38. Shalaby, A.; Farhan, A. Prediction Model of Bus Arrival and Departure Times Using AVL and APC Data. *J. Public Transp.* **2004**, *7*, 41–61. [CrossRef]
39. Zhou, M.; Wang, D.; Li, Q.; Yue, Y.; Tu, W.; Cao, R. Impacts of weather on public transport ridership: Results from mining data from different sources. *Transp. Res. Part C Emerg. Technol.* **2017**, *75*, 17–29. [CrossRef]

40. Julio, N.; Giesen, R.; Lizana, P. Real-time prediction of bus travel speeds using traffic shockwaves and machine learning algorithms. *Res. Transp. Econ.* **2016**, *59*, 250–257. [CrossRef]
41. Hans, E.; Chiabaut, N.; Leclercq, L.; Bertini, R.L. Real-time bus route state forecasting using particle filter and mesoscopic modeling. *Transp. Res. Part C Emerg. Technol.* **2015**, *61*, 121–140. [CrossRef]
42. Zhou, Y.; Yao, L.; Chen, Y.; Gong, Y.; Lai, J. Bus arrival time calculation model based on smart card data. *Transp. Res. Part C Emerg. Technol.* **2017**, *74*, 81–96. [CrossRef]

Article

Reputation System for Increased Engagement in Public Transport Oriented-Applications

David García-Retuerta [1,*], Alberto Rivas [1], Joan Guisado-Gámez [2], Eleni Antoniou [3], Pablo Chamoso [1]

[1] BISITE Research Group, University of Salamanca, Calle Espejo 2, 37007 Salamanca, Spain; rivis@usal.es (A.R.); chamoso@usal.es (P.C.)

[2] Data Management Group, Universitat Politècnica de Catalunya, Jordi Girona 1-3, 08034 Barcelona, Spain; joan@ac.upc.edu

[3] AETHON Engineering Consultants, 25 Em. Benaki Str, 10678 Athens, Greece; e.antoniou@aethon.gr

* Correspondence: dvid@usal.es

Abstract: Increasing user engagement is one of the biggest challenges when a new application is developed. An engaged user is one who finds a product valuable; highly engaged users generate profit. This study focuses on increasing user engagement in a transport application, via a user reputation score feature. The score is to reward application users and activity organisers, as well as to motivate beginners by offering a high reputation score in the first days of use. The algorithms are based on exponential and logarithmic functions, and were first tested on synthetic data. Real-world tests have shown that the algorithms behave as expected, but the COVID-19 pandemic created a disturbance which prevented any user from achieving the maximum score and many users from registering altogether. Data show positive results, although the real number of users is not sufficient to certify a correct behaviour. Further tests will be carried out when transport activities return to normal.

Keywords: reputation algorithm; users' reputation; transport; software application

Citation: García-Retuerta, D.; Rivas, A.; Guisado-Gámez, J.; Antoniou, E.; Chamoso, P. Reputation System for Increased Engagement in Public Transport Oriented-Applications. *Electronics* **2021**, *10*, 1070. https://doi.org/10.3390/electronics10091070

Academic Editor: Claus Pahl

Received: 30 March 2021
Accepted: 28 April 2021
Published: 30 April 2021

Publisher's Note: MDPI stays neutral with regard to jurisdictional claims in published maps and institutional affiliations.

1. Introduction

The vast amount of data generated on the Internet can be converted into highly valuable information if a proper analysis is carried out. Analysing and filtering the information is especially necessary in cases where the user can interact directly with the content offered in the service. Analysis mechanisms, like those applied in recommender systems, are capable of extracting knowledge in systems that manage large volumes of information. This type of system ensures a satisfactory user experience by providing users with the content they are looking for. New innovative solutions have been proposed in recent years to improve urban transport. Mobility services such as bike-sharing, car-sharing, intermodal public transport and the concept of "Mobility as a Service" (MaaS) are effectively shifting demand away from private vehicles [1]. Moreover, smartphone penetration rates are increasing all over the world, facilitating iteration with public transport users via an application. Applications can become an important element of a city, improving citizens' experience and increasing the quality of tourism [2]. As a result, the development of a new app can provide new functionalities and enhancements to a city's infrastructure.

In recent years, mobile apps with user-generated content have become highly popular (TripAdvisor, Amazon, BlaBlaCar, ResearchGate, etc). The trust-building mechanisms of these apps have been enhanced so that a stranger on the internet can be seen as a "trustman" [3], based on the ideal "in truth we trust". Therefore, these apps expand the source of trustworthy information from a few acquaintances to the whole app community, which is of great value to users [4].

Smartphones have the ability to assist users with the completion of tasks (utilitarian), to entertain them (hedonic) and to connect them with others (social) [5]. These three incentives can boost user engagement in a mobile app. Furthermore, a good balance between short-term rewards and medium-term rewards must be found, so that a gradual engagement is achieved. Otherwise, if the user does not perceive an increase in value or perceives a high initial value but no lasting value, then app engagement will lower considerably [6].

When users perceive high value and user engagement is high, the app community can grow on the principles of the gift economy—where valuables are not sold, but given, without an explicit agreement for immediate or future rewards. Webpages like Wikipedia have proven this concept to be highly effective and successful among the internet community [7].

The My-TRAC application has been developed to provide new public-transport-oriented functionalities. Its value lies in presenting practical alternatives to the use of private vehicles by enabling citizens to make better use of public transport. The application creates a healthy user community and a trustworthy source of information.

The main objective of this article is to propose a reputation algorithm to facilitate recommendations on a series of trip-related activities, such as the purchase of tickets, selection of the most appropriate means of transport, tourist activities, etc., which the users will be able to use as a guide while planning their trips.

This work is organised as follows: a review of existing reputation systems is presented in Section 2. Section 3 describes the proposal. Section 4 presents the assessment made with synthetic data and the pilot data. Finally, Section 5 presents the conclusions.

2. Background

Advisory systems provide advice and help solve problems that are normally solved by human experts [8]. In any community, individuals whose opinion is considered more important are normally trusted more. The knowledge of human experts can then be extracted and coded to automatise the process. Reputation systems are a kind of advisory system that allow users to rate each other in online communities so as to build trust through reputation [9].

Numerous proposals for reputation algorithms have been put forward over the years. They are generally quite context-dependent. This is because each problem entails the study of the best solution and, in most cases, it is not enough to have one generic proposal or to apply a specific proposal to a different problem. It is always necessary to adapt the approach to the new problem. From the analysis of the state of the art, it can be inferred that context-dependent solutions generally perform better than those that do not consider the context [10]. Additionally, some platforms have been designed to ease the development of such systems, as an essential part of any developing Smart City [11,12]

The subsections that follow present different existing reputation systems can be divided into groups of academic and commercial proposals.

2.1. Academic Proposals

Among the scientific proposals in the state of the art, two of them stand out (PageRank and EigenTrust). PageRank is the most popular of all the reputation algorithms, presented in [13] and used previously by Google to order the websites in its search engine in an objective and mechanical way. Four years later, researchers from Stanford University proposed an algorithm for reputation management in peer-to-peer (P2P) networks, called Eigen-Trust and described in [14]. With its application, they managed to minimize the impact of malicious peers on the performance of a P2P system.

PathTrust [10], has been presented more recently. It is based on a model that exploits the graph of relationships among the participants of virtual organizations. Its authors indicate that the system is based on the two previous algorithms (PageRank and EigenTrust); however, they are not directly applicable because their personalization is very limited.

Below is a brief description of how each algorithm works, along with its advantages and disadvantages.

- PageRank [13]: **Advantages**: This algorithm converges in about 45 iterations. Its scaling factor is roughly linear in $\log(n)$. It uses graph theory to link the pages. An important component of PageRank is that its calculation can be personalized. PageRank can estimate web traffic and can predict backlinks. **Disadvantages**: PageRank is based on random walks on graphs. This algorithm assumes the behaviour of a "random surfer", but if a real Web surfer ever gets into small loops of web pages, the PageRank will have false positives. This method of random surfer assumes that periodically the surfer "gets bored" and jumps to another random page;
- EigenTrust [14]: **Advantages**: This reputation system is among the most well-known and successful reputation systems. It satisfactorily solves different problems existing in P2P systems, which is the context in which the algorithm was designed. **Disadvantages**: The main drawback of this system is its reliance on a set of pre-trusted peers, which causes nodes to centre around them. As a consequence, other peers are ranked low despite being honest, marginalizing their role in the system [15];
- PathTrust [10]: **Advantages**: This model of reputation (using the trust relationships amongst the participants) is resistant against the attack of faking positive feedback. A group of attackers collaborates to boost their reputation rating by leaving false, positive feedback for each other. In this model of reputation, this will only strengthen the trust relationship among the attackers, but will not necessarily strengthen the path of the attacker to an honest inquirer, such that their reputation does not affect the honest inquirer. Another benefit of exploiting established relationships in member selection is the formation of long-term relationships. **Disadvantages**: The trust relationship between two participants is formed on the basis of past experience with each other. A participant leaves a feedback rating after each transaction, and these ratings are accumulated to a relationship value. Therefore, one user can boost a positive or negative feedback.

2.2. Commercial Proposals

Currently, the most important reputation system proposals are those used by commercial applications. Generally, commercial reputation systems directly focus on assigning users a reputation score within that commercial system (for example, the reputation systems of TripAdvisor, Waze, Amazon and BlaBlaCar).

The conclusion drawn from the review of the state of the art is that all the existing reputation system proposals, especially those of commercial systems, focus exclusively on their context. This implies that a specific algorithm has to be designed to obtain good results. To do this, it is essential to identify the factors and the extent to which they have a direct influence on reputation.

In the same way, although each type of parameter has its weight, which defines its impact on the final score, each occurrence of the parameter may affect the associated factors differently. It is, therefore, necessary to determine how the score assigned to each occurrence of a parameter evolves over time.

Moreover, in the majority of the analysed commercial proposals, the user must know the highest possible reputation level that can be reached in the system. This allows them to understand the relevance of the different scores.

3. Proposal for User and Users' Choices Reputation Algorithm

My-TRAC is an app devoted to the research and development of user-centric services that enhance the passengers' multimodal door-to-door experience. This helps citizens develop greater confidence in, and adhesion to, multimodal transport services. Furthermore, My-TRAC improves adaptation to the users' needs through the provided data, statistics and trends from the passengers' experiences while using the proposed platform. An example of the user interface can be found in Figure 1.

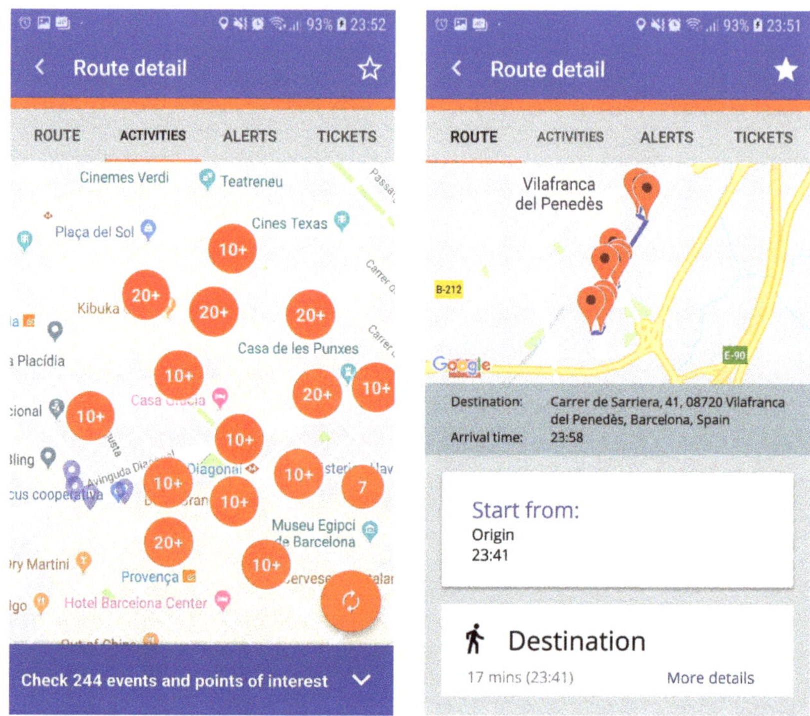

Figure 1. The user interface of the My-TRAC app.

This section describes the algorithms used on My-TRAC to assign a reputation score to each user and each user choice, activity or Point Of Interest (POI), representing their ranking within the system. The two algorithms share a common basis; however, each is used for a different purpose: one calculates the users' reputation and the other one calculates the reputation of the choices made by users. Therefore, each algorithm uses different factors and metrics. As a result, each subsection describes either the part dedicated to the users' reputation algorithm or the users' choices' reputation algorithm.

The proposed model is based on a mixture of exponential and logarithmic functions to create a system of distributed trust, a idea not yet fully explored in the literature. For example, the most common research lines base their mapping functions on the definitions of clever distances [16], graph analysis [13] or the definition of a set of rules affecting to the trust relationships among users [10]. The main advantage of the current proposal is that the mapping functions can be easily adapted or extrapolated to new systems, by just analysing the importance of the considered features and selecting a suitable function for those parameters, resulting in a higher versatility than other works.

This section is structured as follows: the factors identified as essential to determine a user's reputation in the system are presented below. Then, the metrics associated with each of the individual factors are shown, followed by the description of the mechanism that provides the initial score, which is the output of both algorithms. Finally, the proposed adaptive weight mechanisms of both algorithms are described. They adapt the weight of the factors according to the dynamic characteristics of the application where the algorithms are applied. Thus, the role of this mechanism is to re-establish the limits of each factor over time, as the number of users or the number of existing ratings changes with time, providing an adequate maximum score.

3.1. Mathematical Description of the Reputation System

The reputation system is based on a combination of logarithmic and exponential functions to map the inputs onto their corresponding reputation. Metrics are related to each of the identified factors. Therefore, each metric determines the reputation score provided by its corresponding factor and each factor has its own metric. Besides, metrics affect the overall reputation, as it is calculated as the sum of the scores of all the factors. Each metric provides a final score which is calculated as the percentage reached by that user over the total weight of each factor, and these final scores are added to obtain the user/activity reputation.

The number of instances required to reach the maximum score is established for each factor. In addition, the slope of any of the parameter functions determines how fast or how slowly the value for that parameter increases. In this case, the *slope* parameter refers to the steepness, incline, or grade of the function. It has been established that the evolution is not linear, just like ResearchGate's calculation of its "RG Score". Thus, the growth in the score of a specific factor will either be logarithmic or exponential, following Equations (1) and (2).

$$score Parameter_i = y_{maximum} \times \frac{\log(slope \times \frac{x}{x_{maximum}} + 1)}{\log(2 + slope)} \tag{1}$$

The logarithmic equation shown in Equation (1) is useful in cases where the slope should be greater in the initial instances and then gradually decrease in subsequent instances. For example, to encourage new users to rate activities, the first few ratings the user gives will have a considerable effect on their reputation, however, the user will not be able to continue gaining reputation at the same rhythm after producing a considerable amount of ratings. Instead, further ratings will have a smaller impact on the reputation of the user. Logarithmic growth is regulated by the *slope* variable of the equation, whereas the maximum number of instances is regulated by $x_{maximum}$. This factor will be dynamic due to the usage characteristics of the social network. Therefore, in the case of ratings provided by users, the maximum score $x_{maximum}$ can take a value of 200, meaning that a user with more than 200 ratings will obtain a 100% initial score, which will greatly contribute to the final score. In cases where the usage patterns of the application imply that users give a large number of ratings, the factor $x_{maximum}$ is adjusted dynamically, so that $x_{maximum} = 2 \times avg Ratings By User$. Finally, the factor $y_{maximum}$ can reach 1, so that each factor will have a score between 0 and 1.

$$score Parameter_j = y_{maximum} \times \left(\frac{x}{x_{maximum}}\right)^{slope} \tag{2}$$

The exponential equation shown in Equation (2) is useful for factors in which the weight of the initial instances is lesser and becomes more important in the system as the number of instances grows. For example, a user that opens the application three times does not notice a significant increase in their reputation in the system; however, a user who opens the application 200 times is considered a regular user, and therefore obtains a pertinent reputation.

Although the mathematical approach described above is not directly based on any existing work to determine reputation, these types of equation are well known and widely used in the literature for multiple purposes.

Mathematically speaking, the most similar proposed work can be found in [17], where the authors present a trust management system based on reputation mechanisms. The mechanisms proposed in this paper base the evolution of reputation on the number of assessments that follow a logarithmic distribution.

3.1.1. User Reputation Mathematical Model

User reputation is calculated using a mixture of exponential and logarithmic functions. These are selected in order to maximise user engagement, providing them with fast rewards for some easy tasks (logarithmic growth) and slow rewards until they complete a challenging task (exponential growth).

All the equations related to the users can be found in Table 1.

Table 1. Equations related to the user reputation, inputs, outputs and factors.

Factor	Metrics (Equations)	Inputs	Outputs
Days registered	$s_1 = w_{date} \times \frac{c_{date} - r_{date}}{M_{date}}$	c_{date}, r_{date}	s_1
List of valuations	$s_2 = w_{valuations} \times \frac{\log\left(100 \times \frac{n_{valuations}}{M_{valuations}} + 1\right)}{\log(2+100)}$	$n_{valuations}$	s_2
Number of uses of the application	$s_3 = w_{uses} \times \left(\frac{n_{uses}}{M_{uses}}\right)^2$	n_{uses}	s_3
Number of chosen routes	$s_4 = w_{routes} \times \left(\frac{n_{routes}}{M_{routes}}\right)^2$	n_{routes}	s_4

The following variables are used as input:

- c_{date} refers to the current date.
- r_{date} refers to the registration date.
- $n_{valuations}$ refers to the number of valuations of the user.
- n_{uses} refers to the number of times the user opened the app.
- n_{routes} refers to the number of routes the user has chosen to travel.

The following variables are the obtained outputs:

- s_1 is the initial score of days registered.
- s_2 is the initial score of the list of valuations.
- s_3 is the initial score of the number of uses of the application.
- s_4 is the initial score of number of chosen routes.

M represents the number of occurrences of a given parameter to provide the maximum value/weight that it is capable of providing over the total reputation (w). They refer to maximum and weight, respectively. Both are static (but editable) variables obtained from the database. The subscript indicates to which factor they are related.

The final score S is defined as shown in Equation (3):

$$S = \sum_{i=1}^{4} s_i \tag{3}$$

The pseudocode of this procedure can be found below in Algorithm 1.

Algorithm 1 User Reputation Calculation.

```
1: def f_log(x, x_max, slope=100):
2:     return log(slope × (x/x_max)+1)/log(2+slope)
3: def f_exp(x, x_max, slope=2):
4:     return pow((x/x_max), slope)
5: reputation = f_log(User.valuations, Maximum.valuations) × Weight.valuations
       + f_exp(User.uses_app, Maximum.uses_app) × Weight.uses_app
       + f_log(User.tickets_purchased, Maximum.tickets_purchased) ×
       Weight.tickets_purchased + (User.days_registered/Maximum.registration) ×
       Weight.registration + f_log(User.groups, Maximum.groups) × Weight.groups
6: if reputation > 100 then
7:     reputation = 100
8: else if reputation < 0 then
9:     reputation = 0
```

This novel proposal could provide commercial systems with the following advantages: (i) dynamic adaptation of the reputation to the information of the system (non-linear growth), (ii) dynamic adaptation at parameter level, varying the specific weight that each parameter has on the final reputation, (iii) engaging the users with well-modelled changes in its reputation score.

To this end, mechanisms similar to those used in well-known proposals, that have been proven to work well (such as the one presented in [17]), have been integrated with the peculiarities of My-TRAC, which determine the information to be used.

3.1.2. Users' Choices Reputation Mathematical Model

The users' choices (activities and POIs) reputation are calculated using a mixture of linear and logarithmic functions. They are selected in order to maximise users' engagement, providing the users' choices with fast rewards initially, and then basing the rewards on the average star rating received.

All the equations related to users' choices can be found in Table 2.

Table 2. Equations related to the activities' reputation, inputs, outputs and factors.

Factor	Metrics (Equations)	Inputs	Outputs
n-star ratings weighted average	$s_1 = w_{rating} \times \sum_{i=1}^{n} \left(\dfrac{\sum_{k=1}^{n} (re_{user_k} \times ra_{user_{k,i}})}{\sum_{k=1}^{n} re_{user_k}} \times \dfrac{1}{M_{rating}} \right)^2$	re_{user}, ra_{user}	s_1
Number of views of the activity	$s_2 = w_{views} \times \dfrac{\log\left(50 \times \frac{n_{views}}{n_{days} \times M_{views}} + 1\right)}{\log(2+50)}$	n_{days}, n_{views}	s_2

The following variables are used as input:

- $ra_{user_{k,i}}$ is the rating of the k-th user on the i-th activity. The number of users who rated this activity is defined as n.
- re_{user_k} is the reputation of the k-th user who rated the activity. The number of users who rated this activity is defined as n.
- n_{days} is the number of days since the activity was created.
- n_{views} is the number of views of the activity.

The following variables are the obtained outputs:

- s_1 is the initial score of N-star ratings weighted average.
- s_2 is the initial score of the number of views of the activity.

M and w are static (but editable) variables obtained from the database. They refer to maximum and weight, respectively. The subscript indicates which factor they are related to.

The final score, S, is defined as shown in Equation (4):

$$S = \sum_{j=1}^{2} s_j \tag{4}$$

The pseudocode of this procedure can be found below in Algorithm 2.

3.2. Updating the Parameters and Their Weights

The information on My-TRAC is not static; instead, it evolves over time. This obliges the metrics that are part of the reputation algorithms to adapt to the information. For this reason, it is crucial to implement mechanisms that update configurable factors in each of the metrics.

For example, during the pilot stage, when the application begins to obtain real user data, the system will start from zero. In the beginning, a lower number of instances of each factor will be required to obtain a significant final reputation score of a user/activity. The number of instances required will be much higher after a year of system functioning.

Algorithm 2 Activity Reputation Calculation.

```
 1: def f_log(x, x_max, slope=100):
 2:     return log(slope × (x/x_max)+1)/log(2+slope)
 3: def f_exp(x, x_max, slope=2):
 4:     return pow((x/x_max), slope
 5: numerator = 0
 6: denominator = 0
 7: for activity in ratings_database do:
 8:     numerator += activity.rating × activity.reputation_user
 9:     denominator += activity.reputation_user
10: avg_rating_reputation ← numerator / denominator
11: reputation = f_log(Activity.views, days_registered × Maximum.views, 50) ×
    Weight.views + f_exp(avg_rating_reputation, Maximum.average_rating) ×
    Weight.average_rating
12: if reputation > 100 then
13:     reputation = 100
14: else if reputation < 0 then
15:     reputation = 0
```

According to the number of instances of each one of those factors, the metrics that make up the algorithm can be adapted and the values can be updated automatically or manually. Both the weight that the parameter has on the final reputation and number of occurrences that a parameter must have to obtain the maximum score can be updated.

Regarding the weight of the parameter in the final reputation, it is set *a priori* but can be changed at any given point in order to correct certain anomalies or to encourage desired behaviours. On the other hand, there are two ways of updating the number of occurrences that a parameter must have to obtain its maximum score:

- Manually: when an expert administrator/developer decides that it should be changed for some reason.
- Automatically: depending on the evolution of the information on the platform. For example, the rating an activity has in the system will not remain the same; it is going to change over time and, according to its evolution, the maximum weight of this parameter in the system can increase or decrease (if it receives many ratings, its weight will decrease).

In the first version of the model, the system's automatic adaptation has not been evaluated because the data we are using at this stage are not sufficient to test it effectively.

4. Evaluation and Results

The evaluation of the proposed algorithms has been tested using two complementary methods: creation of synthetic data and deployment of a pilot program. Synthetic data are meant to simulate the behaviour of users when the app has gained popularity and is already established, and the pilot phase provides a clear picture of how the algorithms will behave in the beginning of the application deployment phase.

The only way of evaluating the correct functioning of the algorithm with the synthetic data is the following: to analyse whether the obtained output behaves as expected and then draw conclusions as to whether the reputation score assigned to different users corresponds to the initial idea, as a function of the values of each of the parameters affecting the reputation score.

Due to the initial lack of available data of real users, synthetic data were generated in order to evaluate the proposed method. A total of 2000 simulated users were randomly generated considering the following attributes:

1. Gender of users (i.e., male, female).
2. Age group: Three age groups were considered (i.e., 16–30, 31–50, 51–90).

3. Volume of performed actions. Two possibilities: users that performs a small number of actions and users that performs many actions.
4. Type or category of the performed actions. Six major categories of actions were defined for the generated dataset: sports, eating, history, dancing, cinema, shopping.

Considering the above attributes, the generated dataset contained information about the demographics of the users and the number of actions performed for each category. The generation of a synthetic user is carried out by the data generator, which randomly chooses the gender, the age group, and the volume of actions and based on the number of actions performed by the mean users of the category that the user is applied to. The generator randomly calculates (based on a uniform distribution) the number of actions of the generated user for each of the six types of action.

Pilot data were used to analyse the real-world behaviour of the models, in a initial phase. As a result, it is expected that many users register but do not make any usage of the app. As the functionalities of the app are still limited, user engagement is likely to be lower than in the real application.

The evaluation of the obtained results is a subjective task; however, it is important to verify that the algorithms behave as expected. Section 4.1 describes the tests carried out related to the artificially generated users and their results. Section 4.2 describes the real-world experiment and its results.

4.1. Reputation Models Evaluation—Synthetic Data

When creating the synthetic dataset, the aim is to simulate the behaviour of real users and users' choices in the most realistic possible way. This method will provide an *a priori* idea of how well the system works.

4.1.1. Users' Reputation Evaluation

Evaluation methodology. This simulation aims to model the use of the system by users. Therefore, no inactive users will be generated, even though, in a real system, they could become the majority.

In this way, there will be a set of users who use the system a lot, a larger set who use it frequently and an even larger one who use it sporadically. This has involved the creation of three ranges of usage possibilities when creating the data.

This distribution of users is easily observed by analysing the scoreboard of the commercial applications that made their scoreboard public. For example, on Waze [18], one of the tools analysed in Section 2, a user with 100,000 points can reach the maximum level "Waze Royalty", which means they are among the 1% most active users in the country, while the top users listed on the scoreboard have more than one million points.

Results. Figure 2 shows the distribution of reputation among system users. On the x-axis, there are reputation intervals, and on the y-axis, the number of users with a reputation within those intervals.

The resulting scores present a Gaussian distribution which denotes a desirable behaviour—this is the distribution that would be expected from many natural phenomena.

4.1.2. Users' Choices Reputation Evaluation

Evaluation methodology. On the other hand, the users' choices reputation algorithm, which determines the reputation of the activities and POIs included on My-TRAC, has also been evaluated using synthetic data.

In this case, the only case-specific restrictions that have been applied when generating the synthetic dataset are:

- The identifier is a unique integer from 1 to 1000.
- The inclusion date is between 1 September 2017 (start of the project) and 18 December 2018 (the date on which the evaluation was carried out).
- The number of views of an activity is higher than its number of ratings.

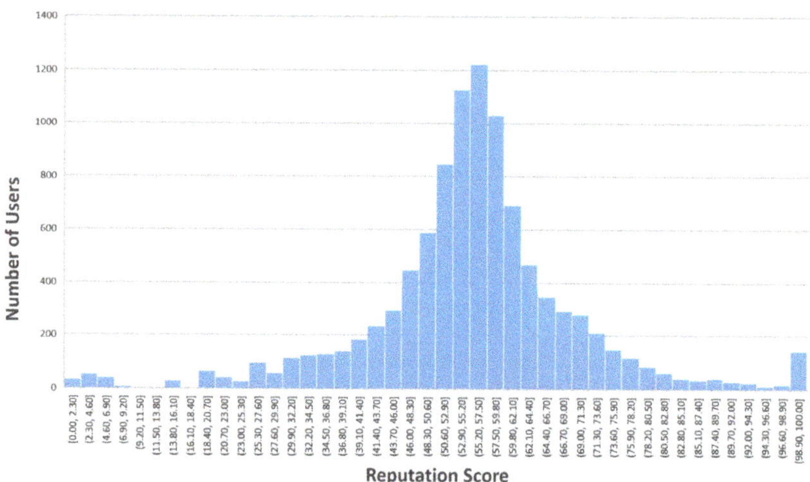

Figure 2. Representation of the results obtained with synthetic data for the evaluation of the user reputation model.

Results. The distribution of the reputation of the 1000 synthetically created activities is shown in Figure 3, which shows, on the x-axis, the reputation values of the activities and on the axis, and the number of activities that there are in the different reputation intervals.

It can be observed that there is no activity with a reputation of less than 21, because the synthetic data were created to test the performance of the models with active users and successful activities. These circumstances are not expected to exist in reality, where it is expected that there may be activities that receive no ratings at all during the pilot stage.

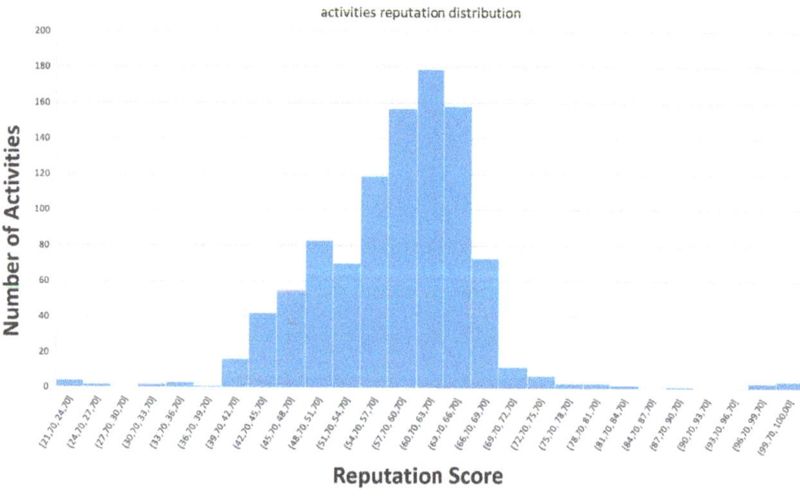

Figure 3. Distribution of the activities' reputation with the generated synthetic data

4.2. Reputation Models Evaluation—Pilot Study

Evaluation methodology. The previously designed reputation models have been evaluated in the pilot phase. A variation in the initial model has been designed and its joint use with the Social Market (another functionality of My-TRAC) is proposed. The Social Market is a means of encouraging use of the application, as it enables the users to exchange the points they have obtained for rewards. The system allows the user to earn free tickets as a reward, in exchange for a set number of points. The number of obtained points is directly related to the user's reputation. It is designed to encourage the user to make more frequent use of the application.

It is necessary to remember that there are reputation models for both users and activities. However, a specific variation in the user reputation model has been designed for the current phase and integrated in the Social Market.

Thus, the version of the reputation model that has undergone major evaluation and been tested by the users in the pilot phase is the original proposed model, with a slight variation. What is different is that the date on which the users register does not affect their reputation.

In the initial version of the model, a very active user who has been registered for a few days would have a greater reputation than a user who has been registered for much longer and who has also used the features of the application (used it sometimes, for example). There are two main reasons for designing a variant for the pilot model:

1. The duration of the pilots is the same for all users and if a user has used the application more times than another user, they should get a higher reward, independently of the date of registration.
2. If the date has a negative effect on the user's reputation, i.e., the more time passes, the less reputation the user will have if they do not participate. This would cause the user's points on the Social Market to decrease even though the user has not spent them. This is an undesirable situation for the evaluation of the model.

Therefore, the score obtained by the users in this phase is a decimal value between 0 and 100, where 0 is the initial reputation value for a user who has just registered, and 100 points can be reached by carrying out repeated interactions with the application. For example, every time an activity or POI is valued, a certain reputation value is assigned according to the previously defined metrics.

These points can be redeemed at the Social Market, where each user's points will be updated periodically at 0:00 (CET) each day. The points on the Social Market have been updated periodically to control possible fraudulent behaviour by users who create multiple accounts, automate actions and obtain rewards illegally at the time. In this way, the development team can act as a moderator if this type of behaviour is detected and proceed accordingly, for example, by deleting the user's account for non-compliance with the terms and conditions of use.

However, although the score that users have been able to visualize throughout the pilot phase is the score that is provided with the user reputation version created for integration with the Social Market, this section of the document also presents the results that would have been obtained with the reputation version not linked to the reward points and the users' choice reputation models version. Thanks to this, it is possible to check how the models operate in the presence of real data, although, after carrying out the evaluation, it can be anticipated that the volume of information that has been collected is again insufficient.

Results. Due to the pandemic, strict mobility restrictions have been implemented, affecting the information that have been collected; this is different from the information we would have expected under normal circumstances.

More specifically, there are 171 valid users out of a total of 206 (which means that 35 decided to delete their account). It can be seen in the results presented below that not all of them have interacted with the tool. This was expected, as it commonly happens in any type of application, as some users download the application and register but never use it.

The pilot was open to everyone who wanted to register, and an advertising campaign was carried out in The Netherlands, Athens (Greece) and Barcelona (Spain) to encourage participation.

The results and evaluation of each of the data models are presented below.

4.2.1. User Points (Social Market Version)

The results of the adapted version of the model for the Social Market reward points calculation, are presented below. They were obtained after carrying out the pilots with different graphs incorporated in the panel of the analysis tool mentioned above.

Figure 4 shows the distribution of the points allocated for the total number of users (206), i.e., both active and non-active users, grouped by ranges of 10 units.

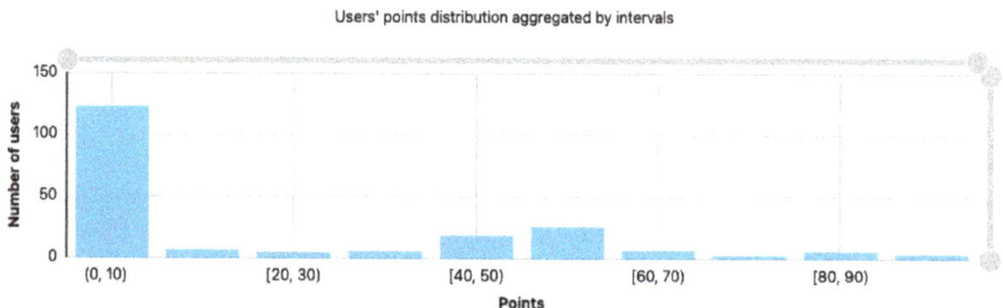

Figure 4. Reward points for all users (206) points grouped ranges of 10 units.

It can be seen that there is a set that encompasses the majority of users (123), and this distorts the results. This is due to the fact that the majority of users have not interacted with the application at all or hardly at all.

To analyse this situation in greater detail, Figure 5 shows the same type of graph as the prior one, but, in this case, the groupings of points are made by unit rather than in groups of 10.

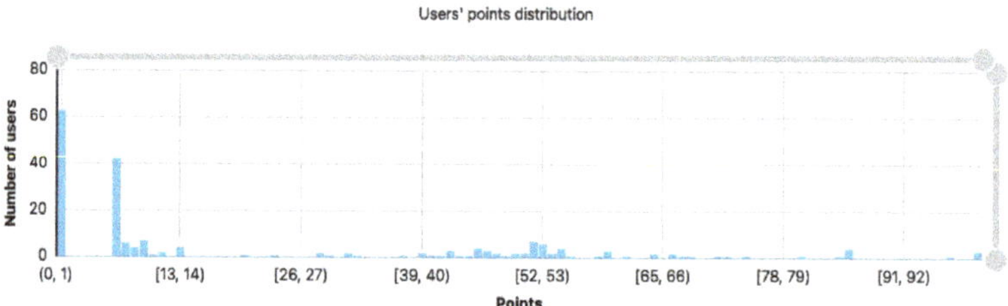

Figure 5. Reward points for all users (206) points grouped by units.

It can be seen that there are 63 users with the minimum value of reputation, which implies that they have registered and have not carried out any more activities, while there are 43 users who have obtained the score that corresponds to a one-time use of the application.

Let us consider the users who have not interacted in any way with the application as non-active users, thus providing 143 active users, and proceed to analyse the results again. Figure 6 again shows a graph with the distribution of users according to their points grouped in ranges of 10.

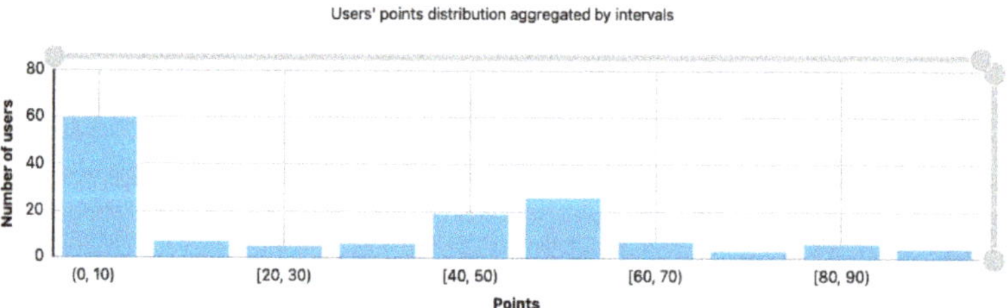

Figure 6. Reward points for active users (143) points grouped in ranges of 10 units.

As with all users, the group of very inactive users still stands out, as they almost have not interacted with the tool, so if we filter the graph by leaving out the first range of values (from 0 to 10), we obtain a graph that is a better representation of the behaviour of the "average" users of the application, as shown in Figure 7.

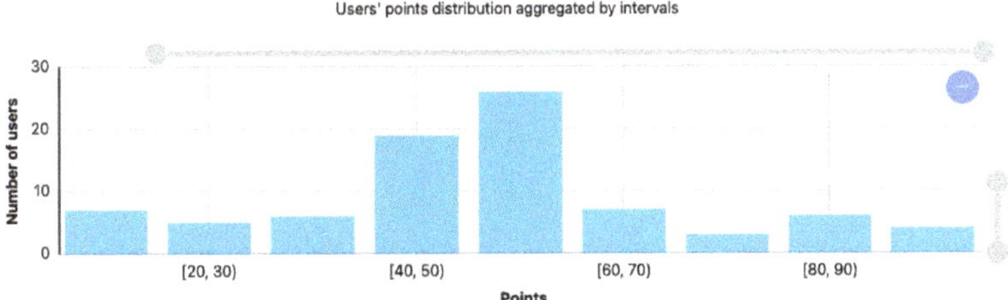

Figure 7. Reward points for active users (143) filtered and grouped in ranges of 10 units.

As mentioned above, the number of users who have participated in the pilots was not significant enough to draw relevant conclusions regarding the functioning of the reputation models; however, a very similar behaviour to the one expected can be observed, which was obtained by generating synthetic data following a series of criteria intended to represent the real behaviour of users. The expected results are represented in the document by the graph shown in Figure 2.

It can be seen that the pursued objective has been achieved: the users who initially participate add points to their reputation score with relative ease until they reach the average values. It becomes more difficult for a user to go above the average reputation values, motivating users to continue to use the app to increase their score, thus increasing their loyalty.

However, a certain number of users were expected to have the highest score and this was not achieved, possibly because users have not been able to travel as much as expected due to the restrictions caused by the COVID-19 pandemic and because 100% of the app's functionality is still not available.

An analysis of user activities was carried out, which provided points to better understand the type of activity carried out by the users of the app. For example, Figure 8 shows the points awarded to users according to the number of times they have used the app.

It can be seen that 67.5% of the users obtained a reward of between 0 and 1 points for using the app, i.e., they were less active, while 20.4% of the users obtained between 9 and 10 points (the maximum) for using the app.

A similar analysis can be made for the score given to users depending on the number of times they have requested a route and followed it. This analysis is shown in Figure 9.

In this case, it can be seen that 80.3% of users were awarded between 0 and 3 points for following suggested routes, while only 2.4% of users obtained between 27 and 30 points (the maximum) for having followed suggested routes. This clearly shows that very few users used this functionality (40 to be exact).

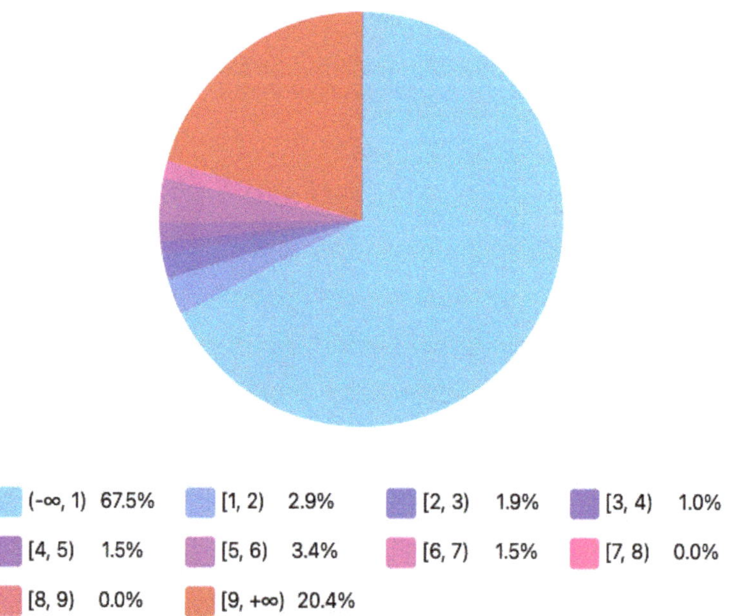

Figure 8. Users (% and size of sectors) who have obtained a certain amount of reward points (colour) depending on how many times they have opened the app.

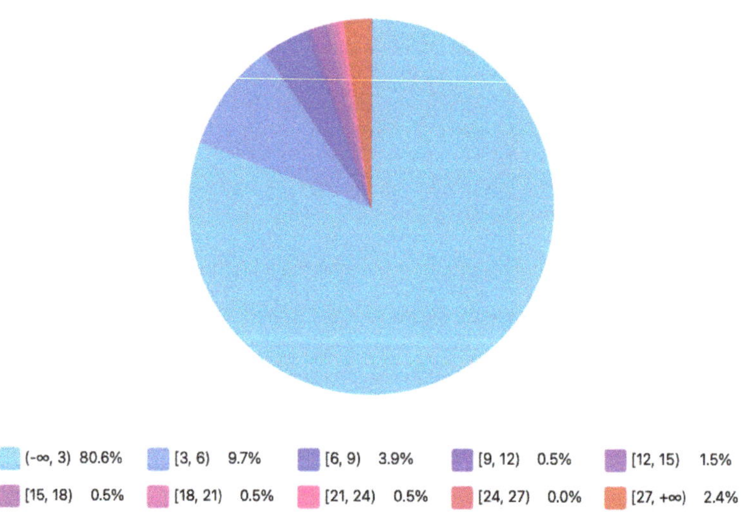

Figure 9. Users (% and size of sectors) who have obtained a certain amount of reward points (colour) depending on how many times they followed a suggested route.

4.2.2. Users' Reputation

Although the first version of the User Reputation Model was not used for the reasons outlined above, it is possible to carry out an assessment to demonstrate how the system would have performed.

In this case, out of the 206 total real-world live users, no one had a reputation of 100, because active users stopped being active before the date of the assessment, and this negatively affected the maintenance of their score at the highest value. The maximum reputation in this case was 86, achieved by two users. To represent this, 10 groupings with equal ranges were created, which are shown in Figure 10.

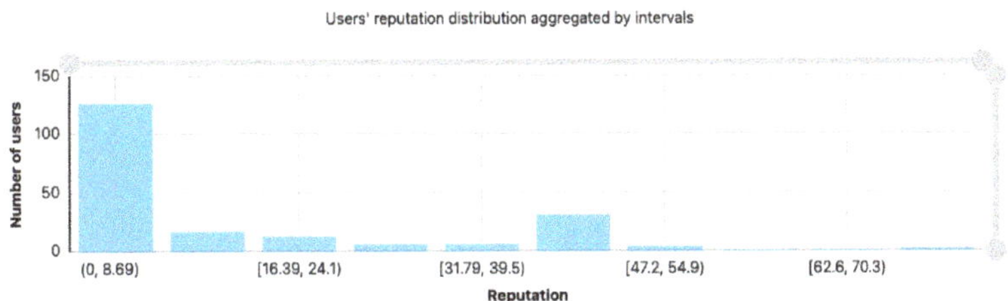

Figure 10. Reputation values for all users (206), scores grouped in 10 ranges.

The distribution is not exactly the same as with the reward points, but, in the same way, the majority continues to remain in low values, mainly due to inactivity, so the results are evaluated by discarding this set of users and focusing again on the 143 real-world live users, who have at least interacted with the app. The distribution of their reputation is shown in Figure 11.

As participation has been lower than expected due to mobility restrictions, it can be seen that the majority of users have a below-average reputation, although the group with the highest number of users is in the intermediate reputation zone, as expected.

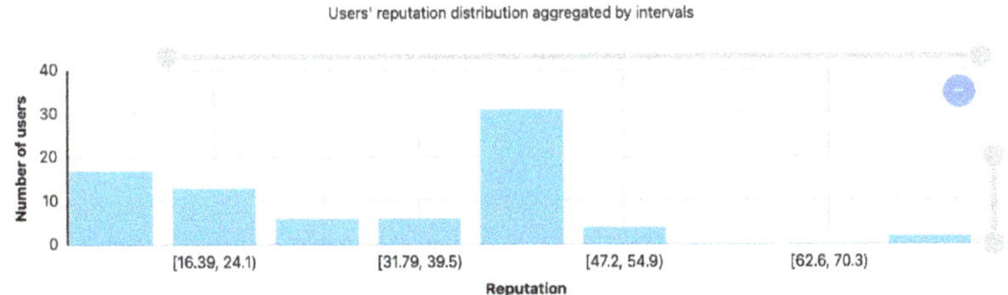

Figure 11. Reputation values for active users (143), scores grouped in 9 ranges.

4.2.3. POIs and Activities' Reputation

As far as the reputation of POIs and activities is concerned, the evaluation that can be made on the basis of the information obtained from the pilots would not truly reflect a real scenario, since the interaction of the real-world live users with this functionality on My-TRAC has not been sufficient. The vast majority of POIs and activities have not been interacted with, so they have no reputation, as can be seen in Figure 12.

If the results are evaluated, leaving aside the activities and POIs that have not been interacted with, the results shown in Figures 13 and 14 are obtained.

Figure 13 shows that users only interacted with two activities, for which they have an average reputation, while Figure 14 shows that users have interacted with a total of 37 POIs.

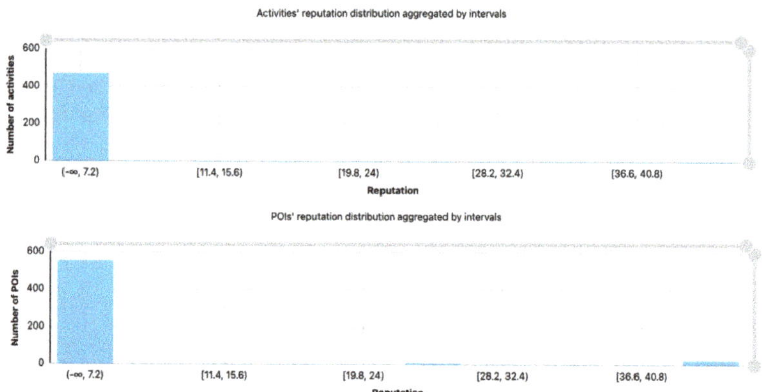

Figure 12. Reputation values for all the activities and all the POIs

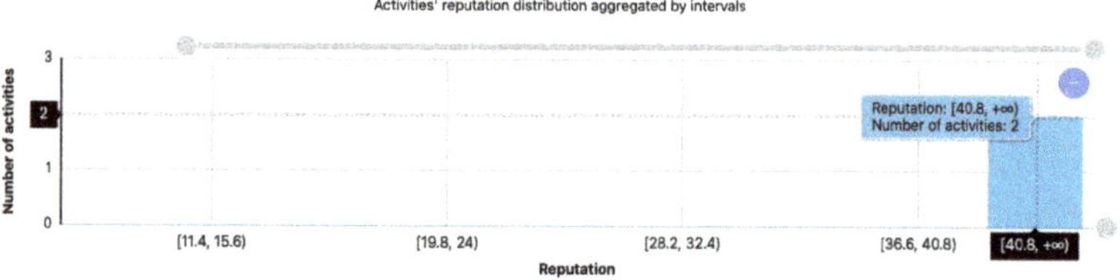

Figure 13. Reputation values for the activities with reputation value > 0.

On the one hand, we can conclude that users interact with POIs more than with the activities offered by the app, despite the fact that there is an even number of options (473 activities and 556 POIs). On the other hand, it can be concluded that users are, in most cases, satisfied with the POIs they visit, as 25 of the 37 POIs they have interacted with have high reputations.

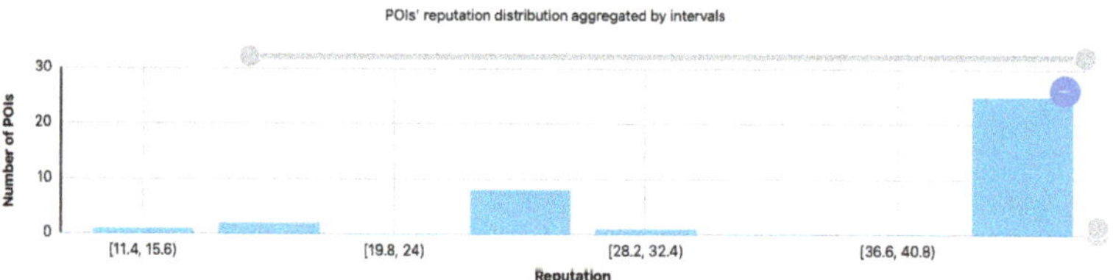

Figure 14. Reputation values for the POIs with reputation value > 0.

5. Conclusions

Following analysis of the results, the conclusion is that, although the results seems to follow the value distribution patterns that were sought with the initially defined models, the number of active users is still not sufficient to certify that, in a real scenario, it will behave as expected.

However, using the data obtained from the pilots and the simulations, the obtained results were satisfactory, as no unexpected behaviours were detected. Moreover, it is clear that the algorithm encourages users to participate more actively by giving them points rapidly, and that reaching the maximum score is such a difficult task that users need to be engaged before achieving it.

The reputation scores seem to form a normal or Gaussian distribution, with peaks on the higher or lower end, resulting from optimal user behaviour in the synthetic data and from a low participation in the pilots, respectively. In general, active real-world users tend to cluster around the reputation value 50 (the maximum reputation value is 100), which is a desirable result. It does not demotivate users by maintaining their low score, and does not cause them to become bored by giving them the maximum score often. Most users will have around 50 points (out of 100), creating healthy competition against similar users, as they try to surpass their equals and not to be left behind.

Activities and POIs also take advantage of having the same basis; therefore, analogous results are obtained and a similar purpose is fulfilled.

It can, therefore, be concluded that, even though there was not enough data, the goal of allowing users to determine the relevance of users and the actions was fulfilled in the case study conducted on the My-TRAC platform.

This research and its results can be taken advantage of by any user who needs to develop a similar system and apply it in a real-world scenario. For example, a new video platform could adapt the developed basic functions (logarithmic and exponential) to assign a reputation to the content creator and content consumers.

The main limitations of this work are related to the limited data gathered during the pilots phase, adversely affected by the effects of the COVID-19 pandemic on mobility. Moreover, user engagement is measured through the distribution of the reputation scores: an indirect measurement instead of a direct one.

Regarding future research on this topic, user engagement will be measured when the application is launched. The parameters' limits will be updated in order to obtain a Gaussian distribution shape, with a moderate number of users obtaining the maximum score. If the resulting distribution has several peaks or is chaotic in any sense, more input will be used to obtain a better modelling of the users' worth.

Author Contributions: Conceptualization, P.C. and A.R.; methodology, D.G.-R.; software, D.G.-R.; validation, J.G.-G., E.A. and P.C.; formal analysis, J.G.-G., E.A. and P.C.; investigation, D.G.-R.; resources, J.G.-G., E.A. and P.C.; data curation, J.G.-G., E.A. and P.C.; writing—original draft preparation, D.G.-R.; writing—review and editing, D.G.-R.; visualization, P.C.; supervision, J.G.-G., E.A. and P.C.; project administration, J.G.-G., E.A. and P.C.; funding acquisition, P.C. All authors have read and agreed to the published version of the manuscript.

Funding: This research has been supported by the European Union's Horizon 2020 research and innovation program under grant agreement No 777640.

Conflicts of Interest: The authors declare no conflict of interest.

References

1. Kamargianni, M.; Li, W.; Matyas, M.; Schäfer, A. A critical review of new mobility services for urban transport. *Transp. Res. Procedia* **2016**, *14*, 3294–3303. [CrossRef]
2. Dickinson, J.E.; Ghali, K.; Cherrett, T.; Speed, C.; Davies, N.; Norgate, S. Tourism and the smartphone app: Capabilities, emerging practice and scope in the travel domain. *Curr. Issues Tour.* **2014**, *17*, 84–101. [CrossRef]
3. Chun, B.N.; Bavier, A. Decentralized trust management and accountability in federated systems. In Proceedings of the 37th Annual Hawaii International Conference on System Sciences, Big Island, HI, USA, 5–8 January 2004; p. 9.
4. Li, N.; Mitchell, J.C.; Winsborough, W.H. Design of a role-based trust-management framework. In Proceedings of the Proceedings 2002 IEEE Symposium on Security and Privacy, Berkeley, CA, USA, 12–15 May 2002; pp. 114–130.
5. Kim, Y.H.; Kim, D.J.; Wachter, K. A study of mobile user engagement (MoEN): Engagement motivations, perceived value, satisfaction, and continued engagement intention. *Decis. Support Syst.* **2013**, *56*, 361–370. [CrossRef]
6. Lalmas, M.; O'Brien, H.; Yom-Tov, E. Measuring user engagement. *Synth. Lect. Inf. Concepts Retr. Serv.* **2014**, *6*, 1–132. [CrossRef]
7. Bruns, A. *Blogs, Wikipedia, Second Life, and beyond: From Production to Produsage*; Peter Lang: New York, NY, USA, 2008; Volume 45.
8. Forslund, G. Toward cooperative advice-giving systems: A case study in knowledge-based decision support. *IEEE Expert* **1995**, *10*, 56–62. [CrossRef]
9. Josang, A.; Ismail, R. The beta reputation system. In Proceedings of the 15th Bled Electronic Commerce Conference, Bled, Slovenia, 17–19 June 2002; Volume 5, pp. 2502–2511.
10. Kerschbaum, F.; Haller, J.; Karabulut, Y.; Robinson, P. Pathtrust: A trust-based reputation service for virtual organization formation. In Proceedings of the International Conference on Trust Management, Pisa, Italy, 16–19 May 2006; Springer: Berlin/Heidelberg, Germany, 2006; pp. 193–205.

11. Corchado, J.M.; Chamoso, P.; Hernández, G.; Gutierrez, A.S.R.; Camacho, A.R.; González-Briones, A.; Pinto-Santos, F.; Goyenechea, E.; Garcia-Retuerta, D.; Alonso-Miguel, M.; et al. Deepint. net: A Rapid Deployment Platform for Smart Territories. *Sensors* **2021**, *21*, 236. [CrossRef] [PubMed]
12. Garcia-Retuerta, D.; Chamoso, P.; Hernández, G.; Guzmán, A.S.R.; Yigitcanlar, T.; Corchado, J.M. An Efficient Management Platform for Developing Smart Cities: Solution for Real-Time and Future Crowd Detection. *Electronics* **2021**, *10*, 765. [CrossRef]
13. Page, L.; Brin, S.; Motwani, R.; Winograd, T. *The PageRank Citation Ranking: Bringing Order to the Web*; Technical Report; Stanford InfoLab: Stanford, CA, USA, 1999.
14. Kamvar, S.D.; Schlosser, M.T.; Garcia-Molina, H. The eigentrust algorithm for reputation management in p2p networks. In Proceedings of the 12th international conference on World Wide Web, Budapest, Hungary, 20–24 May 2003; pp. 640–651.
15. Kurdi, H.A. HonestPeer: An enhanced EigenTrust algorithm for reputation management in P2P systems. *J. King Saud Univ. Comput. Inf. Sci.* **2015**, *27*, 315–322. [CrossRef]
16. Adler, B.T.; De Alfaro, L. A content-driven reputation system for the Wikipedia. In Proceedings of the 16th International Conference on World Wide Web, Banff, AB, Canada, 8–12 May 2007; pp. 261–270.
17. Zacharia, G.; Maes, P. Trust management through reputation mechanisms. *Appl. Artif. Intell.* **2000**, *14*, 881–907. [CrossRef]
18. Waze. Your Rank and Points—Connected Citizens Program. 2019. Available online: https://wiki.waze.com/wiki/Your_Rank_and_Points (accessed on 12 January 2019).

 *electronics*

Article

My-Trac: System for Recommendation of Points of Interest on the Basis of Twitter Profiles

Alberto Rivas [1,2,*], Alfonso González-Briones [1,2,3], Juan J. Cea-Morán [1], Arnau Prat-Pérez [4] and Juan M. Corchado [1,2]

1   BISITE Research Group, University of Salamanca, Edificio I+D+i, Calle Espejo 2, 37007 Salamanca, Spain; alfonsogb@usal.es (A.G.-B.); juanju_97@usal.es (J.J.C.-M.); corchado@usal.es (J.M.C.)
2   Air Institute, IoT Digital Innovation Hub, Carbajosa de la Sagrada, 37188 Salamanca, Spain
3   Research Group on Agent-Based, Social and Interdisciplinary Applications (GRASIA), Complutense University of Madrid, 28040 Madrid, Spain
4   Sparsity-Technologies, 08034 Barcelona, Spain; arnau@sparsity-technologies.com
*   Correspondence: rivis@usal.es

Abstract: New mapping and location applications focus on offering improved usability and services based on multi-modal door to door passenger experiences. This helps citizens develop greater confidence in and adherence to multi-modal transport services. These applications adapt to the needs of the user during their journey through the data, statistics and trends extracted from their previous uses of the application. The My-Trac application is dedicated to the research and development of these user-centered services to improve the multi-modal experience using various techniques. Among these techniques are preference extraction systems, which extract user information from social networks, such as Twitter. In this article, we present a system that allows to develop a profile of the preferences of each user, on the basis of the tweets published on their Twitter account. The system extracts the tweets from the profile and analyzes them using the proposed algorithms and returns the result in a document containing the categories and the degree of affinity that the user has with each category. In this way, the My-Trac application includes a recommender system where the user receives preference-based suggestions about activities or services on the route to be taken.

Keywords: users' profiling; data extraction; natural language processing; recommender system; mapping application

Citation: Rivas, A.; González-Briones, A.; Cea-Morán, J.J.; Prat-Pérez, A.; Corchado, J.M. My-Trac: System for Recommendation of Points of Interest on the Basis of Twitter Profiles. *Electronics* **2021**, *10*, 1263. https://doi.org/10.3390/electronics10111263

Academic Editors: Dimitris Apostolou and Osvaldo Gervasi

Received: 13 April 2021
Accepted: 20 May 2021
Published: 25 May 2021

Publisher's Note: MDPI stays neutral with regard to jurisdictional claims in published maps and institutional affiliations.

1. Introduction

Humans are social beings; we always seek to be in contact with other people and to have as much information as possible about the world around us. The philosopher Aristotle (384–322 B.C.) in his phrase "Man is a social being by nature" states that human beings are born with the social characteristic and develop it throughout their lives, as they need others in order to survive. Socialization is a learning process; the ability to socialize means we are capable of relating with other members of the society with autonomy, self-realization and self-regulation. For example, the incorporation of rules associated with behavior, language, and culture improves our communication skills and the ability to establish relationships within a community.

In the search for improvement, communication, and relationships, human beings seek to get in contact with other people and to obtain as much information as possible about the environment in order to achieve the above objectives. The emergence of the Internet has made it possible to define new forms of communication between people. It has also made it possible to make a large amount of information on any subject available to the average user at any time. This is materialized in the development of social networks. The concept of social networking emerged in the 2000s as a place that allows for interconnection between people, and, very soon, the first social networking platforms appeared on the Internet that served to bring people together. Among the first platforms that emerged were Fotolog, MySpace, Hi5, Buzz, and SecondLife; however, most of them have declined in popularity or disappeared. Today, Facebook, Twitter, and Instagram stand out; they are used by millions of people all over the world. Thanks to these technologies, people from different parts of the world can engage in conversation, post photos of their latest trip, or keep their followers updated by sharing their opinions or experiences.

With regard to writing opinions or experiences, Twitter is the social network par excellence. Twitter is based on the concept of microblogging, i.e., users can post messages about their opinions, preferences, experiences, etc., with a maximum of 280 characters. Twitter allows its users to follow other accounts that interest them, or to comment on events in real time using hashtags. All this translates into one word: information. The information that users provide on social networks can be used in a variety of ways, many of them negative. Exposure on the Internet means that anyone can access the users' data and use it for financial gain. However, it can also be used to make life easier for users who choose to do so, always bearing in mind that there must be express consent on their part. This is precisely the case of the work presented here.

Since the emergence of the first social networks much progress has been made towards the current state of maturity of the social network life cycle. As presented above, they are of vital importance to society, as they fulfill the innate communication function of human beings. In addition to their use as a means of communication, they have begun to be exploited for business purposes in order to profit from the enormous amount of information that is generated on a daily basis. This information, which is generated by the society, is of great value once analyzed and processed correctly.

The data generated by users on social networks allows for the development of commercial and advertising actions that are much more effective than with traditional formats. Advertisement platforms have developed the ability to segment advertising on the basis of the behavior of each user, to show products and services to those who are really interested in those products and services. This increases the effectiveness of advertisements.

Social network users publish practically everything that happens in their daily lives, their opinions, where they are, what they eat, what they would like to buy, where they are going on holiday, and a long list of behaviors that are transformed into valuable information for analysis. The information obtained through the analysis is very interesting as it allows for the elaboration of demographic, socio-economic, and consumer trend profiles. The companies that own these social networks sell high-value information to other companies to enable them to carry out much more powerful and effective marketing and advertising strategies. Another remarkable aspect of data analytics on social networks is the ability to perform real-time analysis of the information to offer products and services according to their characteristics.

Information is a very precious commodity, and, as presented above, Twitter is a great source of data when analyzing human behavior and interactions or when learning about the opinion of certain users on certain topics. This information can be used to improve the multi-modal experience of users when they use the My-Trac application. Therefore, an adaptation of these systems for adoption in mapping applications is proposed.

The European My-TRAC project focuses on providing user-centered services to improve the multi-modal experience of passengers from door-to-door. This helps citizens develop greater confidence in and adherence to multi-modal transport services. In addition, My-TRAC improves customization to users' needs through data, statistics, and trends provided by passengers' experiences when using the proposed platform. Part of the tailoring of services and recommendations to users is determined by the knowledge obtained from their Twitter posts through the use of NLP techniques to classify and understand users.

There are other services that offer similar functionalities to My-Trac, such as ROSE [1], CTRR, and CTRR+ based systems for city-based tourism [2], or to participate in solidarity projects in rural environments [3]. From among the above, ROSE (ROuting SErvice) stands out, which is a mobile phone application that suggests events and places to the user and guides them via public transport. There are many different systems that incorporate both recommendation and navigation. However, there is no system that combines event recommendation and pedestrian navigation with (real-time) public transport. However, it does not employ multi-modal navigation between different public transport modes (bus, train, carpooling, plane, etc.) in different countries and that would use information from the user's social network profile. Instead, current systems utilize a set of information initially entered into the application which is not updated afterwards. Finally, Tables 1 and 2 present a review of similar works.

Table 1. Review of similar works: Part I.

Title / Publication	Functionality	Advantages	Shortcomings
ROSE (ROuting SErvice) [1]	Mobile phone application that suggests events and places to the user and guides them via public transport.	The current systems utilize a set of information initially entered into the application which is not updated afterwards.	There is no system that combines event recommendation and pedestrian navigation with (real-time) public transport. It does not employ multi-modal navigation between different public transport modes (bus, train, carpooling, plane, etc.) in different countries and that would use information from the user's social network profile.
Systems for city-based tourism [2]	A personalized travel route recommendation based on the road networks and users' travel preferences.	The experimental results show that the proposed methods achieve better results for travel route recommendations compared with the shortest distance path method.	It does not use information from public transport services in route recommendations.
Tourism routes as a tool for the economic development of rural areas—vibrant hope or impossible dream? [3]	This paper argues that the clustering of activities and attractions, and the development of rural tourism routes, stimulates co-operation and partnerships between local areas. The paper further discusses the development of rural tourism routes in South Africa and highlights the factors critical to its success.	The article analyzes the realization of routes that include activities and attractions in a way that encourages and enhances rural development in Africa.	Preliminary project that requires public cooperation (institutions, transport, services) for a comprehensive improvement of the proposal.

This article improves on the previous system for the extraction of information regarding Twitter users [4]. The system is capable of obtaining information about a particular user and of elaborating a profile with the user's preferences in a series of pre-established categories. A review of existing reputation systems is presented in Section 2. Section 3 describes the proposal. Section 4 presents the assessment made with synthetic data. Section 5 shows how the system is integrated in My-Trac app. Finally, Section 6 presents the conclusions.

Table 2. Review of similar works: Part II.

Title / Publication	Functionality	Advantages	Shortcomings
Social Recommendations for Events [5]	Outlife recommender assists in finding the ideal event by providing recommendations based on the user's personal preferences.	In addition to the user's preferences, the recommender uses information from the user's group of friends to make event recommendations more satisfactory.	Although it uses information from the user's groups of friends, no use is made of information from the user's social networks to complement the analysis and recommendation.
Smart Discovery of Cultural and Natural Tourist Routes [6]	This paper presents a system designed to utilize innovative spatial interconnection technologies for sites and events of environmental, cultural and tourist interests. The system discover and consolidate semantic information from multiple sources, providing the end-user the ability to organize and implement integrated and enhanced tours.	The system adapts the services offered to meet the needs of specific individuals, or groups of users who share similar characteristics, such as visual, acoustic, or motor disabilities. Personalization is done in a dynamic way that takes place at the time and place of the service.	The very comprehensive system that uses external services, scraping, crawling, geo-positioning but does not include information from social networks to complement the analysis and recommendation of events.
Enhancing cultural recommendations through social and linked open data [7]	Hybrid recommender system (RS) in the artistic and cultural heritage area, which takes into account the activities on social media performed by the target user and her friends	The system integrates collaborative filtering and community-based algorithms with semantic technologies to exploit linked open data sources in the recommendation process. Furthermore, the proposed recommender provides the active user with personalized and context-aware itineraries among cultural points of interest.	The main drawback is the absence of extensive control over the semantics that are not taken into account. It generates difficulties in justifying, explaining, and hence analyzing the resulting scores.
Personalized Tourist Route Generation [8]	Intelligent routing system able to generate and customize personalized tourist routes in real-time and taking into account public transportation.	We have modeled the tourist planning problem, integrating public transportation, as the Time Dependent Team Orienteering Problem with Time Windows (TDTOPTW). We have designed a heuristic able to solve it in real time, precalculating the average travel times between each pair of POIs in a preprocessing step.	Future works consists on extending the system to more cities with a different public transport network topology. The next one consists on integrating an advanced recommendation system in a wholly functional PET. The systema don't use social network capabilities, that allows to store, share and add travel experiences to better help tourists on the destination.

2. Natural Language Processing Techniques Applied to Twitter Profiles

In this section, we review the main techniques applied in the analysis that make it possible to get to know the users preferences through their tweets. This allows for recommendations to be made according to the user profile.

2.1. Word Embedding Techniques

NLP techniques allow computers to analyze human language, interpret it, and derive its meaning so that it can be used in practical ways. These techniques allow for tasks, such as automatic text summarization, language translation, relation extraction, sentiment analysis, speech recognition, and item classification, to be carried out. Currently, NLP is considered to be one of the great challenges of artificial intelligence as it is one of the fields with the highest development activity since it presents tasks of great complexity: how to really understand the meaning of a text, how to intuit neologisms, ironies, jokes, or poetry? It is a challenge to apply the techniques and algorithms that allow us to obtain the expected results.

One of the most commonly used NLP techniques is Topic Modeling. This technique is a type of statistical modeling that is used to discover the abstract "topics" that appear in a series of input texts. Topic modeling is a very useful text mining tool for discovering hidden semantic structures in texts. Generally, the text of a document deals with a particular topic, and the words related to that topic are likely to appear more frequently in the document than those that are unrelated to the text. Topic Modeling collects the set of more frequent words in a mathematical framework, which allows one to examine a set of text documents and discover, on the basis of the statistics of the words in each one, what the topics may be and what the balance is between the topics in each document.

The input of topic modeling is a document-term matrix. The order of words does not matter. In a document-term matrix, each row is a question (or document), each column is a term (or word), we label "0" if that document does not contain that term, "1" if that document contains that term once, "2" if that document contains that term twice, and so on.

Algorithms, such as Bag-of-words or TF-IDF, among others, make it possible to represent the words used by the models and create the matrix defined above, representing a token in each column and counting the number of times that token appears in each sentence (represented in each row).

- **Bag-of-words.** This model allows to extract the characteristics of texts (also images, audios, etc.). It is, therefore, a feature extraction model. The model consists of two parts: a representation of all the words in the text and a vector representing the number of occurrences of each word throughout the text. That is why it is called Bag-of-words. This model completely ignores the structure of the text, it simply counts the number of times words appear in it. It has been implemented through the Genism library [9].
- **Term Frequency - Inverse Document Frequency (TF-IDF).** This is the product of two measures that indicate, numerically, the degree of relevance that a word has in a document within a collection of documents [10]. It is broken down into two parts:

    - *Term frequency:* Measures the frequency with which certain terms appear in a document. There are several measurement options, the simplest being the gross frequency, i.e., the number of times a term t appears in a document d. However, in order to avoid a predisposition towards long documents, the normalized frequency is used:

    $$\mathrm{tf}(t,d) = \frac{\mathrm{f}(t,d)}{\max\{\mathrm{f}(t,d) : t \in d\}}. \tag{1}$$

    As shown in Equation (1), the frequency of the term is divided by the maximum frequency of the terms in the document.

    - *Inverse document frequency:* If a term appears very frequently in all of the analyzed documents, its weight is reduced. If it appears infrequently, it is increased.

    $$\mathrm{idf}(t,D) = \log \frac{|D|}{|\{d \in D : t \in d\}|}. \tag{2}$$

    As shown in Equation (2), the total number of documents is divided by the number of documents containing the term. Term frequency—Inverse document frequency: The entire formula is as shown in Equation (3).

    $$\mathrm{tf} - \mathrm{idf}(t,d,D) = \mathrm{tf}(t,d) \times \mathrm{idf}(t,D). \tag{3}$$

Word embedding is a term used for the representation of words for text analysis, typically in the form of a real-valued vector that encodes the meaning of the word such that the words that are closer in the vector space are expected to be similar in meaning [11]. Word embeddings can be obtained using a set of language modeling and feature learning techniques where words or phrases from the vocabulary are mapped to vectors of real numbers [12].

- **Word2vec.** This technique uses huge amounts of text as input and is able to identify which words appear to be similar in various contexts [13–15]. Once trained on a sufficientñy big dataset, 300-dimensional vectors are generated for each word, forming a new vocabulary where "similar" words are placed close to each other. Pre-trained vectors are used, achieving a wealth of information from which to understand the semantic meaning of the texts.
- **Doc2vec.** This technique is an extension of Word2Vec and is applied to a document as a whole instead of individual words, it uses an unsupervised learning approach to better understand documents as a whole [16]. Doc2Vec model, as opposed to Word2Vec model [17], is used to create a vectorized representation of a group of words taken collectively as a single unit. It does not only give the simple average of the words in the sentence.

2.2. Topic Modeling

As already presented in the previous section, topic modeling is a tool that takes an individual text (or corpus) as input and looks for patterns in word usage; it is an attempt to find semantic meaning in the vocabulary of that text (or corpus).

This set of tools enables the extraction of topics from texts; a topic is a list of words that is presented in a way that is statistically significant. Topic modeling programs do not know anything about the meaning of the words in a text. Instead, they assume that each text fragment is composed (by an author) through the selection of words from possible word baskets, where each basket corresponds to a topic. If that is true, then it is possible to mathematically decompose a text into the baskets from which the words that compose it are most likely to come. The tool repeats the process over and over again until the most probable distribution of words within the baskets, the so-called topics, is established.

The techniques executed by the proposed system are used to discover word usage patterns of each user on Twitter, and they make it possible to group users into different categories. To this end, a thorough review of the main tools for topic modeling has been carried out. Most of the algorithms are based on the paradigm of unsupervised learning. These algorithms return a set of topics, as many as indicated in the training. Each topic represents a cluster of terms that must be related to one of those categories. Precisely for this reason, a large number of tweets have been retrieved as training data. Keywords have been searched for for each category. As part of this research, a total of three algorithms have been evaluated: LDA, LSI, and NMF. In the NMF experiment, the best results were obtained, although the techniques applied in other works have been reviewed in order to contrast their results with this method.

Apart from the comparison itself, there are numerous studies that have made similar comparisons between these techniques so that the decision is supported by similar studies. In the work of Tunazzina Islam, in 2019 a similar experiment was carried out to the one proposed in this paper [18]. In this paper, Apache Kafka is employed to handle the big streaming data from Twitter. Tweets on yoga and veganism are extracted and processed in parallel with data mining by integrating Apache Kafka and Spark Streaming. Topic modeling is then used to obtain the semantic structure of the unstructured data (i.e. Tweets). They then perform a comparison of the three different algorithms LSA, NMF, and LDA, with NMF being the best performing model.

Another noteworthy work is that carried out by Chen et al. [19], in which an experiment is carried out to detect topics in small text fragments.

This is similar to the proposal made in this paper, since tweets can be considered small texts. In this work a comparison is made between the LDA and NMF methods, the latter being the one that provided the best results.

- **Latent Dirichlet allocation (LDA).** Is a generative statistical model that allows sets of observations to be explained by unobserved groups that explain why some parts of the data are similar [20–22]. For example, if observations are collections of words in documents, each document is a mixture of a small number of topics and each word's presence is attributable to one of the document's topics. LDA is an example of a topic model and belongs to the machine learning toolbox and in wider sense to the artificial intelligence toolbox.

- **Nonnegative Matrix Factorization (NMF).** Is an unsupervised learning algorithm belonging to the field of linear algebra. NMF reduces the dimensionality of an input matrix by factoring it in two and approximating it to another of a smaller range. The formula is $V \approx WH$. Let us suppose, observing Equation (4), a vectorization of P documents with an associated dictionary of N terms (weight). That is, each document is represented as a vector of N dimensions. All documents, therefore, correspond to a V matrix

$$V \in \mathbb{R}^{N \times P} = \begin{pmatrix} \square & \cdots & \cdots & \square \\ \square & \ddots & & \vdots \\ \vdots & \ddots & \ddots & \vdots \\ \square & \cdots & \square & \square \end{pmatrix}, \tag{4}$$

where N is the number of rows in the matrix, and each of them represents a term, while P is the number of columns in the matrix and each of them represents a document. Equations (5) and (6) shows matrices W and H. The value r marks the number of topics to be extracted from the texts.

Matrix W contains the characteristic vectors that make up these topics. The number of characteristics (dimensionality) of these vectors is identical to that of the data in the input matrix V. Since only a few topic vectors are used to represent many data vectors, it is ensured that these topic vectors discover latent structures in the text. The H-matrix indicates how to reconstruct an approximation of the V-matrix by means of a linear combination with the W-columns.

$$W \in \mathbb{R}^{N \times r} = \begin{pmatrix} \square & \cdots & \cdots & \square \\ \square & \ddots & & \vdots \\ \vdots & \ddots & \ddots & \vdots \\ \square & \cdots & \square & \square \end{pmatrix}, \tag{5}$$

where N is the number of rows in matrix W, and each of them represents a term (weight), and r is the number of columns in matrix W, where r is the number of characteristics to be extracted.

$$H \in \mathbb{R}^{r \times P} = \begin{pmatrix} \square & \cdots & \cdots & \square \\ \square & \ddots & & \vdots \\ \vdots & \ddots & \ddots & \vdots \\ \square & \cdots & \square & \square \end{pmatrix}, \tag{6}$$

where r is the number of rows in matrix H, r is the number of characteristics to be extracted, and P is the number of columns, with one column for each document. The result of the matrix product between W and H is, therefore, a matrix of dimensions NxP corresponding to a compressed version of V.

The use of Machine Learning techniques for the analysis of information extracted from Twitter is a very common case study today. It is convenient to study what kind of research is being carried out on this subject. One of the main applications is the use of Twitter and Natural Language Processing techniques in order to extract a user's opinion about what is being tweeted at a given time. The article "A system for real-time Twitter sentiment analysis of 2012 U.S. presidential election cycle", written by Hao Wang et al. [23], presents a system for real-time polarity analysis of tweets related to candidates for the 2012 U.S. elections.

The system collects tweets in real time, tokens and cleans them, identifies which user is being talked about in the tweet, and analyzes the polarity. For training, it applies Naïve Bayes, a statistical classifier. It uses hand-categorized tweets as input. Another study similar to this one is the one proposed by J.M.Cotelo et al. from the University of Seville: "Tweet Categorization by combining content and structural knowledge" [24]. It proposes a method to extract the users' opinion about the two main Spanish parties in the 2013 elections. It uses two processing pipelines, one based on the structural analysis of the tweets, and the other based on the analysis of their content.

Another possible line of research is based on categorizing Twitter content. This is the case of the article "Twitter Trending Topic Classification" written by Kathy Lee et al. [25]. It studies the way to classify trending topics (hashtags highlighted) in 18 different categories. To this end, Topic Modeling techniques were used. The key point lies in providing a solution based on the analysis of the network underlying the hashtags and not only the text: "our main contribution lies in the use of the social network structure instead of using only textual information, which can often be noisy considering the social network context".

As it can be seen, there are many studies currently oriented to the analysis of Twitter using Machine Learning tools. The challenge to be faced in this work is to find the optimal way of classifying users according to their tweets. The sections that follow describe the objectives of the project and detail the research and testing that led to the construction of a stable system fit for the purpose for which it has been designed.

3. Proposal

This section proposes a system for the extraction of information about Twitter users. The system is capable of obtaining information about a particular user and of elaborating a profile with the user's preferences in a series of pre-established categories. From an abstract point of view, the proposal could be seen as a processing pipeline, as shown in Figure 1. The different phases of this pipeline contribute to the achievement of the main objective: user classification.

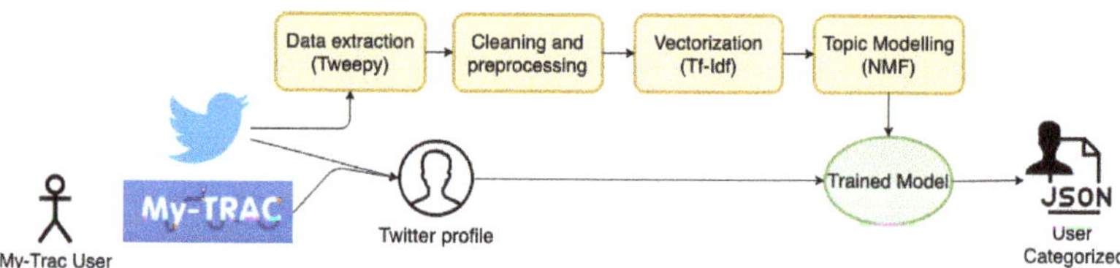

Figure 1. Pipeline representing the system processing steps.

3.1. Category Definition

Matching a given profile to a specific category or topic is one of the objectives of NLP algorithms. As a starting point, it is necessary to prepare the training dataset that is used when investigating the algorithmic model. The strategy followed is based on the model of the Interactive Advertising Bureau (IAB) association [26]. Today, IAB is a benchmark standard for the classification of digital content. In particular, the IAB Tech Lab has developed and released a content taxonomy on which the present categorization is based. This taxonomy proposes a total of 23 categories with their corresponding subcategories covering the main topics of interest. In this way, 8000 tweets from each of these categories have been ingested. As a result, 23 datasets with examples of tweets related to each category were obtained, these datasets have been used to train the system at a later stage. Specifically, the list of topics is shown in Table 3.

Table 3. Categories taxonomy.

Topics
Arts & Entertainment
Automotive
Business
Careers
Education
Family & Parenting
Health & Fitness
Food & Drink
Hobbies & Interests
Home & Garden
Law, Gov't & Politics
News
Personal Finance
Society
Science
Pets
Sports
Style & Fashion
Technology & Computing
Travel
Real Estate
Shopping
Religion & Spirituality

3.2. Twitter Data Extraction

The Twitter data extraction mechanism is a fundamental element of the system. The goal of this mechanism is to recover two types of data.

On the one hand, the system extracts a set of anonymous tweets related to each of the defined preference categories; these tweets are used to train the data classification algorithms.

On the other hand, the mechanism extracts information about the given user for the analysis of their preferences.

Twitter's API enables developers to perform all kinds of operations on the social network. It is, therefore, necessary for our system to use this powerful API. This API could be used by elaborating a module that would make HTTP requests to the API so that the endpoints of interest are executed. However, this involves a remarkably high development cost.

Another option would be to make use of one of the multiple Python libraries that encapsulate this logic and offer a simple interface to developers. The latter option has been chosen for the development of this system, more specifically, library Tweepy [27].

3.3. Preprocessing of Tweets

Once the data has been extracted, it must be prepared for the classification algorithms. Cleaning and preprocessing techniques must be applied, so that the text is prepared for topic modeling algorithms. Libraries, such as NLTK and Spacy, have been used, as can be observed in Listing 1.

The first step involves cleaning tweets, by removing content that does not provide information for language processing. More specifically, this task consists in eliminating URLs, hashtags, mentions, punctuation marks, etc.

Another of the techniques applied to obtain more information from tweets is the transformation of the emojis contained in the text into a format from which it is possible to extract information. To do this, a dictionary of emojis is used as a starting point for the conversion of the data. This dictionary contains a series of values that interpret each of the existing emojis when applying the corresponding analysis. In this way, it has been possible to identify and give a certain value to each emoji for its treatment.

The key activity performed during the preprocessing consist of eliminating stopwords and tokenization. Whether it is a paragraph, an entire document or a simple tweet, every text contains a set of empty words or stopwords. This set of words is characterized by its continuous repetition in the document and its low value within the analysis. These words are mainly articles, determiners, synonyms, conjunctions, and others.

Listing 1. Preprocessing step pseudocode.

```python
from nltk.tokenize import word_tokenize
import~spacy

sp = spacy.load('en_core_web_sm')
stopwords_dict = sp.Defaults.stop_words

def tweet_preprocessing(tweet):
    tweet = hashtag_removal(tweet)
    tweet = mentions_removal(tweet)
    tweet = url_removal(tweet)
    tweet = html_removal(tweet)
    tweet = punctuation_removal(tweet)
    tweet = emojis_removal(tweet)
    tweet = word_tokenize(tweet)
    tweet = [word for word in tweet if not word in stopwords_dict]
    return tweet
```

Table 4 shows the results obtained after the tweets have gone through the preprocessing and preparation process which had been carried out using the tools listed above.

3.4. Vectorization

Vectorization is the application of models that convert texts into numerical vectors so that the algorithms can work with the data. Two algorithms have been considered for the performance of this task, "Bag-Of-Words" and "Tf-Idf". Both are widely used in the field of NLP, but, in general, creation of tf-idf weights from text works properly and is not very expensive computationally. Moreover, NMF expects as input a Term-Document matrix, typically a "Tf-Idf" normalized.

The vectorizer have been tuned manually with some parameters according to the dataset, as can be observed in Listing 2. *Min_df* was set to 100 to ignore words that appear in less than 100 tweets. In the same way, *max_df* was set to 0.85 to ignore words that appear in more than 85% of the tweets. Thanks to that feature, it is possible to remove words that introduce noise in the model. Finally, the algorithm only takes into account single words, so, in order to include bigrams, the parameter *ngram_range* was set to (1, 2)

Table 4. Preprocessing results using NLTK tokenization.

	Text	Nltk_tokenized
0	Read This Before Taking a Road Trip with a Pet	[read, taking, road, trip, pet]
1	@kenwardskorner @Senators @Canucks In addition, does …	[also, name, imply, take, acid, road, trips, I…]
2	Our Art is our Passion \n#apnatruckart #truck	[art, passion, apnatruckart, truck, art, uniqu…]
3	Lelang drop acc budget 40–50 k dong?	[lelang, drop, acc, budget, dong]
4	We agree… and want everyone to know that ou	[agree, want, everyone, know, tours, relaxed, …]
5	Choosing a hotel for a break away with the fam…	[choosing, hotel, break, away, family, special…
6	@_JassyJass Are you,camping?	[camping]
7	How to Pack Your Electronics for Air Travel ht	[pack, electronics, air, travel]
8	Dasar low budget! https://t.co/2YUmUrGjj5	[dasar, low, budget]

Listing 2. Vectorization step pseudocode.

```python
from sklearn.feature_extraction.text import~TfidfVectorizer

def tfidf_vectorization(tweets):
    vectorizer = TfidfVectorizer(
        min_df=100,
        max_df=0.85,
        ngram_range=(1, 2),
        preprocessor=' '.join,
        use_idf=True
    )
    vectorized_tweets = vectorizer.fit_transform(tweets)
    return vectorized_tweets
```

3.5. Topic Modeling

Topic Modeling is a typical NLP task that aims to discover abstract topics in texts. It is widely used to discover hidden semantic structures. In the present work, this technique has been used to discover the main topics of interest of the My-Trac application users based on their Twitter profiles, which should correspond to some of the previously defined categories.

Regarding the features of the model, in Section 3.4, the training tweets were vectorized to create a Term-Document matrix which has been the input of the NMF model. In addition, NMF needs one important parameter, the number of topics to be discovered "*n_components*". In this case, $n\_components$ was set manually to 23, which is the number of topics that were defined initially in the categories taxonomy. Following this approach, the algorithm is trained with 184,000 tweets (8000 per category) with the aim of obtaining as many topics as categories were defined in the taxonomy. Once the model has been trained, it has been possible to determine in which topics a user's profile fits on the basis of their tweets. The implementation of the topic modeling algorithm has been carried out on the basis of NMF using SKLearn library, as is deailed in Listing 3.

Listing 3. NMF Sklearn implementation.

```
from sklearn.decomposition import NMF

def train_model(vectorized_tweets):
    nmf_model = NMF(n_components=23, alpha=.1,
                    l1_ratio=.5, init='nndsvda')
    nmf_model.fit(vectorized_tweets)
    return nmf_model
```

Finally, it is worth mentioning the use of some extra parameters which were set in the implementation of the model. The method used to initialize the procedure was set to "NNDSVa" which works better with the tweet dataset since this kind of data it is not sparse. *Alpha* and *l1_ratio* both are parameters which helps to define regularization.

4. Evaluation and Results

In order to evaluate the results of the algorithm, the most relevant terms have been identified for each resulting topic. Then, by reviewing the main terms for each topic, it is possible to determine if that words really represent the content of the topic. An example is shown in Figure 2, where the most relevant terms have been identified for 4 different topics, proving how well the algorithm identifies the terms associated with each one. As it can be seen, all of them are unambiguously related to their defined categories. Topic 1: Travel. Topic 2: Arts & Entertainment. Topic 9: Personal Finance. Topic 10: Pets.

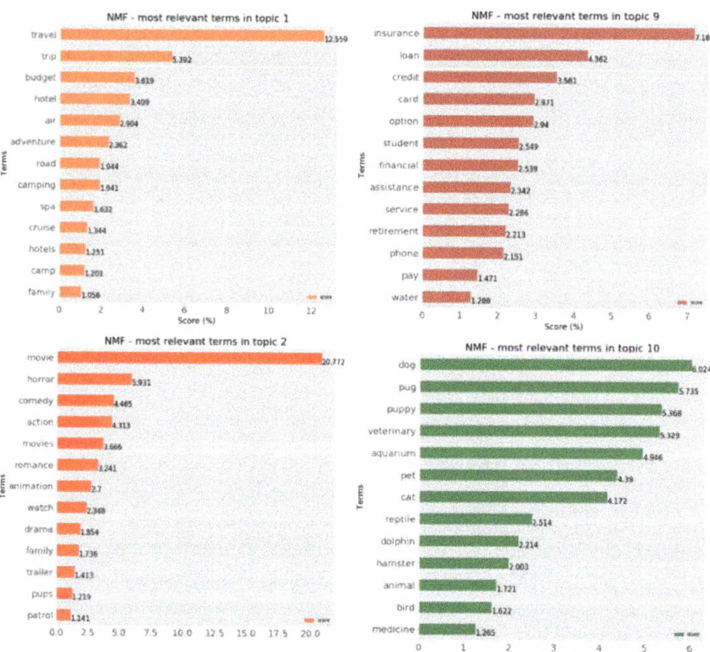

Figure 2. Example topics generated by NMF.

The full list of topics and their top 10 related keywords identified by the algorithm can be seen in Table 5. It should be noted that some of the previously defined categories in Table 3 have been removed during the evaluation of this model. This fact is due to the lack of tweets that would fit into those categories, as well as some topics were quite overlapped amongst them. The initially defined categories that have been removed during training process and evaluation are: "Home & Garden", "Real State", "Society", and "News". In the same way, the algorithm has been able to discover new categories related to the original ones, such as: "Movies", "Videogames", "Music", "Events", and "Medicine & Health", leaving a total of 23 categories in the system.

Table 5. Topics obtained by the algorithm.

	Topic	Top 10 Words
1	Travel	travel, trip, budget, air, hotel, adventure, road, camp, family, day
2	Movies	movie, horror, action, comedy, movies, romance, watch, animation, family, drama
3	Videogames	game, xbox, pc, mmo, nintendo, videogame, esport, rpg, play, console
4	Careers	apprenticeship, internship, job, search, career, interview, vocational, training, remote, advice
5	Events	amusement, concert, cinema, restaurant, birthday, match, holiday, football, funeral, park
6	Health & Fitness	health, nutrition, therapy, physical, fitness, workout, exercise, wellness, medicine, weight
7	Religion & Spirituality	islam, christianity, hinduism, judaism, buddhism, spirituality, religion, astrology, atheism, sikhism
8	Shopping	grocery, lotto, shopping, gift, discount, sale, card, coupon, sales, code
9	Personal Finance	insurance, loan, credit, option, card, student, financial, service, assistance, phone
10	Pets	veterinary, dog, pug, puppy, aquarium, pet, cat, reptile, dolphin, hamster
11	Automotive	truck, car, auto, motorcycle, tesla, van, scooter, pickup, luxury, minivan
12	Science	chemistry, geography, geology, biology, physics, genetic, astronomy, environment, math, science
13	Law, Gov't & Politics	election, political, law, issue, vote, news, state, people, country, trump
14	Education	preschool, college, university, exam, electoral, education, homework, student, school, language
15	Food & Drink	coffee, vegetarian, beer, eat, vegan, cook, drink, wine, tea, dining
16	Family & Parenting	marriage, parent, single, baby, daycare, teen, date, life, toddler, adopt
17	Style & Fashion	wear, beauty, clothing, perfume, deodorant, wallet, casual, fashion, shave, trainer
18	Technology & Computing	software, app, developer, mongodb, database, email, android, internet, computer, ai
19	Hobbies & Interests	meme, draw, puzzle, collect, comic_strip, antique, guitar, art, woodworke, painting
20	Sports	martial, rugby, golf, sport, climb, pool, racing, cricket, skating, basketball
21	Business	industry, agriculture, construction, startup, recall, economy, business, automotive, butterball, turkey
22	Medicine & Health	vaccine, menopause, pregnancy, health, mental, surgery, injury, disease, psychology, substance
23	Music	music, radio, rock, funk, pop, soul, classic, songwriter, listen, classical

Once the resulting model has been evaluated and verified, the next step is to check the effectiveness of the model with real Twitter profiles. The tests have been performed extracting 1200 tweets from different users and predicting for each user the most related topics based on their tweets. The final test results are shown in Table 6, where it can be observed how each profile name match with related topics according to the profile.

As an example, the main topics for the profile "Tesla" are "Automotive", "Technology and computing", and "Travel".

Finally, in order to suggest the main topics of a specific user in the My-Trac app, for each user, the model returns the associated categories, along with the percentage of weight that each category has on the user. The lower the percentage, the less relation the user has with the category. The results of the final classification using some known Twitter accounts are given in Table 7. It should be noted that only the three main categories are shown in the table (together with their associated percentage), as they are the most accurate for categorizing the user.

Table 6. NMF evaluation with data from real Twitter profiles.

	Cat 1	Cat 2	Cat 2
Pontifex	Religion & Spirituality	Family & Parenting	Tech & Computing
Tesla	Automotive	Tech & Computing	Travel
BBCNews	Law, Gov't & Politics	Sports	Family & Parenting
NintendoAmerica	Videogames	Hobbies & Interests	Sports
Theresa_may	Law, Gov't & Politics	Business	Personal Finance
Oprah	Events	Family & Parenting	Sports
SkyFootball	Sports	Events	Hobbies & Interests
ScuderiaFerrari	Sports	Automotive	Law, Gov't & Politics
IMDb	Events	Movies	Hobbies & Interests
ScienceMagazine	Science	Medicine & Health	Tech & Computing
Spotify	Music	Hobbies & Interests	Family & Parenting
Airbnb	Travel	Events	Careers

Table 7. Final results with different accounts.

	First Category	Second Category	Third Category
Tesla	Automotive (70.92%)	Tech & Computing (10.07%)	Travel (3.86%)
RealDonaldTrump	Law, Gov't & Politics (28.02%)	Business (11.36%)	Sports (8.02%)
ScienceMagazine	Science (24.41%)	Medicine & Health (16.73%)	Tech & Computing (12.03%)
NintendoAmerica	Videogames (41.65%)	Hobbies & Interests (12.19%)	Sports (9.74%)

5. Final System Integration in My-Trac Application

Having passed the entire research and evaluation process, a trained algorithm has been obtained capable of classifying different Twitter accounts according to defined and discovered categories. In addition, a reliable data extraction method has been developed. Therefore, the next step consists of applying the algorithm to the My-Trac app to create a system that allows recommendations to My-Trac users based on their Twitter profiles, which is the objective of the present work.

The final system for My-Trac app consists of a mobile app where the user logs in, as it can be seen in Figure 3, and is asked to grant access to their Twitter data. Once the user signs in to the application, My-Trac seeks for the optimal means of transport to reach a specific destination given by the user and suggests the best conveyance for the trip, as Figures 4 and 5 show.

Finally, when the user chooses the route and mean of transport that best fits his trip, based on the present work, My-Trac app recommends some activities and points of interest for the user during the way based on its Twitter information, as can be observed in Figures 6 and 7.

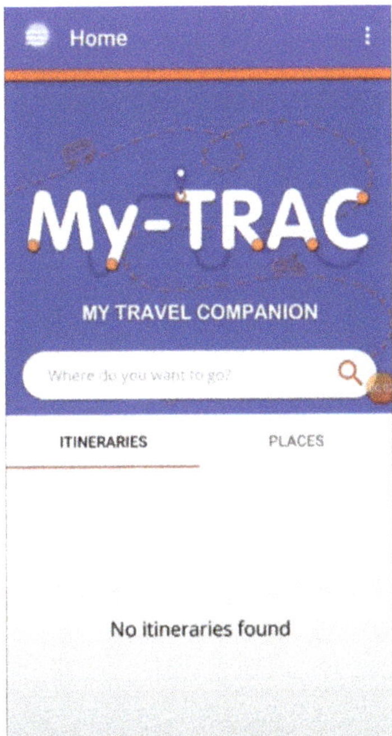

Figure 3. My-Trac application.

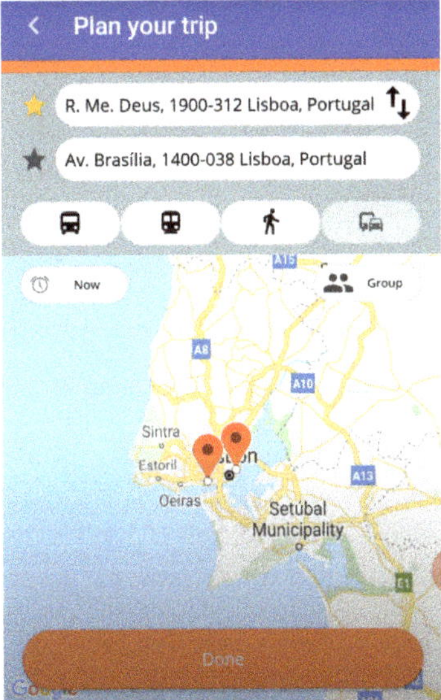

Figure 4. Trip planification using My-Trac app.

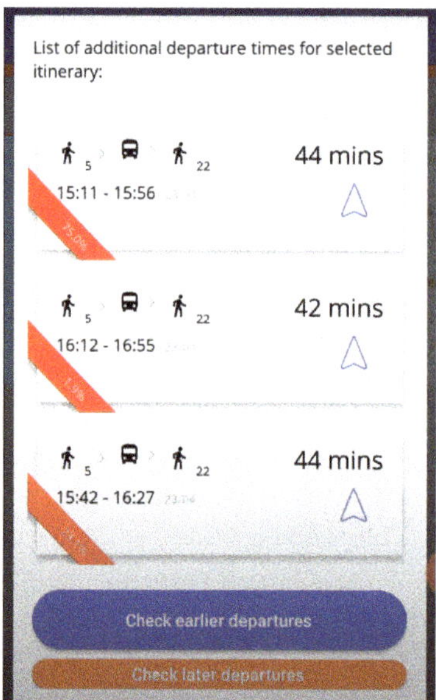

Figure 5. My-Trac suggests optimal means of transport for the destination.

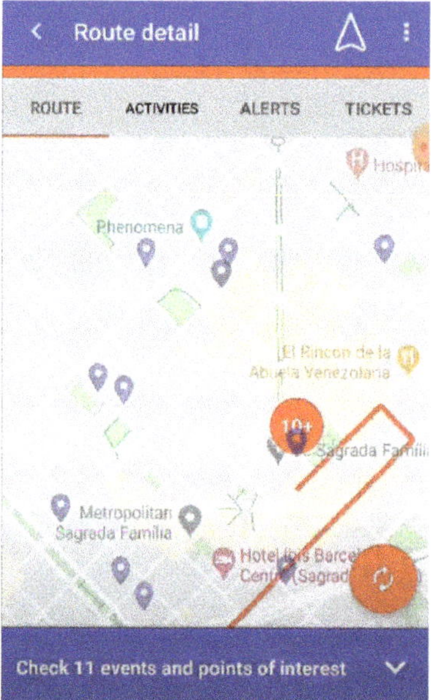

Figure 6. My-Trac recommends activities and point of interest for the user using its Twitter information.

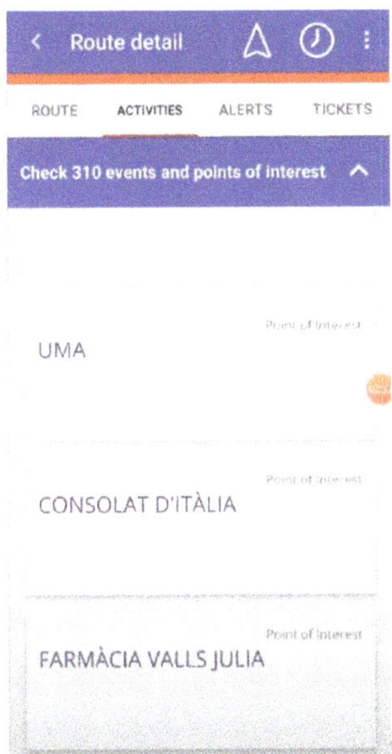

Figure 7. My-Trac suggestions.

Moreover, is possible to get some detailed information for each activity recommended, as Figure 8 shows. In this way, thanks to My-Trac app, the user can improve his experience not only by receiving suggestions for the best conveyance for the trip but also receiving customized activity recommendations and points of interest.

Figure 8. Detailed information about a suggested point of interest.

6. Conclusions and Future Work

This article presents a novel approach to extracting preferences from a Twitter profile by analyzing the tweets published by the user for use in mapping applications. This approach has successfully defined a consistent and representative list of categories, and the mechanisms needed for information extraction have been developed, both for model training and end-user analysis. It is a unique system, with which it has been possible to develop an important feature in the My-Trac app, whereby it is possible to recommend relevant point of interest to the end users.

Regarding future work on this system, many areas of improvement and development have been identified. Tweets are not the only source of information that allows to discern the interests of a profile. It may be the case that a user only writes about football but is the follows many news-related and political accounts. The current system would only be able to extract the sports category. Therefore, one of the improvements would be the implementation of a model that would analyze followed users. This has been started, by extracting the followers and creating wordclouds with the most relevant ones. Similarly, hashtags also provide additional information suitable for analysis. Another line of research is the training of a model that allows to analyze the tweets individually. This would open the doors to performing a polarity analysis that would allow us to know if a user who writes about a certain category does it in a positive, negative, or neutral way.

As for the limitations of the system, it is possible that, in some regions, there may be restrictive regulations on the use of information published on social networks for this type of analysis. Therefore, the user should carry out a study of data protection and the legal framework adapted to each region where the service is to be provided. Furthermore, in terms of performance, it is possible that specific context-dependent systems training an algorithm for each individual user may perform slightly better than the proposed solution.

Author Contributions: Conceptualization, A.R. and A.G.-B.; methodology, A.R. and A.G.-B.; software, A.R. and J.J.C.-M.; validation, A.R., A.G.-B., J.J.C.-M. and A.P.-P.; formal analysis, A.R. and J.J.C.-M; investigation, A.R., A.G.-B. and A.P.-P.; resources, A.R., A.G.-B., J.J.C.-M.; data curation, J.J.C.-M.; writing—original draft preparation, A.R., A.G.-B., J.J.C.-M.; writing—review and editing, A.R., A.G.-B., J.J.C.-M., A.P.-P. and J.M.C.; visualization, A.R. and J.J.C.-M.; supervision, A.G.-B., A.P.-P. and J.M.C.; project administration, A.G.-B., A.P.-P. and J.M.C.; funding acquisition, J.M.C. All authors have read and agreed to the published version of the manuscript.

Funding: This research was funded by the Spanish Ministerio de Ciencia e Innovación under grant number TIN2017-89314-P.

Acknowledgments: This research has been supported by the European Union's Horizon 2020 research and innovation programme under grant agreement No 777,640.

Conflicts of Interest: The authors declare no conflict of interest.

References

1. Ludwig, B.; Zenker, B.; Schrader, J. Recommendation of Personalized Routes with Public Transport Connections. In Proceedings of the International Conference on Intelligent Interactive Assistance and Mobile Multimedia Computing, Rostock-Warnemünde, Germany, 9–11 November 2009; Springer: Berlin/Heidelberg, Germany, 2009; pp. 97–107.
2. Cui, G.; Luo, J.; Wang, X. Personalized travel route recommendation using collaborative filtering based on GPS trajectories. *Int. J. Digit. Earth* **2018**, *11*, 284–307. [CrossRef]
3. Briedenhann, J.; Wickens, E. Tourism routes as a tool for the economic development of rural areas—Vibrant hope or impossible dream? *Tour. Manag.* **2004**, *25*, 71–79. [CrossRef]
4. Cea-Morán, J.J.; González-Briones, A.; De La Prieta, F.; Prat-Pérez, A.; Prieto, J. Extraction of Travellers' Preferences Using Their Tweets. In Proceedings of the International Symposium on Ambient Intelligence, L Aquila, Italy, 17–19 June 2020; pp. 224–235.
5. De Pessemier, T.; Minnaert, J.; Vanhecke, K.; Dooms, S.; Martens, L. Social recommendations for events. In Proceedings of the CEUR workshop Proceedings, Miami, FL, USA, 1 October 2013; Volume 1066.
6. Stathopoulos, E.A.; Paliokas, I.; Meditskos, G.; Diplaris, S.; Tsafaras, S.; Valkouma, E.; Pehlivanides, G.; Riggas, C.; Vrochidis, S.; Votis, K.; et al. Smart discovery of cultural and natural tourist routes. In Proceedings of the IEEE/WIC/ACM International Conference on Web Intelligence-Companion, Thessaloniki, Greece, 14–17 October 2019; pp. 208–214.
7. Sansonetti, G.; Gasparetti, F.; Micarelli, A.; Cena, F.; Gena, C. Enhancing cultural recommendations through social and linked open data. *User Model. User-Adapt. Interact.* **2019**, *29*, 121–159. [CrossRef]
8. Garcia, A.; Arbelaitz, O.; Linaza, M.T.; Vansteenwegen, P.; Souffriau, W. Personalized tourist route generation. In Proceedings of the International Conference on Web Engineering, Vienna, Austria, 5–9 July 2010; Springer: Berlin/Heidelberg, Germany, 2010; pp. 486–497.
9. University, Y. About Yale: Yale Facts. 2017. Available online: https://www.yale.edu/about-yale/yale-facts (accessed on 24 May 2021).

10. Demestichas, K.; Kosmides, P. An offline, statistical method for cost efficient design of experiments and field trials involving electric vehicles. In Proceedings of the 11th ITS European Congress, Glasgow, Scotland, 6–9 June 2016.

11. Ferrari, A.; Donati, B.; Gnesi, S. Detecting domain-specific ambiguities: An NLP approach based on Wikipedia crawling and word embeddings. In Proceedings of the 017 IEEE 25th International Requirements Engineering Conference Workshops (REW), Lisbon, Portugal, 4–8 September 2017; pp. 393–399.

12. Wallace, E.; Wang, Y.; Li, S.; Singh, S.; Gardner, M. Do nlp models know numbers? Probing numeracy in embeddings. *arXiv* **2019**, arXiv:1909.07940.

13. Church, K.W. Word2Vec. *Nat. Lang. Eng.* **2017**, *23*, 155–162. [CrossRef]

14. Rong, X. word2vec parameter learning explained. *arXiv* **2014**, arXiv:1411.2738.

15. Lilleberg, J.; Zhu, Y.; Zhang, Y. Support vector machines and word2vec for text classification with semantic features. In Proceedings of the 2015 IEEE 14th International Conference on Cognitive Informatics & Cognitive Computing (ICCI* CC), Beijing, China, 6–8 July 2015; pp. 136–140.

16. Bilgin, M.; Şentürk, İ.F. Sentiment analysis on Twitter data with semi-supervised Doc2Vec. In Proceedings of the 2017 International Conference on Computer Science and Engineering (UBMK), London, UK, 5–7 July 2017; pp. 661–666.

17. Chen, Q.; Sokolova, M. Word2Vec and Doc2Vec in unsupervised sentiment analysis of clinical discharge summaries. *arXiv* **2018**, arXiv:1805.00352.

18. Islam, T. Yoga-veganism: Correlation mining of twitter health data. *arXiv* **2019**, arXiv:1906.07668.

19. Chen, Y.; Zhang, H.; Liu, R.; Ye, Z.; Lin, J. Experimental explorations on short text topic mining between LDA and NMF based Schemes. *Knowl.-Based Syst.* **2019**, *163*, 1–13. [CrossRef]

20. Resnik, P.; Armstrong, W.; Claudino, L.; Nguyen, T.; Nguyen, V.A.; Boyd-Graber, J. Beyond LDA: Exploring supervised topic modeling for depression-related language in Twitter. In Proceedings of the 2nd Workshop on Computational Linguistics and Clinical Psychology: From Linguistic Signal to Clinical Reality, Denver, CO, USA, 5 June 2015; pp. 99–107.

21. Mehrotra, R.; Sanner, S.; Buntine, W.; Xie, L. Improving lda topic models for microblogs via tweet pooling and automatic labeling. In Proceedings of the 36th international ACM SIGIR conference on Research and development in information retrieval, Dublin, Ireland, 28 July–1 August 2013; pp. 889–892.

22. Tajbakhsh, M.S.; Bagherzadeh, J. Semantic knowledge LDA with topic vector for recommending hashtags: Twitter use case. *Intell. Data Anal.* **2019**, *23*, 609–622. [CrossRef]

23. Wang, H.; Can, D.; Kazemzadeh, A.; Bar, F.; Narayanan, S. A system for real-time twitter sentiment analysis of 2012 us presidential election cycle. In Proceedings of the ACL 2012 System Demonstrations. Association for Computational Linguistics, Jeju Island, Korea, 8–14 July 2012; pp. 115–120.

24. Cotelo, J.M.; Cruz, F.L.; Enríquez, F.; Troyano, J. Tweet categorization by combining content and structural knowledge. *Inf. Fusion* **2016**, *31*, 54–64. [CrossRef]

25. Lee, K.; Palsetia, D.; Narayanan, R.; Patwary, M.M.A.; Agrawal, A.; Choudhary, A. Twitter trending topic classification. In Proceedings of the 2011 IEEE 11th International Conference on Data Mining Workshops, Vancouver, BC, Canada, 11 December 2011; pp. 251–258.

26. IAB Categories | MoPub Publisher UI | MoPub Developers. Available online: https://developers.mopub.com/publishers/ui/iab-category-blocking/ (accessed on 4 May 2021).

27. Tweepy. Available online: https://www.tweepy.org/ (accessed on 4 May 2021).

 *electronics*

Article

A Generic Data-Driven Recommendation System for Large-Scale Regular and Ride-Hailing Taxi Services [†]

Xiangpeng Wan, Hakim Ghazzai * and **Yehia Massoud**

School of Systems and Enterprises, Stevens Institute of Technology, Hoboken, NJ 07030, USA;
xwan6@stevens.edu (X.W.); ymassoud@stevens.edu (Y.M.)
* Correspondence: hghazzai@stevens.edu
† This paper is an extended version of our paper published in IEEE Conference on Vehicular Electronics and Safety (ICVES'19), Cairo, Egypt, 4–6 September 2019.

Received: 6 March 2020; Accepted: 9 April 2020; Published: 15 April 2020

Abstract: Modern taxi services are usually classified into two major categories: traditional taxicabs and ride-hailing services. For both services, it is required to design highly efficient recommendation systems to satisfy passengers' quality of experience and drivers' benefits. Customers desire to minimize their waiting time before rides, while drivers aim to speed up their customer hunting. In this paper, we propose to leverage taxi service efficiency by designing a generic and smart recommendation system that exploits the benefits of Vehicular Social Networks (VSNs). Aiming at optimizing three key performance metrics, number of pick-ups, customer waiting time, and vacant traveled distance for both taxi services, the proposed recommendation system starts by efficiently estimating the future customer demands in different clusters of the area of interest. Then, it proposes an optimal taxi-to-region matching according to the location of each taxi and the future requested demand of each region. Finally, an optimized geo-routing algorithm is developed to minimize the navigation time spent by drivers. Our simulation model is applied to the borough of Manhattan and is validated with realistic data. Selected results show that significant performance gains are achieved thanks to the additional cooperation among taxi drivers enabled by VSN, as compared to traditional cases.

Keywords: intelligent transportation systems; demand prediction; taxi recommendation; vehicle social network; ride-hailing

---

1. Introduction

Modern urbanization has significantly changed people's living arrangements, making public transportation, particularly taxi services, a convenient and affordable means of travel for most people, especially when owning a car and paying parking fees is exorbitant. In New York city, 80% of the residents do not own a car [1]. This leads to an explosive growth of the taxi fleet size (e.g., regular yellow taxis in New York city), and ride-hailing service demand, which results in increasing congestion and inefficient exploitation of the resources. For regular taxi services, like yellow taxis in New York city, the taxi drivers do not know the exact locations of potential customers, while for the ride-hailing taxi services, such as Uber, Lyft, and Didi, customers send requests with their locations to nearby ride-hailing vehicles. In both taxi services, and independently of the level of knowledge about the customers' demand, users experience long waiting time periods before getting a ride. At the same time, taxi drivers are engaged in a tedious customer hunting search, traveling long distances. Indeed, even with the ride-hailing service, customers may find out that the nearest available vehicle needs a long time to pick them up. Therefore, there is a pressing need to improve the utilization of such a means of transportation and enhance the efficiency of both services for the benefits of both customers and drivers.

In regular taxi services, traditional ways for taxi drivers to find potential customers include driving around the city and waiting at some 'hot spots', e.g., taxicab stands. For the first option, taxi drivers usually follow an intuition-based trajectory hoping to find customers as soon as possible, while for the second option, most of the drivers will target the same hot spots since based on their personnel experience, they know when and where customers will be gathered. In the latter case, regular taxi drivers may be subject to an unfair competition since the number of taxis is higher than the demand or vice versa. Hence, traditional solutions for customer hunting are usually exhaustive and inaccurate. On the other hand, for the ride-hailing taxi services, although a central server is dedicated to manage the requests of customers and allocate them to drivers, similar problems that face regular taxi services still exist. Customers' requests might still be raised far away from drivers' locations and high vacant distances are accumulated, resulting in huge and redundant fuel consumption. In Portland, the average waiting times are estimated to be around six and ten minutes for regular and ride-hailing taxi services, respectively, according KGW News [2]. Therefore, it is recommended to enhance the efficiency of such transportation services by tackling the offer/demand problem in both taxi categories.

Thanks to the spread of on-board and infrastructure-based sensors [3], collecting and sharing data have become very common, especially in urban areas, where several novel data-driven applications exist, including Google Navigation, Waze, and parking localization service. This is additionally boosted by the emerging concept of vehicular social network (VSN), which effectively exploits the data availability in transportation networks [4,5]. With the installation and spread of on-board sensors, the data sharing ability has dramatically increased [3]. Mobile apps like Google Navigation and Waze utilize the historical traffic data and human-report accidents to improve the navigation services. The emerging concept of vehicular social network (VSN) has been proposed to better exploit the data availability among road users and transportation networks. A variety of applications and use cases have been discussed in [4–7]. VSN enables interactions between different participants, including human-to-vehicle and vehicle-to-vehicle interactions [8,9]. As an example, the connected vehicle technology in NYC is developed to leverage the safety of road users. It relies on vehicle-to-vehicle (V2V), vehicle-to-infrastructure (V2I) and infrastructure-to-pedestrian (IVP) to share information among them and better assess the transportation network (https://www.cvp.nyc/). Hence applied to our context, VSN can be utilized for improving the communication among taxi drivers and exploit their information to revamp the operation of taxi drivers by enabling efficient and real-time identification and sharing of their locations, as well as knowledge about customers such as pick up time, pick up places, drop off time, and drop off paces, as well as accurate and relevant data about the traffic situation. Such real-time data sharing can provide a clearer vision about the current customers' requests and help continuously predict the future demand at different regions of the navigation map [10,11]. This technological advance significantly contributes to designing novel taxi recommendation solutions for customer hunting [12] or involving highly connected autonomous taxis [13].

In this paper, we propose a combination of data-driven solutions that jointly improve the taxi service efficiency by recommending the operation of both regular taxicabs and ride-hailing taxis [14]. The proposed recommendation systems consists of three phases: (i) a demand prediction phase, (ii) a taxi-to-region matching phase, and (iii) a route planning phase. The proposed system divides the geographical urban area into several sub-regions and predicts the future demand during the next time periods for each region. Afterwards, it assigns taxis to this region based on the predicted demand. The number of taxis associated to each region is determined such that redundant taxi travel is avoided/reduced. This is performed by taking into account the current locations of the taxis and the predicted demand of each region. The problem is modeled as a bipartite graph which is designed such that the total expected traveled distance for taxis during the transition phase (i.e., taxis moving to their assigned locations) is minimized. Finally, the taxi recommendation system employed for realistic maps provide to drivers optimized trajectories to follow given real-time traffic data.

The realistic map is converted into a graph and the Dijkstra's algorithm is applied to determine the fastest paths for each member of the taxi fleet when needed. Three key principal performance indicators, namely total number of pick-ups, total customer waiting time, and total traveled distance for vacant taxis, are evaluated for both regular and ride-hailing taxi services and employed to compare our proposed system versus traditional solutions using realistic data of the area of Manhattan, Borough of New York city.

The main contributions of this paper are summarized as follows:

- We develop demand prediction models to precisely estimate the future demands in each region of the area of interest. One is online learning time series and the other is Long Short Term Memory model (LSTM). Their accuracies are validated using realistic data.
- We convert the taxi-to-region matching problem into a bipartite perfect matching graph, where we evenly assign taxis to different regions based on their future demands and the current locations of the taxis.
- We optimize the routing of each taxi by minimizing the expected time spent from its current location to the guided destination using the Dijkstra's algorithm by considering the real-time traffic data and geodesic distances of the road network.
- We develop a real-time simulated taxi operation based on the recommendation system using realistic maps, which provides evidence that significant performance gains can be achieved as compared to the traditional case.

The rest of the paper is organized as follows. Section 2 provides a literature review. Section 3 presents the system model and the adopted methodology. Section 4 develops the taxi recommendation system. Section 5 describes the proposed simulated taxi operation. Section 6 presents and discusses selected simulation results. Finally, concluding remarks and future directions are drawn in Section 7.

2. Related Work

Over the last few years, researchers have focused on designing solutions to support taxi drivers in enhancing their services. One of the main research directions is the identification of hot spot areas and the prediction of the demand, e.g., using Gaussian process regression [15] or reinforcement learning [16]. The objective is to identify regions with high likelihood of finding potential customers by predicting the spatial distribution of taxi passengers for a short-term time horizon [17,18]. The recommendation system assigns hot spot areas to vacant taxi drivers in order to shorten the waiting time for customers [19]. In [20], the authors proposed a mutual recommendation system that assigns hot spots for both taxi and passengers based on the trajectory of taxis. In [21], the authors developed a route recommendation engine to minimize vacant traveled distance through Monte Carlo tree search algorithm. These studies mainly focus on a single taxi and do not consider the situations where some hot spot areas are attracting a number of taxis larger than the needed demands or the opposite case. Some researchers focused on designing algorithms for ride-sharing services while addressing different research questions including taxi-to-customers assignment, demand and pricing, competition impacts, etc. [22]. In [23,24], the authors proposed Integer Linear Programs (ILP) that can match large groups of riders to a fleet of shared vehicles in real-time with certain capacity size. The algorithms are designed to address the current situation without considering future demands. Moreover, their computational complexity remains high. The adopted routing methods are based on the shortest path algorithm which does not consider traffic data and congestion level.

Spatial-demand prediction was one of the essential topics that are investigated in the context of taxi recommendation systems. In [25], the authors proposed Multi-View Spatial-Temporal Network (DMVST-Net) approach to predict the taxi demand. It is shown that the proposed method achieves a Mean Absolute Percentage Error (MAPE) of $\approx 16\%$. However, the predicted results are daily instead of hourly, which is not suitable for assisting drivers. Moreover, the running time to generate hourly results is also high. In [26], the authors predicted the short-term supply–demand gap of taxis by partitioning the city area into various regular Hexagon lattices-based Convolutional Neural Networks

(H-CNN). However, the proposed model is also computationally expensive compared to traditional methods while achieving slightly better performance. More importantly, it is not necessarily true that cities have uniform partitioning of their area, such as the case of Manhattan, NYC. Therefore, in this study, we use the cities' own region partition to predict the future demand using a faster algorithm in a real-time manner.

Recently, taxi recommendation studies consider more generalized scenarios and are not limited to a specific task. For instance, in [27], the authors developed a recommendation system for taxis by jointly considering the benefits of both drivers and passengers. The driver's utility includes expected revenue, searching time for next passenger, traveled distance, while the passenger's utility includes the waiting time. The authors grouped pick-up locations into clusters and defined them as the hot spot areas, to which it assigns taxis according to their scores. The recommendation system efficiently works for selected hot sport areas but ignores other areas with lower demand. Also, the speed of the vacant taxis is assumed to be constant which is not very practical. In [28], the authors presented a receding horizon control framework to dispatch taxis, with the demand prediction based on the estimated demand distribution. The system is evaluated on a square region without practical road network. In [29], the authors analyzed the dynamic spatial equilibrium of taxis and provided efficient regulation for taxi services in different regions. In [30], the authors presented a two-stage stochastic optimization formulation to consider expected future demand to solve the spatio-temporal matching problem, i.e., taxi matching. Generally, most of the studies discussed earlier do not consider the real-time locations of the taxis.

Furthermore, some other researchers focused on the cruising and matching for the taxi drivers. In [31], the authors provided a data-driven simulation framework for ride-sharing taxis simulated in a simplistic grid map. The proposed approach provides a path for a taxi while optimizing a certain cost function, such as traveled distance or gasoline consumption. In [32], the authors found out that driver's cruising choice is learned from his/her previous experience and his/her interactions with other drivers. In [33], the authors proposed pCruise system to reduce the taxi's cruising miles by providing the shortest cruising route with at least one expected available passengers for this route. In [34], the authors developed efficient algorithms for non-myopic adaptive routing to minimize the collective travel time of all vehicles in the system. In [35], the authors proposed solutions to reduce the number of cruising miles while increasing the number of live miles of taxis by suggesting profitable locations to taxicab drivers. Other research directions have investigated dynamic models to arrange ride-sharing vehicles with discrete simulation environment [36,37]. The authors of [38,39] have proposed data-driven vehicle re-balancing across regions but lack future demands prediction. Some researchers provided a graph partitioning methodology to partition the bipartite graph with lower computational complexity and implemented it in the one-to-one ride-matching problems [40]. Another study has modeled the matching problem as a competition strategy between different ride-hailing companies [41]. Despite the previous studies providing solutions for taxi cruising and matching problem, most of the methods are built in simplistic maps without convincing evidence to show the practicality of their methods. Moreover, they did not take the demand prediction, taxi dispatch, and route selection together into consideration. To the best of our knowledge, the recommendation system that we propose is the first one which jointly takes into account the prediction of future demands, taxi dispatch, and cruising routes selection for both regular and ride-hailing taxi services and is validated using realistic data and map.

3. System Model and Methodology

We propose to design a novel recommendation system for taxis cruising on a large geographical area. The latter is sub-divided into multiple regions for which we aim to predict the demand based on their respective historical data. The demand in the area of interest is estimated and updated in every time period T. In this paper, we focus on both the regular and ride-hailing taxi services. The difference is that regular taxi drivers are supposed to not know the exact locations of the customers as the ride-hailing vehicles, which are informed by the exact locations once they receive the request. Hence, we assume that for regular taxis, the pick-up happens when a taxi driver sees a customer waving his/her hand (e.g., when the distance between the customer and taxi is less than 100 m). In Table 1, we present the summary of the different taxi services managed by the proposed recommendation system.

Table 1. The three taxi services managed by the proposed recommendation system.

Service	Trad. Regular	Smart Regular	Ride-Hailing
Taxi Call	Waving	Waving	Online Request
Information Sharing	Without VSN	With VSN	With VSN
Knowledge level	Taxis locations only	Taxis locations only	Taxis and customers locations
	Future demand every T	Future demand in real-time	Future demand in real-time
Recommendation	Once every T	Continuously	Continuously

Note that the regular taxi services can be split into two categories: the traditional regular taxi services without VSN and the smart regular taxi services with VSN. In both services, taxi drivers are not aware of the locations of customers, but with the smart regular taxi services, when a pick-up happens, all other taxis via the recommendation system are aware about it. Hence, they are instantaneously updated about the changes in the area of interest. In other words, the system can adjust the hunting search locations for vacant taxis during the period T in a real-time manner instead of waiting until the end of the time period T, as it is the case with traditional services. For the ride-hailing taxi service, the taxis are aware of the locations of both users' demand and taxis in real-time and hence, it continuously provides recommendations to vacant taxis.

In Figure 1, we present the overview of the proposed framework for regular and ride-hailing taxi services. There are three major phases: the first phase is the real-time data update phase where information is collected from customers and taxi drivers. The data includes the current locations of customers and taxis in addition to the statuses of taxis (vacant or occupied) and the number of pick-ups already done. The second phase is the demand prediction phase that is executed every period T. In this phase, the historical data is used to predict the demand of the area of interest. Note that for every T time period, the system would predict the demand only once, set at the beginning of that time period T. Then, the demand would be updated by considering the number of pick-ups happening during the entire time period T. The third phase encompasses the process of taxi-to-region matching and taxi routing. For the taxi-to-region matching, the recommendation system assigns vacant taxis to the different regions based on their locations and the potential future demand on that region, e.g., if the system recommends several taxis to some regions, it will only send them to the nearby ones. For the route selection, the system determines the routes for all taxis to reach their destinations by minimizing the expected time spent on their trips by considering the collected real-time traffic data.

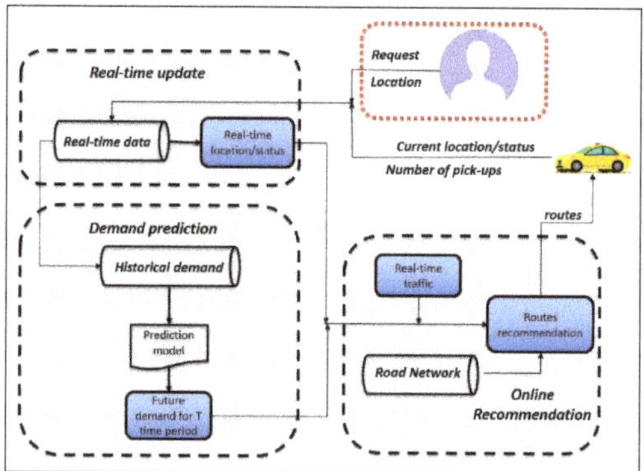

Figure 1. Recommendation framework for regular (without red curved rectangle) and ride-hailing (with red curved rectangle) taxi services.

4. Proposed Taxi Recommendation System

In this section, we introduce the different components and steps of the proposed taxi recommendation system: (1) the taxi demand predictor, (2) the taxi-to-region matching component, and (3) the taxi routing optimizer.

4.1. Taxi Demand Predictor

The first step is to predict the customer demand in the area of interest. We introduce and compare two models that fit the scope of this task. One is Long Short Term Memory (LSTM) model and the other is Autoregressive integrated moving average (ARIMA) model. To illustrate their accuracy, we collect the data about the operation of yellow taxis from the Taxi Limousine Commission (TLC) (https://www1.nyc.gov/site/tlc/about/tlc-trip-record-data.page), which contains the taxi operation information in New York City including the pick-up instants, pick-up locations, drop-off time, drop-off region, trip fare, and trip distance. We then explore the historical demands on the borough of Manhattan which is split into 69 regions as shown in Figure 2. Before feeding the data into our models, we normalize the demands over T ($T = 60$ min in this case) at first.

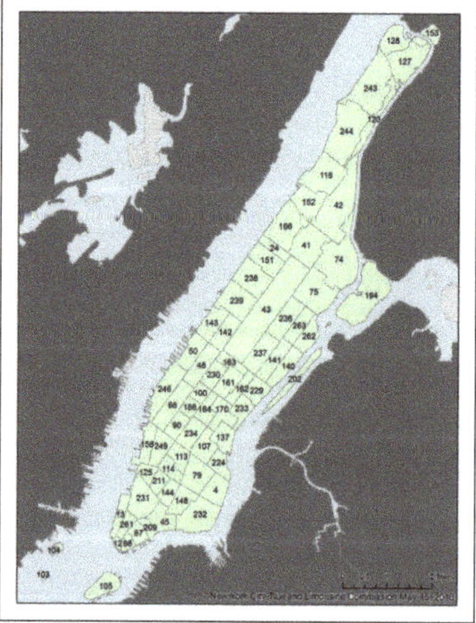

Figure 2. The borough of Manhattan and its taxi region subdivision.

The next step is to predict the future demand for the next period T on each region with ARIMA. In our case, we are using the demand of the previous 168 h to predict the demand of the next hour. In other words, we are using historical data for the previous week to predict the demands in the next hour, which automatically takes into account the weekday and weekends by assessing the trend of its consecutive features (the model could figure out if the date of prediction is a weekend or weekday). To prevent inputting extra information as weekday or weekends, we proceed by predicting the next hour of demands based on previous one-week data inputs. That is, using 168 previous inputs, we predict the next time period's demand, which would take the holidays, weekends, and weekdays into consideration by assessing the trend of its consecutive features. The choice of the demand prediction period is not arbitrary. It considers the objective of the next phase of the proposed recommendation system. Indeed, every hour, the taxi-to-region matching is provided after precisely predicting the hourly demand. Hence, choosing longer values of T may be unadapted with the demand variation in the region and may lead to taxi operation delay. Selecting lower values of T will increase the frequency of executing the taxi-to-region matching algorithm, which may lead to either an excessive re-assignment of taxis, which is not practical, redundant results similar to the ones of the previous time period, as well as extra computational complexity. More importantly, when predicting the traffic demand over the time period T, we aim to estimate the demand of each region at each instant of that period and not only a constant demand. With the help of VSN, the traffic and demand information are shared among the fleet instantaneously. For instance, the predicted remaining customers during the time period are estimated while considering the number of already picked customers.

We set the ARIMA parameter d to zero. In order to find the best model, we test different ARIMA models with different parameter combinations (p, d, q) where we pick the parameters with the lowest Akaike information critera (AIC) value in the end. We find out that ARIMA with $(p = 5, d = 0, q = 3)$ provides the lowest AIC where AIC $= -2.9 \times 10^4$. The ARIMA model fitting is based on the latest updated 168 data time periods before predicting the next time period. In this case, the predicting process is converted into an online learning where the model parameters are updated continuously. The prediction result from the ARIMA model is shown in Figure 3 where the red and blue series correspond to the predicted and actual values, respectively. The resulting mean square error (MSE) is 4.7×10^{-4}. Hence, we can conclude that the online ARIMA model can provide accurate prediction of the future demand, which can be effectively used to provide precise prediction for the taxi recommendation system.

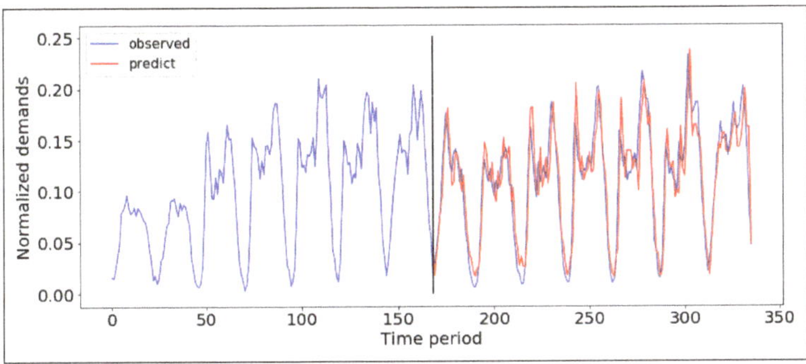

Figure 3. Hourly predicted demand for region 114 using the Autoregressive integrated moving average (ARIMA).

We have compared the performance of the ARIMA model to the LTSM one, which is trained on the historical data first and then employed to predict the future taxi demand. The LSTM model contains two hidden layers and one output neuron. The input shape is 168, which contains previous one-week hourly demand data. It achieves an MSE equal to 6.9×10^{-4} as illustrated in Figure 4.

Unlike the ARIMA model, the LTSM is not trained in an incremental/online manner, which requires a more important amount of data compared to ARIMA. From the comparison results, we find out that the online ARIMA model is more accurate, hence we adopt it in our system.

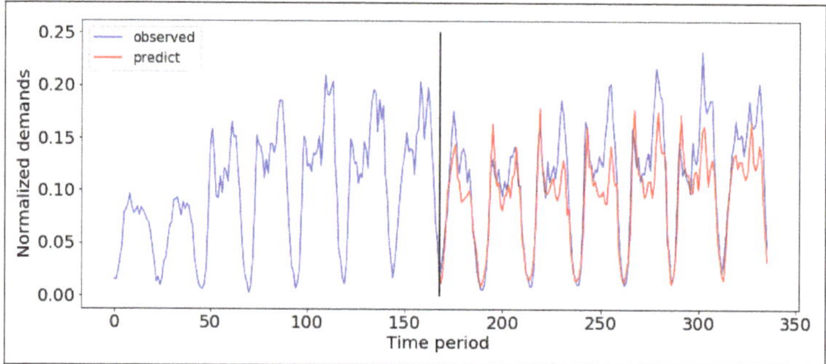

Figure 4. Hourly predicted demand for region 114 using Long Short Term Memory model (LSTM).

4.2. Taxi-to-Region Matching Component

Once accurate future taxi demand is determined for each region, we proceed by assigning vacant taxis to these regions according to the region demands and the taxi current locations. The first metric is used to ensure that the taxi assignment is made proportionally to the demand. In this way, when the expected demand is high, more taxis will be sent to that region and vice versa. The second metric (taxi current locations) is considered in order to minimize the transition phase during which vacant taxis need to reach their assigned regions. This helps in reducing the waiting time of the customers looking for rides at the beginning of the time period. An example is shown in Figure 5 where four regions and eight taxis exist. Obviously, given the demand, we should assign one taxi to region A, two taxis to region B, four taxis to region C, and one taxi to region D based on their respective demand ratios (10, 20, 40, 10).

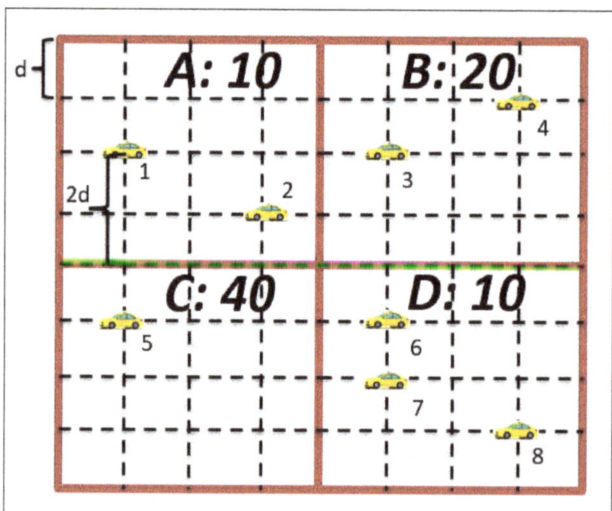

Figure 5. An illustrative example of an $8 \times 8\ d^2$ area composed of four regions A, B, C, and D with different demands 10, 20, 40, and 10, respectively. Eight vacant taxis are circulating around the region. The black dashed lines represent roads.

To ensure an efficient taxi-to-region matching for large-scale problems, we propose to model it by a bipartite weighted graph presented in Figure 6. The weights associated to the graph edges are computed based on the shortest distance needed by the taxi to reach the closest border of the region. To reflect the demand of each region in the graph, we duplicate the ones having higher demand multiple times according to their normalized demand levels with respect to the total number of taxis and total demand in the Borough of Manhattan during that time period. Consequently, the objective of the taxi-to-region matching component is to minimize the sum of the weights while maintaining the perfect matching. In other words, each taxi is assigned to one region. Note that, in practice, the number of taxis is usually higher than the number of regions. We refer to the taxi ID by the index i and the region ID after duplication by the index j. Hence, as shown in Figure 6, $j = 2$ and $j = 3$ refer to the same region B.

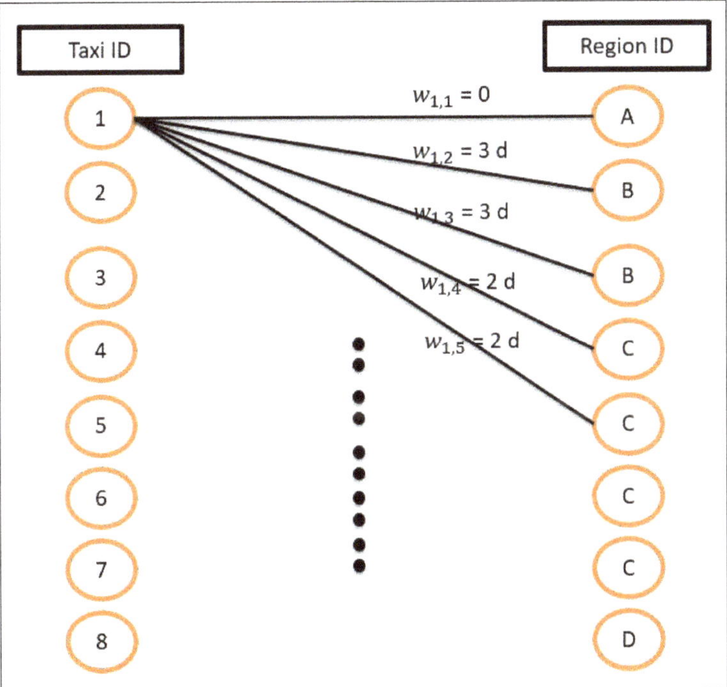

Figure 6. An illustrative example of the bipartite matching graph applied to the scenario presented in Figure 5.

The aforementioned matching procedure can be optimized using an ILP problem formulated as follows:

$$(P):\ \underset{x_{i,j}\in\{0,1\}}{\text{minimize}}\quad \sum_i \sum_j w_{i,j} x_{i,j} \tag{1}$$

subject to:

$$\sum_j x_{i,j} = 1,\ \ \forall i,\ \text{and}\ \sum_i x_{i,j} = 1,\ \ \forall j, \tag{2}$$

where $w_{i,j}$ represents the weights of the edges linking the taxis and the regions while $x_{i,j}$ is a decision variable indicating whether a taxi i is assigned to region j or not. It is equal to 1 if this is the case. In (P), constraints (2) ensure the perfect matching, which forces a taxi to be assigned to only one region.

The matching problem can be also solved optimally using the heuristic minimum weight perfect matching algorithm: the Hungarian method. This algorithm solves the problem in a polynomial time $\approx \mathcal{O}(N)$, much faster than the NP-complete ILP-based solution that adopts the branch-and-bound algorithm, where N is the number of taxis.

4.3. Taxi Routing Optimizer

The road network can be constructed in many ways, such as simple search techniques and complex fuzzy logic theory [42,43]. In this paper, we transform the traffic network of the area of interest into a complex graph composed of intersections and roads. Each road r, where $r \in \{1, \ldots, R\}$ connecting at most two intersections, is divided into multiple small segments with the same length l_r. The graph then has as vertices the connections of different segments and as edges the segments themselves. We define the current location of taxi i as (S_i, Sg_i) and its destination as (D_i, Dg_i), here S_i and D_i represent the ID of the streets and Sg_i, Dg_i represent the ID of their segments. In [44], we propose an optimal solution for route planning problem that takes the real-time traffic into consideration. Integer linear programs are formulated to determine the fastest route given the current locations of vacant taxis and their assigned regions. The fastest paths can then be determined as the real-time traffic feed-back is obtained by the system. With the recurrent updates, ILP is solved regularly to determine the best routes according to the recent data, in other words, the route keeps updated as the new data is received. In order to reduce the complexity of the routing optimizer, we employ the recurrent Dijkstra's algorithm using the metrics evaluating the traffic level at each segment defined in [44,45]. In our approach, routes might be updated every 1 min. The detailed process is provided in Procedure 1. Note that the same routing approach is adopted to determine the trips of occupied taxis after pick-ups.

Procedure 1 Routing Optimizer for Taxi i

1: Inputs $= \{(S_i, Sg_i), (D_i, Dg_i)\}$, time instant t.
2: **while** vehicle does not reach the destination **do**
3: Obtain the latest update traffic data based on the collected information.
4: Update the weights of the road network graph.
5: Run Dijkstra's algorithm to find the fastest route from (S_i, Sg_i) to (D_i, Dg_i).
6: Vehicle follows the proposed route for one minute.
7: Update (S_i, Sg_i).
8: **end while**

---

5. Simulated Taxi Operation and Validation

In this section, we introduce our framework to simulate the operation of taxis in the area of interest. Then, we validate the proposed model with realistic data to ensure that our simulations after determining routes are close to real-world situations.

5.1. Simulation Model

In our simulations, we consider the area of Manhattan, New York city, which is divided into 69 regions. We assume there are N taxis circulating in the area of interest. If it is vacant, we assume that the taxi picks up a customer when the distance separating them is less than 100 m. In the traditional system, where data exchange and knowledge about the customers' demand are absent, we consider that the N taxis move randomly in the whole area when they are vacant, while with the recommendation system, the taxis are always assigned to different regions at the beginning of the time period and will move randomly only within that region. Once a pick-up is made, the status of the taxi is changed to occupied until the customer is dropped off.

For the taxi routing optimization, we extract the parameters of the off-line map from Open Street Map [46]. In total, there are 9070 roads and 4146 intersections in the area of interest. We split each road into segments having length at maximum 100 m. Thus, we obtain a graph of 11,760 edges and 6393 nodes.

Two scenarios are provided to strengthen the persuasive of the model. We consider the one hour demand information on 1 June 2018 from 3 am to 4 am that contains 1813 pick-ups in total (time instants and GPS locations) as the first scenario while the demand information on 1 January 2018 from 5 am to 6 am that contains 2027 pick-ups as the second scenario. We choose these two periods instead of rush hours for tractability and clarity reasons. Indeed, over rush hours, the number of pick-ups is huge and it will be difficult to visualize the results. This also impacts the simulation time, which is expected to be very expensive. Although we have developed low complexity algorithms for both the taxi-to-region matching component and the taxi routing optimizer, simulating the instantaneous operation of a huge number of taxis remains time consuming. It is worth noting that in our simulation results (Section 6) where we compare the different scenarios after simulating the taxi operations, we have investigated the same time periods where identical traffic conditions are experimented with. Since the customer arrival time and waiting time are missing in the dataset, without loss of generality, we assume that their arrival instants are the taxi pick-up times. Three key performance indicators are evaluated in our simulations: (1) the total number of pick-ups, (2) the waiting time of each customer corresponding to the difference between its pick-up time and its arrival time instants, and (3) the vacant traveled distance where no passengers are in the taxis. Precisely, the waiting time of customers corresponds to the period starting from the time instant when the customer arrives on the road for regular taxis or requests the service for ride-hailing taxis. The deadheading or idle distance of drivers is defined as the distance travelled by a taxi without serving any customers either before finding or after dropping a customer. All of these metrics are measured after simulating the taxi operation, as indicated in Section 5. The demands on 69 regions for both scenarios are presented in Figure 7. We notice that the demands mainly exist in mid and lower Manhattan. Although the two scenarios have similar total requests, their distributions in the regions are different. Customers in Scenario 2 are mainly located in regions 48, 68, 246, 230, 249, 79, 148, unlike Scenario 1 where most of them are gathered in regions 48, 186 and 79.

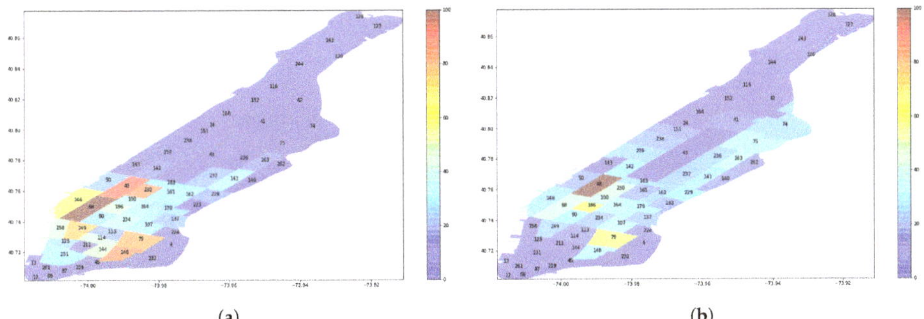

(a)   (b)

Figure 7. Demands percentage heat-map for 69 different regions on 1 January 2018 from 5 am to 6 am and 1 June 2018 from 3 am to 4 am separately. (**a**) Scenario 1. (**b**) Scenario 2.

The detailed algorithm to perform the simulations for regular taxi services without VSN is provided in Algorithm 1. Note that the recommendation occurs at the beginning of every time period T where $T = 1$ h. Hence, the regions assigned to different taxis remain unchanged during this hour. For the next time period, the recommendation system updates its matching procedure for the vacant taxis according to their latest locations and the new demand.

The detailed algorithm to perform the simulations for regular taxi services with VSN is provided in Algorithm 2, where the recommendation occurs on the fly during the time period whenever a pickup is reported to the system. Here, $N_{idle}(t)$ represents the number of vacant taxis at time instant t. In this algorithm, the system continuously provides recommendation during the time period T as the number of pick-ups changes over time. Every $\bar{t}$ minutes, the system sends the vacant vehicles to different regions considering the distance as well as the potential demand for the rest of the time period T. Note that within T the demand in the current step is highly correlated with the one of the next step. Hence, very few re-assignments will occur for vacant taxis.

Algorithm 1 Simulated Taxi Operation for Regular Taxi Services Without VSN

1: Inputs = (S_i, Sg_i) $i \in \{1, \cdots, N\}$.
2: Determine the best assignment of taxi-to-region using the Hungarian method.
3: Send vacant taxis to recommended regions using the Routing Optimizer given in Procedure 1.
4: $t = 0$.
5: **while** $t \leq T$ **do**
6:     **for each** Taxi $i \in \{1, \ldots, N\}$ **do**
7:         **if** Taxi i is vacant **then**
8:             Taxi i circulates towards or within the assigned region during this minute.
9:             Calculate the shortest distance d_{ik} between taxi i and potential nearby customers k's.
10:            **if** $\exists\, k$ such that $d_{ik} < 100$ **then**
11:                Taxi i notices customer k waiving his/her hand and then heads to him/her.
12:                Record the waiting time of customer k.
13:                Change the status of taxi i to occupied.
14:            **end if**
15:         **else**
16:             Taxi i drives towards its destination as per customer request using the Routing Optimizer given in Procedure 1.
17:         **end if**
18:         Update (S_i, Sg_i).
19:     **end for**
20:     $t = t + 1$.
21: **end while**

Algorithm 2 Simulated Taxi Operation for Regular Taxi Services With VSN

1: Inputs = (S_i, Sg_i), $i \in \{1, \cdots, N\}$.
2: $t = 0$.
3: **while** $t \leq T$ **do**
4:     **if** $mod(t, \bar{t}) == 0$ **then**
5:         Update the demand by subtracting the pick-ups happened already.
6:         Find the vacant taxis i, $i \in \{1, \cdots, N_{idle}\}$ at t.
7:         Determine the best assignment of taxi-to-region using the Hungarian method.
8:         Send vacant taxis to recommended regions the Routing Optimizer given in Procedure 1.
9:     **end if**
10:     **for each** Taxi $i \in \{1, \ldots, N\}$ **do**
11:         **if** Taxi i is vacant **then**
12:             Taxi i circulates towards or within the assigned region during this minute.
13:             Calculate the shortest distance d_{ik} between taxi i and potential nearby customers k's.
14:            **if** $\exists\, k$ such that $d_{ik} < 100$ **then**
15:                Taxi i notices customer k waiving his/her hand and then heads to him/her.
16:                Record the waiting time of customer k.
17:                Change the status of taxi i to occupied.
18:            **end if**
19:         **else**
20:             Taxi i drives towards its destination as per customer request using the Routing Optimizer given in Procedure 1.
21:         **end if**
22:         Update (S_i, Sg_i).
23:     **end for**
24:     $t = t + 1$.
25: **end while**

Unlike the regular taxi services, the system for ride-hailing taxi services is aware of the locations for both taxis and customers' requests. Once a customer sends the request to the system, one of the nearby vacant taxis would head to him/her directly. Note that only the vacant vehicles that are within the search range Rg receive the request. In our simulation, we set the minimum search range Rg to 2 km. The detailed algorithm to perform the simulations is provided in Algorithm 3. Here, similar to the regular service with VSN, the system assigns the regions to vacant taxis every $\bar{t}$ minutes as the demand for the rest of T is changing. However, taxis in ride-hailing services do not need to find customers waiving their hands on the street, in other words, the search range of taxis in ride-hailing service Rg is much larger than the regular taxi services. In our algorithm, we collect the location information of every customer and calculate their distance to all vacant vehicles within the search range. The closest available vehicle within that search region will be assigned to the customer. In our simulation, we set $\bar{t} = 5$ min.

Algorithm 3 Simulated Taxi Operation for Ride-Hailing Taxi Services

---

1: Inputs = $(S_i, Sg_i), \quad i \in \{1, \cdots, N\}$.
2: $t = 0$.
3: **while** $t \leq T$ **do**
4:    **if** $mod(t, \bar{t}) == 0$ **then**
5:       Update the demand by subtracting the pick-ups happened already.
6:       Find the vacant taxis $n_i, \quad i \in \{1, \cdots, N_{idle}\}$ at t.
7:       Determine the best assignment of taxi-to-region using the Hungarian method.
8:       Send vacant taxis to recommended regions using the Routing Optimizer given in Procedure 1.
9:    **end if**
10:   **for each** Customer $k \in \{1, \ldots, \mathcal{K}\}$ who shows up before t **do**
11:      Customer k sends its request and its location to the system.
12:      Calculate the shortest distance d_{ik} for customer k with all nearby taxis $i, \quad i \in \{1, \cdots, N_{idle}\}$.
13:      Find the closest taxi i' and the shortest distance $d_{i'k}$.
14:      **if** $d_{i'k} < Rg$ **then**
15:        Taxi i' heads to the customer k to pick him/her up.
16:      **end if**
17:      Record the waiting time of customer k.
18:      Change the status of taxi i' to occupied.
19:   **end for**
20:   **for each** Taxi $i \in \{1, \ldots, N\}$ **do**
21:      **if** Taxi i is vacant **then**
22:        Taxi i cruises towards or within the assigned region during this minutes.
23:      **else**
24:        Taxi i drives towards its destination as per customer request using the Routing Optimizer given in Procedure 1.
25:      **end if**
26:      Update (S_i, Sg_i).
27:   **end for**
28:   $t = t + 1$.
29: **end while**

5.2. Model Validation

Figure 8, we propose to compare the simulation results with the current data to prove the efficiency of the model with respect to real-world scenarios. In the figure, we provide two histograms comparing the gap in terms of trip duration and traveled distance between actual data and simulated data for the different trips. From Figure 8a, we find that the majority of the simulated trips have duration close to the real data with a difference ranging from -3 to 5 min. This is due to the difference between the true traffic status and the simulated one, as well as different drivers' routing preferences, that do not necessarily follow the obtained paths using the approach presented in Section 4. The difference is not huge since the average is close to 0. Moreover, from Figure 8b, we find out that distance differences of the majority of trips are plus or minus 2.5 km from the realistic data since the available dataset only contains the pick-up and drop-off region ID without specifying the exact geographical points. In Figure 8 shows that the simulated model is very close to the real-world case and validates the system model and routing optimization algorithms that we developed.

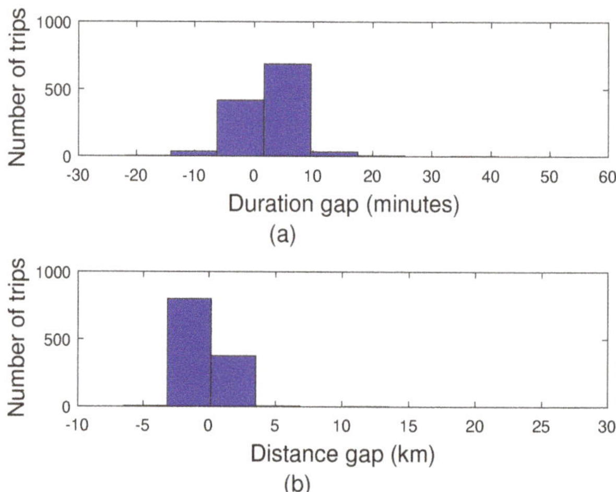

Figure 8. Comparison between the actual and simulated data: (**a**) histogram representing the trip duration and (**b**) histogram representing the distance gap.

6. Performance Evaluation of the Proposed Recommendation System

In this section, we evaluate the performance of our proposed recommendation system and compare it to the traditional case where taxi drivers work individually and based on their own experience for both regular and ride-hailing taxi services. To sum-up, in our simulations, we compare five cases: Two traditional taxi services without recommendations (Regular Trad. and Ride-hailing Trad.) and three taxi services based on our proposed recommendation system (Regular Recom. (w/o VSN), Regular Recom. (VSN), and Ride-hailing Recom.). We start by providing a detailed analysis of the key performance metrics for Scenario 1, followed by a short discussion about Scenario 2.

6.1. Taxi Operation Visualization (Scenario 1)

In Figure 9, we illustrate an example of two selected taxis circulating in the area of interest while considering Scenario 1 (Figure 7a) for regular taxi services without VSN. Two of them, i.e., black and blue trajectories, are moving randomly looking for customers based on their own experience while two other taxis, colored in red and pink, follow the recommendations of the proposed system using Algorithm 1. The starting positions of the black and red taxis as well as the blue and pink taxis are the same, and by comparing the number of pick-ups between those two groups of taxis, we find out that the number of pick-ups increases when the recommendation system is applied. On the other hand, we can notice that the red vehicle spends most of its time cruising within the same region compared to the black vehicle and thus has a greater chance to find customers with lower vacant traveled distance. The starting position of the pink vehicle has lower number of potential customers so it is assigned to other regions that have higher probability to find customers.

In the sequel, we evaluate the performances of the proposed taxi recommendation systems for both regular and ride-hailing taxi services and compare them to the ones of the traditional cases.

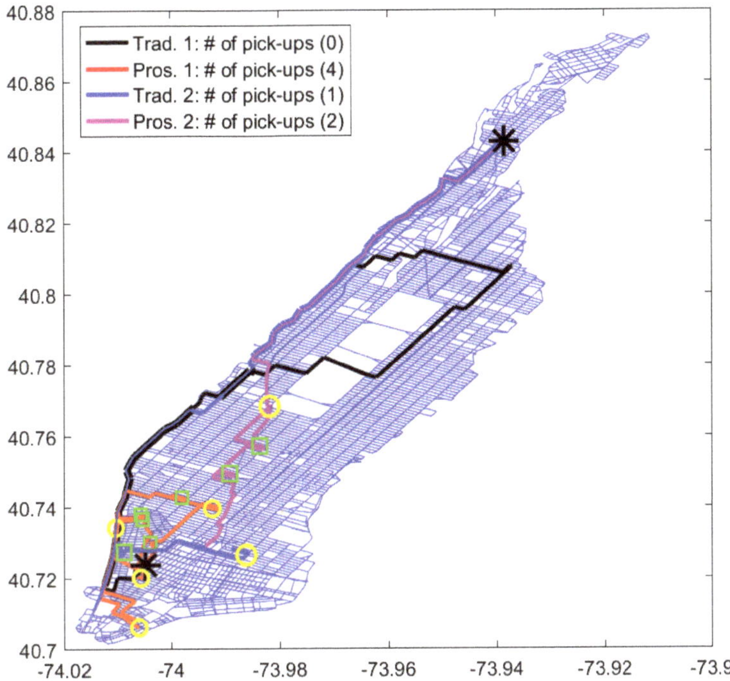

Figure 9. Example of two selected taxis circulating in the area of interest using the traditional and recommendation-based modes for regular taxi services without Vehicular Social Network (VSN). The 'black' and 'blue' trajectories correspond to two taxis moving in a traditional manner. The 'red' and 'pink' trajectories are of the same taxis following the recommendation system instructions (Circles (yellow) = drop off, squares (green) = pick-up locations).

6.2. Number of Pick-Ups (Scenario 1)

In Figure 10, we depict the number of pick-ups achieved by recommendation system in regular (without and with VSN) and ride-hailing taxi services (blue, green, pink) and compare them to the traditional cases of both services (red, black) with different taxi fleet sizes ($N = \{400, 450, 500, 550, 600\}$) for Scenario 1. We can clearly notice that higher performance are achieved with the recommendation system regardless of the taxi fleet size.

For instance, the number of pick-ups with regular taxis increases by around 20% with a fleet size $N = 450$. Adding VSN option also helps in slightly improving the performance. On the other hand, the number of pick-ups in ride-hailing service is higher than those of regular service. For instance, when $N = 600$, with the recommendation system, 100% of the customers have been picked up using ride-hailing taxi service while 95% of the customers have been picked up using regular taxi service. Also, we notice that when $N = 550$ and $N = 600$, the ride-hailing taxis are able to pick up all the customers. In other words, an excess supply is obtained with a taxi fleet of $N = 600$, which corresponds to an unnecessary wasting of fuel and may cause redundant congestion.

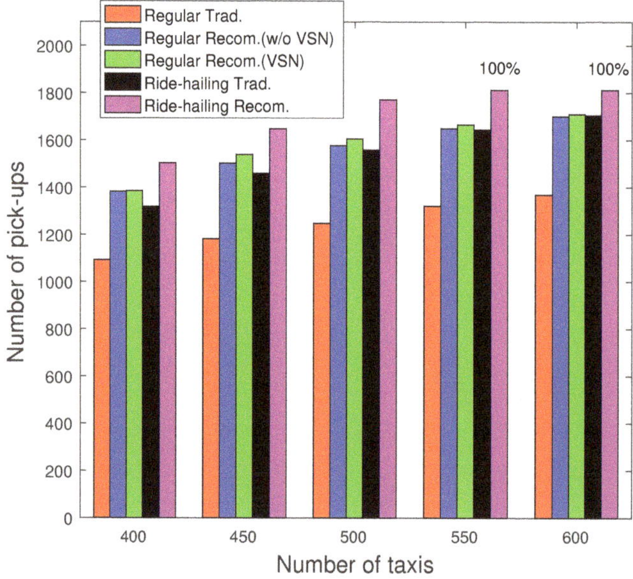

Figure 10. Number of pick-ups with different taxi fleet sizes for Scenario 1.

In order to deeply visualize the number of pick-ups for each region, we plot, in Figure 11, the ratio of number of pick-ups to the total customer's requests with $N = 500$. We notice that the ratio of pick-ups is small in the upper Manhattan since the customers' demands are mainly located in the lower Manhattan. Although we evenly assign the vehicles to different regions according to the expected customers' requests, there is a high probability that a vehicle heading to upper Manhattan from lower Manhattan ends up finding customers on the route before arriving.

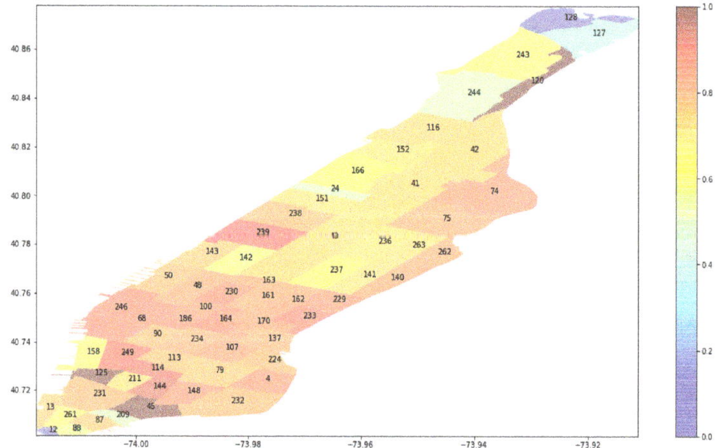

Figure 11. Ratio of pickups with $N = 500$ using the Ride-hailing recommendation system.

6.3. Customer Waiting Time (Scenario 1)

In Figure 12, we evaluate the satisfaction of customers (waiting time) for all the recorded trips during the time period T with $N = 500$ for Scenario 1. We notice that, with the same fleet size of taxis cruising throughout the map, customers wait much less time with the recommendation system. With VSN, the performance of the recommendation system in regular taxi service is improved slightly. It is worth noting that 98% of the customers wait less than 10 min before finding a vacant ride-hailing taxi thanks to the proposed recommendation system compared to 70% with the traditional case. We also notice that without recommendation system, the ride-hailing service provides customers with shorter average waiting time compared to the regular taxi services, which is true in practice. If we apply the recommendation system for both services, then the average waiting time in ride-hailing is much lower than the one obtained with regular taxis.

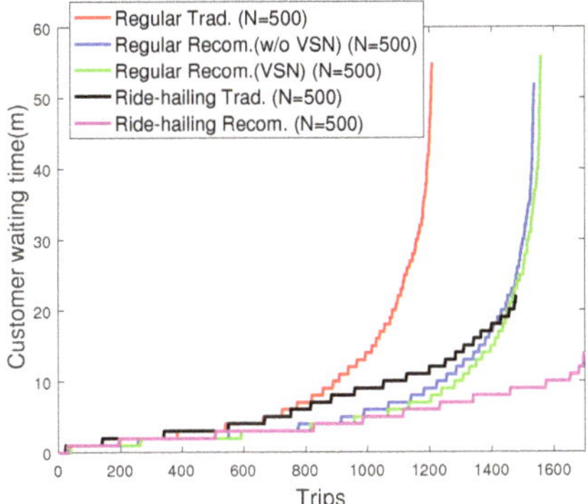

Figure 12. Traditional versus recommendation systems: sorted customer waiting time.

In addition, we present the average customer waiting time per region with $N = 500$ for Scenario 1, as shown in Figure 13. We find out the average waiting time in upper Manhattan is lower than that in lower Manhattan, since the customers are gathered in lower Manhattan and there is competition among customers to find available taxis.

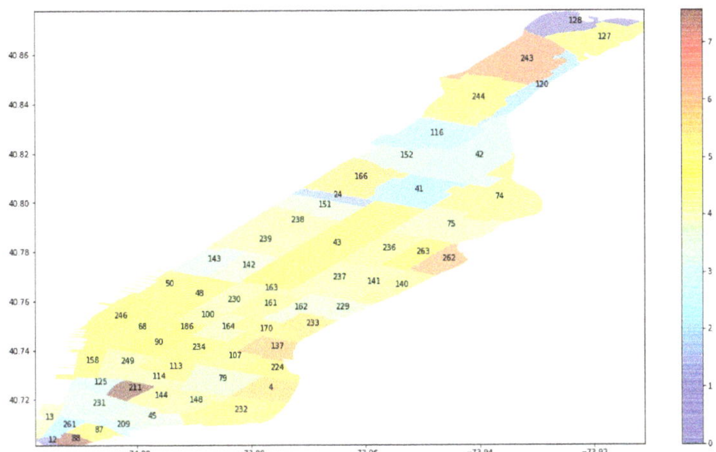

Figure 13. Average customer waiting time with $N = 500$ using Ride-hailing recom.

Finally, in Figure 14, we depict the average customer waiting time including recommendation system in regular and ride-hailing taxi services (blue, green, pink) and traditional case using both services (red, black) with different taxi fleet size ($N = \{400, 450, 500, 550, 600\}$) for Scenario 1. We notice that higher performance is achieved with the recommendation system regardless of the taxi fleet size. For instance, when $N = 600$, on average, with the recommendation system, customers wait 1.66 min lower compared to the traditional case in ride-hailing taxi service and 2.28 min lower for regular taxi service.

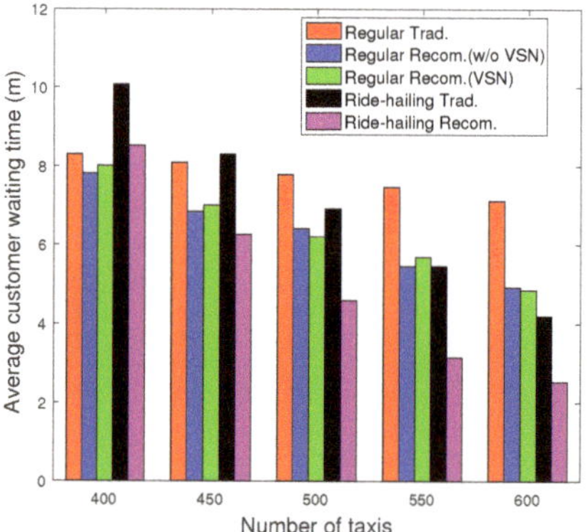

Figure 14. Average customer waiting time with different taxi fleet sizes for Scenario 1.

6.4. Vacant Traveled Distance (Scenario 1)

Similarly, in Figure 15, we evaluate the satisfaction of taxi drivers represented by their idle traveled distance. We notice that with the proposed recommendation system, the taxi drivers have less idle traveled distance, and as expected, drivers in ride-hailing service have less idle traveled distance than those of the regular taxi service. We also notice that with VSN, the performance slightly increases in regular taxi services. It is worth noting that 92% of the taxis idly travel for less than 10 km during one hour when $N = 500$ thanks to the proposed recommendation system. However, with the traditional techniques, only 78% of the fleet achieves a similar result.

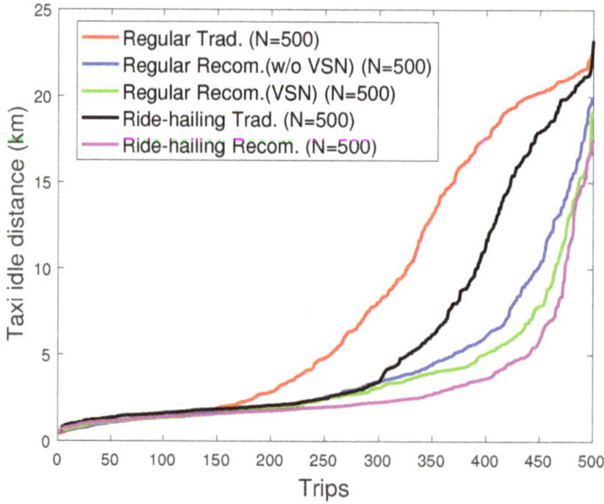

Figure 15. Traditional versus recommendation systems: sorted idle traveled distance.

In addition, we present, in Figure 16, the average idle traveled distance using the recommendation system for regular and ride-hailing taxi services, as well as the traditional cases for Scenario 1. Again, higher performances are achieved with the recommendation system regardless of the taxi fleet size. For instance, when $N = 600$, with the recommendation system, taxis travel 1.7 km less in vacant status compared to the traditional case in ride-hailing taxi service and 4.1 km less compared to the traditional case in regular taxi service. Close performances are achieved with the other fleet sizes.

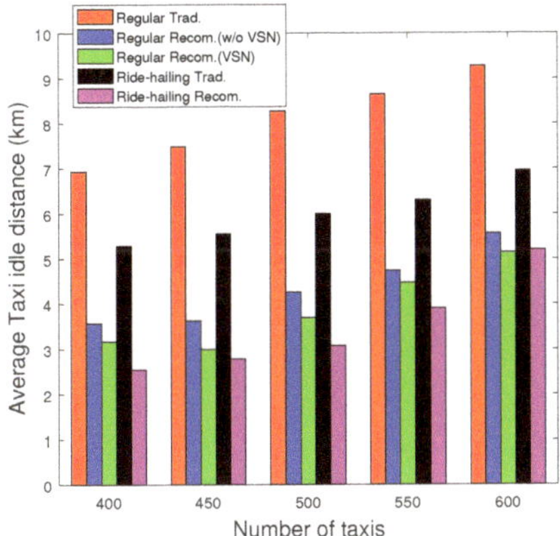

Figure 16. Average idle traveled distance with different taxi fleet sizes for Scenario 1.

In Figure 17, we provide heatmaps for both traditional and proposed schemes illustrating the regions crossed by $N = 500$ regular taxicabs during idle periods when looking for customers. In this figure, we sort the regions according to their geographical locations and place the regions next to each other in both axes where the horizontal axis is the origin region (last dropoff region) and the vertical axis is the destination region (the region where the next customer is found). The figure shows that the taxis in the traditional case are moving from a region to another in a near-uniform pattern where taxi drivers search for customers following their own intuition, while thanks to the recommendation system, taxi drivers are able to reduce their travelling idle distance by searching for customers within the same or nearby regions as it is corroborated by the diagonal pattern given in Figure 17b. In this way, the hunting time is minimized, which allows taxi drivers to save additional time and fuel.

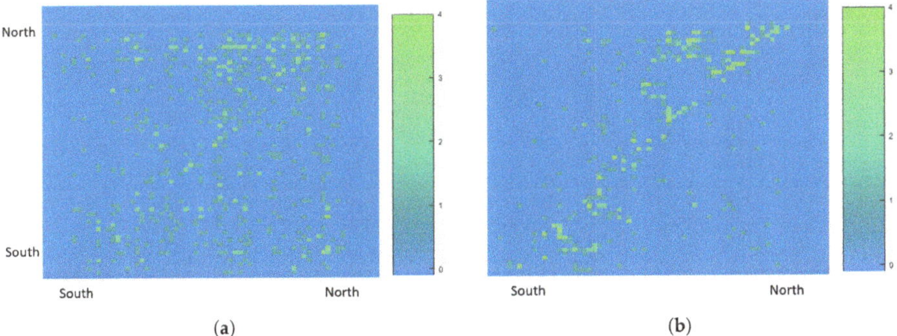

Figure 17. Movement of vacant taxis from a region to another before finding new customers for regular taxicabs. The x-axis represents the origin regions while the y-axis represents the destination. Regions are sorted according to their geographical locations, in Manhattan area, from South to North. (a) Traditional taxi service. (b) Proposed recommendation system.

6.5. Taxi Re-Assignment Frequency (Scenario 1)

We have shown that our recommendation system could achieve outstanding progress for the different key metrics. We then explore whether the system (with VSN) provides excessive re-assignments to taxis during time period T or not and, hence, check the practicality of the system. In other words, we need to pay attention to the number of re-assignments since taxi drivers do not prefer such recommendations in practice. In Table 2, we provide the average number of re-assignments for Scenario 1 based on our simulations. On average, the number of re-assignments is lower than 2 during one hour for vacant taxis. On the other hand, ride-hailing taxis have less re-assignments compared to the those of regular taxi services since the locations of customers are known in ride-hailing taxi services. Also, we notice that when the number of taxis increase, taxi drivers are more likely re-assigned before finding customers since the supply is exceeding the demand.

Table 2. Average taxis re-assignment frequency using Regular Recom. (with VSN) and Ride-hailing Recom.

	Regular Recom. (VSN)	Ride-Hailing Recom.
N = 400	1.62	0.59
N = 450	1.64	1.08
N = 500	1.77	1.09
N = 550	1.83	1.19
N = 600	1.96	1.37

6.6. Summary and Discussion for Scenario 2

Finally, a comprehensive summary of the results for Scenario 2 is shown in Figure 18, which presents similar performance to Scenario 1. For instance, when $N = 600$, customers wait 1.62 min lower compared to the traditional case in ride-hailing taxi service and 2.79 min lower for regular taxi service. Also, with the recommendation system, taxis travel 1.8 km less in vacant status compared to the traditional case in ride-hailing taxi service and 3.9 km less compared to the traditional case in regular taxi service. On the other hand, by comparing the performance between ride-hailing and regular taxi services, we find out the average customer waiting time and the idle traveled distance of drivers are improved when customers' locations are sent to the system. Furthermore, it is worth noting that when the number of taxis increases, the customers' waiting time decreases while the idle traveled distance of taxi drivers increase. It is important to determine the appropriate size of taxi fleet for each time period of the day so that both customer and taxi drivers are satisfied without overloading the region with redundant taxis.

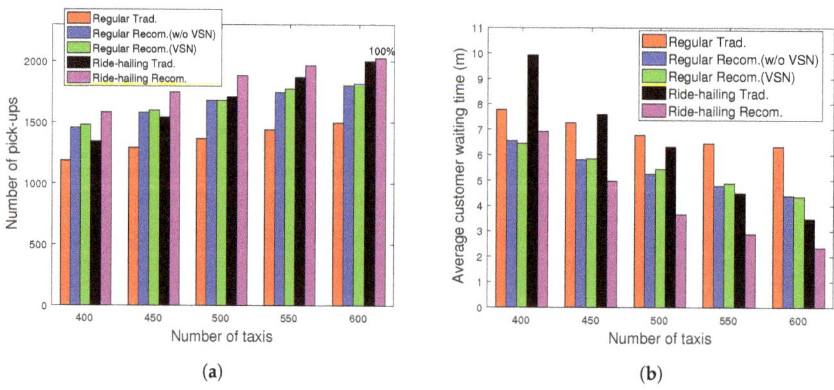

(**a**)　　　　　(**b**)

Figure 18. *Cont.*

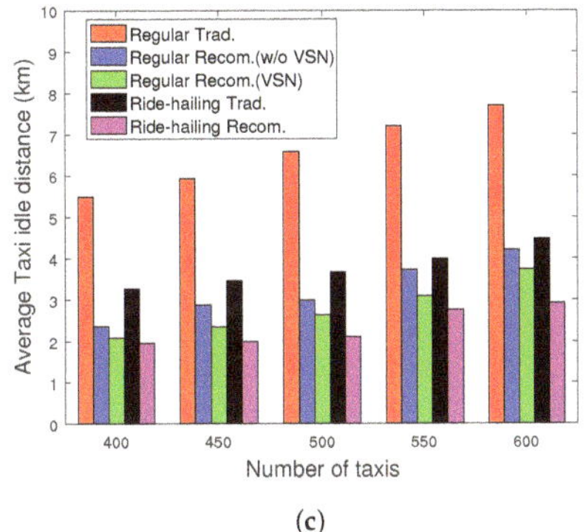

(**c**)

Figure 18. System performance for the three major taxi services for Scenario 2 with 2027 customers in total. (**a**) Number of pick-ups. (**b**) Average customer waiting time. (**c**) Average idle traveled distance.

7. Conclusions

In this paper, we have designed and validated an effective recommendation system for three main taxi services: regular (without VSN), regular (with VSN) and ride-hailing taxi services. The system includes three major components: an incremental predictor of future demands, taxi-to-region matching component, and taxi routing optimizer. By comparing the performance of the proposed recommendation services to the ones of the traditional cases, we found that the proposed approach achieves significant gains in terms of pick-ups efficiency, time, and energy saving for both customers and taxis. The proposed framework can be used as an effective tool for different taxi services by exploiting the power of vehicular social networks and data sharing taxi drivers. Enabling timely and erroneous information exchange of the automatic sensing crowd-sourcing framework will be the scope of our future work in order to ensure efficient operation of the proposed recommendation system.

Author Contributions: Conceptualization, Methodology, Validation, Data Analysis, X.W. and H.G.; Supervision, Project Administration, Funding Acquisition and Writing—review and editing by H.G. and Y.M. All authors have read and agreed to the published version of the manuscript.

Funding: This research received no external funding.

Acknowledgments: The authors would like to acknowledge the NYC DOT government for their open-access traffic data.

Conflicts of Interest: The authors declare no conflict of interest.

References

1. NYCEDC's Blog. New Yorkers and Their Cars. Technical Report. April 2018. Available online: https://edc.nyc/article/new-yorkers-and-their-cars (accessed on 14 April 2020).
2. Roth, S. Report: Taxis Have Longer Wait Times than Uber, Lyft. Technical Report. July 2015. Available online: https://www.google.com.hk/url?sa=t&rct=j&q=&esrc=s&source=web&cd=1&cad=rja&uact=8&ved=2ahUKEwj0i9mxzOnoAhVKQd4KHTjMBowQFjAAegQIAhAB&url=https%3A%2F%2Fwww.its.dot.gov%2Fitspac%2Fdec2014%2Fridesourcingwhitepaper_nov2014.pdf&usg=AOvVaw2BlBGvWScTY3plQNBaxA4d (accessed on 14 April 2020).
3. Kong, X.; Xia, F.; Ning, Z.; Rahim, A.; Cai, Y.; Gao, Z.; Ma, J. Mobility Dataset Generation for Vehicular Social Networks Based on Floating Car Data. *IEEE Trans. Veh. Technol.* **2018**, *67*, 3874–3886, doi:10.1109/TVT.2017.2788441. [CrossRef]
4. Zhao, Y.; Han, Q. Spatial crowdsourcing: Current state and future directions. *IEEE Commun. Mag.* **2016**, *54*, 102–107. [CrossRef]

5. Ning, Z.; Xia, F.; Ullah, N.; Kong, X.; Hu, X. Vehicular Social Networks: Enabling Smart Mobility. *IEEE Commun. Mag.* **2017**, *55*, 16–55. [CrossRef]

6. Vegni, A.M.; Loscrí, V. A Survey on Vehicular Social Networks. *IEEE Commun. Surv. Tutor.* **2015**, *17*, 2397–2419. [CrossRef]

7. Uhlemann, E. Connected-Vehicles Applications Are Emerging [Connected Vehicles]. *IEEE Veh. Technol. Mag.* **2016**, *11*, 25–96. [CrossRef]

8. Ning, Z.; Hu, X.; Chen, Z.; Zhou, M.; Hu, B.; Cheng, J.; Obaidat, M.S. A Cooperative Quality-Aware Service Access System for Social Internet of Vehicles. *IEEE Internet Things J.* **2018**, *5*, 2506–2517, doi:10.1109/JIOT.2017.2764259. [CrossRef]

9. Lucic, M.C.; Ghazzai, H.; Khattab, A.; Massoud, Y. Rapid Management of Unexpected Events in Urban V2I Communications Systems. In Proceedings of the 2019 IEEE International Conference on Vehicular Electronics and Safety (ICVES), Cairo, Egypt, 4–6 September 2019; pp. 1–6, doi:10.1109/ICVES.2019.8906309. [CrossRef]

10. Niu, B.; Li, Q.; Zhu, X.; Cao, G.; Li, H. Achieving K-anonymity in privacy-aware location-based services. In Proceedings of the IEEE Conference on Computer Communications (INFOCOM 2014), Toronto, ON, Canada, 27 April–2 May 2014.

11. Shao, J.; Lu, R.; Lin, X. FINE: A fine-grained privacy-preserving location-based service framework for mobile devices. In Proceedings of the IEEE Conference on Computer Communications (INFOCOM 2014), Toronto, ON, Canada, 27 April–2 May 2014.

12. Yuan, N.J.; Zheng, Y.; Zhang, L.; Xie, X. T-Finder: A Recommender System for Finding Passengers and Vacant Taxis. *IEEE Trans. Knowl. Data Eng.* **2013**, *25*, 2390–2403, doi:10.1109/TKDE.2012.153. [CrossRef]

13. Wan, X.; Ghazzai, H.; Massoud, Y. Online Recommendation System for Autonomous and Human-driven Ride-hailing Taxi Services. In Proceedings of the 2019 31st International Conference on Microelectronics (ICM), Cairo, Egypt, 15–18 December 2019; pp. 351–354.

14. Wan, X.; Ghazzai, H.; Massoud, Y. Incremental Recommendation System for Large-scale Taxi Fleet in Smart Cities. In Proceedings of the 2019 IEEE International Conference on Vehicular Electronics and Safety (ICVES), Cairo, Egypt, 4–6 September 2019; pp. 1–6, doi:10.1109/ICVES.2019.8906451. [CrossRef]

15. Kong, X.; Xia, F.; Wang, J.; Rahim, A.; Das, S.K. Time-Location-Relationship Combined Service Recommendation Based on Taxi Trajectory Data. *IEEE Trans. Ind. Inform.* **2017**, *13*, 1202–1212, doi:10.1109/TII.2017.2684163. [CrossRef]

16. Verma, T.; Varakantham, P.; Kraus, S.; Lau, H.C. Augmenting decisions of taxi drivers through reinforcement learning for improving revenues. In Proceedings of the International Conference on Automated Planning Scheduling (ICAPS'17), Pittsburgh, PA, USA, 18–23 June 2017.

17. Moreira-Matias, L.; Gama, J.; Ferreira, M.; Mendes-Moreira, J.; Damas, L. Predicting Taxi–Passenger Demand Using Streaming Data. *IEEE Trans. Intell. Transp. Syst.* **2013**, *14*, 1393–1402, doi:10.1109/TITS.2013.2262376. [CrossRef]

18. Li, X.; Pan, G.; Wu, Z.; Qi, G.; Li, S.; Zhang, D.; Zhang, W.; Wang, Z. Prediction of urban human mobility using large-scale taxi traces and its applications. *Front. Comput. Sci.* **2012**, *6*, 111–121.

19. Xu, X.; Zhou, J.; Liu, Y.; Xu, Z.; Zhao, X. Taxi-RS: Taxi-Hunting Recommendation System Based on Taxi GPS Data. *IEEE Trans. Intell. Transp. Syst.* **2015**, *16*, 1716–1727, doi:10.1109/TITS.2014.2371815. [CrossRef]

20. Yuan, J.; Zheng, Y.; Zhang, L.; Xie, X.; Sun, G. Where to find my next passenger. In Proceedings of the ACM International Conference on Ubiquitous Computing, Beijing, China, 17–21 September 2011.

21. Garg, N.; Ranu, S. Route Recommendations for Idle Taxi Drivers: Find Me the Shortest Route to a Customer! In Proceedings of the ACM International Conference on Knowledge Discovery and Data Mining (SIGKDD'19), London, UK, 19–23 August 2018.

22. Wang, H.; Yang, H. Ridesourcing systems: A framework and review. *Transp. Res. Part B Methodol.* **2019**, *129*, 122–155. [CrossRef]

23. Alonso-Mora, J.; Samaranayake, S.; Wallar, A.; Frazzoli, E.; Rus, D. On-demand high-capacity ride-sharing via dynamic trip-vehicle assignment. *Proc. Natl. Acad. Sci. USA* **2017**, *114*, 462–467. [CrossRef]

24. Simonetto, A.; Monteil, J.; Gambella, C. Real-time city-scale ridesharing via linear assignment problems. *Transp. Res. Part C Emerg. Technol.* **2019**, *101*, 208–232. [CrossRef]

25. Yao, H.; Wu, F.; Ke, J.; Tang, X.; Jia, Y.; Lu, S.; Gong, P.; Ye, J.; Li, Z. Deep multi-view spatial-temporal network for taxi demand prediction. In Proceedings of the Thirty-Second AAAI Conference on Artificial Intelligence, New Orleans, LA, USA, 2–7 February 2018.

26. Ke, J.; Yang, H.; Zheng, H.; Chen, X.; Jia, Y.; Gong, P.; Ye, J. Hexagon-Based Convolutional Neural Network for Supply-Demand Forecasting of Ride-Sourcing Services. *IEEE Trans. Intell. Transp. Syst.* **2019**, *20*, 4160–4173, doi:10.1109/TITS.2018.2882861. [CrossRef]

27. Wang, X.; Zhang, H.; Wang, L.; Ning, Z. A Demand-Supply Oriented Taxi Recommendation System for Vehicular Social Networks. *IEEE Access* **2018**, *6*, 41529–41538, doi:10.1109/ACCESS.2018.2857002. [CrossRef]

28. Miao, F.; Han, S.; Lin, S.; Stankovic, J.A.; Zhang, D.; Munir, S.; Huang, H.; He, T.; Pappas, G.J. Taxi Dispatch With Real-Time Sensing Data in Metropolitan Areas: A Receding Horizon Control Approach. *IEEE Trans. Autom. Sci. Eng.* **2016**, *13*, 463–478. doi:10.1109/TASE.2016.2529580. [CrossRef]

29. Buchholz, N. Spatial Equilibrium, Search Frictions and Efficient Regulation in the Taxi Industry. Working Paper; Technical Report. December 2017. Available online: https://www.google.com.hk/url?sa=t&rct=j&q =&esrc=s&source=web&cd=1&cad=rja&uact=8&ved=2ahUKEwixg-DrzOnoAhXSP3AKHdq2Ct8QFjAAe gQIAhAB&url=https%3A%2F%2Fscholar.princeton.edu%2Fsites%2Fdefault%2Ffiles%2Fnbuchholz%2Ffil es%2Ftaxi_draft.pdf&usg=AOvVaw3fPzZJAw6mn13xqbuotuLA (accessed on 14 April 2020).

30. Lowalekar, M.; Varakantham, P.; Jaillet, P. Online spatio-temporal matching in stochastic and dynamic domains. *Artif. Intell.* **2018**, *261*, 71–112. [CrossRef]

31. Ota, M.; Vo, H.; Silva, C.; Freire, J. A scalable approach for data-driven taxi ride-sharing simulation. In Proceedings of the 2015 IEEE International Conference on Big Data (Big Data), Santa Clara, CA, USA, 29 October–1 November 2015; pp. 888–897, doi:10.1109/BigData.2015.7363837. [CrossRef]

32. Liu, S.; Wang, S.; Liu, C.; Krishnan, R. Understanding taxi drivers' routing choices from spatial and social traces. *Front. Comput. Sci.* **2015**, *9*, 200–209. [CrossRef]

33. Zhang, D.; He, T. pCruise: Reducing Cruising Miles for Taxicab Networks. In Proceedings of the 2012 IEEE 33rd Real-Time Systems Symposium, San Juan, PR, USA, 4–7 December 2012.

34. Liu, S.; Yue, Y.; Krishnan, R. Non-Myopic Adaptive Route Planning in Uncertain Congestion Environments. *IEEE Trans. Knowl. Data Eng.* **2015**, *27*, 2438–2451, doi:10.1109/TKDE.2015.2411278. [CrossRef]

35. Powell, J.W.; Huang, Y.; Bastani, F.; Ji, M. Towards reducing taxicab cruising time using spatio-temporal profitability maps. In Proceedings of the International Symposium on Spatial and Temporal Databases, Minneapolis, MN, USA, 24–26 August 2011; Springer: Berlin/Heidelberg, Germany, 2011; pp. 242–260.

36. Sayarshad, H.R.; Chow, J.Y. Non-myopic relocation of idle mobility-on-demand vehicles as a dynamic location-allocation-queueing problem. *Transp. Res. Part E Logist. Transp. Rev.* **2017**, *106*, 60–77. [CrossRef]

37. Nourinejad, M.; Ramezani, M. Ride-Sourcing modeling and pricing in non-equilibrium two-sided markets. *Transp. Res. Part B Methodol.* **2020**, *132*, 340–357. [CrossRef]

38. Chen, X.; Miao, F.; Pappas, G.J.; Preciado, V. Hierarchical data-driven vehicle dispatch and ride-sharing. In Proceedings of the 2017 IEEE 56th Annual Conference on Decision and Control (CDC), Melbourne, Australia, 12–15 December 2017; pp. 4458–4463.

39. Ma, T.Y.; Rasulkhani, S.; Chow, J.Y.; Klein, S. A dynamic ridesharing dispatch and idle vehicle repositioning strategy with integrated transit transfers. *Transp. Res. Part E Logist. Transp. Rev.* **2019**, *128*, 417–442. [CrossRef]

40. Tafreshian, A.; Masoud, N. Trip-based graph partitioning in dynamic ridesharing. *Transp. Res. Part C Emerg. Technol.* **2020**, *114*, 532–553. [CrossRef]

41. Pandey, V.; Monteil, J.; Gambella, C.; Simonetto, A. On the needs for MaaS platforms to handle competition in ridesharing mobility. *Transp. Res. Part C Emerg. Technol.* **2019**, *108*, 269–288. [CrossRef]

42. Quddus, M.A.; Ochieng, W.Y.; Noland, R.B. Current map-matching algorithms for transport applications: State-of-the art and future research directions. *Transp. Res. Part C Emerg. Technol.* **2007**, *15*, 312–328. [CrossRef]

43. Quddus, M.A. High Integrity Map Matching Algorithms for Advanced Transport Telematics Applications. Ph.D. Thesis, Imperial College London, London, UK, January 2006.

44. Wan, X.; Ghazzai, H.; Massoud, Y. Real-Time Navigation in Urban Areas Using Mobile Crowd-Sourced Data. In Proceedings of the IEEE International Systems Conference (SYSCON'19), Orlando, FL, USA, 8–11 April 2019.

45. Wan, X.; Ghazzai, H.; Massoud, Y. Mobile Crowdsourcing for Intelligent Transportation Systems: Real-Time Navigation in Urban Areas. *IEEE Access* **2019**, *7*, 136995–137009, doi:10.1109/ACCESS.2019.2942282. [CrossRef]

46. Boeing, G. OSMnx: New methods for acquiring, constructing, analyzing, and visualizing complex street networks. *Comput. Environ. Urban Syst.* **2017**, *65*, 126–139. [CrossRef]

Article

Using a Hybrid Recommending System for Learning Videos in Flipped Classrooms and MOOCs

Jaume Jordán [1,*,†], **Soledad Valero** [1,†], **Carlos Turró** [2,†] and **Vicent Botti** [1,†]

1   Valencian Research Institute for Artificial Intelligence (VRAIN), Universitat Politècnica de València, Camino de Vera s/n, 46022 Valencia, Spain; svalero@dsic.upv.es (S.V.); vbotti@dsic.upv.es (V.B.)
2   Área de Sistemas de Información y Comunicaciones, Universitat Politècnica de València, Camino de Vera s/n, 46022 Valencia, Spain; turro@cc.upv.es
*   Correspondence: jjordan@dsic.upv.es; Tel.: +34-963-877-000
†   These authors contributed equally to this work.

Abstract: New challenges in education require new ways of education. Higher education has adapted to these new challenges by means of offering new types of training like massive online open courses and by updating their teaching methodology using novel approaches as flipped classrooms. These types of training have enabled universities to better adapt to the challenges posed by the pandemic. In addition, high quality learning objects are necessary for these new forms of education to be successful, with learning videos being the most common learning objects to provide theoretical concepts. This paper describes a new approach of a previously presented hybrid learning recommender system based on content-based techniques, which was capable of recommend useful videos to learners and lecturers from a learning video repository. In this new approach, the content-based techniques are also combined with a collaborative filtering module, which increases the probability of recommending relevant videos. This hybrid technique has been successfully applied to a real scenario in the central video repository of the Universitat Politècnica de València.

Keywords: learning recommender system; learning object; learning videos; content-based; collaborative filtering

Citation: Jordán, J.; Valero, S.; Turró, C.; Botti, V. Using a Hybrid Recommending System for Learning Videos in Flipped Classrooms and MOOCs. *Electronics* **2021**, *10*, 1226. https://doi.org/10.3390/electronics10111226

Received: 15 April 2021
Accepted: 17 May 2021
Published: 21 May 2021

Publisher's Note: MDPI stays neutral with regard to jurisdictional claims in published maps and institutional affiliations.

1. Introduction

New challenges in education has raised due to the students' profile changes in the last decade. They demand new ways of learning, better adapted to their way of life and moving away from classical teaching. Academic institutions must be agile in adapting their teaching methodology to the new forms required, taking into account the opportunities offered by the global world [1,2]. In this way, there has been a great increase in the supply of massive online open courses (MOOCs) by academic institutions, as well as in the number of students opting for this type of training [3]. MOOCs mainly relay on learning objects (LOs). As IEEE proposes, a LO is "any entity, digital or non-digital, which can be used, re-used or referenced during technology supported learning" [4]. Thus, MOOCs use different types of LOs, with videos being one of the most commonly used to teach theoretical concepts. In addition, new teaching approaches are emerging in higher education, such as flipped teaching [5–7], in which the theoretical content is studied at home by the students, while the face-to-face sessions are eminently practical, where the knowledge acquired is put into practice by solving problems. To this end, the lecturer instructs the students which LOs they should work on at home before the next face-to-face session. In this way, students are encouraged to acquire the theoretical concepts not only through books or specialised articles, but also through audio–visual material.

Furthermore, in this pandemic context, face-to-face classes have been replaced by online classes in many institutions, increasing the adoption of flipped learning. Lecturers need to plan subjects taking into account possible connectivity problems, as the possibility of students being unable to attend online classes due to health problems (quarantine, hospitalisation) increases. In addition, students may have difficulties in accessing devices during working hours, because they share them with their parents, etc. All these new circumstances make it even more important to provide educators with useful tools to search for and recommend good LOs, which can be accessed by students at any time.

Universitat Politècnica de València (http://www.upv.es, accessed on 14 May 2021) (UPV) is a Spanish Public university which offers undergraduate degrees, dual degrees, masters and doctoral programs. UPV has more than 28,000 students. UPV has been promoting new pedagogical methodologies in their degrees in the last decade, such as flipped teaching [8]. It has also made a great effort in developing MOOCs within the edX platform (https://www.edx.org/, accessed on 14 May 2021), with more than 2 million enrollments and having three courses in Class Central all time top 100 MOOCs in 2019 (https://www.classcentral.com/report/top-moocs-2019-edition/, accessed on 14 May 2021), and two courses top 30 in 2020 (https://www.classcentral.com/report/best-free-online-courses-2021/, accessed on 14 May 2021). In addition, UPV participated in the movement that arose during the first months of the pandemic to offer free certificates for some of the MOOCs offered. This fact, together with the need for new training channels, has led to a significant increase in the number of students on this type of courses (https://www.classcentral.com/report/mooc-stats-2020/, accessed on 14 May 2021).

Students need access to a variety of resources to understand the theoretical concepts required in blended and flipped classroom environments. To facilitate this difficult work, UPV has had a long-standing digital resources project with the aim of producing video content as LOs. This video content is handled in the university central video repository, called mediaUPV (https://media.upv.es/, accessed on 14 May 2021). This portal is not only used in the field of MOOCs, but also in other educational projects.

mediaUPV allows UPV lecturers to upload and manage video content for students. Students access mediaUPV usually through suggestions made by their lecturers through the learning management System (LMS), but they also access the video portal and browse through the content on their own. A relevant feature of mediaUPV against other alternatives (e.g., YouTube) is that the videos have been prepared and recorded by lecturers from the institution, so the quality of the content is guaranteed.

mediaUPV portal has become an essential tool to the institution during this pandemic; however, the size of the mediaUPV content is a growing problem, and hence, it is increasingly difficult to find the most relevant content for both students to view and lecturers to suggest. Thus, UPV determined that both students and lecturers would benefit from the development of a new Learning Recommendation System (LRS), that could recommend relevant and related videos. Therefore, in our previous work [9], we presented the first recommender engine proposed to carry out that purpose, which combined two content-based techniques to recommend useful learning videos to learners and lecturers. However, with that engine it was only possible to recommend videos labelled as high quality, that is, with transcripts. Therefore, it is necessary to apply other techniques to cover the entire mediaUPV catalogue.

In this paper we present an enhancement of the previous work, which extends the proposed recommendation engine also using collaborative filtering techniques. Thus, we describe how we have designed and developed a hybrid recommender system based on both content-based techniques and collaborative filtering. Furthermore, a complete analysis of the results in production of the initial content-based LRS since its application in October 2019 until March 2021 is provided.

This article is structured as follows. In Section 2 related works are described. Following, Section 3 specifies a description about the LOs to recommend and the potential users of the system. In Section 4, the proposed recommender system is explained. Then, Section 5 shows the experimental results of the proposed recommender using the data of the mediaUPV portal. In addition, Section 6 provides an analysis of the results of the LRS used in production. Finally, Section 7 draws the conclusions and future work of this paper.

2. Related Work

Learning Recommender Systems (LRS) should assist learners in discovering relevant LO than keep them motivated and enable them to complete their learning activities [10]. Most of the LRS adopt the same techniques than regular recommender systems [10–13], such as: content-based, in which recommendations are determined considering user profiles and content analysis of the learning objects already visited by the user; collaborative filtering, in which recommendations are based on the choices of other similar user profiles; knowledge-based, in which it is inferred whether a LO satisfies a particular learning need of the user to recommend it; and hybrid, in which recommendations are computed by combining more than one of the above techniques.

In recent years, different approaches have been proposed in order to improve the efficiency and accuracy of the recommendations and retrieval of useful LOs. In this way, in [14], authors provide new metrics for applying collaborative filtering in a learning domain, so users with better academic results have greater weight in the calculation of the recommendations. However, the experiments did not carried out in a learning environment. In another proposal, Zapata et al. provide a tool for filtering the retrieved results from a user query, which uses a combination of different filtering techniques, such as content comparison, and collaborative and demographic searches [15].

Other proposals focus on recommending to students those LOs that can be most useful to them, providing solid arguments. This is the case of [13], which combine content-based, collaborative and knowledge-based recommenders using an argumentation-based module to recommend LOs inside a LMS. In this case, information on student profiles and learning styles is also available. An item-based collaborative filtering method is combined with a sequential pattern mining algorithm to recommend LOs to learners in [16]. In this case, LOs are ranked by the students and it is also possible to obtain the browsing sequences made by them. In a similar way, ref. [17] proposes a hybrid knowledge-based recommender system based on ontology and sequential pattern mining for recommendation of LO. Authors can adequately characterize learners and LOs using an ontology, since they have detailed information about them. Personalized learning paths (sequence of LOs) that maximizes the performance of the learner and effectiveness of learning are provided in [18], hybridizing ant colony optimization with genetic algorithm. In this case, authors use both learners and LOs attributes to determine the appropriate learning paths. In addition, in [19], Dwivedi et al. recommend learning paths using a variable length genetic algorithm. This approach considers learners' learning styles and knowledge levels extracted from the learners' registration process. Finally, in [20], authors propose a method, based in a collaborative filtering approach, for building a unified learner profile which is used to recommend LOs to a group of individuals.

Besides, other works have been done to improve the accuracy of the searches in mediaUPV. For example, in [21] a semi-supervised method is applied to cluster and classify the LOs of mediaUPV, obtaining specific keywords that represent each cluster. In [22], authors applied a custom approach for indexing and retrieving educational videos using their transcripts, which are available in mediaUPV. Videos are classified in different domains using the method described in [21]. In addition, they applied a Latent Dirichlet Allocation algorithm [23] to get a list of topics and their score. User queries are classified in one of the domains, recovering from that cluster those videos whose transcripts are the closest to the query.

As can be seen, most previous work on LRS adopts a hybrid strategy, seeking to harness the strength of each particular technique, overcoming its limitations by using them together. Furthermore, the different strategies that can be applied depend on the data available to describe LOs and users. In our case, a hybrid strategy will also be applied, combining content-based methods. On the other hand, previous experiences on the improvement of searches in the mediaUPV repository show us the usefulness of characterizing LOs using available transcripts and titles.

3. Problem Description

mediaUPV portal started in 2011 and by the end of 2019 it had 55,600 different videos mainly from STEM topics, with more than 10 million views. During 2020 until the end of March 2021, the mediaUPV catalogue increased almost a 44.8%, reaching 80,500. This significant growth may be due to the increased need for the use of online learning resources, not only for new forms of learning, but also for the institution's traditional courses that had to be converted to purely online teaching during the pandemic. From this database, only 13,232 by the end of 2019, and 20,135 by the end of March 2021 are certified as high quality LOs (an increase of 52.2%). All these high quality videos have a transcript (with more than 100 characters), a title, and an author, however the videos are not classified using any taxonomy and no useful keywords are associated to them for all cases. mediaUPV platform generates the transcripts of what is said in the videos using the poli[Trans] service (https://politrans.upv.es, accessed on 14 May 2021), an online platform for automated and assisted multilingual media subtitling offered by UPV. poli[Trans] service is based on transLectures-UPV Platform [24].

Hence, a 75% of the mediaUPV catalogue is composed by videos without transcripts. It was detected that a huge percentage of that videos have a medium quality and could be also interesting to lecturers and students. In fact, some of them are recordings of classes from some subjects, which can be used for reinforcing the learning of students of similar subjects in other degrees. By default, mediaUPV does not obtain the transcripts of these recordings, because is the lecturer who has the permissions to demand them.

mediaUPV portal is mainly used by students and lecturers. The students are mainly formal students of the UPV, but it also receives many visits from anonymous users, who can register in some MOOC offered by the UPV. Moreover, mediaUPV is not connected to the LMS of the UPV, so even though the user is authenticated, it is not possible to know his student profile (e.g., enrolled subjects).

Therefore, the aim of this proposal is to be able to offer recommendations not only to authenticated users but also to anonymous users of the system. In addition, the system should be able to recommend not only students, but also lecturers who want to find quality videos to suggest to their students. Therefore, the main objective is to be able to offer recommendations on learning videos (that is LOs) that meet the quality standards set by the UPV.

It is important to note that all recommendations made on the mediaUPV portal are always through a video. That is, when a user logged into the system is watching a video (which can suggest videos similar to the current video or content that the user has watched previously), or an anonymous user watching a video (which can only suggest content similar to the current video). This implies that there are no recommendations only associated with a logged-in user. Thus, a hybrid system needs to be used in which the current video, and (optionally) the logged-in user are considered.

4. Learning Recommender System Proposal

Collaborative recommendation has some well-known difficulties, such as the necessity of a huge quantity of data for making accurate recommendations. In the same way, content-based approaches also have some problems, such as the lack of serendipity. For this reason, a hybrid approach which combines the previous techniques can improve the accuracy in the provided recommendations and reduce the cold start impact.

In this way, our LRS is based on two different approaches:

- First, a content-based recommender module based on two components. One component recommends taking into account the activity of the user identified in the system, the profile-based module (PB). The other component offers recommendations based on the content of the video being watched at the time, the item-based module (IB).
- Second, a collaborative filtering module (CF), which offers suggestions based on similar other users with a viewing history similar to current user.

Thus, the computed recommendations are based on these two approaches, getting a hybrid recommendation system, so that the user receives recommendations of videos similar to the one she is watching at that moment, or also of videos that may interest her due to her viewing history, and additionally, content considered interesting due to the opinions of other users with a viewing history similar to her own. This fact increases the serendipity, discovering contents to the user that are of her interest, although they are not so similar, a priori, to learning videos already watched by the user so far.

In our proposal, LOs/videos are characterized by their title and their transcript (when available). This characterization is used for the calculation of the similarity between the LOs. As transcripts are in different languages, it is possible to recommend videos from different languages.

Because there is a large set of words in the transcript and title of the videos, it is necessary to have an algorithm that filters out the too common words, which do not serve to differentiate the content. Thus, it is possible to focus on the particular words in the entire collection, which serve to identify the content of a video. We need to use an unsupervised term weighting approach, as our video repository is not categorized. Therefore, the algorithm chosen is the well-known term frequency–inverse document frequency (TF-IDF) [25], as it is a common term weighting scheme used to represent documents. In order to improve the performance of the TF-IDF algorithm, also the stop words from the *nltk Python* (https://www.nltk.org/, accessed on 14 May 2021) package are used. Furthermore, some ad hoc words have been added to this package.

The item-based module takes only the information from the content of the videos, i.e., there is no information about the users. So, we can consider this module as a recommendation by item-item similarity to be used when a (maybe anonymous) user is watching a video. To do this, we take the characteristics (transcript and title) of each of the videos to calculate the item-item matrix with the TF-IDF algorithm. Then, the cosine similarity of two instances of this item-item matrix returns the similarity among the different videos. Cosine similarity calculates the similarity between two n-dimensional vectors by the angle between them in the vector space:

$$cosine\_sim(\vec{p}, \vec{q}) = \frac{\vec{p} \cdot \vec{q}}{|\vec{p}| * |\vec{q}|} \tag{1}$$

The profile-based module considers the content information (the terms extracted by the TF-IDF algorithm from the transcript and title) of the videos viewed by users. In this way, the recommendations to a user are made based on the similarity among the viewed videos of the user in mediaUPV. In this case, the similarity is calculated using the cosine profile-item matrix.

The collaborative filtering module considers the similarity between the users based on their viewed videos. Thus, two users will be more similar depending on how many videos they have both watched. In the opposite way, a user will have no similarity with another user if the sets of videos watched by each of the two users are completely disjoint. Therefore, the similarity is calculated using the cosine similarity between the users based on their watched videos. Then, the prediction values of recommendation for each user are calculated taking the dot product between the similarity of users and the viewed videos matrix, normalizing the data properly. With these process, we have the prediction to recommend each user in the system considering all users similarity (we will call this approach all users). However, we can also obtain a different prediction to recommend if we only consider the top k most similar users to the user to be recommended. To do that, we calculate the top k nearest neighbours (NN) for each user in the system, and then, the prediction values to recommend videos are also calculated (we will call this approach top k NN). We note that the computation cost of this last approach considering the top k NN is higher than the all users approach. The computation cost is also increased as k grows.

Finally, our *hybrid recommendation (HR)* system consists of two components, one content-based (with two modules) and another collaborative-based (with only one module).

Thus, the recommendations are made taking into account the intersection of the videos recommended by the modules (see Equation (2)). The rest of the recommendations are obtained considering first the weight applied to each component, i.e., w_{CB} for the content-base component and w_{CF} for the collaborative filtering module. Furthermore, it is possible to balance the importance given to the two modules of the content-based component by means of other two weights, w_{IB} for the item-based module and w_{PB} for the profile-based one. In this way, in cases where there is no user (anonymous) the w_{PB} and w_{CF} are set to 0. Likewise, when an authenticated user is not yet watching a video, the w_{IB} is set to 0 (however, this case is not currently applicable to mediaUPV portal).

$$HR = (IB \cap PB \cap CF) \cup (w_{CB} \cdot ((w_{IB} \cdot IB) \cup (w_{PB} \cdot PB)) \cup (w_{CF} \cdot CF)) \tag{2}$$

In the next section, we explain the experiments carried out to determine the combination of weights which offers the best performance.

5. Experimental Results

In this section, we explain the experiments carried out with the proposed LRS. We have made tests with data from the videos watched by the users from September 2018 to July 2019, dividing this data into a training set and a test set. We used this data to be able to compare the new collaborative filtering module with the experiments already presented in our previous work [9]. In this section, we first present the general experimental setup in Section 5.1. Then, the experiments of both content-based modules are explained in Section 5.2. Section 5.3 presents the experiments with the collaborative filtering module. The experiments with the hybrid approach, which is the combination of the content-based component and the collaborative filtering module, are explained in Section 5.4. Finally, Section 5.5 contains a discussion of the results of all these experimental results.

5.1. Experimental Setup

As mentioned above, the database is comprised of all the videos available on the mediaUPV as items to be recommended, as well as the usage data (views) of the users logged into the system. Although the platform had more than 55,600 videos by the end of 2019, in the following tests we have filtered out the hidden videos (only available with a direct link), the videos that do not have a transcript, and the videos in which the transcript has less than 100 characters, leaving a total of 13,232 videos suitable for recommendation.

The data set used for these experiments is formed by of learning videos viewed by users during an academic year at the UPV, from September 2018 to July 2019. The training data set consists of the data from September 2018 to April 2019, while the testing data set consists of the data from May to July 2019, i.e., 8 months for training and 3 months for testing. The videos considered are those present on the platform until July 2019, after filtering them as described above. Thus, we try to simulate a real scenario, in which the training data represent the past activity of the users, while the test data is formed by the activity of the following 3 months ("future"). Therefore, any recommendation from the recommender that is among the videos that users have actually watched in the test set is considered a success. Additionally, in these tests we only consider five recommendations since it is the number required by the mediaUPV portal.

We use the well-known precision and recall measures to evaluate the success of the recommender. Precision can be defined as the successful recommendations made (videos that have been viewed by the user in the test set) divided by the number of recommendations made:

$$precision = \frac{success\_recommendation}{recommendations\_made} \tag{3}$$

Recall is defined as the successful recommendations made divided by the number of watched videos in the test set:

$$recall = \frac{success\_recommendation}{watched\_videos} \tag{4}$$

For our tests we considered a set of regular users (*reg_users*) that we define as those who have watched between 10 and 150 videos both in the training and test periods, having 1044 users in this set. We also considered a set of new users that have watched between 1 and 9 videos both in the training and test periods (there are 815 users in this set). If nothing is specified, the regular users set is the one used.

5.2. Content-Based Component

To test the content-based modules we use both of them together to combine their efforts. So this is like using the hybrid recommender engine but without considering the collaborative filtering module. Thus, we focus on the content-based module centred on the video being watched, that is, the item-based module (*IB*); and on the content-based module that considers the user's views, namely the profile-based module (*PB*). The weights for each module in the hybrid recommender engine are set to $w_{IB} = 50\%$, $w_{PB} = 50\%$ and $w_{CF} = 0\%$.

5.2.1. Setting Transcript and Title Features

In this first test, we analyze the success of recommendations for the set of regular users considering different amount of features for the transcript and the title of the video to train the LRS, in order to establish the better amount of both. The graph in Figure 1 shows the precision and recall for different values of the number of features considered for the transcript, while the number of features for the title is kept at zero.

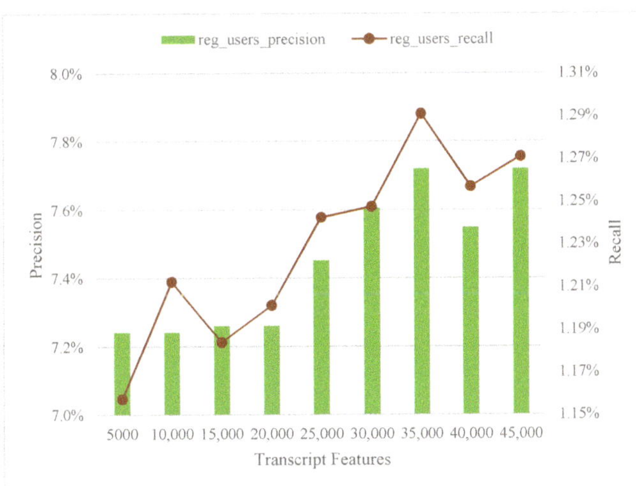

Figure 1. Precision and recall results for the regular users set with different amount of transcript features.

In general, precision and recall increase slightly as the value of the features for the transcript increases (from 7.2% to 7.7% for precision, and from 1.15% to 1.29% for recall), as would be expected when more information is available from the transcript. The best precision values are with 35,000 and 45,000 features for the transcript. However, the best recall value is in the case of 350,00 transcript features. Therefore, the best configuration for the recommender would be to use 35,000 transcript features, since this value achieves higher recall than any other and equals the precision obtained with 45,000 transcript features. In addition, the computation of 35,000 transcript features is computationally less expensive and, in particular, implies a lower memory cost.

It should be noted that, in the experiment made with 35,000 transcript features, the number of users who have been recommended successfully is 254 of 1044, i.e., 24.33%.

Figure 2 shows the precision and recall for different values in the number of features considered for the title having the transcript features fixed to 35,000. The best values of precision and recall are obtained with the number of features of the title at 0. In addition, both values decrease slightly and in a relatively uniform way as the number of title features increases. So, apparently it is better to skip the title features. However, it is interesting to analyze this considering the nature of the LOs of mediaUPV. In this way, we have analyzed the set of words that determines the TF-IDF algorithm. As we mentioned before, the videos on the platform correspond mainly to university courses, so there is a set of terms that are certainly repetitive in the titles of the videos and do not provide any differentiating information with respect to their content. Among these terms, we find the following: {'analysis', 'calculation', 'control', 'creation', 'data', 'design', 'exercise', 'engineering', 'introduction', 'management', 'mechanism', 'model', 'module', 'practical', 'practice', 'presentation', 'simulation', 'system', 'systems', 'theme', 'unit', 'virtual'}.

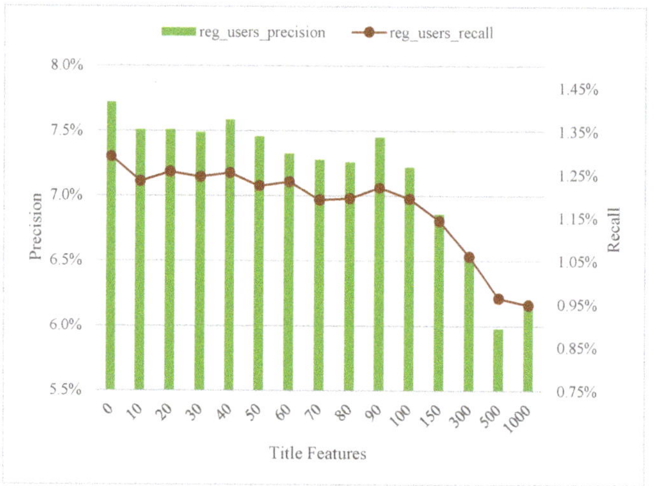

Figure 2. Precision and recall results for the regular users set with different amount of title features.

5.2.2. Filtering Title Features

In order to increase the precision of the recommendations, we decided to filter the terms of the previous list from the titles of the videos by considering them as stop words. The results of applying this correction can be seen in the Figure 3, which shows the precision and recall for 35,000 transcript features and different amount of title features. In this case, it can be seen that the best global values of precision and recall are still obtained with 0 title features. However, it should be noted that with 10 title features, for the case where the specified title terms have been filtered out, the precision and recall almost reach this base case, slightly surpassing the case without filtering (with 10 title features). In addition, for values up to 30 title features, the precision and recall are better for the case with filtered terms. However, from 40 title features on, the effect of filtering is diluted and the results are generally slightly worse, and most of the cases slightly worse than the unfiltered case.

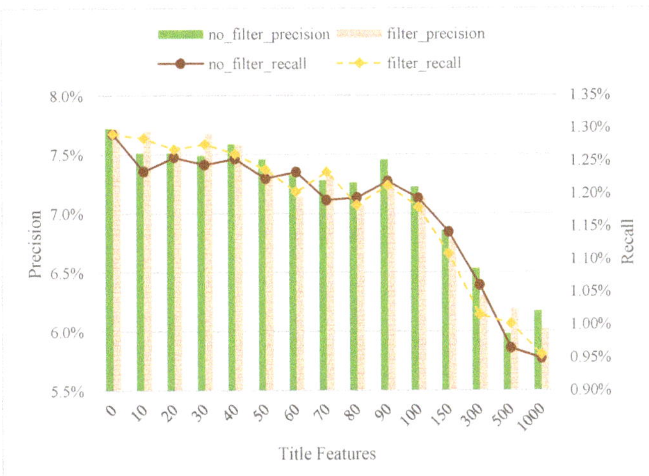

Figure 3. Precision and recall comparison with and without filtering title terms for different amount of title features.

Consequently, we can say that filtering has significantly improved the results but it is not enough to make the title relevant. Perhaps it would be necessary a still greater filtering of terms that we have not considered 'commo' and that the TF-IDF algorithm has not identified as such either. However, since we are considering 35,000 transcript terms, the inclusion of 10 to 50 terms from the title can be considered irrelevant after the analysis. Furthermore, it can also be interpreted as a video being better characterized by its own transcript than by its title.

Although we have already seen that it is better to skip the title in all cases, we will analyze in detail the difference between filtering the title terms and not filtering them for 10 title terms and different amount of transcript features. Figure 4 shows this comparison (precision and recall) with and without filtering. Precision and recall are significantly better if filtering of title terms is performed in all cases, with the only exception of 45,000 transcript features where precision is slightly higher for the unfiltered case (being also the best result for the different transcript values for the unfiltered case of title terms). In this particular case, it could be that by considering only 10 features of the title, but 45,000 for the transcript, the effect of the filtering of the title is diluted. However, this is not a very significant difference. On the other hand, as previously observed, the best results, both in terms of precision and recall, are obtained with 35,000 transcript features in the case of filtering the title terms.

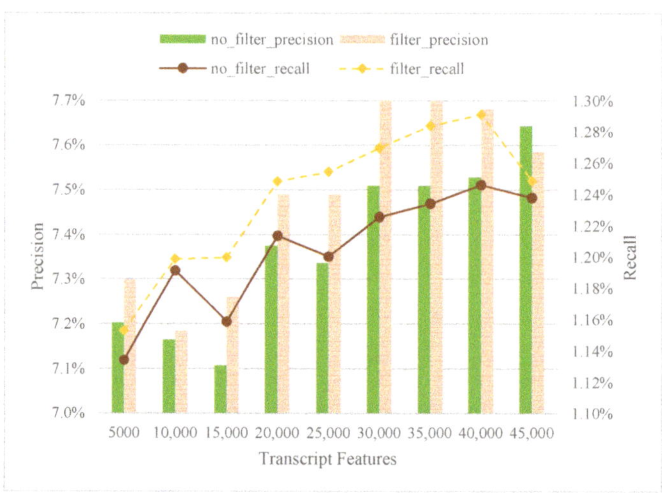

Figure 4. Precision and recall comparison with and without filtering title terms for different amount of transcript features with 10 title features.

5.3. Collaborative Filtering Component

In this subsection we conducted experiments to test the performance of the collaborative filtering module in its two variants, i.e., the one that considers the similarity with all users (which we refer to it as all users) to make recommendations (taking the videos with the highest recommendation values) versus the one that only considers the top k nearest neighbour (NN) users to make the relevant recommendations.

Figure 5 presents the results of precision and recall for the sets of regular users (watched from 10 to 150 videos) and new users (watched from 1 to 9 videos). The values shown correspond to the collaborative filtering module considering all users recommendations in the first column, and the subsequent columns consider the top k NN users for recommending. Generally, for both the regular and new users sets, the best results in precision and recall are obtained by the CF all users approach, which in fact doubles the values for almost all the top k NN approaches. In this way, it is clear that the best approach for collaborative filtering with mediaUPV data is the one that uses CF all users instead of any number of the top k NN. Additionally, this approach is less costly computationally. If we compare the different k values of the top NN, it seems that the best are 25 for the regular users set, and 50 for the new users set, slightly decreasing in both cases as k grows. All in all, the difference with the CF all users approach is significant enough to avoid any top k NN approach.

In the case of the regular users, the CF all users approach achieves almost 9.5% precision and 1.5% of recall. However, with the new users set, the precision is almost 3.7% and the recall over 10%. This difference between both users sets is mainly due to the amount of historical data of the users. In the case of the new users set, data about them is limited, as they have only watched 1 to 9 videos. This causes the precision of the recommendation to be significantly lower than the precision obtained in the regular users set, in which the amount of data is larger to build a more solid recommender. This difference is commonly known in literature as the cold start problem. However, since our recommender engine has different modules, we can leverage them to obtain satisfactory results in any case.

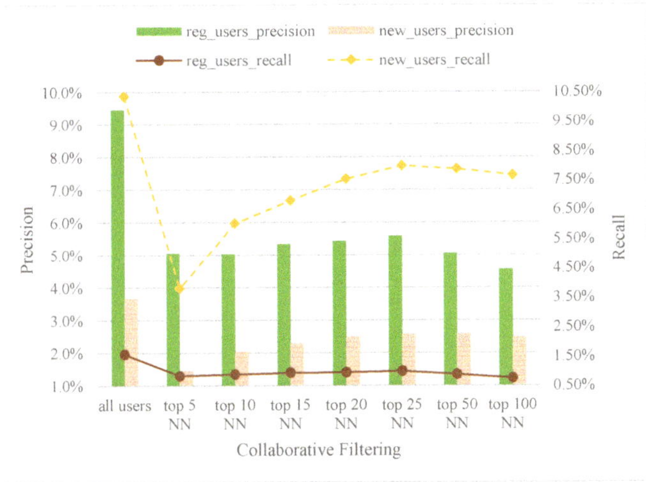

Figure 5. Precision and recall comparison of regular users and new users for collaborative filtering module with different values of top N users obtained by nearest neighbour (NN) technique.

5.4. Hybrid Weights Setting

Our LRS is a hybrid approach that considers, on the one hand, a content-based component with two modules named profile-based module and item-based module, and on the other hand, a collaborative filtering component (with a unique module) as it is specified in Equation (2). In this subsection, we first analyse the best combination of weights for the content-based modules. Once found them, we study how to tune the weights of the content-based component and the collaborative filtering component.

In Figure 6 we make a comparison of precision and recall of two different sets of users, using different weights for the content-based modules of our hybrid LRS. We first consider the set of new users of which we have little knowledge as they have only watched 1 to 9 videos in the data set (815 users). The second set are the regular users, formed by users that have already watched between 10 and 150 videos in the data set (1044 users). For regular users, the best precision and recall is obtained with balanced weights, i.e., $\{w_{PB} = 40\%; w_{IB} = 60\%\}$ and $\{w_{PB} = 60\%; w_{IB} = 40\%\}$. However, for new users, the best precision and recall values are obtained with low w_{PB}, with 15.83% of new users receiving successful recommendations.

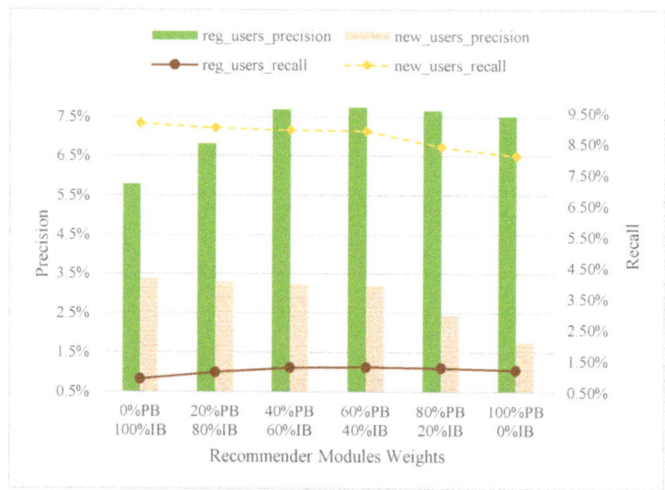

Figure 6. Precision and recall comparison of regular users and new users, using different content-based modules weights.

Having set the weights of the content-based modules to $w_{PB} = 60\%$ and $w_{IB} = 40\%$, since they obtain the best results, we can now set the weights of the combination of the content-based component (w_{CB}) with the collaborative filtering component (w_{CF}) as specified in Equation (2). For this, Figure 7 shows the results of precision and recall for both groups of regular and new users with different weights for the recommender modules, the content-based modules (CB) and the collaborative filtering module (CF). In this case, the results for the different values of weights are the same except of the extreme values of $\{w_{CB} = 0\%; w_{CF} = 100\%\}$, and $\{w_{CB} = 100\%; w_{CF} = 0\%\}$, that are slightly lower. The reason behind this is that the intersection of recommendations of both the content-based modules and the collaborative filtering module already gives the best results, and hence, any combination of the weights (except the extremes) can be considered.

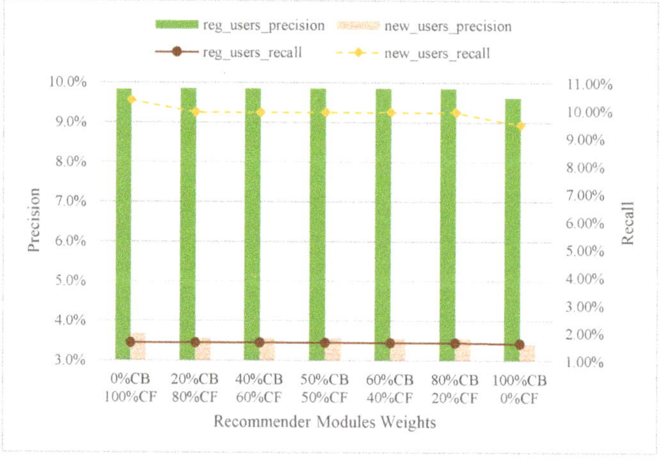

Figure 7. Precision and recall comparison of regular users and new users, using different content based and collaborative filtering modules weights.

The incorporation of the collaborative filtering module improves the precision from 7.7% to 9.85% for regular users (for new users is increased from 3.3% to 3.56%) with respect to the version of the LRS with only the content-based modules proposed in [9]. Recall is also slightly improved from 1.3% to 1.6% for regular users (9.1% to 9.9% for new users). In this case, 28.5% of the regular users and 16.4% of the new users received a useful recommendation. We also note that the results of the collaborative filtering module alone (see Figure 5) when compared with the combined version with the content-based modules also improve from 9.5% to 9.85% of precision for the regular users. Therefore, all of these are positive results that justify the need of applying the collaborative filtering component to our LRS to improve its accuracy.

Since there is no significant difference between the weights of the content-based component and the collaborative filtering component, we propose balanced weights to apply in production $\{w_{CB} = 50\%; w_{CF} = 50\%\}$.

5.5. Discussion

For new users from which the system has few information is harder to make successful recommendations; however, this could improve in the future during the application of the LRS as the users get more engaged to mediaUPV portal. In the case of the users from which there is more historical data available, the accuracy of our LRS improves significantly, specially when using the new developed collaborative filtering module combined with the content-based modules. Furthermore, the collaborative filtering module will be able to recommend videos that do not have transcript (we remind that only 20,135 from the 80,500 videos have transcript), which will suppose a lot of more possibilities to recommend to the logged-in users when applied to production.

We emphasize that even though our hybrid LRS obtains a precision of 9.85% simulating a real environment (we improved previous results of 7.7% precision without the collaborative filtering module of [9]), 28.5% regular users and 16.4% of new users received some good recommendation. In addition, the precision and recall obtained are significantly better than a random recommendation.

In conclusion, after this analysis we established the parameters of the LRS to be applied in production in the mediaUPV portal as 35,000 transcript features, 0 title features, $w_{CB} = 50\% \cdot (w_{PB} = 60\%, w_{IB} = 40\%), w_{CF} = 50\%$.

6. Production Results

In this section we show the results of the application of the LRS described in [9] to the mediaUPV portal. That proposal did not contain a collaborative filtering module. It should be noted that mediaUPV did not have a LRS until the application in October 2019. The parameter configuration of the recommender system applied to production was the one that worked best in the experiments of the previous work [9], that is, 35,000 transcript features, 0 title features, $w_{PB} = 60\%$ and $w_{IB} = 40\%$ (without the collaborative filtering module which has been developed for the present work).

The graph of Figure 8 presents the global precision and recall results of the recommendations made to the users for the period from October 2019 until March 2021. We compare different time ranges, from 10 to 480 min, in which the user can watch the recommended videos. Hence, the column of 10 min represents the precision and recall of the video recommendations that are watched within the 10 min after the recommendation is made. So the precision increases as the time range grows. As it can be seen in Figure 8, the precision is significantly lower for the cases below 120 min, and then, it only increases slightly. The reason behind that might be that the users of mediaUPV portal usually watch long videos, or they even do other tasks (like homework in the case of students, or preparing other classes or new material in the case of lecturers) between watching a video and the next. This may explain why the precision increases significantly if we consider a time after recommendation of at least 120 min instead of 10 or 30 min.

Overall, a precision of around 9.5% for 120 min after the recommendation, or higher than 10% if we consider more minutes is a suitable result for our recommender system. We note that most of the users that enter in mediaUPV portal do not seek for recommendations, since they only watch the video that they need to (i.e., a student who must watch the corresponding lesson). Additionally, the production results are significantly better (around 2–3% higher precision) than the experimental results with the training and test sets of September 2018 to July 2019 originally used in our previous work [9].

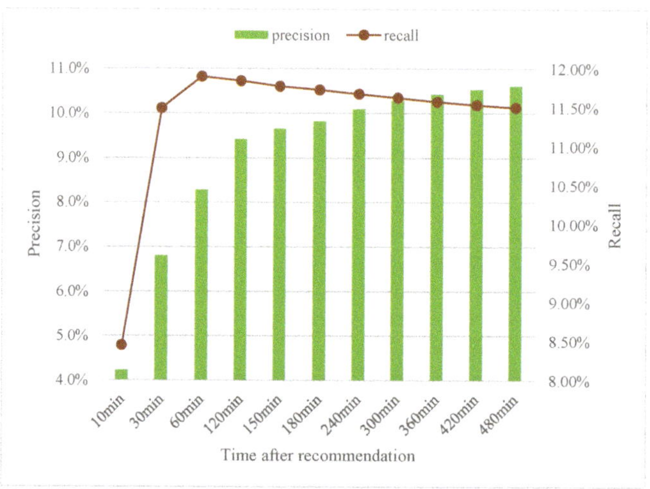

Figure 8. Production results of precision and recall considering different time range for the user to watch the recommended video after the recommendation.

Figure 9 presents the results of precision and recall for different type of users considering at most 120 min after the recommendation to watch the video. All users (all_users) is the set that includes the total amount of users that watched any video and received any recommendation in our system from October 2019 to March 2021. The set of all users is divided in other three sets, namely: new users (new_users), which are those that watched between 1 and 9 videos in the period of application of our recommender system; regular users (reg_users), the ones that watched 10 to 150 videos in the period; and top users (top_users), those who watched more than 150 videos in the period. According to this classification, we have 2297 new users, 8202 regular users, and 634 top users. The results in precision and recall for all users and regular users are reasonably similar since this last set is the larger, so it influences more the all users set, but also it is the middle point between the new users and top users set.

The main results of Figure 9 show that the precision of our recommender engine for new users is 6.3% with a recall of 12.74%. This precision is significantly lower than the precision for both the regular users and the top users, that is 9.23% and 9.8%, respectively, with a lower recall for the top users of 10.6%. From these results, we can confirm that as the known data from the user increases, the accuracy of the recommendations also increase. However, it is even hard to increase precision with top users due to the nature of the recommendations of mediaUPV portal, in which most users only come to watch the specific videos they have to. This scenario differs significantly from platforms like YouTube where most users enter looking to spend some leisure time.

Globally, we make some successful recommendation to 35% of all users. Particularly, we have successful recommendations for 24% of new users, 51% in the case of regular users, and 78,6% for the top users. This results clearly show that a successful or useful recommendation for a user is totally related to the amount of (historical) data that the system has of the user. It is important to note that almost the 80% of the top users (the most informed set of users we defined) received any successful recommendation, which is a very positive result for the recommender system. We can assume that our recommender would be more accurate as the users are more engaged with it.

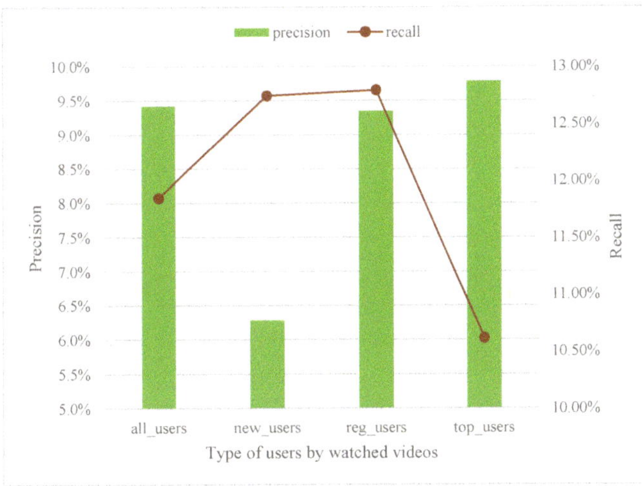

Figure 9. Production results of precision and recall by different type of users within the 120 min after the recommendation. All users include the total amount of users that received any recommendation. New users are those that watched between 1 and 9 videos, regular users watched 10 to 150 videos, and top users watched more than 150 videos.

In Figure 10, we show a violin plot (https://towardsdatascience.com/violin-plots-explained-fb1d115e023d, accessed on 14 May 2021) depicting the distribution of daily accesses to the videos of the mediaUPV platform during the year 2019, with no recommender (before October) and with the proposed one. As can be seen clearly, the existence of the LRS is correlated with an increase of the number of accesses to the videos by the users.

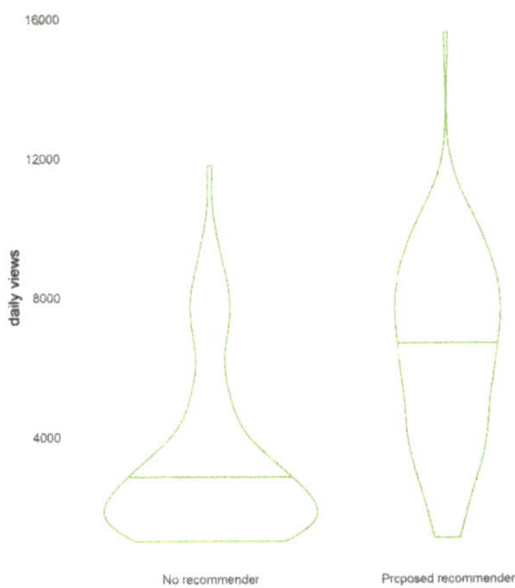

Figure 10. Video access with and without a recommender in production.

A common way in industry to measure relative quality of a recommender system is the Click-Through Rate (CTR) (https://en.wikipedia.org/wiki/Click-throughrate, accessed on 14 May 2021), that measures the percentage of clicks in the recommender per number of views. As the CTR is used by the ads industry, there is an ongoing interest in CTR prediction techniques [26]. In the case of a generic recommender system, anything above 0.35% means you are doing a good job (www.acquisio.com/blog/agency/what-is-a-good-click-through-rate-ctr/, accessed on 14 May 2021). As can be seen in Figure 11, the CTR is 1.28% on average, with notable peaks over 4%. These results can be considered quite satisfactory, since they imply that the recommendations made to users generate interest in them.

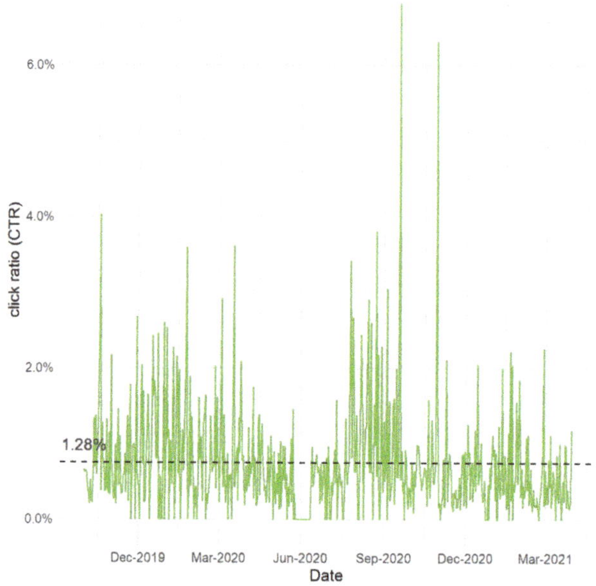

Figure 11. Clicks in recommender (CTR) in production.

Finally, we point out the percentage of clicks on each of the recommended videos according to their order in the list, demonstrating the relevance of this order to users (see Figure 12). About 28% of users click on the first video, and more than half of the users click on the first two recommendations with a distribution that seems to be heavy tailed. So the most relevant video by far for users is the first one and then the rest of the videos follow a decreasing order of importance, which also points to a reasonable work of the presented LRS.

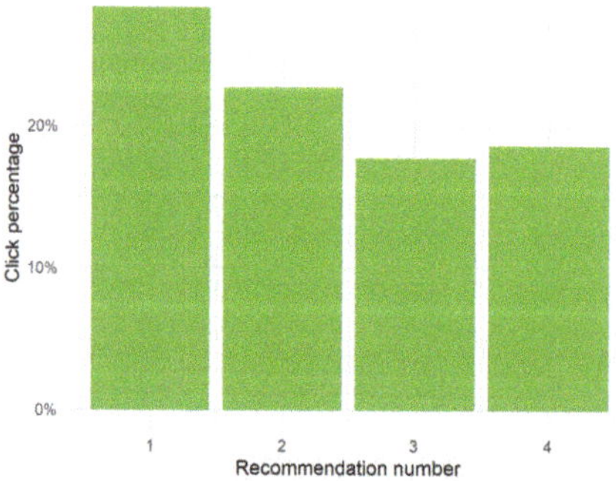

Figure 12. Percentage of clicks in recommendations per list position.

7. Conclusions

This work proposes a new hybrid LRS based on collaborative filtering and content-based components capable of recommend learning videos based on viewing history and current video content. Thus, the LRS proposed is able to recommend not only to authenticated users but also to anonymous users from the mediaUPV portal, independently if they are lecturers or students. In fact, mediaUPV portal has not information about learners' profiles or needs, as it is not connected with any LMS.

The hybrid LRS has been applied to a simulated environment, using a data set of learning videos and user profiles from the 2018-2019 academic year at UPV. The best hybrid LRS configuration obtained 9.85% of precision and 1.6% of recall, where 28.5% of regular users received some useful recommendation.

Furthermore, the content-based component of the approach has been applied in a real scenario, the mediaUPV portal of the UPV from October 2019 to March 2021. This portal is mainly used by learners and trainers to access to useful LOs for their MOOCs and flipped classrooms. We can state that the application of this LRS to the mediaUPV portal was positive as it improved the precision of the original experimental results of [9] (from 7.7% to 9.85%), it brought an increase in visits to the videos, and it had a significant CTR of 1.28% on average, with notable peaks of over 4%.

The results of our LRS must be seen in the context of its application. In this respect, it should be remembered that we are dealing with videos of university lectures or subject-specific lessons. This means that users of the system do not usually enter for leisure purposes as on YouTube or Netflix, or to make purchases as on platforms such as Amazon. Thus, users of this system usually enter to watch a specific lesson of the subjects they are studying, or to search for a specific video about a particular topic. Therefore, it is difficult for a recommender to obtain better results than the ones we show. In addition, it is important to highlight what has been observed with the results in production with respect to the time that elapses between the recommendation and the moment in which the users watch one of the recommended videos. In this sense, it has been shown (Figure 8) that the precision of the recommender almost doubles if instead of considering the 30 min after the recommendation we consider 120 min or more.

As future work, we want to analyse the results in production of both components working together, the collaborative filtering and content-based one. Thus, the collaborative filtering component presented in this work is able to recommend videos that have no transcript, which opens up more possibilities to increase the serendipity of the recommendations. In addition, it would be interesting to evaluate whether to add the classification of the mediaUPV videos obtained by [21] to the current characterization of the videos used by our proposal, which is currently based on the video transcript and collaborative filtering.

We also want to test if we achieve better results by changing the term frequency–inverse document frequency (TF-IDF) algorithm in the content-based module using other techniques such as delta TF-IDF [27] or TF.IDF.ICF [28], which try to avoid the TF-IDF problem of not considering intraclass or interclass distributions. In addition, we need to further study if it will be possible to apply other variant weighting approaches of TF-IDF, such as the presented in [29], in which the number of occurrences of a term, the number of documents that include the term, and the number of classes in which the term appears, are used to obtain a more accurate set of characteristic features.

Finally, we would like to conduct random user surveys to analyse the precision of the recommender in a less automatic way and to get direct feedback from users. In this way, we could know better the precision of the recommendations as the users could answer if these recommendations are useful instead of basing them on whether they have seen the video or not. This would avoid the uncertainty of the current analysis in which we do not know for sure if users do not find the recommendations useful or if they only enter the platform to watch the content they need without paying attention to any recommendation, whether it is useful or not.

Author Contributions: Conceptualization, J.J., S.V. and V.B.; Data curation, J.J. and C.T.; Formal analysis, J.J. and S.V.; Funding acquisition, V.B.; Investigation, J.J., S.V. and C.T.; Methodology, V.B.; Project administration, V.B.; Resources, C.T.; Software, J.J.; Supervision, C.T. and V.B.; Validation, J.J., S.V. and C.T.; Writing—original draft, J.J. and S.V.; Writing—review & editing, J.J., S.V., C.T. and V.B. All authors have read and agreed to the published version of the manuscript.

Funding: This research was partially supported by MINECO/FEDER RTI2018-095390-B-C31 and TIN2017-89156-R projects of the Spanish government, and PROMETEO/2018/002 project of Generalitat Valenciana.

Conflicts of Interest: The authors declare no conflict of interest.

Abbreviations

The following abbreviations are used in this manuscript:

CB	Content-Based Recommender Modules
CF	Collaborative Filtering Module
CTR	Click-Through Rate
HR	Hybrid Recommender
IB	Item-Based Recommender Module
LMS	Learning Management System
LRS	Learning Recommender System
LO	Learning Object
MOOC	Massive Online Open Course
PB	Profile-Based Recommender Module
TF-IDF	term frequency—inverse document frequency
UPV	Universitat Politècnica de València

References

1. Maassen, P.; Nerland, M.; Yates, L. (Eds.) *Reconfiguring Knowledge in Higher Education*; Higher Education Dynamics, 50; Springer International Publishing: Cham, Switzerland, 2018.
2. Zajda, J.; Rust, V. *Globalisation and Higher Education Reforms*; Springer: Cham, Switzerland, 2016; Volume 15.
3. van Dijck, J.; Poell, T. Higher Education in a Networked World: European Responses to U.S. MOOCs. *Int. J. Commun. IJoC* **2015**, *9*, 2674–2692.
4. Institute and Committee of Electrical and Electronics Engineers; Learning Technology Standards. *IEEE Standard for Learning Object Metadata*; IEEE Standard 1484.12.1; IEEE: Piscataway, NJ, USA, 2002.
5. O'Flaherty, J.; Phillips, C. The use of flipped classrooms in higher education: A scoping review. *Internet High. Educ.* **2015**, *25*, 85–95. [CrossRef]
6. Roehl, A.; Reddy, S.L.; Shannon, G.J. The flipped classroom: An opportunity to engage millennial students through active learning strategies. *J. Fam. Consum. Sci.* **2013**, *105*, 44–49. [CrossRef]
7. Tucker, B. The Flipped Classroom. Online instruction at home frees class time for learning. *Educ. Next* **2012**, *2012*, 82–83.
8. Turró, C.; Morales, J.C.; Busquets-Mataix, J. A study on assessment results in a large scale Flipped Teaching Experience. In Proceedings of the 4th International Conference on Higher Education Advances (HEAD'18), Valencia, Spain, 20–22 June 2018; pp. 1039–1048.
9. Jordán, J.; Valero, S.; Turró, C.; Botti, V. Recommending Learning Videos for MOOCs and Flipped Classrooms. In *Advances in Practical Applications of Agents, Multi-Agent Systems, and Trustworthiness*; Demazeau, Y., Holvoet, T., Corchado, J.M., Costantini, S., Eds.; The PAAMS Collection; Springer International Publishing: Cham, Switzerland, 2020; pp. 146–157.
10. Klašnja-Milićević, A.; Ivanović, M.; Nanopoulos, A. Recommender systems in e-learning environments: A survey of the state-of-the-art and possible extensions. *Artif. Intell. Rev.* **2015**, *44*, 571–604. [CrossRef]
11. Burke, R. Hybrid Recommender Systems: Survey and Experiments. *User Model. User-Adapt. Interact.* **2002**, *12*, 331–370. [CrossRef]
12. Herlocker, J.; Konstan, J.; Terveen, L.; Riedl, J. Evaluating collaborative filtering recommender systems. *ACM Trans. Inf. Syst.* **2004**, *22*, 5–53. [CrossRef]
13. Rodríguez, P.; Heras, S.; Palanca, J.; Duque, N.; Julián, V. Argumentation-Based Hybrid Recommender System for Recommending Learning Objects. In *Multi-Agent Systems and Agreement Technologies*; Springer International Publishing: Cham, Switzerland, 2016; pp. 234–248.
14. Bobadilla, J.; Serradilla, F.; Hernando, A. Collaborative filtering adapted to recommender systems of e-learning. *Knowl.-Based Syst.* **2009**, *22*, 261–265. [CrossRef]
15. Zapata, A.; Menéndez Domínguez, V.; Prieto, M.; Romero, C. A Hybrid Recommender Method for Learning Objects. *Int. J. Comput. Appl. (DEDCE)* **2011**, *1*, 1–7.

16. Chen, W.; Niu, Z.; Zhao, X.; Li, Y. A Hybrid Recommendation Algorithm Adapted in E-Learning Environments. *World Wide Web* **2014**, *17*, 271–284. [CrossRef]
17. Tarus, J.K.; Niu, Z.; Yousif, A. A hybrid knowledge-based recommender system for e-learning based on ontology and sequential pattern mining. *Future Gener. Comput. Syst.* **2017**, *72*, 37–48. [CrossRef]
18. Vanitha, V.; Krishnan, P.; Elakkiya, R. Collaborative optimization algorithm for learning path construction in E-learning. *Comput. Electr. Eng.* **2019**, *77*, 325–338. [CrossRef]
19. Dwivedi, P.; Kant, V.; Bharadwaj, K.K. Learning Path Recommendation Based on Modified Variable Length Genetic Algorithm. *Educ. Inf. Technol.* **2018**, *23*, 819–836. [CrossRef]
20. Dwivedi, P.; Bharadwaj, K.K. e-Learning recommender system for a group of learners based on the unified learner profile approach. *Expert Syst.* **2015**, *32*, 264–276. [CrossRef]
21. Stoica, A.S.; Heras, S.; Palanca, J.; Julian, V.; Mihaescu, M.C. A Semi-supervised Method to Classify Educational Videos. In *Hybrid Artificial Intelligent Systems*; Springer International Publishing: Cham, Switzerland, 2019; pp. 218–228.
22. Turcu, G.; Heras, S.; Palanca, J.; Julian, V.; Mihaescu, M.C. Towards a Custom Designed Mechanism for Indexing and Retrieving Video Transcripts. In *Hybrid Artificial Intelligent Systems*; Springer International Publishing: Cham, Switzerland, 2019; pp. 299–309.
23. Blei, D.M.; Ng, A.Y.; Jordan, M.I. Latent Dirichlet Allocation. *J. Mach. Learn. Res.* **2003**, *3*, 993–1022.
24. MLLP Research Group (Universitat Politècnica de València). TLP: The transLectures-UPV Platform. Available online: https://www.mllp.upv.es/tlp/ (accessed on 14 May 2021).
25. Salton, G.; Buckley, C. Term-weighting approaches in automatic text retrieval. *Inf. Process. Manag.* **1988**, *24*, 513–523. [CrossRef]
26. Richardson, M.; Dominowska, E.; Ragno, R. Predicting clicks: Estimating the click-through rate for new ads. In Proceedings of the 16th International Conference on World Wide Web, Banff, AB, Canada, 8–12 May 2007; pp. 521–530.
27. Martineau, J.; Finin, T. Delta tfidf: An improved feature space for sentiment analysis. In Proceedings of the International AAAI Conference on Web and Social Media, San Jose, CA, USA, 17–20 May 2009; Volume 3.
28. Ren, F.; Sohrab, M.G. Class-indexing-based term weighting for automatic text classification. *Inf. Sci.* **2013**, *236*, 109–125. [CrossRef]
29. Kansheng, S.; Jie, H.; Liu, H.T.; Zhang, N.T.; Song, W.T. Efficient text classification method based on improved term reduction and term weighting. *J. China Univ. Posts Telecommun.* **2011**, *18*, 131–135.

 *electronics*

Article

Forecasting Energy Consumption of Wastewater Treatment Plants with a Transfer Learning Approach for Sustainable Cities

Pedro Oliveira *, Bruno Fernandes , Cesar Analide and Paulo Novais

ALGORITMI Centre, Department of Informatics, University of Minho, 4710-057 Braga, Portugal;
bruno.fmf.8@gmail.com (B.F.); analide@di.uminho.pt (C.A.); pjon@di.uminho.pt (P.N.)
* Correspondence: poliveira199208@gmail.com

Abstract: A major challenge of today's society is to make large urban centres more sustainable. Improving the energy efficiency of the various infrastructures that make up cities is one aspect being considered when improving their sustainability, with Wastewater Treatment Plants (WWTPs) being one of them. Consequently, this study aims to conceive, tune, and evaluate a set of candidate deep learning models with the goal being to forecast the energy consumption of a WWTP, following a recursive multi-step approach. Three distinct types of models were experimented, in particular, Long Short-Term Memory networks (LSTMs), Gated Recurrent Units (GRUs), and uni-dimensional Convolutional Neural Networks (CNNs). Uni- and multi-variate settings were evaluated, as well as different methods for handling outliers. Promising forecasting results were obtained by CNN-based models, being this difference statistically significant when compared to LSTMs and GRUs, with the best model presenting an approximate overall error of 630 kWh when on a multi-variate setting. Finally, to overcome the problem of data scarcity in WWTPs, transfer learning processes were implemented, with promising results being achieved when using a pre-trained uni-variate CNN model, with the overall error reducing to 325 kWh.

Keywords: deep learning; energy consumption; sustainable cities; transfer learning; wastewater treatment plants

Citation: Oliveira, P.; Fernandes, B.; Analide, C.; Novais, P. Forecasting Energy Consumption of Wastewater Treatment Plants with a Transfer Learning Approach for Sustainable Cities. *Electronics* **2021**, *10*, 1149. https://doi.org/10.3390/electronics10101149

Academic Editors: Juan M. Corchado, Josep L. Larriba-Pey, Pablo Chamoso and Fernando De la Prieta

Received: 31 March 2021
Accepted: 3 May 2021
Published: 12 May 2021

Publisher's Note: MDPI stays neutral with regard to jurisdictional claims in published maps and institutional affiliations.

1. Introduction

Over the years, there has been an increase in global urbanisation through a greater concentration of people in small spaces. According to the World Urbanisation Perspectives report carried out in 2017 by the United Nations on the number of people living in urban and rural areas worldwide, it was found that 4.1 billion people already lived in urban areas [1]. In fact, cities have a fundamental role in sustainable development, namely related to economic and environmental concerns.

With the increase in energy consumption, concerns about the energy sector have expanded substantially. Although there has been a greater awareness of the impact of non-renewable energy sources on the planet and the high emission of greenhouse gases, if concrete and imperative measures are not applied, this problem will only worsen. Thus, over the years, the term energy efficiency has become increasingly important and indispensable. Energy efficiency can help reduce energy production and, consequently, reduce greenhouse gas emissions and preserve fossil fuel resources, ensuring a notable contribution to reducing environmental problems on our planet [2].

There are several infrastructures where energy consumption is high in a city, with Wastewater Treatment Plants (WWTPs) being one of them. In a WWTP, achieving a high energy efficiency level has become an increasingly important topic [3]. WWTPs, with the execution of their functions, demand high levels of energy, reflecting about 7% of all energy consumed worldwide [4]. In Portugal, about 4% of the consumed electricity is urban water cycle's responsibility, with approximately 25% of that energy being used in WWTPs [5].

Reducing energy consumption, emission of greenhouse gases and operating costs has been one of the main concerns of WWTP managers, who have been adopting more efficient equipment and technologies [6,7]. Hence, a WWTP must always consider the efficient management of all its resources, including energy.

Currently, in most WWTPs, low levels of energy efficiency performance are found. In fact, several factors influence the consumed energy in this type of facilities, depending on their characteristics and the types of treatments being applied. In general, the lack of energy efficiency is due to [8]:

- A growing need for water recycling due to the scarcity of this resource;
- Types of motors and pumps that are used;
- Higher requirements on discharge parameters in the treated effluent;
- Water pumping processes, which require high energy consumption;
- Absence of energy recovery mechanisms;
- Low efficiency in operations, mixing, and aeration systems;
- Influent flow.

1.1. State of the Art

A study carried out by Li et al. [9] aimed at predicting energy consumption in a WWTP through the use of a Radial Basis Function (RBF) neural network. To evaluate the conceived models, they compared these with a Multi-variate Linear Regression (MLR) model. The data were based on a WWTP located in China, with daily periodicity. The data collected corresponded to 360 records, between December 2015 and December 2016, with six invalid records removed. To decide which features were given as input to the model, the authors used the Fuzzy C-Means (FCM) method. This method identified three indicators: the influential charge, the Chemical Oxygen Demand (COD), and the total nitrogen removed. The authors defined the FCM hyperparameters without any search for the best value for each of them, such as the number of iterations or clusters. Each of these selected indicators was used, one at a time, as input to the RBF model. The authors used min relative error, max relative error, and mean absolute percentage error (MAPE) for performance measurement metrics. In total, the authors developed four models with different inputs, three of them for each set of selected indicators and another with the total data. Using only data from each indicator's subset, the RBF model performed better than the MLR model. On the contrary, the MLRM model performed better when using the total dataset as input. Overall, both models performed better when using only the data subset of the indicators.

Harrou et al. [10] conducted a study to make short-term forecasts of energy consumption in a WWTP, using statistical and Deep Learning (DL) models. The data used in this study are between 2010 and 2017, belonging to a WWTP in Saudi Arabia. In total, the authors used six statistical models, such as the Auto-regressive Integrated Moving Average (ARIMA) or the Ordinary Least Square (OLS). Two types of networks were based on DL models, Long Short-Term Memory (LSTM) and Gated Recurrent Units (GRU). The models conceived by the authors used a uni-variate approach, where only the feature they intend to forecast, the energy consumption, is given as input to the different candidate models. The data were normalised between 0 and 1 for all conceived models. There was no particular attention to the case of LSTM networks working internally with a hyperbolic tangent. Throughout the manuscript, no cross-validation or overfitting control techniques are mentioned in the conceived models. Regarding the evaluation metrics of the models, the authors used four, i.e., MAPE, Mean Absolute Error (MAE), Root Mean Square Error (RMSE), and Root Mean Squared Log Error (RMSLE). By observing the obtained results, the authors verified that the statistical-based models slightly outperformed the DL models, with ARIMA getting a MAPE of 2.29%, while the best DL model, LSTMs, presented a MAPE of 2.42%. The authors also verified that the models' parameters were updated recursively, given a better performance than the models with no updates. However, they concluded that the DL models could provide forecast results with more significant performance when applying more data. No reference was made regarding fitting times.

The study carried out by Huang et al. [11] had as objective the construction of an energy consumption model in a WWTP based on Elman Neural Network-Energy Consumption Model (ENN-ECM) to identify the relationship between energy consumption and the quality of the effluent. The benchmark simulation model (BSM1) was used to compare the authors' model results. Both models were based on data related to an activated sludge model, being obtained from BSM1, which provided data for a period of two weeks in 15-minutes time intervals. Firstly, the authors used the energy consumption model to verify which effluent characteristics had a more significant relationship with the characteristics related to energy consumption. Then, they implemented the ENN-ECM with five characteristics of the effluent obtained from the energy consumption model to forecast four energy consumption parameters. The network architecture, namely the number of layers, was obtained through empirical formulas and the Kolmogorov theorem. The authors concluded that the ENN-ECM model obtained better performance concerning energy consumption with the analysis of the obtained results.

Ramli et al. [12] conducted a study to forecast energy consumption in a WWTP in Malaysia using an ARIMA model. To compare the obtained results, the authors used a linear regression method. The data used in this study were based on four years of active power in the WWTP. To achieve the best ARIMA model, the authors used the Time Series Modeler, incorporated in the SPSS software, obtaining the values (0, 1, 0) for ARIMA's parameters. The results allowed the authors to verify that the ARIMA model obtained better performance than the linear regression with an RMSE of 55.59, compared to 67.51, respectively. The authors further concluded that it was possible to increase energy efficiency by 10% of energy recovery, which could reduce the cost of electricity in the studied WWTP.

Another study carried out by Maki et al. [13] aimed to forecast the total energy consumption of a WWTP and the consumption in different processes, using a Markov switching model. The data were collected by applying several sensors connected to a WWTP energy distribution network in Japan and transmitted over a 3G line. The data collection was carried out between March 2015, and March 2017, with a 1-min periodicity. The authors then grouped the data into an hourly periodicity. In addition to the forecast of total energy consumption at the WWTP, the authors also forecast the energy consumption in the water treatment, sludge treatment, and auxiliary facilities processes. Additionally, as the sum of the three identified processes' energy consumed did not coincide with the total energy consumed in the WWTP, they made the forecast for the remaining operations, marked as "others". An analysis was made of energy consumption over time, where it was possible to verify that there is greater energy consumption in summer than winter. In addition to the data collected by the sensors, the authors added six more features to be used in the conceived model: holidays, office hours, temperature, humidity, wind speed, and the previous five hours of energy consumption. Only 1 week was considered as input. With the obtained results, the authors found that, except for the sludge treatment and auxiliary facilities, the values were below 10%. Besides, the relationships between the variables that affect the energy consumption forecast equation were verified in each process. The authors then concluded that an increase in the WWTP's energy consumption, together with the increase in seasonal temperatures, leads to a rise between 0.1% and 0.2% for each 1 °C in temperature.

Oulebsir et al. [14] conducted a study where they conceived an Artificial Neural Network (ANN) to create an energy consumption model in a WWTP using the active sludge process. The authors used data provided by a WWTP in Algeria between January 2006 and March 2016. In this study, the authors use four parameters: (1) the Biological Oxygen Demand (BOD_5), (2) the COD, (3) suspended solids and (4) ammonium. In addition, they also use the water temperature, and flow of the influent, the flow of recirculated sludge, and the total consumed energy. The authors applied a set of methods to clean the dataset, keeping 318 days of observations even though the original dataset had 10 years of data. The different ANNs had six hidden layers with a total of 200 neurons each. The architecture of the models was established using the trial-and-error method. In each conceived model, data were divided into 80% for training and 20% for testing, without using a time series cross-validator. The authors confirmed that the pollution load contributes more significantly to forecasting energy consumption than the removal efficiency. The authors also applied the k-means algorithm, observing three clusters. The authors were thus able to verify three classes of energy consumption: under-consumption, over-consumption, and optimal consumption.

As an overall conclusion, it can be said that some studies have already considered the use of DL models to forecast energy consumption in a WWTP. Typically, studies follow a single-step approach, i.e., they only forecast consumption value for the next day. Furthermore, it is usual to find studies that do not consider certain aspects of time series problems, such as using an appropriate cross-validator, not breaking the time series when removing missing values or missing timesteps, or even when searching the best hyperparameters. In addition, it is not easy to understand the existence of overfitting as learning curves are not analysed. All this may lead to significant problems when deploying the best candidate model in a real-life scenario.

1.2. Goals, Research Questions, and Paper Structure

This work aims to conceive, tune, and evaluate a set of candidate DL models to forecast energy consumption in a WWTP, going from recurrent to convolutional candidates. In addition, the goal is to implement a recursive multi-step approach to forecast the next two days, providing a stronger understanding of future patterns. We also aim to experiment two different methods for outliers' handling and the performance of the candidates in uni- and multi-variate settings. Then, as last goal, we aim to evaluate the best candidate model in a WWTP with a low volume of data. For that, we are required to apply transfer learning processes, overcoming the problem of data scarcity.

This study uses data provided by a Portuguese water company. The elicited goals can be translated into the following research questions:

1. Do recurrent neural networks have a better performance when forecasting energy consumption in a WWTP than convolutional ones?
2. Which features facilitate the process of forecasting energy consumption in a WWTP?
3. Is it possible to apply transfer learning processes, with the goal being to use a pre-trained model to forecast the energy consumption of a WWTP with low volumes of data?

The remainder of this manuscript is structured in three more sections. Section 2 describes the materials and methods, namely the collection, exploration, and pre-processing of data, the developed DL models, and the conducted experiments. Section 3 is responsible for summarising the obtained results, as well as their interpretation. Finally, Section 4 discusses the obtained results and gathers the conclusions drawn from this study.

2. Materials and Methods

The following lines describe the materials and methods used throughout this study, including collecting, exploring, and treating data. Additionally, the models used throughout the work are described, as well as the evaluation metrics, the used technologies, and the designed experiments.

2.1. Dataset

The data used in this study took into account three different datasets. Dataset one was related to energy consumption while the second dataset described the volume of the flow of water at the entrance of a WWTP. The third dataset described the climatological conditions. The first two datasets were made available by a Portuguese wastewater company and were related to a single WWTP. Regarding the energy consumption value, which is the target feature, there is an intrinsic relationship between the different processes present in a WWTP and the required energy (typically, the larger the WWTP, the greater its energy consumption). However, this relation was captured and described in the time series in itself as the values were a snapshot of the state of the WWTP. The third dataset was collected using the Open Weather Map API, and contains climatological data regarding the same city where the WWTP was located. All datasets contained observations belonging to the period between January 2016 to May 2020.

Figure 1 illustrates the WWTP layout used in this study. This WWTP was based on four main stages: preliminary, primary, secondary and tertiary treatments. In addition, there was also a line responsible for the sludge treatment. The preliminary treatment, which included bar screening, was accountable for removing solids and materials of greater volume, an essential step in the WWTP process since some of these objects could damage some equipment in the following steps. The primary treatment, which included the primary classifier, aimed to remove the smaller volume solids, namely the suspended solids, from the previous stage and the organic matter present. In the secondary treatment, two processes were included, the aeration tank and the secondary classifier. This stage aimed to remove biodegradable organic matter from wastewater, in addition to suspended solids and nutrients, such as nitrogen. Finally, the tertiary treatment was responsible for removing the remaining suspended solids resulting from the previous stages. The sludge produced in the primary and secondary treatment was inserted in the sludge treatment line. This line was responsible for dewatering and disinfecting the sludge, reusing it as an energy source.

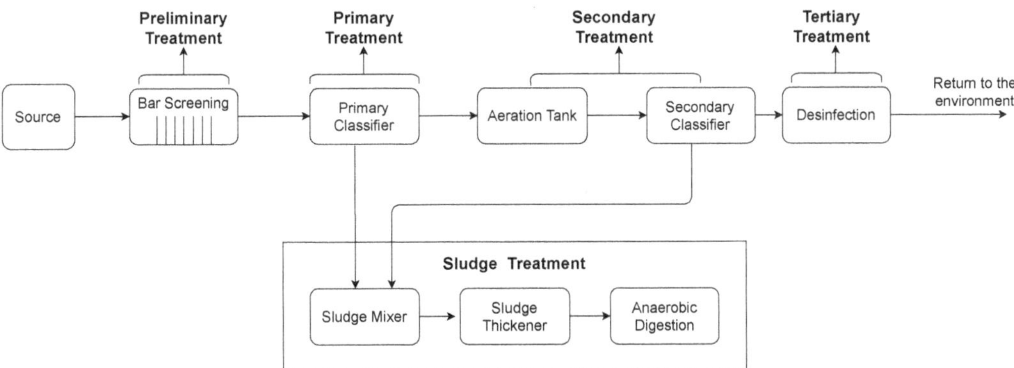

Figure 1. WWTP (Wastewater Treatment Plants) layout.

2.1.1. Data Exploration

The energy consumption dataset comprised two features: the energy consumption value (in kWh) and the corresponding timestamp, making 1522 records with a daily periodicity. The influent flow dataset also contained two features, i.e., the value of the influent flow (in m^3) and the timestamp, with a total of 1535 records, again with a daily periodicity. Finally, the climatological dataset had a total of 25 features, including the timestamp, air temperature, and humidity, among others, with a total of 38,651 hourly timesteps. Table 1 presents the different features available in the three datasets, detailing its characteristics and presenting the corresponding units of measure.

None of the three datasets had missing values. However, as in its genesis the problem identified in this study was based on a time series problem, it was essential to pay attention to missing timesteps. In the case of the climatological dataset, there were no missing timesteps. On the contrary, both the energy consumption and the influent inflow datasets contained missing timesteps. In the former, there were 88 missing timesteps, while in the latter 75 missing timesteps were identified. In a subsequent section, it is explained how to overcome the missing timesteps problem.

As the main goal of this study was to forecast energy consumption, data exploration emphasized the *value_energy* feature of the energy consumption dataset. Firstly, it is worth mentioning that this feature presented an accumulated value. Hence, it was necessary to subtract, from each observation, the value of the previous one, in order to obtain its real value. Since the first observation had no previous one, it was removed. A box plot analysis allowed us to identify the existence of some extreme outliers that were derived from an incorrect insertion of values by the operators of the WWTP.

Table 1. Features available in the used datasets. Only the main features of the climatological dataset are presented.

#	Features	Description	Unit
		Energy Consumption Dataset	
1	*date*	Timestamp	date and time
2	*value_energy*	Total energy consumption	kWh
		Influent Flow Dataset	
1	*date*	Timestamp	date and time
2	*flow*	Accumulated influent flow value	m^3
		Climatological Dataset	
1	*dt_iso*	Timestamp	date and time
2	*temp*	Temperature	°C
3	*feels_like*	Human perception of climate	°C
4	*temp_min*	Minimum temperature	°C
5	*temp_max*	Maximum temperature	°C
6	*pressure*	Atmospheric pressure	hPa
7	*humidity*	Humidity percentage	%
8	*wind_speed*	Wind speed value	m/s
9	*wind_deg*	Wind direction	Degrees
10	*rain*	Rain volume	mm
11	*clouds_all*	Cloudiness percentage	%

A statistical analysis of the energy consumption values was performed, being described in Table 2. It was possible to verify that the mean energy consumption value in the dataset presents a value of 8050.96 kWh, with a standard deviation of 3736.359 kWh. The skewness was 3.172, representing an asymmetric distribution, i.e., the positive value indicates a positive inclination in the distribution of the data, in which the tail size of the right hand is larger than that of the left. Regarding the kurtosis value, it was 28.101. A kurtosis value greater than 1 indicates that the distribution of energy consumption has a very high peak (a leptokurtic distribution).

Table 2. Descriptive statistics for energy consumption.

Number of Items	Mean	Median	Std. Deviation	Skewness	Kurtosis
1522	8050.960	7689	3736.359	3.172	28.101

We then explored the energy consumption over the months of a year, during the 5 years present in the dataset. In Figure 2 it is possible to verify a pattern in all the explored years, with a constant drop in energy consumption between July and August.

Another analysis took into account the variation in energy consumption over the different days of the week. This analysis was based on the mean value of the days of the week for each year. As shown in Figure 3, it is possible to verify that Sunday and Monday were the days when there was less energy consumption in the WWTP. In conclusion, it appears that the traditional working days had a higher energy consumption on average, while on weekends there was a decrease.

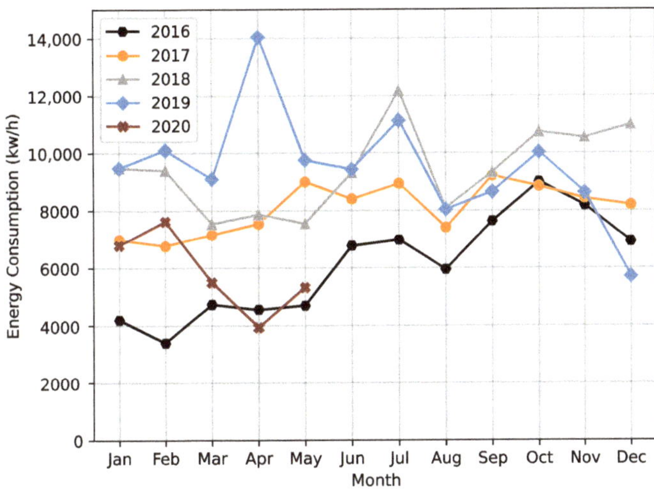

Figure 2. Monthly variation of energy consumption over the years present in the dataset.

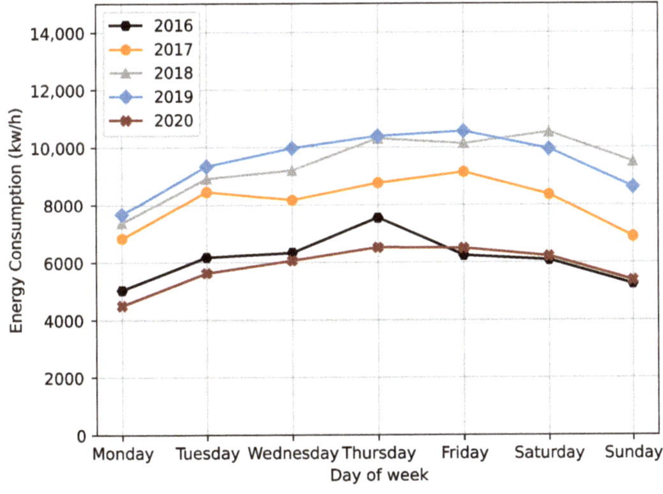

Figure 3. Day of the week variation of energy consumption over the years.

To understand seasonality, we performed two different analyses on the energy consumption data between 2016 and 2019: the first relative to the average consumption by season and the second related to the energy consumption per trimester. Figure 4 depicts the first analysis, being possible to verify that, typically, more energy was consumed during the autumn. Interestingly, in 2019, autumn was the season with the lowest average energy consumption value. In general, it was also possible to see that over the years, energy consumption was rising in different seasons. Despite a higher number of average consumption values, it was not in the autumn that the highest average peak was reached, but in the spring of 2019 with a value of 10,912 kWh. Regarding the lowest peak, it occurred in the winter of 2016, with a value of 4398 kWh. Additionally, it was possible to verify that, in general, winter was the season with less consumption of energy.

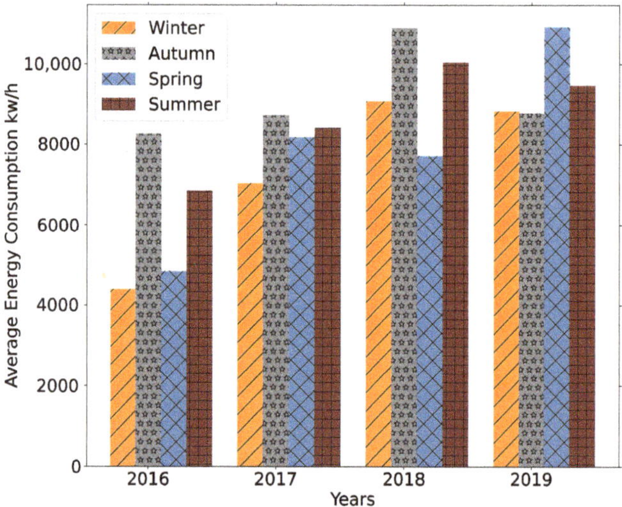

Figure 4. Mean energy consumption per seasons of the year.

The trimesters analysis showed that the fourth trimester had the highest energy consumption values over the first three years. Despite this, the highest value was verified in the second trimester of 2019, with 11,072 kWh. As demonstrated in the seasons' analysis, in general, the average values increased during the first three years. In 2019, there was an increase in the first and second trimester and a decrease in the third and fourth ones.

Regarding the influent flow, an analysis was carried out considering the average for each year, described in Table 3. As can be seen, 2019 was the year with the highest volume of influent flow on the WWTP (1155.33 m^3). Interestingly, checking the year of 2019 concerning the energy consumption (Figure 2), we verified that this year also obtained, in general, the highest average of energy. On the other hand, looking at 2016, excluding the incomplete year of 2020, this was where the lowest average influent flow value occurred, this being, in general, the year with the lowest energy consumption value.

Table 3. Average influent flow per year.

Year	Value (m^3)
2016	910.69
2017	1025.23
2018	981.26
2019	1155.33
2020	849.19

2.1.2. Data Preparation

The first step to prepare the data were to carry out a feature engineering process in the three datasets, thus creating three new features from the timestamps (i.e., *year*, *month*, and *day*). The dataset related to climatological data, as mentioned, had an hourly periodicity, so to match the same periodicity as the other datasets, these were grouped by day, month and year, aggregating the mean value per feature.

As referred above, as both the energy consumption and influent flow datasets presented accumulated values, a method was applied to obtain the value that would correspond to each specific day. The identified extreme outliers, which corresponded to miss insertions of values by the operators of the WWTP (for example, extra digits), were also solved. The remainder of the data treatment is specified in the following lines.

Handling Missing Timesteps

To deal with the missing timesteps verified in the energy consumption and the influent flow datasets, a dataset was created comprising all days (i.e., timesteps) that should have been present in the dataset. In both cases, the start date was 2nd January 2016 and the end date 28 May 2020. The datasets were joined, with missing timesteps being added and having its features filled with the −99 value. Solving the missing timesteps problem created a new one, missing values, i.e., timesteps that were missing were now present but all their features had the −99 value.

Handling Missing Values

To fill the missing values, a queue-based approach was followed. Each record was read for each of the two datasets with missing values, saving its value (energy consumption or influent flow) in the mentioned structure, with a maximum size of eight values. Whenever reading a record, if the queue was full, a push operation would be performed at the beginning of the queue. When a timestep had a feature with the −99 value, its value would be computed based on the average of the last eight records, i.e., the previous 8 days, present in the queue. Once calculated, this value would then be pushed to the queue, eliminating the oldest record. By the end of this process, no dataset had missing values neither missing timesteps.

Joining Datasets

When reaching this point, each one of the three datasets was made of 1609 observations. However, we were required to join the three datasets into a single one. This was performed using the features *year*, *month*, and *day*. In the end, a single dataset was created, having 1609 observations with 30 features each.

Correlation Analysis

To verify which features had a more significant correlation with the target feature (*value_energy*), it was first necessary to check whether the data followed a normal distribution. Using a $p < 0.05$ and the Kolmogorov–Smirnov test, it was possible to verify that all features assumed a non-Gaussian distribution. Hence, it was necessary to use the non-parametric Spearman's rank correlation coefficient, being possible to verify that the features that had a more significant correlation with the target were the year, month, temperature, and *flow_value*. Since the other features had a low correlation with the target, they were removed. After this treatment, the final dataset had 1609 observations with a shape (1609, 5). Table 4 shows an example of a record in the final dataset.

Table 4. Features present in the final dataset.

#	Features	Observation Example
1	*year*	2018
2	*month*	5
3	*temperature*	11.96
4	*flow_value*	829
5	*value_energy*	5155

Handling Outliers

Extreme outliers were above 14,000 kWh. Only six observations were below 2000 kWh. Since the range between the maximum and minimum values for the feature *value_energy* was large, and considering the reduced amount of observations that were causing it, two different methods were experimented to handle outliers. These two methods provided a comparative term for the different experiments, causing slight modifications to the input data that were fed to the models. The two methods were as follows:

- Method 1—to further reduce the amplitude of the target feature, the few timesteps with *value_energy* greater than 10,000 kWh or lower than 2000 kWh had their value updated, using the queue-based approach described above. The goal was to use interpolation to replace the outliers;
- Method 2—to further reduce the amplitude of the target feature, the few timesteps with *value_energy* greater than 10,000 kWh or lower than 2000 kWh had their value truncated. The goal was not to use interpolation to update the target value.

Normalisation

With the data prepared, the next step was to normalize them. Since LSTMs work internally with the hyperbolic tangent, we decided that the applied normalization would be in the range $[-1, 1]$, according to the following equation:

$$\frac{x_i - min(x)}{max(x) - min(x)} \tag{1}$$

Supervised Problem

The final step was to go from an unsupervised problem to a supervised one, with the respective inputs (X) and corresponding labels (y). Thus, it was necessary to create sequences of data, which depend on the number of timesteps used as input for the models. A sliding window was used over the initial dataset to create the different sequences and the respective labels, thus creating a set of sequences that can be fed to the models. As an example, if the shape of a model's input was (1601, 7, 5), the first element set the number of samples, the second the number of input timesteps, and the last the number of features. In this example, the labels would have the shape (1601, 1). A similar algorithm can be seen in the work of Fernandes et al. [15].

2.2. Model Conception

To achieve the objective of forecasting energy consumption in a WWTP, three different DL models were conceived and evaluated, namely LSTMs, GRUs, and uni-dimensional Convolutional Neural Networks (CNNs). Regarding the choice of models, concerning the LSTM and GRU models, these were selected since they belong to the set of Recurrent Neural Networks (RNNs), which has shown an outstanding performance in time series problems. While traditional ANNs cannot remember what they learned in previous iterations, RNNs can learn from earlier timesteps [16–19]. Regarding the choice of CNNs as the third model to be used, despite its greater use in image processing, it has shown promising results in terms of time series problems when using uni-dimensional convolutions [20–23].

To find the best combination of hyperparameters, two error metrics were used. The RMSE is an error measure, as it measures the difference between the values predicted by the model ($\hat{y}$) and the true values observed (y). RMSE equation is as follows:

$$\text{RMSE} = \sqrt{\frac{\sum_{i=1}^{n}(y_i - \hat{y}_i)^2}{n}} \tag{2}$$

The second metric, the MAE, is the mean of the differences between predicted and observed values. Its use is mainly to complement and strengthen the confidence on the obtained values. Its equation is as follows:

$$\text{MAE} = \frac{1}{n}\sum_{i=1}^{n}|y_i - \hat{y}_i| \tag{3}$$

2.2.1. LSTMs

One of the models used in this study was based on a particular RNN, i.e., LSTMs. RNNs are a type of network that, unlike ANNs, can have as input the current input and pay attention to past inputs [24,25]. In other words, the decision taken on the timestep $t-1$ will affect the timestep t. LSTMs, introduced in 1997 by Hochreiter and Schmidhubber [26], can learn temporal dependencies over a long period, in addition to the short term. These networks came to fill an existing problem in RNNs, where there was an exponential drop in the backpropagated error in long periods. Nowadays, LSTMs are widely used in forecasting problems, such as in road traffic or weather, and their use in detecting anomalies in time series problems [27–31].

Regarding the architecture of LSTMs, it consists of multiple memory cells. There are two states in each of these memory cells: the hidden state and the cell state. The hidden state, already present in RNNs, is responsible for short-term memory, while on the other hand, the cell state (not present in RNNs) has the capacity for long-term memory. Additionally, each memory cell has internal gates, which allow a LSTM to forget (f_t), include (i_t), and output (o_t) information [26]. The following equations describe the calculation performed on each of the gates.

$$i_t = \sigma(w_i[h_{t-1}, x_t] + b_i) \tag{4}$$

$$f_t = \sigma(w_f[h_{t-1}, x_t] + b_f) \tag{5}$$

$$o_t = \sigma(w_o[h_{t-1}, x_t] + b_o) \tag{6}$$

where σ represents the sigmoid function, w_x the weight for the respective gate, h_{t-1} the output of the previous block, x_t input at current timestep and b_x the biases for the respective gate.

First, through the sigmoid layer, it is necessary to decide which information will leave the cell state (forget gate) and remain the same. The action on what will keep information is divided into two stages, the first deciding which values should be updated through another sigmoid layer (input gate) and the second creating a vector of new deals that can add to the state through a hyperbolic tangent layer. The next cell state update is obtained through a point multiplication operation on the two previous steps results. Finally, the output is decided using a sigmoid layer (output gate) followed by a hyperbolic tangent one [26]. Figure 5 provides a graphical view of such a memory cell. The following equations describe the calculation of the cell state, the candidate cell state and the final output.

$$\tilde{c}_t = tanh(w_c[h_{t-1}, x_t] + b_c) \tag{7}$$

$$c_t = f_t \times c_{t-1} + i_T \times \tilde{c}_t \tag{8}$$

$$h_t = o_t \times tanh(c^t) \tag{9}$$

where c_t represents the cell state at timestep t and $\tilde{c}_t$ represents the candidate for cell state at timestep t.

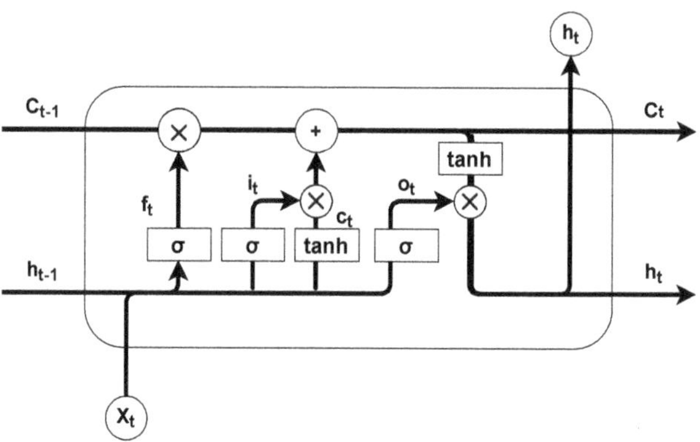

Figure 5. Architecture of a LSTM (Long Short-Term Memory) cell.

2.2.2. GRUs

Another model used in this study was the GRU. These networks are a subtype of RNNs, introduced in 2014 by Kyunghyun Cho [32]. Like LSTMs, GRUs were developed to solve the vanishing gradient problem of RNNs. GRUs are a simpler version of LSTMs, and they can be faster than these, obtaining similar performance. Unlike LSTMs, GRU cells only have the hidden state, which can maintain long and short term dependencies, thus eliminating the LSTM cell state. Another difference is that GRUs only have two layers of neural networks and have only two gates: reset (r_t) and update (z_t) [33]. The following equations describe the calculation performed on each of the gates.

$$z_t = \sigma(w_z.[h_{t-1}, x_t]) \tag{10}$$

$$r_t = \sigma(w_r.[h_{t-1}, x_t]) \tag{11}$$

The first step performed in a GRU cell is to represent the information removed by a sigmoid layer, from the previous hidden states, through the reset gate, working in a very similar way to the LSTM forget gate. Then, through the update gate, the amount of information from the previous timesteps is decided to be transmitted to the next state through a sigmoid layer. The next step uses the reset gate, applying a hyperbolic tangent layer, to introduce a new memory content, called the hidden state candidate. Finally, the update gate effect is incorporated to create the new hidden state [33]. GRUs are, like LSTM, widely used in forecasting problems in time series [34–36]. Figure 6 provides a graphical view of a GRU cell. The following equations describe the calculation of the current memory content and the final memory at current time step.

$$h_t = tanh(w.[r_t \times h_{t-1}, x_t]) \tag{12}$$

$$h_t = (1 - z_t) \times h_{t-1} + z_t \times \tilde{h}_t \tag{13}$$

where $\tilde{h}_t$ represents the current memory cell and h_t the vector which holds information for the current unit.

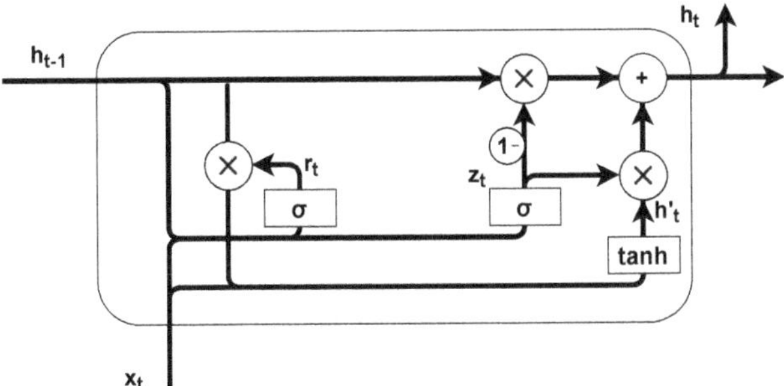

Figure 6. Architecture of a GRU (Gated Recurrent Units) cell.

2.2.3. CNNs

The last model used in this study was a CNN, a type of neural network developed a few decades ago [37,38]. Its appearance was based on a survey carried out by Hubel and Wiesel, in 1962, on the visual cortex of cats [39]. Over the past few years, CNNs has been closely linked to the classification of images and object detection [40,41]. In general, CNNs have a set of essential aspects: the convolutional layer, the pooling layer, and the fully connected one. Based on an image as an input, the convolutional layer is responsible for dividing the image's features, while the fully connected layer uses the output of the convolutional layer to classify. The pooling layer is used to reduce the amount of information coming from the convolutional one.

Recent times came with the use of CNNs for time series problems, mainly using uni-dimensional ones [21–23]. In the context of a time series problem, a significant aspect that needs to be taken into account is the approach being followed in terms of the data format, i.e., whether channels' last or channels' first. Concerning channels' last, this approach aims to reduce the number of timesteps while keeping the number of filters intact. On the other hand, the channels' first approach does just the opposite, i.e., reduces the number of filters and keeps the number of timesteps intact. Depending on the followed approach, this will always cause differences in the convolutional layer, which has the format $(timesteps, filters)$. The kernel size is yet another parameter responsible for defining the timesteps window length that is affected by each filter. An illustrative example of a channels' last approach can be seen in work of Oliveira et al. [23]. Finally, the form of calculating the shape of the output follows the following equation:

$$(Timesteps - KernelSize) + 1 \tag{14}$$

2.3. Experiments

Several experiments were carried out, taking into account different scenarios as shown in the next lines. The same random seed (91195003) was used in all conducted experiments.

2.3.1. Technologies

For data exploration, the *Knime* platform was used as well as the Python programming language, version 3.7. Python was also used for data pre-processing and for the development and evaluation of the DL models. *Pandas, NumPy, scikit-learn,* and *matplotlib* were the used libraries. In addition to these, *TensorFlow v2.0.0* was used to develop the models. Regarding the hardware, all of it was made available by Google's Colaboratory.

2.3.2. Experimental Setup

To achieve the goal of forecasting the energy consumption of a WWTP, it was necessary to evaluate multiple candidate models. All candidates were designed to follow a recursive multi-step approach, i.e., to forecast energy consumption for the next 2 days.

For each type of DL model used in this study, candidate models were designed based on an uni-variate and multi-variate approach. In the case of being uni-variate, the models would only receive, as input, the *value_energy* feature. In the multi-variate approach, three distinct scenarios were defined, with each scenario consisting in a different set of features. Table 5 summarises the features that each scenario contains. These scenarios are useful to understand the importance of temporal and climatological context data in the energy consumption of WWTPs. The influent flow is included in all multi-variate scenarios, since it had the highest correlation coefficient with the target feature.

Table 5. Uni- and multi-variate data scenarios.

	Uni-Variate
Scenario 1	*value_energy*
	Multi-Variate
Scenario 1	*value_energy, year, month, temperature, flow_value*
Scenario 2	*value_energy, temperature, flow_value*
Scenario 3	*value_energy, flow_value*

Two distinct datasets were built, one for each outliers' method. For each method, two approaches were followed: uni- and multi-variate. Then, for each approach, a set of scenarios were defined. Figure 7 sets the different combinations of data used to fit and evaluate the candidate models.

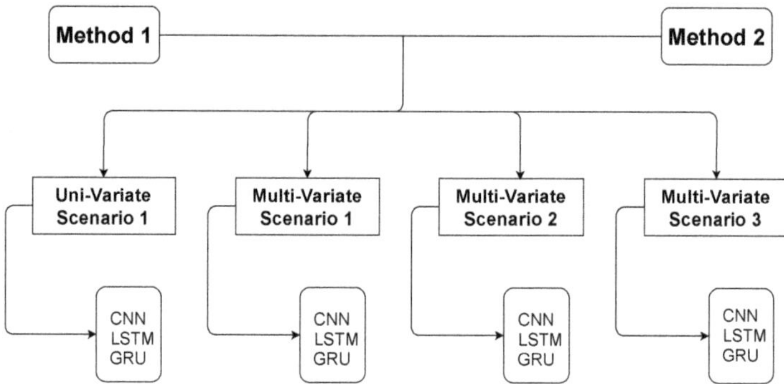

Figure 7. Different combinations for the conception of the candidate models.

The search for the best hyperparameters' configuration was performed using grid search. This method was applied to tune parameters such as the model architecture, batch size, or the number of timesteps that make an input sequence. Table 6 describes the hyperparameters' searching space considered for each model type. Besides, two callbacks were defined over the validation's loss. One aimed to automatically reduce the learning rate, while the other stopped the training when the RMSE stopped improving.

To prevent overfitting and underfitting situations, learning curves were plotted, stored, and analyzed. It should also be noted, taking into account that we were facing a time series problem, that a time series cross-validator was used ($k = 3$), namely the *TimeSeriesSplit* API of scikit-learn. This cross-validator, unlike traditional ones, had successive training sets as supersets of those that came before. Each of these training sets was further split into training and validation sets.

Table 6. Hyperparameters searching space.

Parameter	LSTM and GRU	CNN
Layers	[3, 4, 5]	[3, 4, 5]
Neurons	[32, 64, 128]	-
Activation	[ReLU, Tanh]	[ReLU, Tanh]
Timesteps	[14, 21, 28]	[14, 21, 28]
Batch Size	[5, 10, 20]	[5, 10, 20]
Dropout	[0.0, 0.5]	[0.0, 0.5]
Kernel Size	-	[3, 4, 5]
Filters	-	[16, 32]
Pool Size	-	[2, 3]

3. Results

Several hundred experiments were run in order to evaluate all possible candidate models. The candidates were evaluated considering their RMSE and MAE.

3.1. Method 1

The first method had the outliers updated as per the conceived queue-based approach. Table 7 presents the best hyperparameter configurations for each combination in this method. Within these combinations, it was possible to verify that the best one concerned CNNs for the third multi-variate scenario, with a MAE of 630 and a RMSE of 690 kWh.

Table 7. Best results, per scenario, for Method 1. The letters stand as follows: a. timesteps; b. batch size; c. number of layers; d. number of neurons/filters; e. pool size; f. kernel size; g. dropout; h. activation; i. RMSE; j. MAE; k. time (s).

Model	a.	b.	c.	d.	e.	f.	g.	h.	i.	j.	k.
Uni-Variate-Scenario 1											
CNN	14	10	3	32	3	3	0.0	tanh	702.71	645.70	33
LSTM	21	20	3	32	-	-	0.5	tanh	779.13	714.58	36
GRU	21	20	5	64	-	-	0.0	ReLU	715.42	653.75	78
Multi-Variate-Scenario 1											
CNN	21	20	5	32	3	3	0.5	ReLU	737.47	677.48	32
LSTM	21	5	4	64	-	-	0.0	tanh	788.46	720.77	175
GRU	14	10	3	32	-	-	0.5	tanh	755.56	693.78	73
Multi-Variate-Scenario 2											
CNN	21	20	4	16	3	3	0.0	ReLU	742.35	684.92	19
LSTM	21	5	3	64	-	-	0.0	ReLU	760.75	699.45	136
GRU	28	20	3	64	-	-	0.0	ReLU	727.07	670.53	50
Multi-Variate-Scenario 3											
CNN	28	20	4	32	3	3	0.5	ReLU	690.00	630.63	27
LSTM	21	20	4	128	-	-	0.5	ReLU	729.73	668.62	91
GRU	21	20	3	32	-	-	0.5	ReLU	746.98	683.02	38

Regarding the uni-variate approach, it was possible to verify some differences between some hyperparameters between the RNN-based models and the CNN-based model. Concerning the number of timesteps and the value of the batch size, it appeared that the CNN-based had the lowest value of both models, 14 and 10, respectively. Regarding the number of layers and the number of neurons/filters, the GRU-based model presented the highest values of 3, 5 and 64, respectively.

Overall, CNN candidates showed better results in all uni- and multi-variate scenarios, except for the second scenario, where GRUs presented a better performance. Regarding the training times, CNNs candidate models demonstrated lower values than the other two, with LSTM-based models being the ones taking more time to fit. It was also notable that the number of timesteps given as input increased, in general, with the number of features provided to the model. Concerning the activation function, it was also possible to verify that there was a tendency to use *tanh* in the uni-variate approach, while in the multi-variate, the best candidate models tended to use *ReLU*.

The best multi-variate scenario is the one that added, to the *value_energy* feature, the *flow_value*, i.e., the influent flow value combined with energy consumption value. In this approach, it was possible to verify that approach in terms of the number of timesteps, in Scenario 1 and Scenario 2, the model based on LSTM and the model based on CNN presented the same value in both cases, 21 timesteps (3 weeks). Regarding the batch size, note that the LSTM-based model in Scenarios 1 and 2 had a lower value than the others, while in Scenario 3, both models had the same value (20). It was also possible to verify that the best candidate models had a better performance with climatological context and without temporal context, except for CNN-based models. On the other hand, GRU-based models had their the best performance in the uni-variate approach, while the other two models presented their best performance in the multi-variate approach, more specifically in Scenario 3 (*value_energy* and *flow_value* features).

Figure 8 plots eight multi-step forecasts for the best candidate model in this method (the best CNN candidate in the third multi-variate scenario). These forecasts describe a set of 28 timesteps (i.e., days) given as input, making a successive two-day forecast for a total of 8 days.

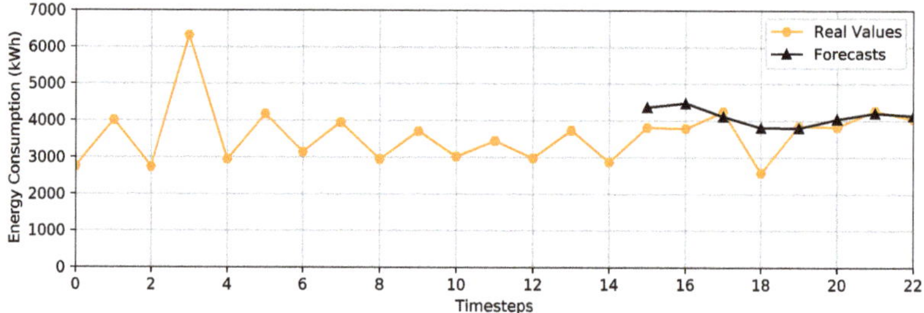

Figure 8. Eight multi-step forecasts for the best candidate model in Method 1.

3.2. Method 2

The second method used a dataset that had the outliers truncated. Table 8 depicts the best hyperparameter configuration for each combination of this method, with the best candidate, a CNN, following a uni-variate approach and presenting a MAE of 784 and a RMSE of 869 kWh. This meant that when truncating the outliers, a uni-variate approach presented better results than a multi-variate one.

As in Method 1, the CNN-based models presented a training time shorter than the others. It was also possible to verify that the CNN-based models had better performance. These models show an interesting uniformity in the cardinality of timesteps, while in the other models there was a higher fluctuation. Regarding the number of layers, it was possible to verify a constant value in most models (three layers), except for two CNN-based models.

In the uni-variate approach, it was possible to verify that the model based on CNN presented a lower value of timesteps given as input to the model (14) than models based on RNN. On the other hand, regarding the batch size value, the CNN-based model presented a higher value than the others (30).

Table 8. Best results, per scenario, for Method 2. The letters stand as follows: a. timesteps; b. batch size; c. number of layers; d. number of neurons/filters; e. pool size; f. kernel size; g. dropout; h. activation; i. RMSE; j. MAE; k. time (s).

Model	a.	b.	c.	d.	e.	f.	g.	h.	i.	j.	k.
Uni-Variate-Scenario 1											
CNN	14	30	4	32	2	4	0.5	tanh	869.78	784.23	13
LSTM	21	20	3	32	-	-	0.5	tanh	913.90	828.48	72
GRU	28	20	3	64	-	-	0.5	ReLU	869.85	798.94	83
Multi-Variate-Scenario 1											
CNN	21	10	5	16	3	4	0.0	ReLU	926.23	845.11	27
LSTM	21	5	3	64	-	-	0.0	ReLU	961.34	881.92	186
GRU	21	10	3	32	-	-	0.5	ReLU	950.09	863.89	46
Multi-Variate-Scenario 2											
CNN	14	20	3	16	3	4	0.0	tanh	885.90	796.07	21
LSTM	28	20	3	128	-	-	0.0	ReLU	913.90	845.28	51
GRU	21	20	3	64	-	-	0.0	ReLU	887.41	808.75	43
Multi-Variate-Scenario 3											
CNN	14	20	3	16	3	3	0.0	tanh	916.27	831.90	12
LSTM	14	30	3	128	-	-	0.5	tanh	946.63	854.81	31
GRU	21	20	3	32	-	-	0.5	ReLU	898.67	816.27	39

Regarding the models conceived over the multi-variate approach, it was possible to verify that the best performance was again obtained by a CNN-based model but now in the second scenario. This scenario had, as input features, the *value_energy*, *temperature*, and *flow_value*. In this approach, it was possible to verify that in the scenario with the most significant number of features given with input to the models, all three models presented the same value of timesteps (21). In the remaining scenarios, where there was a decrease in the number of features, in general, the CNN-based model requires a lower timestamp value than the rest. It should also be noted that, for the most part, all DL models required an equal value of layers in each of the scenarios. It is also interesting to note that this scenario held the best multi-variate candidates for CNNs, LSTMs, and GRUs.

Figure 9 illustrates several multi-step forecasts made by the best candidate model in this method. Here, the input sequence was made of 14 timesteps (i.e., days).

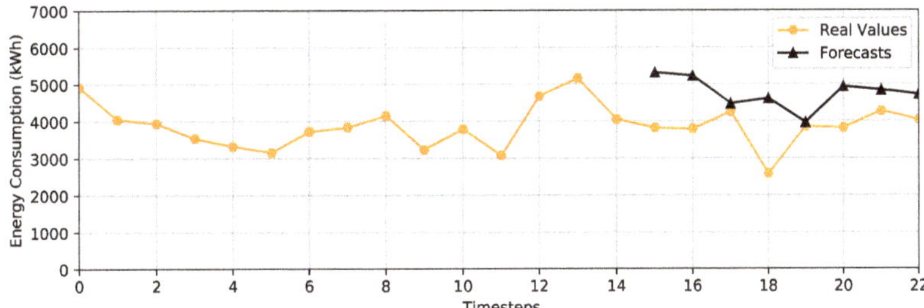

Figure 9. Eight multi-step forecasts for the best candidate model in Method 2.

3.3. Transfer Learning

It is usual to find situations where an WWTP has insufficient data. Hence, a goal of this study was to understand the applicability of transfer learning processes in this domain. To achieve such a goal, data were obtained from a second WWTP. However, no influent flow data were available. Hence, we were limited to apply transfer learning processes over the uni-variate approach since it only considers the *value_energy* feature, which was only available in a daily periodicity for the years of 2016 and 2017. The best uni-variate candidate model, a CNN, was conceived over the first method, i.e., the one that had the outliers interpolated. Hence, the data from the second WWTP were treated similarly. Finally, 2016 data were used for training and 2017 for testing.

To carry out the transfer learning process, it was necessary to store several parameters of the best uni-variate CNN including its architecture, hyperparameters, and weights (the pre-trained model). Two different settings were tried. The first one re-trained the entire pre-trained CNN model, while the second one only re-trained the layers after the last *Conv1D/AveragePooling1D* pair, inclusive. This is achieved by enabling, or disabling, the *trainable* property of each layer. Table 9 describes the results achieved by the pre-trained uni-variate CNN model, in each setting.

Table 9. Results of the pre-trained CNN (Convolutional Neural Networks) model on the second WWTP (Wastewater Treatment Plants).

Setting	RMSE	MAE
1-All model re-trained	357.98	324.69
2-Re-train after the last pair, inclusive	367.72	334.18

It was possible to verify that the method with better performance was the one that re-trained the entire model. This method had a MAE of 324 and a RMSE of 357 kWh. Figure 10 illustrate eight multi-step forecasts for the best model. A total of 14 timesteps were used as input, with successive two-day forecasts encompassing the next 8 days.

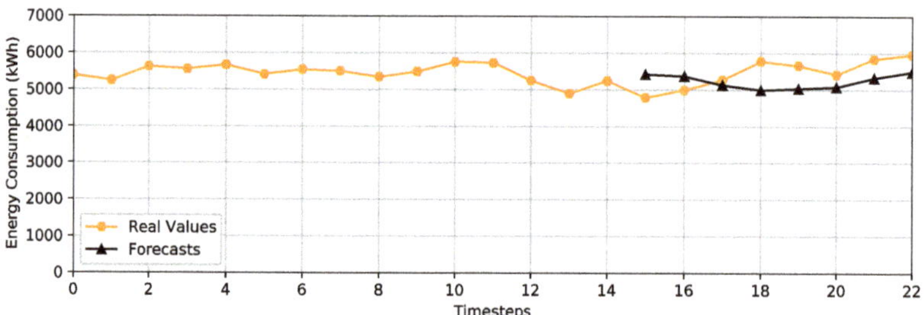

Figure 10. Eight multi-step forecasts when re-train the entire model.

4. Discussion and Conclusions

Energy consumption forecasting in a WWTP can significantly impact these installations, making them increasingly sustainable, obtaining greater energy efficiency, and reducing costs. After a diversity of experiments being carried out, from all the candidate models, the one achieving a better performance was a multi-variate CNN over the dataset created by Method 1, with a RMSE and MAE of 690 and 630 kWh, respectively.

Another interesting result was the differences in performance concerning the uni- and multi-variate approaches, for the two methods. If in Method 1 the best candidate model was a multi-variate one, in Method 2 it was uni-variate. Regarding both methods, it can be said that the method in which interpolations are made (Method 1) allowed all candidate models to achieve better performances when compared to the method that truncated the outliers (Method 2). Overall, CNN models presented a better performance than the remaining models. Table 10 summarises the obtained results.

Table 10. Ordered list of best candidate models.

Method	Approach	Scenario	Best Candidate	RMSE	MAE
1	Multi-Variate	3	CNN	690.00	630.63
1	Uni-Variate	1	CNN	702.71	645.70
1	Multi-Variate	2	GRU	727.07	670.53
1	Multi-Variate	1	CNN	737.47	677.48
2	Uni-Variate	1	CNN	869.78	784.23
2	Multi-Variate	2	CNN	885.90	796.07
2	Multi-Variate	3	GRU	898.67	816.27
2	Multi-Variate	1	CNN	926.23	845.11

Within the different scenarios in the multi-variate approach, there were some differences between both methods. In Method 1, the best multi-variate scenario was found when combining the influent flow with the energy consumption values (Scenario 3). However, in Method 2, the best multi-variate scenario was found when adding the climatological context to the influent flow and energy consumption values (Scenario 2). In both methods, it was possible to verify that the temporal context (*year* and *month*) worsened the energy consumption forecasts.

Regarding the cardinality of timesteps required as input by the models, in CNN-based models, the increase in the number of features usually led to an increase in the number of timesteps. On the other hand, GRU-based models showed that more features led to a lower number of timesteps. LSTM candidates had their results varying significantly.

Finally, an analysis was carried out to compare the three models' performance. A critical difference diagram was developed to represent the results of a two-tailed Nemenyi post-hoc test, with a $p < 0.05$, as depicted in Figure 11. When the average ratings of two models differ by, at least, the critical difference, we can say that the performance between the two is statistically significant. Considering the mean MAE as measure, it is possible to verify that CNNs have better performance than LSTMs and GRUs, being this difference statistically significant.

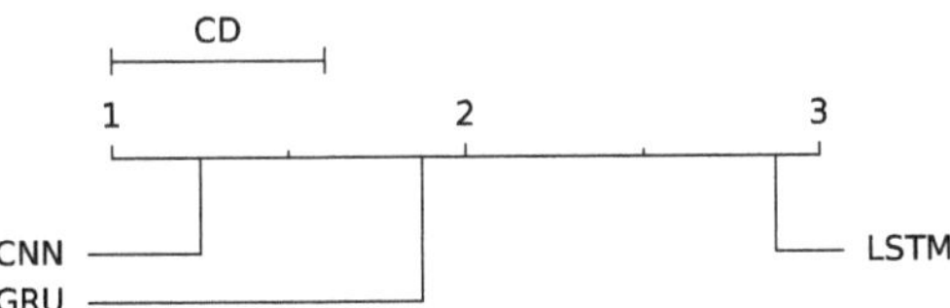

Figure 11. Critical difference diagram showing pairwise comparison of the average ranks in terms of MAE (Mean Absolute Error) ($p < 0.05$).

In regard to the applied transfer learning processes, promising results were achieved using a pre-trained uni-variate CNN model. The best performance was achieved when re-training the whole model. To answer the research questions raised at the beginning of the study, it can be said that (RQ1) CNNs performed better than RNNs, with CNN-based models being the best in practically the whole set of experiments; (RQ2) that the feature that most facilitated the process of forecasting energy consumption in a WWTP was the influent flow; and (RQ3) it was found that it is viable to use transfer learning processes in WWTP with a low volume of data and still present promising results.

However, it is known that other factors can be correlated with energy consumption in a WWTP, such as the concentration of certain pollutants in water like BOD_5. Nevertheless, to obtain this data, laboratory analysis of WWTP waters is required. Thus, it can take us several days to know the BOD_5 value, among many others. Hence, from a data exploration perspective, it is interesting to understand the impact of such pollutants on energy consumption. Although, from an engineering point of view, this is a significant limitation as the goal of this study is to deploy the best DL model to have real-time forecasts of energy consumption. If we were expected to include the concentration of such pollutants, it would only be possible to predict the value of energy consumption in the WWTP for tomorrow after obtaining the results from the laboratory, and this would only be available the day after tomorrow. In this way, we would not be able to implement the model to predict the value of energy consumption for tomorrow due to some input parameters of the model would be unknown and would only be available in a few days.

Considering that we are handling a real-life scenario and that the goal is to deploy the best candidate model in a WWTP, future work and research will focus on the use of more extensive sets of data, as well as the conception and evaluation of hybrid models to forecast energy consumption. An additional goal is to conceive a dashboarding platform for Machine Learning Operations (MLOps) to improve the process of monitoring the execution and performance of the deployed models.

Author Contributions: Conceptualization, P.O. and B.F.; methodology, P.O. and B.F.; software, P.O. and B.F.; validation, P.O. and B.F.; formal analysis, P.O. and B.F.; investigation, P.O. and B.F.; resources, P.N. and C.A.; data curation, P.O. and B.F.; writing—original draft preparation, P.O. and B.F.; writing—review and editing, P.N. and C.A.; visualization, P.O. and B.F.; supervision, P.N.; project administration, P.N. and C.A.; funding acquisition, P.N. and C.A. All authors have read and agreed to the published version of the manuscript.

Funding: The work of Paulo Novais and Cesar Analide has been supported by FCT—Fundação para a Ciência e Tecnologia within the R&D Units Project Scope: UIDB/00319/2020. The work of Pedro Oliveria and Bruno Fernandes is also supported by National Funds through the Portuguese funding agency, FCT—Fundação para a Ciência e a Tecnologia within project DSAIPA/AI/0099/2019.

Data Availability Statement: The data presented in this study are available on request from the corresponding author. The data are not publicly available due to having been made available by a multi-municipal water systems company.

Conflicts of Interest: The authors declare no conflict of interest.

Abbreviations

The following abbreviations are used in this manuscript:

ANN	Artificial Neural Networks
ARIMA	Autoregressive Integrated Moving Average
BSM1	Benchmark Simulation Model
BOD_5	Biological Oxygen Demand
COD	Chemical Oxygen Demand
CNN	Convolutional Neural Network
DL	Deep Learning
ENN-ECM	Elman Neural Network-Energy Consumption Model
FCM	Fuzzy C-Means
GRU	Gated Recurrent Units
LSTM	Long Short-Term Memory
MAE	Mean Absolute Error
MAPE	Mean Absolute Percentage
MLOps	Machine Learning Operations
MLR	Multi-variable Linear Regression
OLS	Ordinary Least Square
RBF	Radial Basis Function
RMSE	Root Mean Square Error
RMSLE	Root Mean Squared Log Error
RQ	Research Question
WWTP	Wastewater Treatment Plant

References

1. World Urbanization Prospects-Population Division-United Nations. 2018. Available online: https://population.un.org/wup/ (accessed on 21 January 2021).
2. Omer, A.M. Energy, environment and sustainable development. *Renew. Sustain. Energy Rev.* **2018**, *12*, 2265–2300. [CrossRef]
3. Daw, J.; Hallett, K.; DeWolfe, J.; Venner, I. *Energy Efficiency Strategies for Municipal Wastewater Treatment Facilities*; National Renewable Energy Lab.(NREL): Golden, CO, USA, 2012. [CrossRef]
4. Liu, F.; Ouedraogo, A.; Manghee, S.; Danilenko, A. *A Primer on Energy Efficiency for Municipal Water and Wastewater Utilities*; World Bank: Washington, DC, USA, 2012.
5. Frade, J.; Lacasta, N.; Mendes, P.; Cardoso, P.; Trindade, I.; Newton, F.; Franco, P.; Serra, A.; Póvoa, C.; Narciso, F. PENSAAR 2020–Uma Estratégia ao Serviço da População: Serviços de Qualidade a um Preço Sustentável. Available online: https://www.apambiente.pt/index.php?ref=16&subref=7&sub2ref=9&sub3ref=1098 (accessed on 22 January 2021).
6. Rajaeifar, M.; Ghanavati, H.; Dashti, B.; Heijungs, R.; Aghbashlo, M.; Tabatabaei, M. Electricity generation and GHG emission reduction potentials through different municipal solid waste management technologies: A comparative review. *Renew. Sustain. Energy Rev.* **2017**, *79*, 414–439. [CrossRef]
7. Zeng, S.; Chen, X.; Dong, X.; Liu, Y. Efficiency assessment of urban wastewater treatment plants in China: Considering greenhouse gas emissions. *Resour. Conserv. Recycl.* **2017**, *120*, 157–165. [CrossRef]
8. De Haas, D.; Foley, J.; Marshall, B.; Dancey, M.; Vierboom, S.; Bartle-Smith, J. Benchmarking Wastewater Treatment Plant Energy Use in Australia. Available online: https://www.researchgate.net/profile/David-De-Haas-2/publication/276921977_Benchmarking_Wastewater_Treatment_Plant_Energy_Use_in_Australia/links/5599093e08ae793d137e2735/Benchmarking-Wastewater-Treatment-Plant-Energy-Use-in-Australia.pdf (accessed on 25 January 2021).
9. Li, Z.; Zou, Z.; Wang, L. Analysis and forecasting of the energy consumption in wastewater treatment plant. *Math. Probl. Eng.* **2019**, *2019*. [CrossRef]
10. Harrou, F.; Cheng, T.; Sun, Y.; Leiknes, T.O.; Ghaffour, N. A Data-Driven Soft Sensor to Forecast Energy Consumption in Wastewater Treatment Plants: A Case Study. *IEEE Sens. J.* **2020**, *21*, 4908–4917. [CrossRef]
11. Huang, X.; Han, H.; Qiao, J. Energy consumption model for wastewater treatment process control. *Water Sci. Technol.* **2013**, *67*, 667–674. [CrossRef] [PubMed]
12. Ramli, N.A.; Abdul, M.F. Analysis of energy efficiency and energy consumption costs: A case study for regional wastewater treatment plant in Malaysia. *J. Water Reuse Desalin.* **2017**, *7*, 103–110. [CrossRef]
13. Maki, S.; Chandran, R.; Fujii, M.; Fujita, T.; Shiraishi, Y.; Ashina, S.; Yabe, N. Innovative information and communication technology (ICT) system for energy management of public utilities in a post-disaster region: Case study of a wastewater treatment plant in Fukushima. *J. Clean. Prod.* **2019**, *233*, 1425–1436. [CrossRef]
14. Oulebsir, R.; Lefkir, A.; Safri, A.; Bermad, A. Optimization of the energy consumption in activated sludge process using deep learning selective modeling. *Biomass Bioenergy* **2020**, *132*, 105420. [CrossRef]
15. Fernandes, B.; Silva, F.; Alaiz-Moreton, H.; Novais, P.; Neves, J.; Analide, C. Long Short-Term Memory Networks for Traffic Flow Forecasting: Exploring Input Variables, Time Frames and Multi-Step Approaches. *Informatica* **2020**, *31*, 723–749. [CrossRef]
16. Jin, X.; Yang, N.; Wang, X.; Bai, Y.; Su, T.; Kong, J. Integrated predictor based on decomposition mechanism for PM2.5 long-term prediction. *Appl. Sci.* **2019**, *9*, 4533. [CrossRef]
17. Zhang, T.; Song, S.; Li, S.; Ma, L.; Pan, S.; Han, L. Research on gas concentration prediction models based on LSTM multidimensional time series. *Energies* **2019**, *12*, 161. [CrossRef]
18. Mbatha, N.; Bencherif, H. Time series analysis and forecasting using a novel hybrid LSTM data-driven model based on empirical wavelet transform applied to total column of ozone at Buenos aires, Argentina (1966–2017). *Atmosphere* **2020**, *11*, 457. [CrossRef]
19. Chatterjee, A.; Gerdes, M.W.; Martinez, S.G. Statistical explorations and univariate timeseries analysis on covid-19 datasets to understand the trend of disease spreading and death. *Sensors* **2020**, *20*, 3089. [CrossRef]
20. Zhang, W.; Yu, Y.; Qi, Y.; Shu, F.; Wang, Y. Short-term traffic flow prediction based on spatio-temporal analysis and CNN deep learning. *Transp. Transp. Sci.* **2019**, *15*, 1688–1711. [CrossRef]
21. Dong, X.; Qian, L.; Huang, L. Short-term load forecasting in smart grid: A combined CNN and K-means clustering approach. In Proceedings of the International Conference on Big Data and Smart Computing (BigComp), Jeju, Korea, 13–16 February 2017; pp. 119–125. [CrossRef]
22. Hussain, D.; Hussain, T.; Khan, A.A.; Naqvi, S.A.A.; Jamil, A. A deep learning approach for hydrological time-series prediction: A case study of Gilgit river basin. *Earth Sci. Informatics* **2020**, *13*, 915–927. [CrossRef]
23. Oliveira, P.; Fernandes, B.; Aguiar, F.; Pereira, M.A.; Analide, C.; Novais, P. A Deep Learning Approach to Forecast the Influent Flow in Wastewater Treatment Plants. In Proceedings of the International Conference on Intelligent Data Engineering and Automated Learning, Guimarães, Portugal, 4–6 November 2020; pp. 362–373. [CrossRef]
24. Schmidhuber, J. Deep learning in neural networks: An overview. *Neural Netw.* **2015**, *61*, 85–117. [CrossRef] [PubMed]
25. Medsker, L.; Jain, L. *Recurrent Neural Networks: Design and Applications*; CRC Press: Boca Raton, FL, USA, 1999.
26. Hochreiter, S.; Schmidhuber, J. Long short-term memory. *Neural Comput.* **1997**, *9*, 1735–1780. [CrossRef]
27. Kang, D.; Lv, Y.; Chen, Y. Short-term traffic flow prediction with LSTM recurrent neural network. In Proceedings of the 20th International Conference on Intelligent Transportation Systems (ITSC), Yokohama, Japan, 16–19 October 2017; pp. 1–6. [CrossRef]

28. Yang, B.; Sun, S.; Li, J.; Lin, X.; Tian, Y. Traffic flow prediction using LSTM with feature enhancement. *Neurocomputing* **2019**, 320–327. [CrossRef]
29. Fente, D.N.; Singh, D.K. Weather forecasting using artificial neural network. In Proceedings of the 2018 Second International Conference on Inventive Communication and Computational Technologies (ICICCT), Coimbatore, India, 20–21 April 2018; pp. 1757–1761. [CrossRef]
30. Kim, T.; Cho, S. Web traffic anomaly detection using C-LSTM neural networks. *Expert Syst. Appl.* **2018**, *106*, 66–76. [CrossRef]
31. Feng, C.; Li, T.; Chana, D. Multi-level anomaly detection in industrial control systems via package signatures and LSTM networks. In Proceedings of the 47th Annual IEEE/IFIP International Conference on Dependable Systems and Networks (DSN), Denver, CO, USA, 26–29 June 2017; pp. 261–272. [CrossRef]
32. Cho, K.; Van Merriënboer, B.; Bahdanau, D.; Bengio, Y. On the properties of neural machine translation: Encoder-decoder approaches. *arXiv* **2014**, arXiv:1409.1259.
33. Chung, J.; Gulcehre, C.; Cho, K.; Bengio, Y. Empirical evaluation of gated recurrent neural networks on sequence modeling. *arXiv* **2014**, arXiv:1412.3555.
34. Niu, Z.; Yu, Z.; Tang, W.; Wu, Q.; Reformat, M. Wind power forecasting using attention-based gated recurrent unit network. *Energy* **2020**, *196*, 117081. [CrossRef]
35. Wang, R.; Li, C.; Fu, W.; Tang, G. Deep learning method based on gated recurrent unit and variational mode decomposition for short-term wind power interval prediction. *IEEE Trans. Neural Netw. Learn. Syst.* **2019**, *31*, 3814–3827. [CrossRef] [PubMed]
36. Wang, Y.; Liao, W.; Chang, Y. Gated recurrent unit network-based short-term photovoltaic forecasting. *Energies* **2018**, *11*, 2163. [CrossRef]
37. Fukushima, K.; Miyake, S. Neocognitron: *A Self-Organizing Neural Network Model for a Mechanism of Visual Pattern Recognition*; Springer: Berlin/Heidelberg, Germany, 1982; pp. 267–285. [CrossRef]
38. LeCun, Y.; Bottou, L.; Bengio, Y.; Haffner, P. Gradient-based learning applied to document recognition. *Proc. IEEE* **1998**, *86*, 2278–2324. [CrossRef]
39. Hubel, D.H.; Wiesel, T.N. Receptive fields, binocular interaction and functional architecture in the cat's visual cortex. *J. Physiol.* **1962**, *160*, 106–154. [CrossRef] [PubMed]
40. Li, Q.; Cai, W.; Wang, X.; Zhou, Y.; Feng, D.D.; Chen, M. Medical image classification with convolutional neural network. In Proceedings of the 13th International Conference on Control Automation Robotics & Vision (ICARCV), Singapore, 10–12 December 2014; pp. 844–848. [CrossRef]
41. Chen, C.; Liu, M.; Tuzel, O.; Xiao, J. R-CNN for small object detection. In Proceedings of the 13th Asian Conference on Computer Vision, Taipei, Taiwan, 20–24 November 2016; pp. 214–230. [CrossRef]

electronics

Article

Extending a Trust model for Energy Trading with Cyber-Attack Detection

Rui Andrade *, Sinan Wannous, Tiago Pinto and Isabel Praça *

GECAD—Knowledge Engineering and Decision Support Research Centre, School of Engineering, Polytechnic of Porto (ISEP/IPP), 4050-535 Porto, Portugal; sinai@isep.ipp.pt (S.W.); tcp@isep.ipp.pt (T.P.)
* Correspondence: rfaar@isep.ipp.pt (R.A.); icp@isep.ipp.pt (I.P.)

Abstract: This paper explores the concept of the local energy markets and, in particular, the need for trust and security in the negotiations necessary for this type of market. A multi-agent system is implemented to simulate the local energy market, and a trust model is proposed to evaluate the proposals sent by the participants, based on forecasting mechanisms that try to predict their expected behavior. A cyber-attack detection model is also implemented using several supervised classification techniques. Two case studies were carried out, one to evaluate the performance of the various classification methods using the IoT-23 cyber-attack dataset; and another one to evaluate the performance of the developed trust mode.

Keywords: cyber-attack detection; IoT; trust; energy trading; trusted negotiations

Citation: Andrade, R.; Wannous, S.; Pinto, T.; Praça, I. Extending a Trust model for Energy Trading with Cyber-Attack Detection. *Electronics* 2021, *10*, 1975. https://doi.org/10.3390/electronics10161975

Academic Editor: Myung-Sup Kim

Received: 15 July 2021
Accepted: 6 August 2021
Published: 17 August 2021

1. Introduction

The energy market and electric grid play a major role in everyday life. Most areas in modern society require electric energy to operate properly. The electric grid has become indispensable for life in modern society. Due to these reasons, it is important to maintain and improve the stability and reliability of the energy grid.

Currently, energy grids tend to follow a very strict and somewhat inefficient structure. A high number of entities that desire to consume energy are connected to a single centralized energy supplier entity. Traditional energy markets, such as wholesale or retail markets, were not designed to support the rising in distributed energy generation coming from Renewable Energy Sources (RES) in households, small commerce and small industry. Such facts raise questions about different ways of structuring energy markets to deal with these challenges.

One of the possible proposals to answer to this problem is the creation and implementation of local energy markets (LEMs). LEMs are structured in such a way as to enable small-scale negotiations and energy exchanges between participants who traditionally would only be final consumers. These markets are designed to operate within a regional area, such as a neighborhood or a city. Participants in this market are the local households, small commerce and small industry, that may be regular consumers or consumers with some type of local energy generation, being referred to as prosumers. Furthermore, local small-scale power plants can also participate in the LEM. The LEM is better designed to deal with distributed energy generation from RES because the surplus in generation from local energy producers and prosumers can be purchased and utilized by local consumers. This flexibility of response makes LEM an attractive proposition for the future of energy markets.

In order to guarantee the success and desired operation of the LEM, it is necessary to ensure security and trust in negotiations. While security is focused on the traditional measures of cyber-security, such as security in network communications, trust is focused on ensuring that the LEM participants and their proposals in the negotiations are viable and trustworthy.

The objective in this work is to create a LEM simulation, and incorporate a trust model. The trust model should be able to score participants' trust level during negotiations, allowing the untrustworthy participants (low trust score) to be prohibited from participating in the LEM. Furthermore, the goal is to also create a cyber-attack detection model utilizing supervised classification techniques.

A MAS is a system that combines several agents, which are software entities that have the capacity to interact among themselves. For this reason it is ideal to simulate the LEM, as each market participant can be simulated individually, and by the means of their interactions, it is possible to simulate a far more complex environment, as is the case of the LEM.

After this introductory Section, the document is organized as follows: Section 2 contextualizes the work, describing the concepts of the LEM, and of trust models for MAS. Section 3 presents possible approaches to obtain the cyber-security guaranties needed for the safe operation of the LEM. Section 4 describes the LEMMAS system. Section 5 describes the developed cyber-attack detection system. Section 6 presents the analyses done to Cyber-Attack detection models. Lastly, Section 7 presents the conclusions of this work.

2. Local Energy Market

The local energy market (LEM) is a novel energy market model. There is no unique definition on what a LEM is; however, many authors have addressed this issue, and among their work a general idea of the LEM begins to emerge. Authors exploring this topic tend to define three key aspects: (i) market structure; (ii) advantages; and (iii) challenges.

The structure of the LEM is generally defined as a group of local participants (such as a neighborhood) [1–3], which are capable of trading energy among themselves. Participants in the local market are separated into three kinds [3–5]:

- Consumers: who wish to buy energy;
- Producers: who wish to sell energy;
- Prosumers: (consumers with some source of energy generation) who wish to buy and sell energy.

Both the market participants and the underling electrical grid, which serves as a basis for the LEM, are defined as having monitoring sensors for consumption, generation, energy storage and other data sources; and network communication technologies to share this information [3,4]. Such an energy grid is referred to as a Smart-Grid [1].

The LEM brings several potential advantages when compared to traditional energy markets. Some authors [1,5] claim that the LEM would make a more efficient use of electrical grids. Simultaneously, it is believed that the shift to local energy markets (LEMs) could reduce the greenhouse effect [1] and create a more sustainable environment [4,5]. Participants in the LEM (especially traditional consumers) take a much more involved role in the market when compared to traditional markets. These participants gain the ability to directly negotiate and can achieve cost reductions or even profits with their participation [1,4,5]. Lastly, the versatility of the LEM makes it possibly for the coexistence with traditional markets [4,5], that being the case, the local market can adapt to the needs of each specific community.

Currently, the LEM is facing some challenges that prevent its adoption at a large scale. Abidin et al. [1] identify security concerns as one of these challenges. The local market, and consequently the underling Smart-Grid, deal with a lot of sensitive information that needs to be properly secured from unauthorized access; and from malicious entities who may tamper with data in order to have some financial gain. The former also emphasizes the need for trust in negotiations in the local energy market (LEM). Interest from the community and an economic upfront investment by investors are also seen as one of the current challenges to the LEM adoption [2,5]. From a technical point of view, the implementation of Smart-Grids capable of providing the support needed for the LEM is still a challenge that needs further research [5]. Lastly, the support from governments and creation of adequate legislation is a must for the success of the LEM [5].

2.1. Trust in Multi-Agent Systems

Trust and reputation systems (TRS) are designed with the objective of predicting the reliability in the behavior of an entity by analyzing data from past interactions [6]. By performing such analyzes, TRS are able to associate a reputation to each user. Good reputation indicates that the user is trustworthy in its negotiations, and vice versa.

In [6], several trust models are identified, some of which are specific for the marketplace area of applicability. Two of these trust models seem interesting for this project since they are targeted at a marketplace, but apply different strategies. These models are the e-commerce model and ReGreT [7].

Reference [8] views the e-commerce trust model from the perspective of eBay. eBay operates as an online auction web site. Users of this platform can propose their sale offers and/or place bids on other users' offers. In online auction web sites such as eBay, the participants in the transactions are humans, and these platforms implement mechanisms for participants to review their experience in the transaction. This feedback provided by the users is then used to feed the TRS with the data necessary to access the reputation of the users [6].

ReGreT is a trust and reputation model proposed by [7]. This model is different from the eBay model because it does not consider trust as a global value. ReGreT has a focus towards modularity [9]. Modules might be used or not depending on the needs of each context. ReGreT considers three kinds of information for trust: the agent's own experiences, information from other agents and the social structure among agents. These types of information coincide with the three dimensions used in ReGreT to calculate trust, which are the following:

- Individual dimension: Considers the outcomes observed directly by the agent when in negotiation with another agent.
- Social dimension: Considers information provided by other entities. Something that can be useful when direct information is not available.
- Ontological dimension: Considers the contextual information that can be gained by the reputation.

In another work [10], the authors identify three distinct approaches that can be followed when developing a trust mechanism for a MAS. Each approach considers a different dimension of trust. These approaches are:

- Security Approach: Is focused on the traditional security mechanisms [11]: confidentiality, availability, authentication, integrity, non-repudiation. This dimension aims to prevent cyber-security threats;
- Institutional Approach: Considers the idea of a centralized entity that acts as an overseer in the MAS. This centralized entity, takes the place of an institution that must evaluate all agents and ensure that each one of them is trustworthy. This is the case for the e-commerce trust model;
- Social Approach: Is similar to the way humans interact in the real world. With this trust model, each agent decides who it considers trustworthy. Agents can make this decision based on their interactions with other agents, and/or by considering others' opinions. This is the case for the ReGreT trust and reputation model.

2.2. Security Risks

In a LEM, part of the physical layer corresponds to the network and sensor infrastructure that is necessary for collecting data and allowing communication, and this is what makes the grid be called a smart-grid. However, this infrastructure can be a vector of cyber-attacks [12]. The sensor infrastructure in the smart-grid is often composed of IoT devices. IoT devices can have a potential risk of being tied to a company or cloud network and having access to the data collected by the sensor. A security breach in the cloud network would also expose the data related to the sensor and intern the LEM where this sensor is [13].

Traditional cyber-attacks to the LEM's smart-grid are also a security risk [13]. These can be attacks that aim at gathering private information, such as Man In The Middle attacks, or can even be attacks that try to tamper with the communications in the network. There might even be financial incentives to try such an attack by an ill intending participant in the LEM since he might be able to change the final market price in order to have a financial gain. Unknown sensor hardware malfunctions can also be problematic since they can leave the system working with incorrect data. In this work, the aim is to provide tools to help detect both malicious data tampering and hardware malfunctions.

In this paper, we present a first approach towards including mechanisms to detect attacks that can be made on devices of market participants, with the aim of making LEMMAS a system that provides trusted and secured negotiation.

3. Cyber-Attack Detection

In order to obtain a secure environment for local energy market negotiations, traditional cyber-security cannot be forgotten. It is as important to trust in the participants as it is to have a secure network and computer systems. One option to create this environment is to combine intrusion detection systems with artificial intelligence algorithms as an anomaly/attack detection tool by analyzing network data.

In [14], the authors developed a system to perform intrusion detection of smart meters. They combined support vector machine (SVM) and temporal failure propagation graph (TFPG) techniques with a pattern recognition algorithm. The study showed that the system provided good results.

The authors in [15] tried to detect false data injection in a smart grid using deep learning techniques. Their approach combined a Convolutional Neural Network (CNN) with a Long Short Term Memory (LSTM) network. The system was able to achieve an accuracy result above 90% for certain kinds of attacks. They conclude that their approach can be combined with a different technique to obtain a highly accurate attack detection system for all kinds of attacks.

In [16], the authors also tackled the problem of false data injection in power systems. Their approach used an autoencoder network with 4 hidden layers. A case study was performed, and their system was able to detect the kinds of attacks the study was focusing on, and the system also outperformed the techniques currently used in that scenario.

In [17], the authors researched the problem of face spoofing attacks. Instead of using the traditional methods for such a problem, the authors opted to use ensemble based technique by combining multiple one-class classifiers. A case study was conducted to evaluate the performance of their approach using three face anti-spoofing datasets. Their proposed solution showed a good performance for the problem.

In [18], the authors developed a cyber-attack detection system for network based attacks. In this work, the methods of random forest, multi-layer perceptron and long-short term memory were implemented and experimented using the CIDDS-001 dataset [19,20]. The results of this study showed that the long-short term memory technique was the best, achieving an accuracy score above 99%.

The authors in [21] studied the application of unsupervised learning techniques in order to perform cyber-attack anomaly detection. The authors experimented with six different techniques: Isolation Forest, K-Means, 1-Nearest Neighbor, Autoencoder, Scaled Convex Hull, Support Vector Machines; combined with the best pre-processing steps for each. A case study was performed with the NSL-KDD [22] and the ISCX [23] datasets in order to evaluate the algorithms. Based on the results, the authors concluded that all detection algorithms showed a good performance for the cyber-attack anomaly detection problem.

4. LEMMAS System

The developed MAS follows the agent structure as proposed in [24]. In that work, a computational model of a LEM is separated into three kinds of agents:

- Market Interactions Manager: Is at the center of every LEM and is responsible for managing all negotiations within the market. The task of ensuring trust in the negotiations is also a responsibility of the market interaction manager (MIM).
- Participant agent: Acts on behalf of consumers, producers or prosumers. This agent assumes the role of a negotiator that seeks to best satisfy the needs of the respective market participants (home owners, local commerce and small industry owners). The Participant Agent will have Sensor Agents that report to them the information needed for the negotiations. The Participant Agent is then able to make proposals to buy or sell energy in the market;
- Sensor agent: Has the single responsibility of acquiring one type of data and reporting said data to their respective Participant Agent. A Sensor agent can be, for example, connected to a meter measuring energy consumption in a household, while another sensor agent can be connected to a web service in order to obtain the weather forecast.

With these three kinds of agents, it is possible to create a reasonably complete representation of a LEM, which includes: consumers, producers and prosumers. The Sensor Agents allow the cyber-physical system, such as the ones of smart houses and other connected environments. A complete representation of the proposed LEM model is presented in Figure 1.

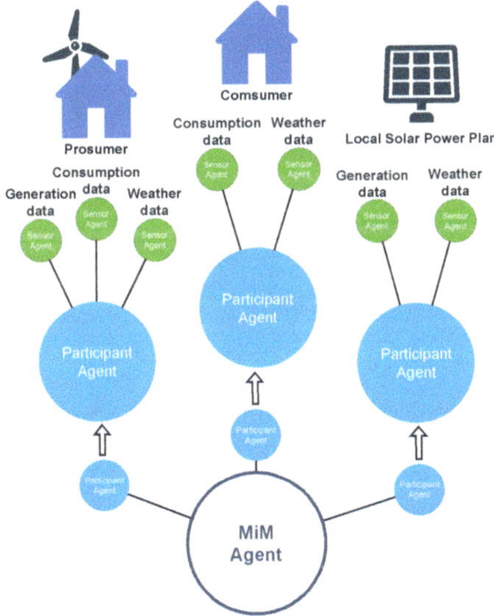

Figure 1. Proposed LEM model diagram [24].

The LEM is composed of several participants, represented by their respective Participant Agent, and all of these agents are connected to the MIM. In Figure 1, three participants are further detailed as examples of how real participants might be structured in a realistic scenario. These participants are the following:

- Consumer: Represents a household without self-generation that participates in the market.
- Prosumer: Represents an household that participates in the market and has its own energy generation, with a small wind generator.
- Local Solar Power Plant: Exemplifies a small photovoltaic power plant that is part of the LEM.

4.1. Trust Model

To support the market, an institutional based trust model is proposed to be used by the MIM, capable of evaluating the behavior of participants and detecting faulty or malicious activities. This trust model was chosen over the social model because with a social model, participants might need access to sensitive (consumption, generation, etc.) data from other participants in order to make their own trust evaluation.

The idea for the trust mechanism is that with information such as weather, historical consumption and generation data, and other contextual data, it is possible to use forecasting methods to try to predict what the participant's consumption, generation or proposals should be in the coming market negotiation period.

Using such forecasted values, it is possible to obtain an idea if the participant is trustworthy over time. Since forecasting methods always have a certain degree of uncertainty, a single proposed value that does not match the forecasted value does not provide a reliable metric. So, by using an evaluation over time, it is thought that incorrectly forecasted values become negligible.

Figure 2 presents a diagram of the proposed trust evaluation process. As shown, the trust evaluation process takes three values as input: the participant's proposed values for the current market negotiation period, the participant's trust value from the previous negotiation period, and the forecasted value based on the participants historical and contextual data.

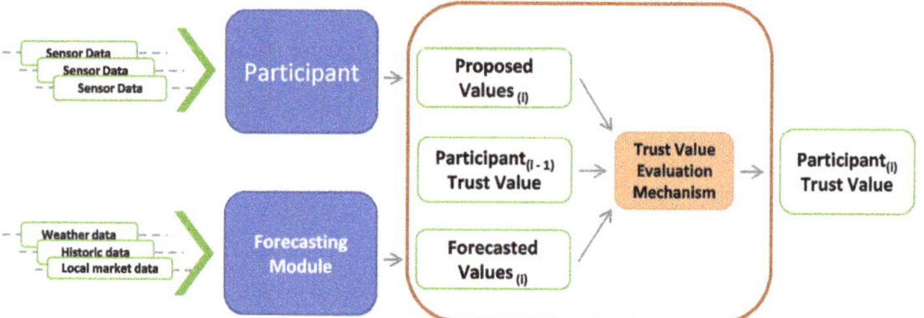

Figure 2. Trust evaluation process.

The definition of the proposed trust model is the following:

The trust value ranges from $[0, 1]$ where 1 is the highest trust and 0 lowest trust value. The trust value for a participant p in negotiation period i is represented as t_{pi}.

The evaluation formula takes several variables into consideration that can be configured to obtain the best possible results, these variables are:

- vr: represents the *variable acceptance range*, which is a percentage value;
- fr: represents the *fixed acceptance range*, which is a static value;
- tiv: represents the *trust increase value*, a value by which the participant's trust is increased;
- tdv: represents the *trust decrease value*, a value by which the participant's trust is decreased;
- sv_{pin}: represents the *submitted value* by participant p, from the data source of sensor n in market negotiation period i;
- fv_{pin}: represents the *forecasted value* for participant p, for sensor n in negotiation period i;
- t_{p0}: represents the *default trust value*, the trust value which all participants start with.

Equation (1) shows how the trust evaluation is calculated by being combined with either the Equation (2) for the asymmetric acceptance range or the Equation (3) for the symmetric acceptance range.

The difference between the asymmetric and the symmetric acceptance range is that the asymmetric has a higher acceptance range when the forecasting mechanism overestimates the value, since a percentage from a higher value results in a higher range.

$$t_{pi} = t_{p(i-1)} + trust\_eval(sv_{pi}, fv_{pi}) \tag{1}$$

$$trust\_eval_{asym}(sv_{pi}, fv_{pi}) = \begin{cases} tiv & \text{if } sv_{pi} > fv_{pi} * (1 - vr) \text{ AND } sv_{pi} < fv_{pi} * (1 + vr) \\ tiv & \text{if } sv_{pi} > fv_{pi} - fr \text{ AND } sv_{pi} < fv_{pi} + fr \\ tdv & \text{otherwise} \end{cases} \tag{2}$$

$$trust\_eval_{sym}(sv_{pi}, fv_{pi}) = \begin{cases} tiv & \text{if } fv_{pi} > sv_{pi} * (1 - vr) \text{ AND } fv_{pi} < sv_{pi} * (1 + vr) \\ tiv & \text{if } fv_{pi} > sv_{pi} - fr \text{ AND } fv_{pi} < sv_{pi} + fr \\ tdv & \text{otherwise} \end{cases} \tag{3}$$

There needs to be some consideration of how each participant's trust value is interpreted. Two things need to be taken into account: a participant that always submits real and true values should be fully trustworthy and so should be evaluated with a 1.0 trust value; on the other hand, a participant that always submits false values should not be trusted and should have a trust evaluation of 0.0.

There is, however, some subjectivity in considering these trust evaluations. For example, a participant that always submits real and true values and is evaluated with a 0.9 trust value, or a participant that always submits false values and is evaluated with a 0.1 trust value, also seem like acceptable evaluations. Given this subjective nature of the trust evaluation, three trust ranges are proposed:

- **Trustworthy**: range where the trust value is $[h_t, 1]$, and any participant in this range is fully trusted;
- **Unsure**: range where the trust value is $[m_t, h_t[$, and any participant in this range is considered to be a possible malicious or faulty participant and should, for example, be further evaluated by the market authority;
- **Untrustworthy**: range where the trust value is $[0, m_t[$, and any participant in this range is considered a malicious or faulty participant and should be prevented from participating in the market.

The values of h_t minimum threshold for high trust and m_t minimum threshold for medium trust are variable values that can be configured accordingly to the needs of the LEM.

4.2. Cyber-Security Model

Having a trust model capable of correctly analyzing the trust evaluation of participants supports the LEM negotiations; however, this kind of analysis leaves an important aspect neglected, the origin of a malicious proposal.

To fully understand the safety of negotiations, it is also necessary to consider the traditional security aspects. The idea behind the security model is to analyze the data coming from the sensor agents to discover potential security intrusions.

Discovering a security intrusion would also influence the ability of a participant to negotiate in the market. Regarding the cyber-security aspect we consider a simple binary classification for participants: *Secure, Insecure*. This leaves us with the final possible participants classifications in Table 1.

Table 1. Possible participant classifications.

	Trustworthy	**Unsure**	**Untrustworthy**
Secure	Secure Trustworthy	Secure Unsure	Secure Untrustworthy
Insecure	Insecure Trustworthy	Insecure Unsure	Insecure Untrustworthy

Considering the classifications from Table 1 the participants with classifications of: *Insecure Trustworthy*, *Insecure Unsure* and *Insecure Untrustworthy* should be prevented from negotiating because they can be under a cyber-attack. Participants with a classification of *Secure Untrustworthy* will also be prevented from negotiating as their trust score does not allow it. Participants classified as *Secure Unsure* are in a grey area where they can be allowed to participant, but further investigation is required in order to ensure if the negotiations are at risk. Furthermore, lastly, the participants classified as *Secure Trustworthy* will be allowed to negotiate.

5. LEMMAS Case Study

The idea for this case study is to simulate a LEM with several participants that vary in the amount and intensity of false proposals and observing how the proposed trust model evaluates these participants. Since the trust model is based on forecasting, forecasting methods are simulated as a normal standard distribution based on what the real proposal value should be, this way forecasting methods with distinct levels of accuracy and precision can be estimated, and it is possible to see how the performance of the forecasting method influences the trust model performance.

The LEM was simulated for a 24 h period and with 15 min market negotiation period duration, which results in a total of 96 market negotiation periods. Each simulation was performed 10 times, and its results were averaged. The 24 h simulated were of a Monday, simulated from hour 00:00 to hour 24:00. The LEM aggregates 4 participants using real consumption data from private homes publicly available in [25]. Each participant has their own bias in the proposals it submits:

- TP—True Proposer: Is the only who does not have a bias and always sends the real value;
- LUaUP—Low Under and Over Proposer: Sends a real value 80% of the times and a value between 30% under to 30% over the real value the rest of the time;
- MUaOP—Medium Under and Over Proposer: Sends a real value 50% of the times and a value between 60% under to 60% over the real value the rest of the time;
- HUaOP—High Under and Over Proposer: Sends a real value 30% of the times and a value between 90% under to 90% over the real value the rest of the time.

With these participants configurations, the expected result is a correlation between the trust value of the participant and the amount of false submissions. The True Proposer acts as a base line showing if trustworthy participants are being correctly identified.

As for the estimated forecasting methods, four were simulated, in decreasing levels of accuracy and precision. The estimated forecasting methods have the following mean $\bar{x}$ and standard deviation σ:

- Perfect Predictor: $\bar{x} = 1.0\ \sigma = 0.0$;
- Low Center Predictor: $\bar{x} = 1.0\ \sigma = 0.2$;
- High Center Predictor: $\bar{x} = 1.0\ \sigma = 0.4$.

The simulations are preformed with both the symmetric and asymmetric acceptance methods. Lastly, the trust formula variables are configured as such: $ar = 0.5$, $tiv = 0.01$, $tdv = -0.08$ and $t_{p0} = 0.8$; and the trust ranges are: *Trustworthy* $[0.8, 1]$, *Unsure* $[0.5, 0.8[$ and *Untrustworthy* $[0, 0.5[$.

These values were chosen after some experimentation, as they proved to be adequate values for the specific scenario in study.

Case Study Results and Discussion

To present these results in a clear way, each simulation was divided into 2 graphs showing the trust value for each participant over time, separated by the forecasting method and acceptance method.

Looking at Figure 3, there is a clear distinction in the trust evaluation of each participant. Analyzing Figure 4, the results have some changes from the asymmetric model. All participants obtained a higher trust value compared to the results of the asymmetric acceptance.

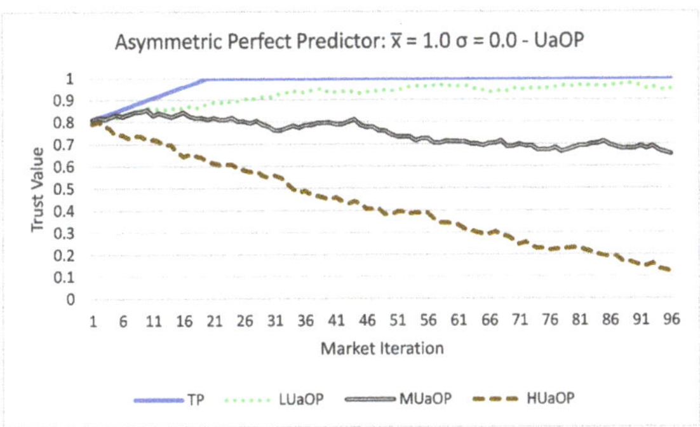

Figure 3. Asymmetric perfect predictor.

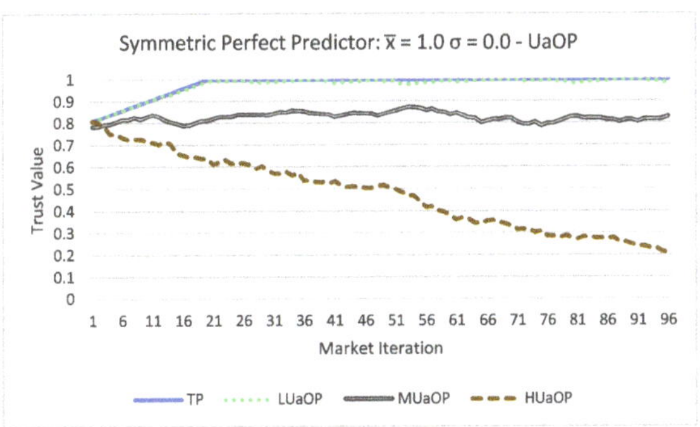

Figure 4. Symmetric Perfect Predictor.

Figures 5 and 6 show different result. In the previous estimator the trust value for the TP participant was always 1.0, but now with the uncertainty in the estimated forecasting method the trust value oscillates; however, it remains close to 1.0. In Figure 6, the results are very similar to the ones obtained with the Perfect Predictor. The biggest difference is in Figure 5 where the TP and LUaOP participants obtained trust values very similar to the ones obtained with the Perfect Predictor, and the MUaOP and HUaOP participants obtained evaluations significantly lower. This demonstrates that the acceptance formula used can make a big difference in the results.

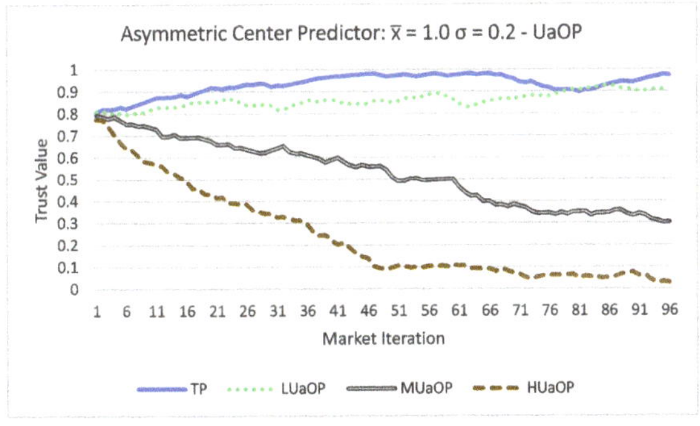

Figure 5. Asymmetric Low Predictor.

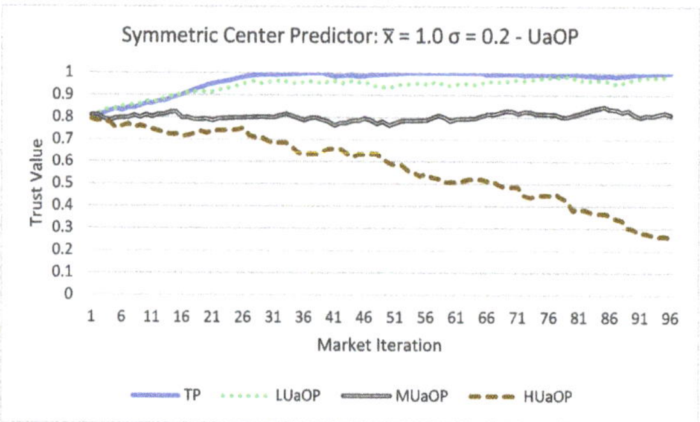

Figure 6. Symmetric Low Predictor.

Lastly both Figures 7 and 8 show low trust evaluations for all participants. There are some differences in the way the trust value changed over time between the asymmetric and symmetric acceptance formulas; however, at the end, the values are very similar (all below 0.2). Even the TP participants obtained a low trust evaluation, and this result shows that with a low performing forecasting method the trust evaluation is also low performing.

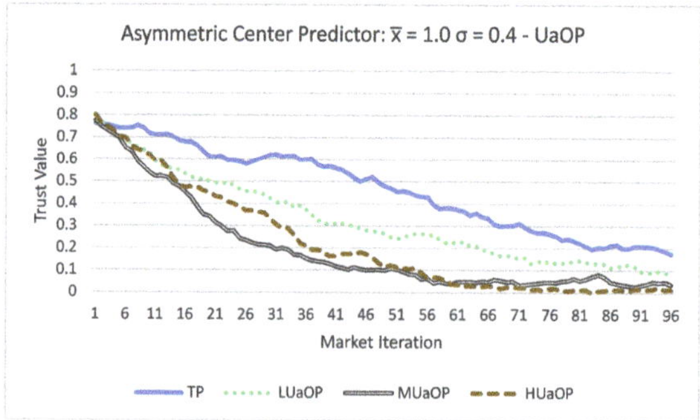

Figure 7. Asymmetric High Predictor.

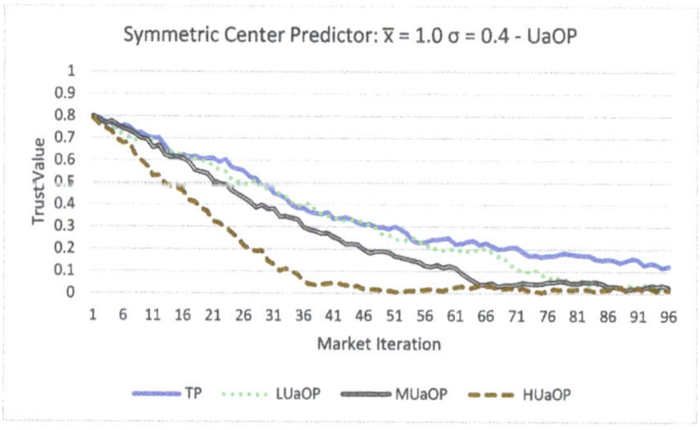

Figure 8. Symmetric High Predictor.

Finally these results lead us to conclude that:

1. Using the proposed trust methodology, it is possible to dynamically update the trust value of a participant;
2. The MIM agent is able to use the proposed trust methodology to access the trust value of a participant;

3. The performance of forecasting methods has a direct impact on the trust evaluation;
4. The acceptance formula can have an impact on the trust evaluation;
5. The higher the amount of false values a participant submits, the lower their trust value will be.

6. Developed Security Model and Analysis

As our proposed LEM architecture is based on IoT sensors and their representation as sensor agents, it is fundamental that the information and network communications coming from the sensor agents is secured. With this goal in mind, a cyber-security module is needed to classify participants as described in Table 1. A cyber-attack detection system was developed to complement LEMMAS, as is shown in Figure 9, and the objective is to achieve a negotiation environment with only secured and trusted data.

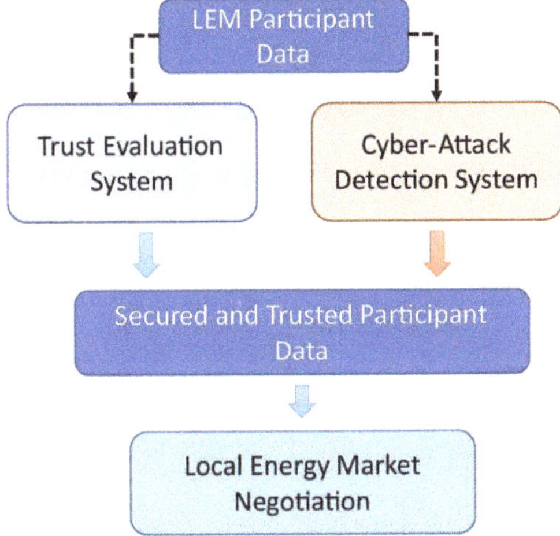

Figure 9. Trust and security modules.

In order to create the necessary security model the python library, Scikit-Learn was used. The goal is to train an artificial intelligent supervised classification model that can analyze the sensor data and classify it as malicious or not.

Six classification models were selected and implemented in order to evaluate which ones were the best for this application. The models are the following.

- Nearest Neighbors;
- Decision Tree;
- Random Forest;
- Neural Network;
- AdaBoost;
- Naive Bayes.

6.1. Dataset

Aposemat IoT-23 [26] is a publicly available dataset containing Internet of Things (IoT) network traffic data. This dataset is labeled, including both benign and malicious data entries, and subdivided into 23 sub datasets, 20 containing malicious cyber-attack samples and 3 containing only benign data samples. The data was collected between 2018 and 2019 in three kinds of IoT network devices, namely a Philips HUE smart LED lamp, an Amazon Echo home intelligent personal assistant and a Somfy smart doorlock. These kinds reflect some of the devices that would be part of a smart-home in a smart-grid and as so are aliened with the data generated and collected by LEM participants.

Table 2 presents the different datasets showing which ones include malicious samples, the IoT devices involved, the duration of the attack, the number of packets recorded, the information flows and the size in (GB).

The dataset structure contains the following fields:

- ts—Time of the connection;
- uid—Unique identifier for he connection;
- id.orig_h, id.orig_p, id.resp_h, id.resp_p—Ports and addresses used in the communication;
- proto—Communication protocol used;
- service—Service protocol used;
- duration—Duration of the communication;
- orig_bytes, resp_bytes—Bytes sent in the communication;
- conn_state—State of the connection;
- local_orig, local_resp—Origin of the communication;
- missed_bytes—Bytes missed in the communication;
- history—History of the state of the connection;
- orig_pkts—Number of packets sent in the original message;
- orig_ip_bytes—Number of IP level bytes sent in the original message;
- resp_pkts—Number of packets sent in the response message;
- resp_ip_bytes—Number of IP level bytes sent in the response message;
- tunnel_parents—UID of tunnel parents connections;
- label—Label of the sample (malicious or benign);
- detailed-label—Detailed label of the sample (specific attack used in case of a malicious sample).

Table 2. IoT-23 dataset description.

#	Type	Capture Name	Malware Device	Duration	Number of Packets	Total Flows	Total Size
1	Malicious	Capture-34-1	Mirai	24	233,000	23,146	121 MB
2	Malicious	Capture-43-1	Mirai	1	82,000,000	67,321,810	6 GB
3	Malicious	Capture-44-1	Mirai	2	1,309,000	238	1.7 GB
4	Malicious	Capture-49-1	Mirai	8	18,000,000	5,410,562	1.3 GB
5	Malicious	Capture-52-1	Mirai	24	64,000,000	19,781,379	4.6 GB
6	Malicious	Capture-20-1	Torii	24	50,000	3210	4 MB
7	Malicious	Capture-21-1	Torii	24	50,000	3287	4 MB
8	Malicious	Capture-42-1	Trojan	8	24,000	4427	3 MB
9	Malicious	Capture-60-1	Gagfyt	24	271,000,000	3,581,029	21 GB
10	Malicious	Capture-17-1	Kenjiro	24	109,000,000	54,659,864	7.8 GB
11	Malicious	Capture-36-1	Okiru	24	13,000,000	13,645,107	992 MB
12	Malicious	Capture-33-1	Kenjiro	24	54,000,000	54,454,592	3.9 GB
13	Malicious	Capture-8-1	Hakai	24	23,000	10,404	2 MB
14	Malicious	Capture-35-1	Mirai	24	46,000,000	10,447,796	3.6 GB
15	Malicious	Capture-48-1	Mirai	24	13,000,000	3,394,347	1.2 GB
16	Malicious	Capture-39-1	IRCBot	7	73,000,000	73,568,982	5.3 GB
17	Malicious	Capture-7-1	Linux Mirai	24	11,000,000	11,454,723	897 MB
18	Malicious	Capture-9-1	Linux Hajime	24	6,437,000	6,378,294	472 MB
19	Malicious	Capture-3-1	Muhstik	36	496,000	156,104	56 MB
20	Malicious	Capture-1-1	Hide & Seek	112	1,686,000	1,008,749	140 MB
21	Benign	Capture-7-1	Soomfy Doorlock	1.4	8276	139	2 MB
22	Benign	Capture-4-1	Phillips HUE	24	21,000	461	4 MB
23	Benign	Capture-5-1	Amazon Echo	5.400	398,000	1383	364 MB

6.2. Dataset Pre-Processing

In order to utilize this dataset to train an evaluate models, first a pre-processing step was needed.

The dataset was divided into X and Y, with Y being the target column "label" and X being the remaining data. The columns of "UID" and "ts" were dropped as they do not provide any valuable information. The column "detailed-label" was also dropped since the current objective is only to classify as "Malicious" or "Benign", meaning s binary classification. All columns containing IPs were converted to the corresponding integer number. The columns of "proto", "service", "conn_state" and "history" were also converted to a numeric value. Regarding missing values, all are imputed and replaced by the median corresponding value. Lastly, the data was randomly split in 80% train data and 20% test data.

6.3. Train, Test and Results

Due to the specifications of our test machine, 64 GB of RAM, 20 cores CPU and a GPU with 8 GB, and the large size of parts of the datasets only some datasets were used, namely: Capture-1-1, Capture-8-1, Capture-3-1, Capture-42-1 and Capture-34-1. Within these datasets, only Capture-1-1 was balanced in the percentage of malicious an benign samples, so we decided to create 3 sub datasets: Capture-1-1 SMALL, Capture-1-1 MEDIUM and Capture-1-1 LARGE, created by randomly selecting samples of each category in a balanced way. The original Capture-1-1 dataset was also used with the name Capture-1-1 FULL. We decided to use this approach to analyze how the performance of the algorithms changes with more data. Table 3 presents in detail the information about the datasets used, including the time it took for each one to be processed. To train and test the model, we used a 80/20 data split, 80% for training and 20% for testing. The analyses were performed with a 5 fold cross validation.

Table 3. Datasets used.

File	Total Time	Total Samples	Benign%	Malicious%
1-1 SMALL	0:00:22	20,000	50.00%	50.00%
1-1 MEDIUM	0:23:09	200,000	50.00%	50.00%
1-1 LARGE	1:33:50	400,000	50.00%	50.00%
1-1 FULL	5:31:42	1,008,748	46.52%	53.47%
8-1	0:00:10	10,403	20.96%	79.03%
3-1	0:06:36	156,103	2.90%	97.09%
42-1	0:00:07	4426	99.86%	0.13%
34-1	0:00:29	23,145	8.30%	91.69%

Looking at the results from training and testing presented in Figures 10 and 11, we can see how each technique performed with each dataset.

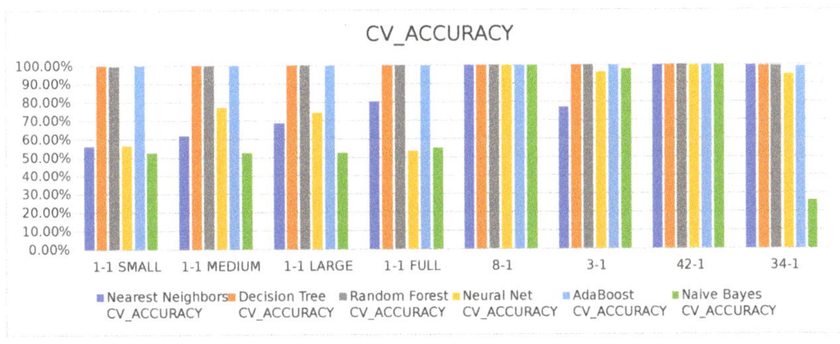

Figure 10. Accuracy of techniques per dataset.

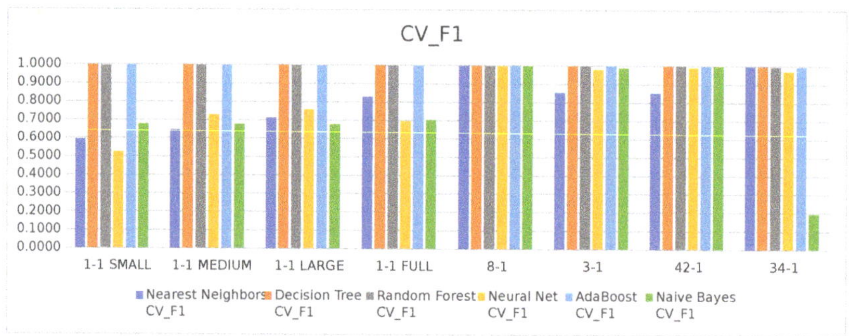

Figure 11. F1 score of techniques per dataset.

Regarding the unbalanced datasets, Capture-8-1, Capture-3-1, Capture-42-1 and Capture-34-1, almost all techniques achieved great results with the exception of Naive Bayes when using the dataset Capture-34-1, where the results were low performing. However, these datasets are unbalanced, and the technique might just be over fitting one of the results.

When looking at the results for the balanced datasets, Capture-1-1 SMALL, Capture-1-1 MEDIUM, Capture-1-1 LARGE and Capture-1-1 FULL we see different results, only Decision Tree, Random Forest and AdaBoost were capable of maintaining the strong results of both Accuracy and F1 score. The different amounts of data did not change the results, it only increased the processing time, with the 1-1_FULL dataset taking more than five and a half hours to process.

Lastly, it is necessary to analyze the percentage of false positives in these algorithms, this is because a model that generates a high percentage of false positives is unpractical and will generate more confusion rather than help find and stop cyber-attacks. This metric is presented in Table 4, where each algorithm's false positive rate is shown for each sub-dataset tested. Looking at the results, we can see that once again Decision Tree, Random Forest and AdaBoost are the best options since they obtained a false positive scores below 1% on all sub-datasets, while the other algorithms reached more than 20% on some occasions.

Table 4. False positive percentage of each algorithm per sub-dataset.

File	Nearest Neighbors	Decision Tree	Random Forest	Neural Net	AdaBoost	Naive Bayes
1-1 SMALL	25.43%	0.09%	0.59%	25.46%	0.00%	47.34%
1-1 MEDIUM	23.24%	0.00%	0.13%	8.13%	0.00%	47.39%
1-1 LARGE	20.09%	0.01%	0.08%	16.32%	0.00%	47.38%
1-1 FULL	13.99%	0.00%	0.04%	46.52%	0.00%	44.93%
8-1	0.02%	0.02%	0.02%	3.35%	0.02%	0.02%
3-1	1.12%	0.06%	0.00%	2.32%	0.00%	2.13%
42-1	0.25%	0.00%	0.00%	0.27%	0.00%	0.00%
34-1	0.29%	0.46%	0.38%	4.24%	0.28%	0.34%

7. Conclusions

The local energy market (LEM) is an emergent market model that is aimed towards solving the challenges currently faced in the energy landscape. One of the requirements for the success of LEM is trust in its negotiations. The main goals in this work are the development of a multi-agent system (MAS) for simulation and modeling LEM; and the proposal of a trust model capable of ensuring trust the LEM negotiations.

A MAS was developed with three types of agents, namely: (i) the Sensor Agent, (ii) the Participant Agent and (iii) the Market Interaction Manager (MIM) Agent, each with their own responsibilities, thus facilitating, the process of modeling the market.

To give a response to the needs of trust in the LEM, a formulation was proposed to calculate a trust value for each participant based on the analysis of the participant's historical data, contextual data, such as weather data, and by using forecasting methods to predict the participants expected behavior. The trust value given to participants evolves over time and takes into consideration its market submissions to the LEM, the forecasting of those submissions and considers the disparity between those values.

A case study was carried out in which several simulations were made with four participants using realistic consumption data and with different biases towards submitting false values. Each simulation used a different estimated forecasting mechanism with distinct levels of accuracy and precision.

The LEM was simulated for a 24 h period and 15 min market negotiation period duration, which resulted in a total of 96 market negotiation periods. This case study's aim was to evaluate the ability of the proposed trust formulation to respond to market needs by evaluating each participant with an appropriate trust value. The realization of the case study made it possible to conclude that: (i) The forecasting methodology used has a big impact on the performance of the trust formulation, but the acceptance formula also needs to be considered; (ii) a bad forecasting method, will provide a bad trust evaluation; and (iii) the higher the amount of false values a participant submits the lower their trust value will be, which is the desired outcome.

A study was carried out to evaluate the six supervised classifications techniques implemented. The training and testing of these classifications techniques were done using the IoT-23, a dataset containing IoT device data under malicious cyber-attacks. The classifications techniques were evaluated using the Accuracy and F1 score metrics. The results showed that the techniques of Decision Tree, Random Forest and AdaBoost provided excellent results. With these results in mind we believe that further studying is necessary with larger datasets and using multi-class classification in order to improve this cyber-attacks detection model. None the less, these results show that such an approach seems viable for the local energy market.

Lastly, one aspect we want to further improve is to develop the LEMMAS system in order to make use of the security and trust models at the same time, and developing a case study to evaluate how these models perform when working together.

Author Contributions: Conceptualization, R.A. and I.P.; methodology, R.A., T.P. and I.P.; software, R.A. and S.W.; validation, R.A., S.W., I.P., and T.P.; investigation, R.A.; resources, I.P.; data curation, R.A. and S.W.; writing—original draft preparation, R.A. and S.W.; writing—review and editing, I.P. and T.P.; supervision, I.P.; project administration, I.P.; funding acquisition, I.P. All authors have read and agreed to the published version of the manuscript.

Funding: This work has received funding from FEDER Funds through COMPETE program and from National Funds through FCT under the project SPET–PTDC/EEI-EEE/029165/2017 and UIDB/00760/2020.

Conflicts of Interest: The authors declare no conflict of interest.

References

1. Abidin, A.; Aly, A.; Cleemput, S.; Mustafa, M.A. Towards a Local Electricity Trading Market Based on Secure Multiparty Computation. 2016. Available online: https://www.esat.kuleuven.be/cosic/publications/article-2664.pdf (accessed on 6 August 2021).
2. Bremdal, B.A.; Olivella, P.; Rajasekharan, J. EMPOWER: A network market approach for local energy trade. In Proceedings of the 2017 IEEE Manchester PowerTech, Manchester, UK, 18–22 June 2017; pp. 1–6. [CrossRef]
3. Ampatzis, M.; Nguyen, P.H.; Kling, W. Local electricity market design for the coordination of distributed energy resources at district level. In Proceedings of the 2014 IEEE PES Innovative Smart Grid Technologies Conference Europe (ISGT-Europe), Istanbul, Turkey, 12–15 October 2014; pp. 1–6.
4. Teotia, F.; Bhakar, R. Local energy markets: Concept, design and operation. In Proceedings of the 2016 National Power Systems Conference (NPSC), Bhubaneswar, India, 19–21 December 2016; pp. 1–6.
5. Mendes, G.; Nylund, J.; Annala, S.; Honkapuro, S.; Kilkki, O.; Segerstam, J. Local Energy Markets: Opportunities, Benefits, and Barriers. 2018. Available online: https://www.cired-repository.org/handle/20.500.12455/1265 (accessed on 6 August 2021).

6. Rahimi, H.; Bekkali, H.E. State of the art of Trust and Reputation Systems in E-Commerce Context. *arXiv* **2017**, arXiv:1710.10061.

7. Sabater, J.; Sierra, C. Regret: A reputation model for gregarious societies. In Proceedings of the Fourth Workshop on Deception Fraud and Trust in Agent Societies, Barcelon, Spain, 4 June 2001; Volume 70, pp. 61–69.

8. Houser, D.; Wooders, J. Reputation in auctions: Theory, and evidence from eBay. *J. Econ. Manag. Strategy* **2006**, *15*, 353–369. [CrossRef]

9. Sabater, J.; Sierra, C. Reputation and Social Network Analysis in Multi-agent Systems. In *Proceedings of the First International Joint Conference on Autonomous Agents and Multiagent Systems: Part 1*; ACM: New York, NY, USA, 2002; pp. 475–482. [CrossRef]

10. Pinyol, I.; Sabater-Mir, J. Computational trust and reputation models for open multi-agent systems: A review. *Artif. Intell. Rev.* **2013**, *40*, 1–25. [CrossRef]

11. Yan, Y.; Qian, Y.; Sharif, H.; Tipper, D. A Survey on Cyber Security for Smart Grid Communications. *IEEE Commun. Surv. Tutor.* **2012**, *14*, 998–1010. [CrossRef]

12. Ghirardello, K.; Maple, C.; Ng, D.; Kearney, P. Cyber security of smart homes: Development of a reference architecture for attack surface analysis. In Proceedings of the Living in the Internet of Things: Cybersecurity of the IoT, London, UK, 28–29 March 2018.

13. Hall, F.; Maglaras, L.; Aivaliotis, T.; Xagoraris, L.; Kantzavelou, I. Smart Homes: Security Challenges and Privacy Concerns. *arXiv* **2020**, arXiv:2010.15394.

14. Sun, C.C.; Cardenas, D.J.S.; Hahn, A.; Liu, C.C. Intrusion Detection for Cybersecurity of Smart Meters. *IEEE Trans. Smart Grid* **2020**, *12*, 612–622. [CrossRef]

15. Niu, X.; Li, J.; Sun, J.; Tomsovic, K. Dynamic detection of false data injection attack in smart grid using deep learning. In Proceedings of the 2019 IEEE Power & Energy Society Innovative Smart Grid Technologies Conference (ISGT), Washington, DC, USA, 18–21 February 2019; pp. 1–6.

16. Wang, C.; Tindemans, S.; Pan, K.; Palensky, P. Detection of false data injection attacks using the autoencoder approach. In Proceedings of the 2020 International Conference on Probabilistic Methods Applied to Power Systems (PMAPS), Liege, Belgium, 18–21 August 2020; pp. 1–6.

17. Fatemifar, S.; Awais, M.; Arashloo, S.R.; Kittler, J. Combining multiple one-class classifiers for anomaly based face spoofing attack detection. In Proceedings of the 2019 International Conference on Biometrics (ICB), Crete, Greece, 4–7 June 2019; pp. 1–7.

18. Oliveira, N.; Praça, I.; Maia, E.; Sousa, O. Intelligent cyber attack detection and classification for network-based intrusion detection systems. *Appl. Sci.* **2021**, *11*, 1674. [CrossRef]

19. Ring, M.; Wunderlich, S.; Grüdl, D.; Landes, D.; Hotho, A. Flow-based benchmark data sets for intrusion detection. In Proceedings of the 16th European Conference on Cyber Warfare and Security; Dublin, Ireland, 29–30 June 2017; pp. 361–369.

20. Ring, M.; Wunderlich, S.; Grüdl, D.; Landes, D.; Hotho, A. Creation of flow-based data sets for intrusion detection. *J. Inf. Warf.* **2017**, *16*, 41–54.

21. Meira, J.; Andrade, R.; Praça, I.; Carneiro, J.; Bolón-Canedo, V.; Alonso-Betanzos, A.; Marreiros, G. Performance evaluation of unsupervised techniques in cyber-attack anomaly detection. *J. Ambient Intell. Humaniz. Comput.* **2020**, *11*, 4477–4489. [CrossRef]

22. Tavallaee, M.; Bagheri, E.; Lu, W.; Ghorbani, A.A. A detailed analysis of the KDD CUP 99 data set. In Proceedings of the 2009 IEEE Symposium on Computational Intelligence for Security and Defense Applications, Ottawa, ON, Canada, 8–10 July 2009; pp. 1–6.

23. Shiravi, A.; Shiravi, H.; Tavallaee, M.; Ghorbani, A.A. Toward developing a systematic approach to generate benchmark datasets for intrusion detection. *Comput. Secur.* **2012**, *31*, 357–374. [CrossRef]

24. Andrade, R.; Pinto, T.; Praça, I. Trust Model for a Multi-agent Based Simulation of Local Energy Markets. In *International Conference on Practical Applications of Agents and Multi-Agent Systems*; Springer: Berlin, Germany, 2020; pp. 183–194.

25. Open Data Sets IEEE PES Intelligent Systems Subcommittee. Available online: https://site.ieee.org/pes-iss/data-sets/ (accessed on 8 May 2012).

26. Garcia, S.; Parmisano, A.; Erquiaga, M.J. IoT-23: A Labeled Dataset with Malicious and Benign IoT Network Traffic (Version 1.0.0). Available online: https://zenodo.org/record/4743746#.YR3cFt8RVPY (accessed on 6 August 2021).

 *electronics*

Article

SPD-Safe: Secure Administration of Railway Intelligent Transportation Systems

George Hatzivasilis [1,2,*], Konstantinos Fysarakis [3], Sotiris Ioannidis [4], Ilias Hatzakis [2], George Vardakis [2], Nikos Papadakis [2] and George Spanoudakis [3]

[1] Institute of Computer Science, Foundation for Research and Technology—Hellas, Vassilika Vouton, GR-70013 Heraklion, Greece

[2] Electrical and Computer Engineering, Hellenic Mediterranean University (HMU), Estavromenos, GR-71410 Heraklion, Greece; hatzakis@cs.teicrete.gr (I.H.); gvardakis@cs.hmu.gr (G.V.); npapadak@cs.hmu.gr (N.P.)

[3] Sphynx Technology Solutions AG, Innovation Department, 6300 Zug, Switzerland; fysarakis@sphynx.ch (K.F.); spanoudakis@sphynx.ch (G.S.)

[4] Electrical and Computer Engineering, Akrotiri Campus, Technical University of Crete, GR-73100 Chania, Greece; sotiris@ece.tuc.gr

* Correspondence: hatzivas@ics.forth.gr; Tel.: +30-2810-391600

Abstract: The railway transport system is critical infrastructure that is exposed to numerous man-made and natural threats, thus protecting this physical asset is imperative. Cyber security, privacy, and dependability (SPD) are also important, as the railway operation relies on cyber-physical systems (CPS) systems. This work presents SPD-Safe—an administration framework for railway CPS, leveraging artificial intelligence for monitoring and managing the system in real-time. The network layer protections integrated provide the core security properties of confidentiality, integrity, and authentication, along with energy-aware secure routing and authorization. The effectiveness in mitigating attacks and the efficiency under normal operation are assessed through simulations with the average delay in real equipment being 0.2–0.6 s. SPD metrics are incorporated together with safety semantics for the application environment. Considering an intelligent transportation scenario, SPD-Safe is deployed on railway critical infrastructure, safeguarding one outdoor setting on the railway's tracks and one in-carriage setting on a freight train that contains dangerous cargo. As demonstrated, SPD-Safe provides higher security and scalability, while enhancing safety response procedures. Nonetheless, emergence response operations require a seamless interoperation of the railway system with emergency authorities' equipment (e.g., drones). Therefore, a secure integration with external systems is considered as future work.

Keywords: intelligent transportation; railway; CPS; security; safety; critical infrastructure

Citation: Hatzivasilis, G.; Fysarakis, K.; Ioannidis, S.; Hatzakis, I.; Vardakis, G.; Papadakis, N.; Spanoudakis, G. SPD-Safe: Secure Administration of Railway Intelligent Transportation Systems. *Electronics* **2021**, *10*, 92. https://doi.org/10.3390/electronics10010092

Received: 18 November 2020
Accepted: 29 December 2020
Published: 5 January 2021

Publisher's Note: MDPI stays neutral with regard to jurisdictional claims in published maps and institutional affiliations.

1. Introduction

Railways continue to be one of the main transport systems nowadays [1,2], covering public, private, and military needs over a wide operational area. Thus, railway assets are an attractive target for malicious actors and are exposed to various threats, from natural events to man-made ones, such as terrorism or vandalism (e.g., [3–6]).

The associated risks are exacerbated by the fact that railway infrastructure assets are typically placed along the route, including remote areas where physically protecting them is challenging. Moreover, railway premises have a large attack surface (due to their numerous electronic and electrical parts, such as power supply, switches, scheduling, and other subsystems), but often reside far from the main stations. While auditing for physical threats is quite important [7], the premises are usually inspected remotely through cameras. Sensory equipment is also deployed to monitor environmental parameters. The goal is to prevent potential intruders [8–10], avoid machinery overheating, and detect fires. Since the interconnection of this monitoring equipment is, at least partly, wireless, it can become a target of several types of attacks.

In this context and considering that a successful attack could damage the railway's operation or even cause severe injuries and deaths, cybersecurity is an important consideration for such interconnected critical systems [11,12]. Attackers can disrupt communications (e.g., through jammers) or even infiltrate the networks and take control of critical equipment [13]. Cyber-attacks on the command-and-control centers (C&C) and the information systems are also feasible [13,14]. Thus, the secure interconnection of all the deployed elements and platforms is important, and the cyber and physical security of the critical infrastructure becomes imperative [9,10].

As sketched above, safety is another design factor and one that is closely related to security. Cargo and passengers are transported in high volumes each day, covering long distances. In the past, railway accidents have caused a number of deaths, along with significant financial losses [15]. While the introduction of electronic controllers reduced the occurrence of such situations [16], safety risks cannot be ignored, considering the wide railway coverage that still includes aspects such as uninspected car-crossings, system malfunctions (like signal loss), and, of course, the human factor [17,18].

Within the ever-changing technological landscape, there is currently a move from automated to intelligent cyber-physical systems (CPS), motivated by the speedy infiltration of the Internet of Things (IoT) and cloud computing and enabled by wireless networking [19–22]. Wireless sensor networks (WSNs) [23] can cover the wide railway operational territory, gathering and processing pieces of ambient knowledge, while gateways can be used to transmit the data to the controlling center or a cloud service. The railway controlling software at the backend can, then, collect and integrate the spatial information and manage the underlying subsystems [24,25]. Therefore, WSNs are an ideal solution for covering the railway operating area, including the railway routes and various scattered shelters.

However, the railway cyber infrastructure and networks currently only adopt rudimentary defenses (e.g., cryptography), which provide protection against the most basic threats, forfeiting effective ways of detecting advanced cyber-attacks [26]. While initially designed as closed systems, current infrastructure networks are vulnerable to various network layer attacks, like blackhole, badmouthing, and jamming attacks [27].

Motivated by the above, this work presents "SPD-Safe", (security, privacy, and dependability (SPD)), an administration framework for railway CPS, aiming to enhance the security, privacy, dependability, and safety of the intelligent railway infrastructure, while enabling services for monitoring and managing the overall setting. The framework integrates mechanisms for mitigating cyber-attacks attempting to disrupt communications or compromise infrastructure assets, and periodic malfunctioning of assets is also taken into consideration. SPD-Safe can act as an intelligent communications-based train control (CBTC) system for railway CPS, leveraging artificial intelligence (AI) to manage the system at runtime. The system uses standardized solutions, and its building blocks can be easily retrofitted in current deployments.

In addition to the detailed description of the proposed framework, a preliminary implementation is described and evaluated, concentrating on the management of: (a) In-carriage, and (b) on-route sub-systems. WSNs are deployed inside the carriage and by the railway tracks to safeguard carriages that transfer dangerous freight and to help avoid crashes with objects blocking the train's route (like stuck vehicles on rail track crossings), respectively. Furthermore, smart cameras are installed to improve the physical security of the critical infrastructure. In the context of the two use cases (a) and (b) above, through SPD-Safe the railway CPS is configured in real-time to tackle ongoing cyber-attacks and control safety-related incidents. This hands-on validation was developed and demonstrated under the EU-funded project new embedded Systems arcHItecturE for multi-Layer Dependable solutions (nSHIELD) [28], with the cooperation of major industrial partners in the railway and defense domains, including Ansaldo STS (http://www.railway-technology.com/contractors/signal/ansaldo-sts/), Selex ES (now Leonardo

S.p.A.: https://www.leonardocompany.com/en/home), and HAI (http://www.haicorp. com/en/). Simulation analysis was also conducted during the design phase, utilizing the security-aware Cyber-Physical Systems (CPS) Simulator Framework (COSSIM) [29], paving the way for the final installation of the proposed system, as presented in the following sections.

The rest of the paper is structured as follows: In Section 2, related work on railway signaling systems is reviewed. In Section 3, the middleware platform and intelligent agent technologies that manage the underlying equipment are presented. In Section 4, the network layer protection mechanisms are detailed. In Section 5, the implementation details of SPD-Safe are provided and the application in the railway setting is demonstrated. The proposed system is also compared with relevant systems in Section 6, while Section 7 features the concluding remarks.

2. Materials and Methods—Related Work

Smart transportation ecosystems involve, among others, passenger services as well as critical infrastructure-related applications and the associated safeguards. The fundamental goals in this context include "green" (i.e., environment-friendly) operation, improved performance and efficacy, as well as enhanced security and safety.

Railways, in specific, rely on signaling systems that direct the trains' traffic. Infrastructure control and management is achieved via various telecommunication means that are installed on carriages and tracks. Communication between track equipment and trains is achieved via CBTC signaling systems [30–32] enabling the railway's management and infrastructure control. For the European Union (EU), the international wireless communications standard for railways includes the European Train Control System (ETCS) [33]. The communication baseline is implemented by the Global System for Mobile Communications—Railway (GSM-R) [34], which is further enhanced with the General Packet Radio Service (GPRS) [35] and forms the base of an intelligent transportation application. ETCS utilizes trackside equipment that transmits information regarding the route to unified controlling equipment within the train cab. Thus, all lineside data are passed wirelessly to the driver, without requiring the direct observation of lineside visual signals, as was the case in legacy railway settings. The adoption of ETCS results in more and longer running trains, with increased traffic and railway management capabilities.

In addition to the signaling developments, WSNs can now cover a wide railway operational area, gathering ambient data. Embedded systems implement intelligence solutions encompassing the underlying critical assets as well the interlinked smart city ecosystems. Related frameworks for intelligent monitoring of the critical infrastructure have already been proposed in the literature (e.g., [36,37]). The Integrated System for Transport Infrastructure surveillance and Monitoring by Electromagnetic Sensing (ISTIMES) project [36] implements a transport infrastructure surveillance and monitoring system with electromagnetic sensing. Distributed and local sensory equipment (e.g., optic fiber sensors, infrared thermography, low-frequency geographical techniques, etc.) are utilized to perform non-destructive electromagnetic sensing and monitoring of the critical infrastructure. The Cloud to Infrared Thermography (Cloud2IR) [37] deploys an infrared and environmental Structural Health Monitoring (SHM) information system. The software architecture enables multi-sensor connection and the interplay with cloud computing services (e.g., data aggregation, system management, etc.). However, the heterogeneity of the deployed equipment and diverse demands of the various applications make the administration of the underlying infrastructure a challenging task.

In parallel, as Service-oriented Architectures (SoAs) increase in popularity, a continuous effort to deploy SoAs within the Industrial IoT (IIoT) domain and the smart railway CPSs can be observed. Several technologies are proposed that support the required functionality, ranging from agent frameworks and middleware platforms, to communication protocols and data representation standards. Such state-of-the-art solutions are presented in the subsections that follow.

2.1. Management Platfroms and Reasoning Systems

Agent technologies constitute the typical option for modeling ambient intelligent systems that exchange information with the environment and user [38,39]. Intelligent agents inspect the surrounding setting and react to upcoming events at normal operation. Their AI modules process context-aware data, as collected from the surrounding environment by the attached devices.

Regarding the various agent technologies, 24 frameworks were analyzed in Kravari and Bassiliades [40], including the popular Java Agent DEvelopment framework (JADE), Agent Globe (A-GLOBE), and Jason (the hero's name from Greek mythology). JADE implements the relevant standards for Semantic Web and the Foundation for Intelligent Physical Agents (FIPA: http://www.fipa.org/) (e.g., the Agent Communication Language (ACL) [41]). The platform is easy to learn and user-friendly, while offering portability and compatibility with all Java Virtual Machines (JVMs). The open-source and stable developer versions operate with several programming languages, such as Java, Jess, and Prolog. The agent communication is fast, and the overall framework is efficient and scalable. Moreover, JADE supports strong user authentication and cryptographic solutions—i.e., JADE security (JADE-S)—along with Hypertext Transfer Protocol Secure (HTTPS). The framework is widely-used and is deployed in several fields, including reasoning in multiple domains, general purpose applications, mobile computing, and e-commerce. The study of Kravari and Bassiliades [40] also infers that JADE is the most popular framework due to the pure Java design and the co-operation with several web systems. In addition, five respectable organizations (France Telecom, Motorola, Profactor, TILAB, and Whitestein TEchnologies AG) supervise the framework [40].

Regarding middleware systems and messaging protocols, a comparative analysis of relevant IoT solutions (Constrained Application Protocol (CoAP), Message Queuing Telemetry Transport (MQTT), and Devices Profile for Web Services (DPWS)) was conducted in Fysarakis et al. [42]. DPWS [43], by the Organization for the Advancement of Structured Information Standards (OASIS: https://www.oasis-open.org/), constitutes the benchmark in terms of ease-of-design. The framework is flexible and robust in terms of service eventing, discovery, and subscription; following initialization, the underlying devices can discover the provided services and communicate in a seamless manner.

Finally, concerning the deductive rule engines that enable the AI reasoning features, National Aeronautics and Space Administration (NASA) examines the capabilities of several related approaches (Jess, Drools, Microsoft Business Rule Engine, Official Production System Java (OPSJ), and Intelligence Logiciel (ILOG)), as described in [44,45]. Jess is efficient and excels in many categories. It works with dynamic facts and dynamism in variables and rules and is appropriate for NASA's mission-critical applications as well as many other research areas [46,47].

2.2. Intelligent Railway Systems

To the best of the authors' knowledge, only a few multi-agent systems (MAS) have been developed and tested on actual railway environments [48,49]. In addition, despite strong industrial involvement, not all developed agent capabilities are fully used in practice and thus, the full potential of agent technologies is not exploited fully. Three indicative installations on railway settings are: Train Integrity (TrainIntegrity) [50], Condition Monitoring of a Light Rail Vehicle and Track (CMLRVT) [51], and Sensor Networks for Railways (SENSORAIL) [52].

TrainIntegrity utilizes WSNs to check the integrity of cargo trains [50]. The nodes consist of the RCM 3400 RabbitCore module and sense environmental parameters. The WSN raises an alarm if it infers that an unexpected change has occurred in the train's composition. CMLRVT is built and tested on a tramway operation in Poland. The system consists of a dispersed sensor network that is installed on the vehicles and railway infrastructure, along with the data acquisition component and a data server that maintains the artifacts of the management and analysis procedures. At first, the system collects data during normal operation, which is stored in the server. Then, the new pieces of knowledge that are sensed by the devices are compared with the nominal values. The detected variations are further analyzed by the system, revealing safety-related incidents (e.g., rail cracks) that are presented to the user through a dedicated application. However, neither of the two systems considers security issues at the network link, nor do they integrate and manage the heterogeneous underlying embedded systems.

SENSORAIL [52] is an early warning system for the railway monitoring infrastructure. WSNs collect and integrate data to enable the detection of structural failures and security threats. Sensor clusters communicate information towards distant controlling centers through GSM-R/GPRS mobile equipment. The integration of heterogeneous sensors is managed by a component referred to as "scalable software architecture for the integration of heterogeneous sensor systems" (SeNsIM) [53], while the detection of events is made by a model-based data correlation component, called "novel framework for the detection of attacks to critical infrastructure" (DETECT) [54]. SENSORAIL specifies the examined threats in the Event Description Language (EDL) [55] and maintains them within a scenario repository. Upcoming events are stored in a history database and model-checking is performed at runtime. Regarding the middleware and agent platform, SeNsIM does not utilize any semantic technologies and does not support related standards, thus lacking in terms of interoperability and ease of integration with existing setups. Moreover, SENSORAIL does not include any protection mechanisms, solely focusing on the detection of threats.

2.3. Network Layer Protection

Several schemes are suggested in the literature for protecting communication in WSNs, attempting to address pertinent security concerns (e.g., [56–59]).

The Reputation-based Framework for Sensor Networks (RFSN) [56] authenticates underlying nodes with the Timed Efficient Stream Less Tolerant Authentication protocol (μTESLA) [60], implementing a beta Bayesian formulation for fading and evaluating the reputation of the routing operation and the legitimacy of the reported sensed variables. Ariadne [57] also utilizes a TESLA variant for authentication, collecting feedback regarding the successful delivery of packets to choose optimum communication paths and avoid malicious behavior. The Cooperative Secure routing protocol based on ARAN (CSRAN) [58] integrates digital certificates and asymmetric cryptography for authentication. As in the case of RFSN, it uses a Bayesian distribution for fading and, when a node detects malicious activity, it automatically re-routes communication from that point on. The Secure Resilient Reputation-based Routing (SR3) [59] adopts lightweight cryptography (LWC) and symmetric modules for security and authentication. Fading is accomplished by a First In, First Out (FIFO) finite list. The system combines reputation with a reinforced random walk algorithm, producing enhanced load-balancing at the cost of a high intermediate forwarding node count.

Despite the plethora of proposed solutions, and while most can tackle basic security attacks and malfunctioning cases, there are still various open avenues for attackers, including flooding attacks in congested periods, topology-related attacks, and jamming [61,62].

3. Administration of IoT Deployments

Considering the landscape sketched above, this section presents the proposed SPD-Safe solution, and more specifically the deployed platform and the reasoning process of each SPD-Safe agent. The core reasoning engine has been previously presented by the authors in [63]. This version enhances the network layer security and is applied in a mobile setting, forming an intelligent transportation system that complies with the real-time requirements of CBTC for railways.

SPD-Safe comprises a framework that integrates variants of the aforementioned primitives (i.e., agents, middleware, rule engine, and network layer protection) across all system layers to implement an efficient, scalable, practical, and easy-to-deploy and maintain solution, with adequate reasoning and management capabilities. From top-to-bottom, the system consists of four layers: (i) An overlay with intelligent agents that control distinct subsystems; (ii) a middleware platform that enables communication between the agents and underlying networks; (iii) the network layer that consists of interconnected IoT devices; and (iv) the node layer that represents the devices themselves. The core technological building blocks will be detailed in the subsections that follow.

3.1. Agent Technologies & Middleware Solutions

SPD-Safe utilizes JADE [64] as the top-layer multi-agent system. It adopts and implements standardized approaches to agent deployment, such as the ACL [41] by FIPA. The JADE-S add-on [65,66] safeguards communication at the overlay and offers built-in security functionality for confidentiality, integrity, authentication, and authorization.

Then, each agent is ported as a bundle in the middleware platform Open Service Gateway initiative (OSGi) [67]. Through it, an agent can monitor the underlying subsystem, enhancing real-time management. Network gateways also deploy a controlling bundle in the same platform, defining the offered functionality as a service in the DPWS standard [43]. The agent and the related network bundles interchange well-structured semantic data, as defined in the OASIS standardized Common Alerting Protocol (CAP) [68]. The OSGi platform also provides its own built-in security features for the inner-platform communication, limiting bundle functionality to pre-defined capabilities and protecting both the agent and controller bundles.

Here, other than these built-in features that are provided by the deployed platforms at the overlay and middleware layers, SPD-Safe integrates an additional defense mechanism for the network and node layers, namely Secure Route (SecRoute) [61], a security protocol that protects the wireless ad hoc communication of the underlying embedded devices. This protocol counters several types of threats and attacks at the network link, protects the nodes' assets and their resource consumption, and acts as an intrusion detection module for the upper layers. When a security incident is recorded, the network gateway bundle will send related CAP messages to the responsible agent, which may take further action. SecRoute is detailed in the next section.

Metrics that evaluate the various system aspects are now an integral feature of the development cycle. They offer a quantitative indication regarding the compliance with the targeted requirements of the application domain. An evaluation method for the estimation of the security, privacy, and dependability (SPD) properties for configurable embedded systems is presented by the authors in Hatzivasilis et al. [63]. For every configuration option, the metrics derive a triple vector of <Security, Privacy, Dependability>, whereby the vector's factors are assigned a value from 0–100, representing no to full protection respectively. SPD-Safe adopts this methodology to enable a metric-driven SPD- and safety-aware administration, where the reasoning procedure triggers runtime system adaptations to reach specific SPD goals [69].

3.2. Reasoning Capabilities & Conflict Resolution

The artificial intelligence (AI) behavior of each agent is developed in the rule engine Jess [70,71]. For knowledge representation and reasoning, the Jess-EC [72] is used. The latter is an Event Calculus (EC) [73] implementation in Jess, offering the required semantics modeling. SPD-Safe's software layers are illustrated in Figure 1.

Each agents' AI procedure implements automated temporal, casual, and epistemic reasoning with real-time events, action preconditions, rule priorities, indirect effects, context-sensitive side-effects, as well as the common law of inertia. Moreover, the reasoning capabilities can cope with the requirements of dynamic and partially known or uncertain domains.

However, as agents exchange information, contradictory reasoning results may occur due to the local viewpoint of each entity and the lack of global knowledge. Thus, for resolving conflicts, SPD-Safe introduces the epistemic mechanism of share theories [63]. The participating entities send the involved theory rules to a mediator agent, along with the recently sensed local events. The mediator combines these elements and performs a reasoning operation that determines the final outcome and the state of the conflicting assets.

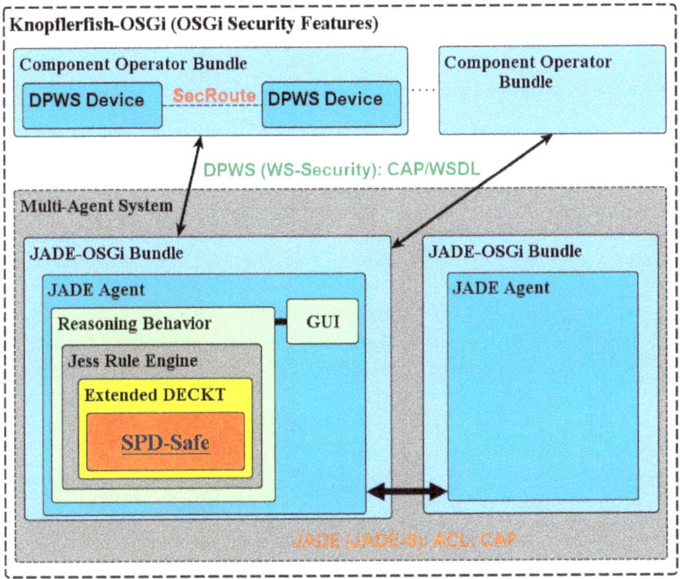

Figure 1. The software layers of the proposed security, privacy, and dependability (SPD)-Safe framework.

Nevertheless, if an agent utilizes protected data that must be maintained locally and not distributed (e.g., confidential information regarding user policies or system settings), it will not be able to contribute in the share theory with its full knowledge. For this occasion, an alternative relational grading mechanism, called certainty degree [63], resolves the affair quickly and efficiently. The mechanism utilizes subjective criteria as well as the agents' roles and hierarchy, marshaling the problem without constructing the related share theory and retaining the system's coherency. Thus, the certainty degree is applied in affairs where reasoning with locally protected data is involved, otherwise a share theory is constructed.

3.3. SPD Measurement

The SPD multi-metric methodology [69] measures the provided protection level of a system and its various configurations. The system's perimeter is identified and the data sources, entry, and exit points are recorded. Then, the mechanisms that protect each of these elements are assessed based on the standardized Criteria Evaluation Methodology (CEM) [69]. This involves the attack potential risk analysis that evaluates the attacker's motive to misuse specific system elements, expertise, and the resources that they are willing to devote for an attack. Henceforth, five parameters are examined for the analysis of a potential threat:

-   Required time: The time that it is required to perform a specific attack (e.g., in days or weeks);
-   Expertise: The technical skills and knowledge that the attacking group can exhibit (such as copy-cat, advanced, or expert);
-   Knowledge of the target: Familiarity with the targeted system and its operation (e.g., public, sensitive, or critical information concerning some subsystems, etc.);
-   Window of opportunity: The attacker may require appreciable access to the system in order to exploit a vulnerability and avoid detection;
-   Resources: The software, hardware, or other equipment that is necessary to perform an attack (such as specialized or common resources).

The method does not investigate every possible attack but educes a good indication of the defense status in accordance with standard ratings. The protection level for each of the three SPD properties is calculated by integrating the risk analysis with the efficacy of the installed defenses against known attacks and/or other limitations (e.g., based on the latest reports from Computer Emergency Response Teams (CERTs) or Common Vulnerabilities and Exposures (CVE) repositories). The result is a value in the range of 0–100, where 0 represents the absence of defense mechanisms and 100 represents full protection. The final outcome is a vector of <Security, Privacy, Dependability>, which represents the total SPD value of the currently composed setting of the system. The SPDs of different system configurations can be estimated either in advance or at runtime. The first option is leveraged by the AI units of SPD-Safe in order to perform proactive and/or automated changes in the state architecture when a safety or security event occurs. The second option provides indications to the human operator in order to take decisions and make manual interventions.

Therefore, the protection status of all mechanisms and their integration in the demonstration examples are pre-calculated based on this method, as described in Sections 4 and 5. Then, automated administration policies are triggered in response to real-time events, as presented in Section 5.

These features enable the implementation of a relative novel protection strategy, called Moving Target Defenses (MTDs) [74]. When a system is stable, it is seen as a "sitting duck" by the attacker, who has plenty of time to analyze it, detect potential vulnerabilities, and exploit them. With MTD, a system that is aware of the defense level of its various components, their configurations, and the integration of all of them, can alter the setting automatically or semi-automatically in a periodic fashion. The AI modules are always keeping the system in a secure state, while the different configuration and architectural sets increase the system states that have to be analyzed by the attacker. In addition, the time that a specific setting remains active is determined by the time required for an average hacker to analyze it (i.e., based on the "Required Time" factor of the attack potential risk analysis). Performing attacks is becoming quite hard, while the window of opportunity for the malicious entities has significantly decreased.

3.4. AI Processing & Performance

The reasoning component of Jess implements the RETE algorithm (Latin word for net, meaning network in this domain) [75]. This is the most widely-used pattern matching technique for rule-based systems and is optimized for speed. Scalability and performance are affected by the three factors of: (i) The rules' volume (R), (ii) the average number of patterns in the left-hand-side of each rule (P), and (iii) the facts in the working memory (F). Computational complexity is linear to the working memory size and in the order of O (RPF). For each SPD agent in the railway mission-critical applications that are examined in the following sections, the theory rules volume (R) is very low (around 30 rules per scenario). In order to reduce the pattern-matching space, unique identifiers are assigned to every modeled entity, and therefore, occurring events affect specifically defined parameters, keeping the pattern-matching ration low (P) and in the order of 1–3. Performance is mostly influenced by the number of facts (F). In the demonstrated cases, it requires 10–20 facts per scenario. The computational overhead for an SPD agent is in the range of a nanosecond with additionally 50 bytes in memory.

For a central agent that collects information from the whole railway system, it requires around 500 facts and 40 rules to model the underlying setting. At boot time, the reasoning engine takes to run around 1.6 s, 87 MB for code, and 45 MB in RAM. Then, a reasoning process for a theory and a few hundreds of facts would require 0.002 s on average, representing the actual delay that affects the applications.

3.5. Relevant Methodologies for Secure IoT Modeling

Over time, several solutions have been proposed that try to resolve the open issues of capturing the security posture of an IoT or other system and facilitate its administration [76–80]. Eby et al. [76] integrated the Simple Modeling Language for Embedded Systems (SMoLES) with the Security Model Analysis Language (SMAL) [76]. SMAL provides security extensions to the composition meta-model of the Domain Specific Modeling Language (DSML) [77] and can express access control policies for IoT applications. The resulting framework is called SMoLES Security (SMoLES-SEC). However, its reasoning capabilities are bounded due to the constrained expressiveness of the underlying SMAL. Furthermore, SMoLES-SEC cannot deduce which security characteristics hold after the compositions of two components or the final security status of the composed system.

Service Dependency Trees (SDTs) [78] support the verification of service secure composition in IoT ecosystems. The IoT devices/nodes construct their own SDT. For each provided service, the relevant SDT defines the potential external service nodes that the service is depending on. The nodes are also aware of all recursive SDTs for their composed services. Thus, secure service composition is performed by enabling integration only with SDTs where all paths and involved entities are trusted. On the other hand, creating a SDT for a real IoT application is not trivial, while trustworthiness and consistency in an actual complex and dynamic environment may be challenging.

Albanese et al. [79] utilize attack surface metrics in order to evaluate the security aspects of system and materialize MTDs strategies. This solution calculates the distance of the security surface of the various system states. The goal is to administrate responses against ongoing attacks as well as to deduce a system setting that exhibits specific desirable parameters. Techniques for assessing and reducing the cost for the defender are also included.

Savola and Sihvonen [80] propose a MTD approach based on a multi-metric-driven management framework. The overall solution has been applied in an e-health digital environment for chronic diseases [80], where three metric types are considered. Risk-driven security assurance and engineering metrics are defined at deployment-time to offer an early assessment on the deployed defense mechanisms and their effectiveness. Continuous security monitoring metrics are determined at operational-time, enabling the security correctness assessment, enhanced systematization, and traceability of the various product requirements and involved metrics. Thereupon, automated adaptive decision-making metrics are assigned at operational-time and accomplish a higher quality security effectiveness understanding in operational security auditing and future versioning of the system. The method supports continuous security monitoring and automated metric-driven security-related actions.

Table 1 presents the outcomes of the qualitative analysis. The modeling expressiveness of SPD-Safe is quite general and can also be utilized in complex and dynamic systems. Moreover, it assesses all three security, privacy, and dependability properties and can evaluate their status both before a composition is performed and after the integration of the system. As with the other relevant approaches, the MTD features are driven by metrics and SPD-Safe provides a concrete implementation of this modern defense type. The overall solution fits with the distributed nature of IoT ecosystems and can resolve conflicts that may arise due to knowledge sharing between the various entities.

Table 1. AI (artificial intelligence) modeling features.

Feature	SPD-Safe [This Paper]	SMoLES-SEC [76]	SDT [78]	Attack Surface MTD [79]	Multi-Metric-Driven MTD [80]
System composition					
Expressiveness generality	Y	N	Y	N	N
Dynamicity	Y	Y	Y	N	N

Table 1. *Cont.*

Feature	SPD-Safe [This Paper]	SMoLES-SEC [76]	SDT [78]	Attack Surface MTD [79]	Multi-Metric-Driven MTD [80]
Validation					
Pre-composition	Y	Y	Y	N	N
Post-composition	Y	N	N	N	N
Evaluated Properties					
Security	Y	Y	Y	Y	Y
Privacy	Y	N	N	N	N
Dependability	Y	P	P	N	P
Artificial Intelligence					
Distributed reasoning/processing	Y	N	Y	N	Y
Conflict resolution	Y	N	N	N	N
MTD	Y	N	N	Y	Y

Y(es), N(o), P(atrial). Service Dependency Trees (SDTs); Moving Target Defenses (MTDs); Simple Modeling Language for Embedded Systems Security (SMoLES-SEC).

4. Network Layer Security

The protection of the network link is essential in order to safeguard the underlying systems of critical railway infrastructure (e.g., WSN, signaling equipment, surveillance, etc.). For this purpose, as mentioned, SecRoute [61] is developed; a novel defence primitive that provides the core security properties for authentication, integrity, and confidentiality, along with energy-aware secure routing and authorization.

The secure routing protocol protects the involved entities from malicious operations while improving performance and offering load-balancing. It consists of three main primitives:

- The cryptographic service with the Timed Efficient Stream Less Tolerant Authentication protocol (μTESLA) [60], which implements message authentication, confidentiality, and integrity;
- The efficient secure routing service with the Self-Channel Observation Trust and Reputation System (SCOTRES) [62] that safeguards the communication link against ad-hoc routing attacks and network layer vulnerabilities;
- The authorization service with the Policy-Based Access Control framework (PBAC) [42], which offers authorization and access control based on policies.

Table 2 summarizes the overall security properties that are provided by the integrated network layer defense mechanism and the relevant threats and attacks that are countered, while a brief analysis is presented in the subsections that follow. More details regarding the three services are presented in the relevant papers for μTELSA [60], SCOTRES [62], and PBAC [42], respectively.

Figure 2 presents the block diagram of the main SPD-Safe modules and their connection.

Table 2. Protection aspects of the SPD-Safe's network layer security.

Primitive	System Property	Countered Threats
μTESLA	Authentication	Impersonation, Sybil attacks
	Integrity	Data tampering, modification, interruption
	Forward security	Replay attacks
	Confidentiality (optional)	Disclosure

Table 2. *Cont.*

Primitive	System Property	Countered Threats
	Topology-awareness	Attacks on topology-significant entities
	Energy-awareness & Load-balancing	Energy dissipation, overloading attacks on congested periods
	Channel health	Jamming attacks
	Reputation	Malicious or selfish activity on the network operations of: - Routing (link spoofing, routing table poisoning, HELLO flooding, false link break, loops, nonexistent paths), - Forwarding (blackhole, grayhole, sleep deprivation), - Or making recommendations (badmouth, ballot-stuffing)
SCOTRES	Trust	Overall misbehavior on the previously mentioned networking perspectives
	Secure routing	General attacks on the pure routing protocol (e.g., Denial of Service (DoS), inject arbitrary packets)
PBAC	Authorization based on policies	Unauthorized access

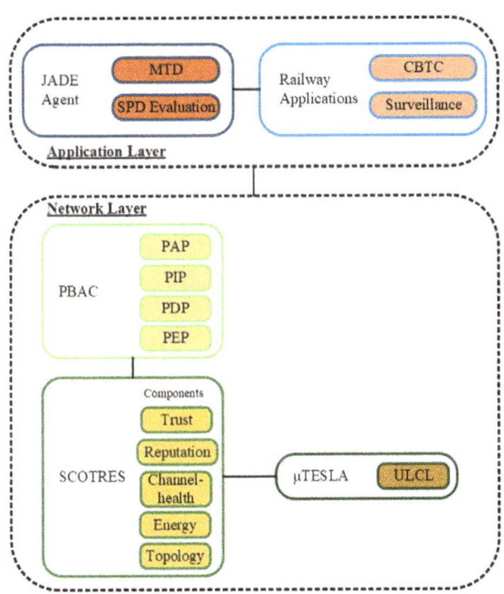

Figure 2. The building blocks of the SPD-Safe framework.

4.1. Cryptographic Service—μTESLAs

μTELSA is a building-block for the Sensor Protocols for Information via Negotiation (SPIN) [81]. Loose time synchronization is required between the receiver and sender, with μTESLA utilizing broadcast messages and symmetric cryptography to implement the aforementioned core cryptographic properties. The security functionality of asymmetric cryptography is achieved by utilizing keyed Message Authentication Code (MAC) operations. In brief, the sender includes a keyed MAC on every transmitted packet, where this key is initially known only to this entity. Receivers maintain the received packets without authenticating the sender at this point. Shortly after, the key is revealed by the sender and then the receiver authenticates the packet and proceeds to further processing. Otherwise, the receiver discards the unauthenticated packets after a time-slot.

The protocol μTESLA is efficient and exhibits low computational and communicational overheads. It also tolerates packet loss and scales well for large networks. We use the Ultra-Lightweight Cryptographic Library (ULCL) [61] in order to develop the cryptographic functionality of μTESLA, adopting the Secure Hash Algorithm (SHA) with 256-bits message digest (SHA-256) for the MAC computations and the Advanced Encryption Standard (AES) with 256-bits cryptographic keys (AES-256) for the encryption/decryption.

4.2. Secure Routing Service—SCOTRES

After authenticating a package with μTESLA, SCOTRES evaluates the sender's trustworthiness and its contribution to the network [62]. SCOTRES is a secure routing system for wireless ad-hoc systems that is based on trust computing and is designed around the intricacies of CPS solutions. It maximizes the information that is inferred regarding the network state, based on the knowledge that a node already processes. It safeguards communication against Internet-originating attacks or compromised equipment and jammers. The overall setting is utilized for real-time monitoring of IoT and CPS applications and their management through the cloud.

SCOTRES consists of five components that rate different aspects of the networking operation: (i) The topology-aware component improves the traffic load-balancing and defends distant entities from being isolated; (ii) the energy-aware component estimates the remaining energy of each node, defending the network against energy dissipation and other relevant threats; (iii) the channel-health component identifies jamming in the wireless medium, constraining its effects by routing communication through unaffected paths; (iv) the reputation component ranks a node's fair use of the network resources for routing, forwarding, and recommending activities; and (v) finally, the trust component aggregates all these pieces of knowledge and evaluates the trustworthiness and overall cooperativeness of network entities. Performance and security analyses for the five components have been conducted in [62].

4.3. Authorization Service—PBAC

After verifying the message's legitimacy, the receiver node must decide if it will perform the requested action or not. The PBAC framework is used to implement this authorization functionality. The framework manages direct access to a smart device's resources as determined by a pre-defined collection of policies and rules that are modeled on the OASIS standards DPWS [43] and the eXtensible Access control Markup Language (XACML) [82]. PBAC consists of four components that are placed between the backend infrastructure and devices: The Policy Administrator Point (PAP) and Policy Information Point (PIP) that maintain the attribute values for creating and managing policies in a central repository, the Policy Decision Point (PDP) that runs on a trusted gateway node with sufficient computational capabilities, evaluates the request, and renders the authorization decision, and the Policy Enforcement Point (PEP) that enforces authorization at the end device and makes decision requests. These are combined to provide fine-grained, policy-based access control on assets from remote endpoints (like control stations, sensors, or cameras). Therefore, the specification of an active policy set can be used to define the rights to access to acquired resources (e.g., sensed data and video/audio streams), the rights to update the settings, and even the rights to push notifications of emergency alerts (e.g., blocked routes and train crashes).

4.4. Performance Evaluation

To assess the performance and validate the feasibility of the proposed approach, SecRoute is deployed on an embedded system which features BeagleBone (http://beagleboard.org/bone) devices and is integrated with the Distance Source Routing (DSR) protocol (DSR Uppsala University: http://dsruu.sourceforge.net/). BeagleBone is a low-cost and credit-card-sized device with ARM architecture, executing compact Linux operating systems (ARM Cortex-A8 processor at 720 MHz, 256 MB RAM, Ubuntu Linux). The devices sense environmental conditions, like humidity and temperature, and exchange data wirelessly with a central processing unit via a USB-WiFi.

We measure the processing overhead for SecRoute under normal operation without attacks taking place. Table 3 details the resource consumption of the proposed network layer defense. As indicated in the results, the calculation of reputation is the most compute-intensive part, as it maintains a history with previous interactions, which also increases the overall resource demands for trust computations. The requirements of authentication, routing, forwarding, as well as policy check are low. For the end-to-end interaction, the network latency is also low, ranging between 0.2–0.6 s on average. In an in-carriage setting where the distance among the nodes is short (a few meters) the transmission overhead is minimal, while the maximum delay is recorded for outdoor deployment where the nodes are placed hundreds of meters away from each other. The two scenarios are detailed in the next section.

Table 3. Resource allocation for SecRoute on BeagleBone devices.

Component	ROM (KB)	RAM (KB)	CPU (ms)
Cryptographic service			
Authentication	3.7	2.22	0.0020
Encryption	25.0	10.41	0.0028
Authenticated encryption	28.7	12.63	0.0048
Secure routing service			
Direct trust	5.6	4899.00	677.52
Reputation evaluation	20.0	1621.00	108.97
Indirect trust (recommendations)	30.0	185.00	37.90
Total trust	2.9	45,756.00	9.48
Accept route request	2.0	0.00	104.23
Suitable route selection	15.0	40.00	33.17
Authorization service			
PBAC policy check	24.7	36.00	7.50
Total resource consumption			
Total SecRoute	210.4	46,000.00	1652.50
DSR	310.0	90,000.00	2300.00
SecRoute_DSR	520.4	136,000.00	3952.50

4.5. Comparison with Other Protocols

Efficiency and security analysis of the proposed network layer solution and five relevant systems (RFSN, Ariadne, CSRAN, and SR3) have been presented by the authors in [62]. Table 4 summarizes the main features of the examined secure routing protocols.

Table 4. Secure routing protocols.

Property	SecRoute	RFSN	Ariadne	CSRAN	SR3
Authentication	μTESLA	μTESLA	TESLA	Certificates	LWC
Routing method	DSR	DSR	DSR	ARAN	Random walk
Reputation Fading	Bayesian	Bayesian	NO	Bayesian	FIFO *
Load-balancing	YES	NO	NO	NO	Partially
Energy-aware	YES	NO	NO	NO	NO
Anti-jamming	YES	NO	NO	NO	NO
Authorization	YES	NO	NO	NO	NO

* FIFO: First In First Out finite list.

The secure network communication link of SPD-Safe is compared with the five most relevant proposals for protecting WSNs. Simulation analysis has been performed in the Network Simulator 2 (NS2: http://www.isi.edu/nsnam/ns/), analyzing the performance of each scheme and the provided protection level on a medium-size WSN with 50 nodes [62]. Four attack cases are considered for blackhole, ballot-based attacks for link-spoofing, topology- and energy-aware attacks, and jamming. For each setting, several experiments have been conducted, with the attackers' participation in the network ranging from 10–50%. Figure 3 presents the evaluation of the simulation results. SecRoute counters the attacks and outperforms the relevant schemes, providing the highest level of security and demonstrating the best energy- and load-balancing characteristics.

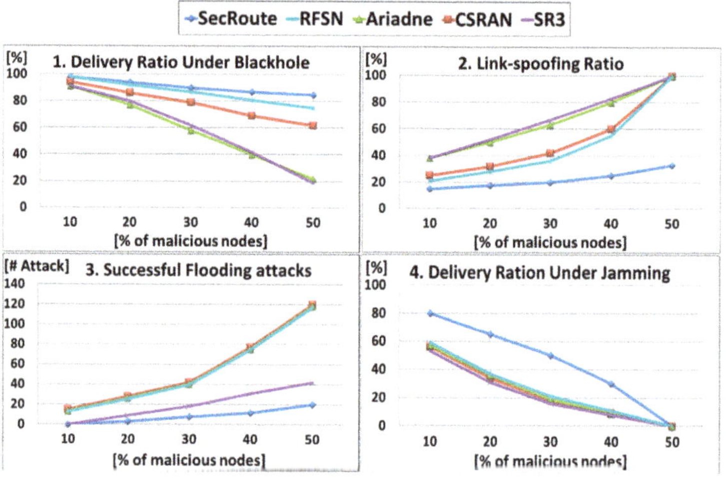

Figure 3. Simulation results for the evaluation of the network layer security solutions against four attack scenarios.

5. SPD-Safe Demonstration

5.1. Railway CPS Architecture

This section details the demonstration and evaluation of the whole SPD-Safe framework in the context of protecting and managing a railway CPS. In the proof-of-concept setting, our proposal assesses and manages the system and ambient ecosystem with the goal of safeguarding the trains' carriages and railway's routes. The hardware platforms incorporate embedded devices that control smart equipment (e.g., cameras and electronic doors), inspect environmental conditions, and exchange information wirelessly. Furthermore, the PBAC framework is applied for the control of the physical access for personnel, determined by access rights that are specified in XACML policies. Every agent manages a smart subsystem, like a train or a station. Backend agents can also run at the cloud in order to gather high level information, perform big data analysis, and enable interaction with external systems and actuators. These agents run on virtual machines deployed on the research cloud platform GRNET Virtual MAchines (ViMA: http://vime.grnet.gr/about/info/en/). Figure 4 illustrates the railway system architecture. The whole setting is administered by a master agent (MA) at the C&C. At the edge, simple and more lightweight agents (SAs) protect the local subsystems (applying access control, lightweight data analysis, incident detection, etc.) and exchange information with the MA (i.e., security/safety events and response strategies). The MA can, optionally, forward data to a cloud SA for storing or in-depth analysis. The cloud SA also presents high-level knowledge to end-users as well as the current SPD status of the railway infrastructure.

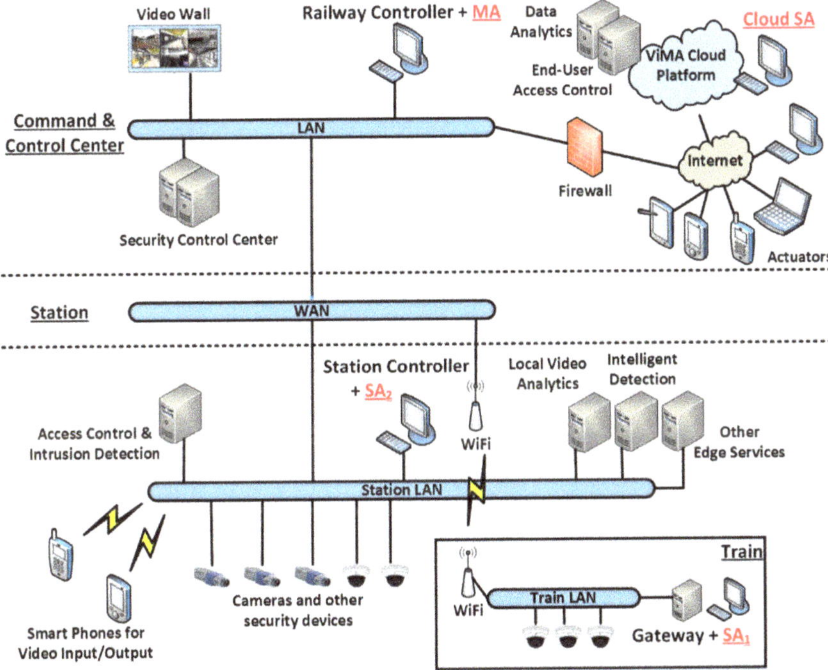

Figure 4. The smart railway use case architecture.

For this demonstration, the MA and the C&C services are deployed on a laptop. Both MA and cloud SA are installed on machines witha 2.1 GHz Intel Core i-7 processor, 8 GB of RAM, and the Ubuntu Linux Operating System (OS). The SAs are deployed on the BeagleBone devices at the edge systems.

As a case study, two deployments are evaluated. In the first indoor setting, which emulates in-carriage or shelter equipment, we test the system under normal operation and the aforementioned attacks on routing. In the second outdoor scenario, which emulates the on-route equipment, we examine the system's response to safety-related incidents. Both networks run the SecRoute protocol [61] to enable communication, protect the network layer against cyber-attacks, and act as an intrusion detection and incident response system for the upper layers.

5.2. Indoor Setting—Cyber-Security

The demonstration setting includes a carriage/shelter inspecting application, which is equipped with a surveillance system and WSNs. Those components are sensitive to network layer threats, like jamming and blackhole attacks. The deployed network is depicted in Figure 5, where these devices are deployed in a shelter [28]:

- At the entrance, the smart camera inspects for physical intrusion;
- Two WSNs are deployed in the shelter. WSN_{1-1} (green color) monitors light and temperature, and WSN_{1-2} (red color) senses temperature. WSN_{1-1} and WSN_{1-2} utilize different hardware to enhance diversity and ensure redundancy for the monitored factors;
- A gateway interconnects the rest of the components with the C&C.

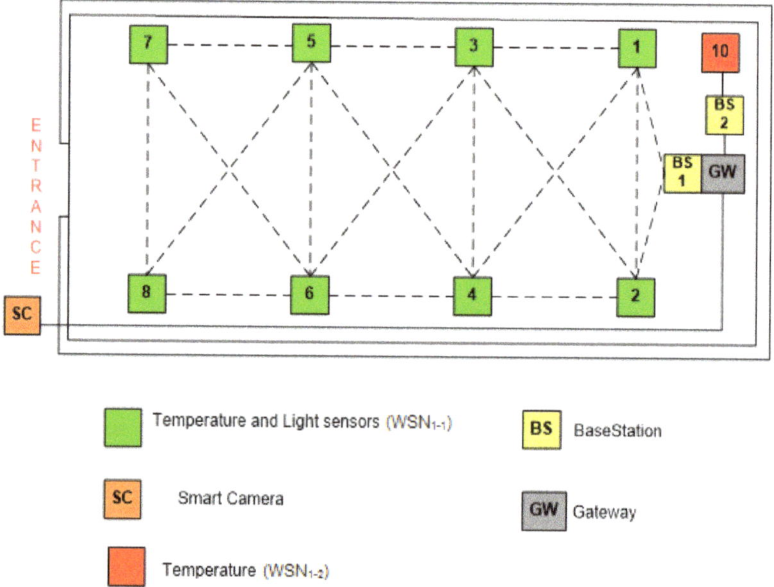

Figure 5. The internal wireless sensor network (WSN) for the carriage setting.

WSN_{1-1} consists of eight Memsic Iris sensor nodes (16 MHz Atmel ATMega 1281 processor, 8 KB RAM, Contiki OS). The devices are battery powered and measure light and temperature. Furthermore, the smart camera is controlled by the node at the carriage's entrance. WSN_{1-2} is installed for redundancy and is comprised of Zolertia Z1 sensor nodes (16 MHz MSP430 processor, 8 KB RAM, and Contiki OS) that collect temperature data. The two WSNs are monitored by two relevant simple agents (SA_{1-1} and SA_{1-2} respectively). Every device executes the PEP module of the PBAC framework. The devices also exchange data with the gateway, which runs the access policies for PBAC, and communicates with an MA which administrates the whole network.

The devices gather environmental information and send data to the relevant base station (laptop with WiFi connectivity). This component integrates and processes the received information. It also runs an application with which the user accesses and manages the overall testbed.

The different components are evaluated by the corresponding agents, who also estimate the aggregate SPD value of the whole system. The agents inspect their underlying domain, managing it based on an SPD-aware reasoning operation. Furthermore, the system is re-configurable at runtime according to the SPD protection and performance goals defined in the activated policy. Affected agents configure their subsystem's settings to raise the SPD value when attacks are performed and then return to normal when the attacks are over (to save resources). Regarding the adaptation capabilities integrated within the proof-of-concept, the cryptographic service provides three communication states: Plaintext, authenticated, as well as authenticated encryption. Additionally, the trust scheme supports two trust evaluation states: Direct trust only, as well as a combination of direct and indirect trust.

The system begins with a moderate SPD configuration to conserve resources (i.e., authenticated communication and direct trust). If SPD-Safe observes malicious activity, it informs the system entities to raise their protection level. The relevant response actions are specified in a security policy (applicable to the specific device type), such as applying authenticated and encrypted communication with combined direct and indirect trust information. The SPD value and status of each system component is then altered as a response to the launched attacks, so as to achieve a sufficient level of protection. The WSNs comply with the current policies, becoming stricter to misbehavior and isolating the compromised nodes. The main protection mechanism against cyber-attacks (i.e., blackhole or link-spoofing) is provided by SecRoute, while the smart camera enhances physical protection. In the same way, the system returns to the previous (initial) state when the triggering conditions are over.

For WSN_{1-1}, we emulate scenarios where: (i) A node is malfunctioning due to low battery, and (ii) a compromised node launches a badmouth attack. In (i) the node is protected when a low energy level is observed by not including in traffic forwarding operations. The administrator gets notified accordingly. When the issue is fixed, the trust level is restored and the nodes' operational status returns back to normal. In case (ii), the compromised entity is detected when the attack rate reaches a threshold and it is blocked from routing operations. For WSN_{1-2}, we launch blackhole and jamming attacks against congested or topology significant components. The secure routing mechanism successfully identifies both attacks and mitigates them. Table 5 presents in detail the above-mentioned scenario phases. The SPD levels are depicted with: (i) Red for values of 0–50—i.e., a situation where the provided protection is low, the proper functionality may not be available, and the operator must take immediately the related countermeasures; (ii) yellow for values of 51–70—i.e., moderate protection but still safe operation; and (iii) green for values of 71–100—i.e., high levels of protection.

Table 5. Scenario steps of the smart transportation use case.

Event	Description	SPD State	Total <S, P, D> Value	SPD Visualization
1	Start of all components and services.Discovery/registration operations.	Initial State	<80, 70, 65>	
2	Bad-mouthing attack to WSN_{1-1}. MA (master agent) is alerted for the attack and commands the rest agents to increase security.	Security level decreases	<60, 70, 65>	
3	Security status is enhanced on all SAs (simple agents). MA is notified.	Security level increases	<85, 70, 65>	
4	WSN_{1-1} counters bad-mouthing and SA_{1-1} informs the MA.The MA requests from the SAs to restore the normal state (to conserve resources).	Security level returns to initial state	<80, 70, 65>	
5	Blackhole attack to WSN_{1-2}. MA is alerted for the attack and commands the rest agents to increase security.	Security level decreases	<50, 70, 65>	
6	Security status is enhanced on all SAs. MA is notified.	Security level increases	<85, 70, 65>	

Table 5. *Cont.*

Event	Description	SPD State	Total <S, P, D> Value	SPD Visualization
7	WSN$_{1-2}$ counters the blackhole attack and SA$_{1-2}$ informs the MA. The MA requests from the SAs to restore the normal state (to conserve resources).	Security level returns to initial state	<80, 70, 65>	
8	A node has died in WSN$_{1-2}$. MA is informed.	Dependability level decreases	<80, 70, 30>	
9	The dead node is replaced by the personnel. Dependability is restored.SA$_{1-1}$ reports the new status to MA.	Dependability level returns to initial state	<80, 70, 65>	
10	Simulated jamming attack against the network layer of WSN$_{1-2}$. MA is informed.	S & D levels decrease	<40, 70, 40>	
11	The trust-based routing component counters the attack. SA$_{1-2}$ reports new state to MA.	S & D levels return to initial state	<80, 70, 65>	

5.3. Outdoor Setting—Safety Scenario

For outdoor on-route defense, a similar WSN with four BeagleBone nodes is installed. The nodes are connected with a mains power supply and control a smart camera as well as weather sensors. In the emulated use-case, the nodes and related SAs are deployed on: (i) The passenger's station; (ii) the track; (iii) the carriage departure, and; (iv) all bridges and tunnels along the track. Figure 6 illustrates the on-route WSN$_2$ [28] along with the central MA and underlying SA$_{2-1}$–SA$_{2-4}$.

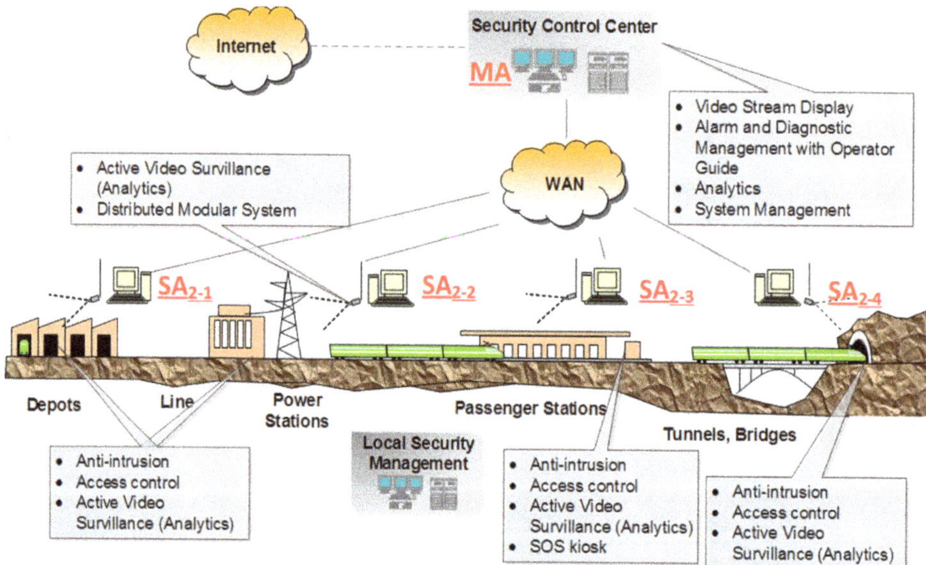

Figure 6. The outdoor WSN for the on-route scenario. The MA is deployed in the security control center and the four SAs are installed in the edge system.

Through the responsible SA, the networking components (e.g., sensors and cameras) send real-time information to a security control center and the related master agent. Figure 7 depicts the graphical user interface and the visualization of the information that is collected by the on-route equipment, as developed by Ansaldo STS.

Figure 7. The railway on-route WSN graphical user interface.

In case of an emergency, the agents manage the system components to advise the personnel and assist the passengers. The demonstrated incident emulates the response strategy for a fire-alarm, where decisions concerning both safety and security must be taken. In Appendix A, the code sample Figure A1 describes the CAP message that indicates the fire alarm.

Normally, for the indoor setting, the personnel and passengers are allowed to open doors based on their access rights (as determined by safety and security rules). When fire is detected by the sensors, an alarm is triggered, and the associated agent is notified. The agent takes the decision to degrade the security status by unlocking all doors, therefore enabling the unhindered evacuation of the train. Furthermore, via GSM, the agent automatically transmits an SMS to the responsible authorities concerning this incident (including situation's severity, GPS coordinates) and alerts the neighboring entities to be aware (e.g., agents on nearby trains). The train agents that cross the area are also notified to perform related actions (such as stop to the nearest station or change route). Moreover, it is assumed that during normal operation the smart cameras capture frames at a low rate to preserve bandwidth. When the alarm is raised, the setting is reconfigured at runtime, offering a high framerate and continuous monitoring of the affected area. As the fire is extinguished and the damaged components are restored, the normal status is restored. The code shown in Figure A2 summarizes the main processing flow and the emergency response rules that perform the described actions (for more information regarding EC, please refer to Mueller [73]).

6. Discussion

6.1. Comparison

This subsection compares SPD-Safe with the related works presented in Section 2.2 (i.e., TrainIntegrity, CMLRVT, and SENSORAIL) in terms of features. Table 6 summarizes the comparison results.

Table 6. Smart railway systems.

Property	SDP-Safe	TrainIntegrity	CMLRVT	SENSORAIL
AI technologies	JADE/Jess	NO	NO	SeNsIM
Reasoning & processing	Distributed	Centralized	Centralized	Centralized
Conflict resolution	YES	NO	NO	NO
Security management	YES	NO	NO	NO
Safety management	YES	YES	YES	YES
Middleware	OSGi	NO	NO	NO
Network layer protection	SecRoute	NO	NO	NO
Cloud management	ViMA	NO	NO	NO

All the related smart railway systems identified adopt semantic representation and reasoning. The service-oriented approaches conform to the specific application aspects and, therefore, in all relevant systems the agents are uniquely responsible for specific operations. The conflicting patterns are also not examined in most of these designs, limiting their applicability to specific deployments.

Furthermore, the three related systems do not use any management middleware for embedded devices. This approach is quite limiting in the IoT era, where high volumes of heterogeneous equipment have to be deployed and co-function. The systems also neglect the popular agent frameworks which, among others, provide efficient agent-related functionality and implement relevant standards. The reasoning operation is developed with general purpose programming languages, ignoring the advantages offered by the deductive rule-based techniques. Mechanisms for resolving conflicts, when implemented, are based either on epistemic or relational reasoning. More importantly, these related systems do not safeguard security, privacy, and dependability, and do not utilize any built-in protection technologies.

Conversely, SPD-Safe is a solution focusing on the SPD management of IoT and CPS settings. The SPD modeling is based on well-structured metrics that analyze the various configuration options of a multi-layered system. The AI process adjusts the railway CPS and counters attacks at runtime. SPD-Safe integrates state-of-the-art technological building blocks and platforms for the implementation of reasoning, as well as the management of devices and agents. Epistemic and relational reasoning are incorporated for resolving conflicts. Furthermore, the proposed framework adopts standardized technologies, from semantic standards to communication protocols and authorization schemes.

6.2. Future Work

SPD-Safe integrates several technologies in a secure manner. It preserves the SPD properties, enables active defenses and countermeasures, and can facilitate emergency response operations.

Active and offensive types of defenses are proposing nowadays, as the next step to enhance protection and mitigate threats, that the mainstream passive mechanisms (e.g., cryptography, network slicing, anti-viruses, etc.) cannot tackle. MTD is such an approach. It is becoming harder to analyze the system and exploit its vulnerabilities. Furthermore, in conjunction with other intrusion detection techniques, it can mitigate or even block some type of ongoing attacks. Nevertheless, more research is needed in order to make guidelines for the implementation of effective MTD policies as well as strategies to mitigate more advance attacks.

Moreover, safety-related events require the participation of relevant authorities. In modern settings, emergency authorities possess their own equipment, which is utilized during safety incidents. The cooperation of the involved systems becomes vital when it comes to rescuing lives. For the effectiveness of the response services, the systems must authenticate and authorize the various participants and exchange information (e.g., sensors' data, surveillance video, etc.) in real time. The seamless interoperation will be examined in future extensions of SPD-Safe.

7. Conclusions

This paper introduced SPD-Safe, an administration framework for IoT settings in ambient secure and safety-critical domains, applied to protect a railway CPS. For secure connectivity, an innovative secure routing protocol was integrated in the network layer. The protocol covers all core security properties (confidentiality, integrity, and authentication) and features policy-based authorization. It was found to be energy efficient and could effectively counter a variety of attacks, providing defense against several threats that are not mitigated by existing solutions. For smart monitoring and automatic adaptation, smart agents were deployed at the edge systems and backend infrastructure, and performed the required AI processes. A multi-agent system was developed in the JADE platform and integrated on the OSGi middleware for the management of DPWS-enabled equipment, also utilizing various built-in protection mechanisms. The core reasoning process was implemented in Event Calculus. The SPD validation and metric-driven administration were modeled as a heuristic framework in a security-related theory. The implementation of MTDs was enabled, providing extra protection against attacks that were not mitigated by passive defenses. Furthermore, the system models a safety-related theory and implemented associated AI ambient strategies and plans. The two features were incorporated to administrate the underlying components, considering both the SPD and safety aspects. To validate the proposed approach, SPD-Safe was deployed to administrate WSNs on a complex railway CPS testbed, where the underlying components were successfully configured at runtime and mitigate security-related attacks, while AI reactive plans preserved the safety of personnel and passengers in emergency situations. The average delay in real equipment was around 0.2–0.6 s.

In terms of future work, advances in MTD solutions and integration with emergency response services were considered. MTDs are coming to the foreground nowadays and are expected to play a significant role in future defense strategies as AI becomes an integral part of new generation systems. Safety critical systems, such as the railway ones, must provide an adequate means to collaborate with emergency authorities and support their operations. Facilitating emergency response and a rapid restoration of service must also be considered by modern smart railway installations.

Author Contributions: Conceptualization, G.H. and K.F.; methodology, G.H. and S.I.; software, G.H.; validation, K.F., N.P., G.V., and I.H.; writing—original draft preparation, G.H.; writing—review and editing, K.F. and S.I.; Supervision and Project administration, S.I. and G.S. All authors have read and agreed to the published version of the manuscript.

Funding: This work has received funding from the European Union Horizon's 2020 research and innovation program under the grant agreements No. 786890 (THREAT-ARREST), No. 830927 (CONCORDIA), and No. 269317 (nSHIELD).

Institutional Review Board Statement: Not applicable.

Informed Consent Statement: Not applicable.

Data Availability Statement: Data sharing not applicable.

Conflicts of Interest: The authors declare no conflict of interest. The funders had no role in the design of the study; in the collection, analyses, or interpretation of data; in the writing of the manuscript, or in the decision to publish the results.

Appendix A

The code sample Figure A1 describes the CAP message that indicates the fire alarm.

```
<soap:Envelope xmlns:soap="http://schemas.xmlsoap.org/soap/envelope/">
 <soap:Body>
  <ns3:Notify xmlns="http://docs.oasis-open.org/wsrf/bf-2"
   xmlns:ns2="http://www.w3.org/2005/08/addressing"
   xmlns:ns3="http://docs.oasis-open.org/wsn/b-2"
   xmlns:ns4="http://protectrail.eu/model/events/resource"
   xmlns:ns5="urn:oasis:names:tc:emergency:cap:1.2"
   xmlns:ns6="http://docs.oasis-open.org/wsn/t-1">
   <ns3:NotificationMessage>
    <ns3:Topic Dialect="http://docs.oasis-open.org/wsn/t-1/TopicExpression/Full">FireDetection</ns3:Topic>
    <ns3:Message>
     <ns5:alert>
     <ns5:identifier>urn:rixf:com.tuc.SPD-Safe:id/FireAlert_01</ns5:identifier>
      <ns5:sender>WSN2</ns5:sender>
      <ns5:sent>2018-10-19T10:43:09.000+02:00</ns5:sent>
      <ns5:status>Actual</ns5:status>
      <ns5:msgType>Alert</ns5:msgType>
      <ns5:source>urn:rixf:com.tuc.fireprotection/devices/WSN</ns5:source>
       <ns5:scope>Public</ns5:scope>
      <ns5:info>
        <ns5:category>Fire</ns5:category>
        <ns5:event>FireDetection</ns5:event>
        <ns5:responseType>Evacuate</ns5:responseType>
        <ns5:urgency>Immediate</ns5:urgency>
        <ns5:severity>Extreme</ns5:severity>
        <ns5:certainty>Observed</ns5:certainty>
        <ns5:parameter>
          <ns5:valueName>Area</ns5:valueName>
          <ns5:value>51.468928,16.858863 0.01</ns5:value>
        </ns5:parameter>
       </ns5:info>
      </ns5:alert>
     </ns3:Message>
    </ns3:NotificationMessage>
   </ns3:Notify>
  </soap:Body>
</soap:Envelope>
```

Figure A1. Simple Object Access Protocol (SOAP) message that contains the CAP alert for the fire alarm.

The code shown in Figure A2 summarizes the main processing flow and the emergency response rules that perform the described actions (for more information regarding EC, please refer to Mueller [73]).

```
Rule 1: Happens(EventDitection(WSN,Fire),t) ∧ HoldsAt(ControllingAgent(SA,WSN),t)
=>
Happens(InformAgent(SA,WSN,Fire),t+1)


Rule 2: Happens(InformAgent(SA,WSN,Fire),t)
=>
Happens(InformTrainOperator(SA,SA_TrainOperator,Fire),t+1) ∧ Happens(UnlockDoors(SA, SA_Train),t+1) ∧
Happens(InformMA(SA,MA,Fire),t+1) ∧
Happens(InformNearbyAgents(SA,Fire),t+1)


Rule 3: Happens(InformNearbyAgents(SAᵢ,Fire),t) ∧ HoldsAt(LocatedIn(SAᵢ,area),t) ∧ HoldsAt(NearbyAgents(SAᵢ, SAⱼ),t)
=>
Happens(AvoidArea(SAⱼ, SAⱼ_TrainOperator, area),t+1)


Rule 4: Happens(InformMA(SA,MA,Fire),t)
=>
Happens(InformSystemOperator(MA,SystemOperator,Fire),t+1) ∧ Happens(InformAuthorities(MA,FireBrigate,Fire),t+1)
```

Figure A2. EC rules that perform the main safety reasoning behavior of each intelligent agent.

References

1. Xu, S.; Zhu, G.; Ai, B.; Zhong, Z. *A Survey on High-Speed Railway Communications: A Radio Resource Management Perspective. Computer Communications*; Elsevier: Amsterdam, The Netherlands, 2016; Volume 86, pp. 12–28.
2. Chamoso, P.; González-Briones, A.; Rodríguez, S.; Corchado, J.M. Tendencies of Technologies and Platforms in Smart Cities: A State-of-the-Art Review. *Wirel. Commun. Mobile Comput.* **2018**, *2018*, 3086854.
3. Boudi, Z.; El Koursi, E.M.; Ghazel, M. The New Challenges of Rail Security. *J. Traffic Logist. Eng.* **2016**, *4*, 56–60. [CrossRef]
4. Kour, R.; Thaduri, A.; Karim, R. Railway Defender Kill Chain to Predict and Detect Cyber-Attacks. *J. Cyber Secur. Mobil.* **2019**, *9*, 47–90. [CrossRef]
5. Luxton, A.; Marinov, M. Terrorist Threat Mitigation Strategies for the Railways. *Sustainability* **2020**, *12*, 3408. [CrossRef]
6. Zhang, J.; Hu, F.; Wang, S.; Dai, Y.; Wang, Y. Structural vulnerability and intervention of high speed railway networks. *Phys. A Stat. Mech. Appl.* **2016**, *462*, 743–751. [CrossRef]
7. González-Briones, A.; Garcia-Martin, R.; de AlbaJuan, F.L.; Corchado, M. Agent-Based Platform for Monitoring the Pressure Status of Fire Extinguishers in a Building. In *International Conference on Practical Applications of Agents and Multi-Agent Systems (PAAMS)*; Springer: Berlin/Heidelberg, Germany, 2020; Volume 1233, pp. 373–384.
8. Catalano, A.; Bruno, F.A.; Galliano, C.; Pisco, M.; Persiano, G.V.; Cutolo, A.; Cusano, A. An optical fiber intrusion detection system for railway security. *Sens. Actuators A Phys.* **2017**, *253*, 91–100. [CrossRef]
9. Fraga-Lamas, P.T.; Fernández-Caramés, M.; Castedo, L. Towards the Internet of Smart Trains: A Review on Industrial IoT-Connected Railways. *Sensors* **2017**, *17*, 1457. [CrossRef]
10. Wang, Y.; Zhu, L.; Yu, Z.; Guo, B. An adaptive track segmentation algorithm for a railway intrusion detection system. *Sensor* **2019**, *19*, 2594. [CrossRef]
11. Gai, K.; Qiu, M.; Hassan, H. Secure Cyber Incident Analytics Framework using Monte Carlo Simulations for Financial Cybersecurity Insurance in Cloud Computing. In *Concurrency and Computation: Practice and Experience*; Wiley: Hoboken, NJ, USA, 2017; Volume 29, issue 7.
12. Chang, S.E.; Liu, A.Y.; Lin, S. Exploring privacy and trust for employee monitoring. *Ind. Manag. Data Syst.* **2015**, *115*, 88–106. [CrossRef]
13. Paganini, P. Modern Railroad Systems Vulnerable to Cyber Attacks. Security Affairs. 2016. Available online: http://securityaffairs.co/wordpress/43196/hacking/railroad-systems-vulnerabilities.html (accessed on 18 November 2020).
14. Bababeik, M.; Khademi, N.; Chen, A.; Nasiri, M.M. Vulnerability analysis of railway networks in case of multi-link blockage. *Transp. Res. Procedia* **2017**, *22*, 275–284. [CrossRef]
15. Khanmohamadi, M.; Bagheri, M.; Khademi, N.; Ghannadpour, S.F. A security vulnerability analysis model for dangerous goods transportation by rail–Case study: Chlorine transportation in Texas-Illinois. *Saf. Sci.* **2018**, *110*, 230–241. [CrossRef]
16. Salmane, H.; Khoudour, L.; Ruichek, Y. A Video-Analysis-Based Railway–Road Safety System for Detecting Hazard Situations at Level Crossings. *IEEE Trans. Intell. Transp. Syst.* **2015**, *16*, 596–609. [CrossRef]
17. Chernov, A.V.; Savvas, I.K.; Butakova, M.A. Detection of Point Anomalies in Railway Intelligent Control System Using Fast Clustering Techniques. In Proceedings of the 3rd International Scientific Conference Intelligent Information Technologies for Industry, Sochi, Russia, 17–21 September 2018; pp. 267–276.
18. Coppola, P.; Silvestri, F. Assessing travelers' safety and security perception in railway stations. *Case Stud. Transp. Policy* **2020**, *8*, 1127–1136. [CrossRef]
19. Mrazovic, P.; Eser, E.; Ferhatosmanoglu, H.; Larriba-Pey, J.L.; Matskin, M. Multi-vehicle Route Planning for Efficient Urban Freight Transport. In Proceedings of the 2018 International Conference on Intelligent Systems (IS), Funchal, Madeira, Portugal, 25–27 September 2018; pp. 744–753.
20. Zhu, C.; Shu, L.; Leung, V.C.M.; Guo, S.; Zhang, Y.; Yang, L.T. Secure multimedia Big Data in trust-assisted sensor-cloud for smart city. *IEEE Commun. Mag.* **2017**, *55*, 24–30. [CrossRef]
21. Chamoso, P.; De La Prieta, F.; De Paz, F.; Corchado, J.M. Swarm Agent-Based Architecture Suitable for Internet of Things and Smartcities. In *Distributed Computing and Artificial Intelligence, 12th International Conference*; Springer: Berlin/Heidelberg, Germany, 2015; Volume 373, pp. 21–29.
22. Zhang, Q.; Chen, Z.; Leng, Y. Distributed fuzzy c-means algorithms for big sensor data based on cloud computing. *Int. J. Sens. Networks* **2015**, *18*, 32–39. [CrossRef]
23. Tsaramirsis, G.; Karamitsos, I.; Apostolopoulos, C. Smart Parking: An IoT application for Smart City. In Proceedings of the 10th INDIACom-2016 International Conference, New Delhi, India, 16–18 March 2016; pp. 2271–2275.
24. Yin, W.; He, S.; Zhang, Y.; Hou, J. A Product-Focused, Cloud-Based Approach to Door-to-Door Railway Freight Design. *IEEE Access* **2018**, *6*, 20822–20836. [CrossRef]
25. Dong, Q.; Hayashi, K.; Kaneko, M. An Optimized Link Layer Design for Communication-Based Train Control Systems Using WLAN. *IEEE Access* **2017**, *6*, 6865–6877. [CrossRef]
26. Fanian, F.; Rafsanjani, M.K. Cluster-based routing protocols in wireless sensor networks: A survey based on methodology. *J. Netw. Comput. Appl.* **2019**, *142*, 111–142. [CrossRef]
27. Khanna, N.; Sachdeva, M. Study of trust-based mechanism and its component model in MANET: Current research state, issues, and future recommendation. *Int. J. Commun. Syst.* **2019**, *32*, 1–23. [CrossRef]
28. Cesena, M. SHIELD Technology Demonstrators. In *Measurable and Composable Security, Privacy, and Dependability for Cyberphysical Systems*; CRC Press: Boca Raton, FL, USA, 2017; pp. 381–434.

29. Brokalakis, A.; Tampouratzis, N.; Nikitakis, A.; Andrianakis, S.; Papaefstathiou, I.; Dollas, A. An Open-Source Extendable, Highly-Accurate and Security Aware CPS Simulator. In Proceedings of the 2017 13th International Conference on Distributed Computing in Sensor Systems (DCOSS), Ottawa, ON, Canada, 5–7 June 2017; pp. 81–88.

30. Farooq, J.; Soler, J. Radio Communication for Communications-Based Train Control (CBTC): A Tutorial and Survey. *IEEE Commun. Surv. Tutor.* **2017**, *19*, 1377–1402. [CrossRef]

31. Sun, W.; Yu, F.R.; Tang, T.; Bu, B. Energy-Efficient Communication-Based Train Control Systems with Packet Delay and Loss. *IEEE Trans. Intell. Transp. Syst.* **2016**, *17*, 452–468. [CrossRef]

32. Garcia-Loygorri, J.M.; Val, I.; Arriola, A.; Briso-Rodriguez, C. 2.6 GHz Intra-Consist Channel Model for Train Control and Management Systems. *IEEE Access* **2017**, *5*, 23052–23059. [CrossRef]

33. Fotso, S.J.T.; Frappier, M.; Laleau, R.; Mammar, A. Modeling the Hybrid ERTMS/ETCS Level 3 Standard Using a Formal Requirements Engineering Approach. In *International Conference on Abstract State Machines*; Alloy, B., Ed.; Springer: Berlin/Heidelberg, Germany, 2018; pp. 262–276.

34. Chetty, K.; Chen, Q.; Woodbridge, K. Train monitoring using GSM-R based passive radar. In Proceedings of the 2016 IEEE Radar Conference (RadarConf), Philadelphia, PA, USA, 1–6 May 2016; pp. 1–4.

35. Bates, R.J. GPRS: General Packet Radio Service. In *Book GPRS: General Packet Radio Service*; McGraw-Hill, Professional Telecom: New York, NY, USA, 2001.

36. Proto, M.; Bavusi, M.; Bernini, R.; Bigagli, L.; Bost, M.; Bourquin, F.; Cottineau, L.-M.; Cuomo, V.; Della Vecchia, P.; Dolce, M.; et al. Transport Infrastructure Surveillance and Monitoring by Electromagnetic Sensing: The ISTIMES Project. *Sensors* **2010**, *10*, 10620–10639. [CrossRef] [PubMed]

37. Crinière, A.; Dumoulin, J.; Mevel, L.; Andrade-Barroso, G. Cloud2IR an Infrared and Environmental SHM Information System. In Proceedings of the 13th Quantitative Infrared Thermography Conference (QIRT), Gdansk, Poland, 4–8 July 2016; pp. 226–235.

38. Xie, J.; Liu, C.-C. Multi-agent systems and their applications. *J. Int. Counc. Electr. Eng.* **2017**, *7*, 188–197. [CrossRef]

39. De la Prieta, F.; Rodríguez-González, S.; Chamoso, P.; Corchado, J.M.; Bajo, J. Survey of agent-based cloud computing applications. *Future Gener. Comput. Syst.* **2019**, *100*, 223–236. [CrossRef]

40. Kravari, K.; Bassiliades, N. A Survey of Agent Platforms. *J. Artif. Soc. Soc. Simul.* **2015**, *18*, 11–29. [CrossRef]

41. FIPA, "FIPA ACL Message Structure Specification," Foundation for Intelligent Physical Agents. 2002. Available online: http://www.fipa.org/specs/fipa00061/SC00061G.html (accessed on 18 November 2020).

42. Fysarakis, K.; Askoxylakis, I.; Soultatos, O.; Papaefstathiou, I.; Manifavas, C.; Katos, V. Which IoT Protocol? Comparing Standardized Approaches over a Common M2M Application. In Proceedings of the 2016 IEEE Global Communications Conference (GLOBECOM), Washington, DC, USA, 4–8 December 2016; pp. 1–6.

43. OASIS. "Devices Profile for Web Services Version 1.1," Organization for the Advancement of Structured Information Standards. 2009. Available online: http://docs.oasis-open.org/ws-dd/dpws/1.1/os/wsdd-dpws-1.1-spec-os.pdf (accessed on 18 November 2020).

44. Thirumalainambi, R. Pitfalls of Jess for dynamic systems. In Proceedings of the International Conference on Artificial Intelligence and Pattern Recognition (AIPR), Orlando, FL, USA, 9–12 July 2007; Volume 1, pp. 491–494.

45. Kumar, S.; Prasad, R. Importance of expert system shell in development of expert system. *Int. J. Innov. Res. Dev.* **2015**, *4*, 128–133.

46. Semmel, G.; Davis, S.; Leucht, K.; Rowe, D.; Kelly, A.; Boloni, L. Launch commit criteria monitoring agent. In Proceedings of the 4th International Joint Conference on Autonomous Agents and MultiAgent Systems (AAMAS), Utrecht, The Netherlands, 25–29 July 2005; pp. 3–10.

47. Goseva-Popstojanova, K.; Tyo, J. Experience Report: Security Vulnerability Profiles of Mission Critical Software: Empirical Analysis of Security Related Bug Reports. In Proceedings of the 28th International Symposium on Software Reliability Engineering (ISSRE), Toulouse, France, 23–26 October 2017; pp. 152–163.

48. Leitao, P.; Karnouskos, S. *Industrial Agents: Emerging Applications of Software Agents in Industry*, 1st ed.; Elsevier Science: Amsterdam, The Netherlands, 2015; pp. 1–476.

49. Ghadimi, P.; Wang, C.; Lim, M.K.; Heavey, C. Intelligent sustainable supplier selection using multi-agent technology: Theory and application for Industry 4.0 supply chains. *Comput. Ind. Eng.* **2019**, *127*, 588–600. [CrossRef]

50. Scholten, H.; Westenberg, R.; Schoemaker, M. Sensing Train Integrity. In Proceedings of the IEEE Sensors Conference, Christchurch, New Zealand, 25–28 October 2009.

51. Firlik, B.; Chudzikiewicz, A. Condition monitoring of a light rail vehicle—From concept to implementation. *Key Eng. Mater.* **2012**, *518*, 66–75. [CrossRef]

52. Flammini, F.; Gaglione, A.; Ottello, F.; Pappalardo, A.; Pragliola, C.; Tedesco, A. Towards wireless sensor networks for railway infrastructure monitoring. In Proceedings of the Electrical Systems for Aircraft, Railway and Ship Propulsion (ESARS), Bologna, Italy, 19–21 October 2010.

53. Casola, V.; Gaglione, A.; Mazzeo, A. A reference architecture for sensor networks integration and management. In Proceedings of the 3rd International Conference on Geosensor Networks, Oxford, UK, 13–14 July 2009; pp. 158–168.
54. Flammini, F.; Gaglione, A.; Mazzocca, N.; Pragliola, C. DETECT: A novel framework for the detection of attacks to critical infrastructures. In *Safety, Reliability and Risk Analysis: Theory, Methods and Applications*; Taylor & Francis: Abingdon, UK, 2008; pp. 105–112.
55. Chakravarthy, S.; Mishra, D. Snoop: An expressive event specification language for active databases. *Data Knowl. Eng.* **1994**, *14*, 1–26. [CrossRef]
56. Ganeriwal, S.; Balzano, L.; Srivastava, M. Reputation-based framework for high integrity sensor networks. *ACM Trans. Sen. Netw.* **2008**, *4*. [CrossRef]
57. Hu, Y.-C.; Perrig, A.; Johnson, D.B. Ariadne: A secure on-demand routing protocol for ad hoc networks. *Wirel. Netw.* **2005**, *11*, 21–38. [CrossRef]
58. Zhang, Y.; Xu, L.; Wang, X. A Cooperative Secure Routing Protocol based on Reputation System for Ad Hoc Networks. *J. Commun.* **2008**, *3*, 43–50. [CrossRef]
59. Altisen, K.; Devismes, S.; Jamet, R.; Lafourcade, P. SR3: Secure resilient reputation-based routing. In Proceedings of the 2013 IEEE International Conference on Distributed Computing in Sensor Systems (DCOSS), Cambridge, MA, USA, 20–23 May 2013; pp. 258–265.
60. Dhaheri, A.A.; Yeum, C.Y.; Damiani, E. New Two-Level μTESLA Protocol for IoT Environments. In Proceedings of the 2019 IEEE World Congress on Services (SERVICES), Milan, Italy, 8–13 July 2019; pp. 84–91.
61. Hatzivasilis, G.; Papaefstathiou, I.; Askoxylakis, I.; Fysarakis, K. SecRoute: End-to-end secure communications for wireless ad-hoc networks. In Proceedings of the 22nd IEEE Symposium on Computers and Communications (ISCC), Heraklion, Crete, Greece, 3–6 July 2017; pp. 558–563.
62. Hatzivasilis, G.; Papaefstathiou, I.; Manifavas, C. SCOTRES: Secure Routing for IoT and CPS. *IEEE Internet Things J.* **2017**, *4*, 2129–2141. [CrossRef]
63. Hatzivasilis, G.; Papaefstathiou, I.; Plexousakis, D.; Manifavas, C.; Papadakis, N. AmbISPDM: Managing embedded systems in ambient environment and disaster mitigation planning. In *Applied Intelligence*; Springer: Berlin/Heidelberg, Germany, 2017; pp. 1–21.
64. Java Agent Development (JADE) Framework. Available online: http://jade.tilab.com/ (accessed on 18 November 2020).
65. Tilab, S.P.A. JADE Security Add-On Guide. 2005. Available online: http://jade.tilab.com/doc/tutorials/JADE_Security.pdf (accessed on 18 November 2020).
66. Ali, B.; Manzoor, U.; Zafar, B. eJADE-S: Encrypted JADE-S for Securing Multi-Agent Applications. In Proceedings of the International Conference on Artificial Intelligence (ICAI), Athens, Greece, 27–30 July 2015; pp. 548–554.
67. Open Services Gateway Initiative (OSGi). Available online: http://www.osgi.org/ (accessed on 18 November 2020).
68. OASIS. Common Alerting Protocol Version 1.2, Organization for the Advancement of Structured Information Standards. 2010. Available online: http://docs.oasis-open.org/emergency/cap/v1.2/CAP-v1.2-os.pdf (accessed on 18 November 2020).
69. Hatzivasilis, G.; Papadakis, N.; Hatzakis, I.; Ioannidis, S.; Vardakis, G. AI-driven composition and security validation of an IoT ecosystem. *Appl. Sci.* **2020**, *10*, 4862. [CrossRef]
70. Friedman-Hill, E.J. Jess: The Rule Engine for Java Platform. Sandia National Laboratories. 2008. Available online: http://www.jessrules.com/docs/71/ (accessed on 18 November 2020).
71. Lu, Y.; Li, Q.; Zhou, Z.; Deng, Y. Ontology-based knowledge modeling for automated construction safety checking. *Saf. Sci.* **2015**, *79*, 11–18. [CrossRef]
72. Patkos, T.; Plexousakis, D.; Chibani, A.; Amirat, Y. An event calculus production rule system for reasoning in dynamic and uncertain domains. In *Theory and Practice of Logic Programming*; Cambridge University Press: Cambridge, UK, 2016; Volume 16, pp. 325–352.
73. Mueller, E.T. *Commonsense Reasoning*, 2nd ed.; Kaufmann, M., Ed.; Elsevier: Amsterdam, The Netherlands, 2015; pp. 1–516.
74. Lei, C.; Zhang, H.-Q.; Tan, J.-L.; Zhang, Y.-C.; Liu, X. Moving Target Defense Techniques: A Survey. *Secur. Commun. Netw.* **2018**, *2018*, 1–25. [CrossRef]
75. Berstel, B. Extending the RETE algorithm for event management. In Proceedings of the 9th International Symposium on Temporal Representation and Reasoning, Manchester, UK, 7–9 July 2002; pp. 49–51.
76. Eby, M.; Werner, J.; Karsai, G.; Ledeczi, A. Integrating security modeling into embedded system design. In Proceedings of the 14th Annual IEEE International Conference and Workshops on the Engineering of Computer-Based Systems (ECBS), Tucson, AZ, USA, 26–29 March 2007; pp. 221–228.
77. Kelly, S.; Tolvanen, J.-P. *Domain-Specific Modeling: Enabling Full Code Generation*; Wiley-IEEE Computer Society Pr.: Hoboken, NJ, USA, 2008; pp. 1–444.
78. Ko, H.; Jin, J.; Keoh, S.L. Secure Service Virtualization in IoT by Dynamic Service Dependency Verification. *IEEE Internet Things J.* **2016**, *3*, 1006–1014. [CrossRef]
79. Albanese, M.; Battista, E.; Jajodia, S.; Casola, V. Manipulating the Attacker's View of a System's Attack Surface. In Proceedings of the IEEE Conference on Communications and Network Security, San Francisco, CA, USA, 29–31 October 2014; pp. 472–480.

80. Savola, R.M.; Sihvonen, M. Metrics driven security management framework for e-health digital ecosystem focusing on chronic diseases. In Proceedings of the MEDES '12: International Conference on Management of Emergent Digital EcoSystems, Addis Ababa, Ethiopia, 28–31 October 2012; pp. 75–79. [CrossRef]
81. Ayyappan, B.; Kumar, P.M. Security protocols in WSN: A survey. In Proceedings of the 2017 Third International Conference on Science Technology Engineering & Management (ICONSTEM), Chennai, India, 23–24 March 2017; pp. 301–304.
82. Parducci, B.; Lockhart, H. *eXtensible Access Control Markup Language (XACML) Version 3.0*; OASIS Standard: Burlington, MA, USA, 2013; pp. 1–154.

 *electronics*

Article

Exploratory Data Analysis and Data Envelopment Analysis of Urban Rail Transit

Guillermo L. Taboada [1,*] and **Liangxiu Han** [2]

1   Computer Architecture Group, CITIC, Universidade da Coruña, Campus de Elviña, 15071 A Coruña, Spain
2   Department of Computing and Mathematics, Manchester Metropolitan University, Manchester M1 5GD, UK;
    l.han@mmu.ac.uk
*   Correspondence: guillermo.lopez.taboada@udc.es

Received: 6 July 2020; Accepted: 5 August 2020; Published: 7 August 2020

Abstract: This paper deals with the efficiency and sustainability of urban rail transit (URT) using exploratory data analytics (EDA) and data envelopment analysis (DEA). The first stage of the proposed methodology is EDA with already available indicators (e.g., the number of stations and passengers), and suggested indicators (e.g., weekly frequencies, link occupancy rates, and CO_2 footprint per journey) to directly characterize the efficiency and sustainability of this transport mode. The second stage is to assess the efficiency of URT with two original models, based on a thorough selection of input and output variables, which is one of the key contributions of EDA to this methodology. The first model compares URT against other urban transport modes, applicable to route personalization, and the second scores the efficiency of URT lines. The main outcome of this paper is the proposed methodology, which has been experimentally validated using open data from the Transport for London (TfL) URT network and additional sources.

Keywords: urban rail transit (URT); exploratory data analysis (EDA); data envelopment analysis (DEA); sustainable transport systems; intelligent transportation systems (ITS); big-data applications

---

1. Introduction

Rail is one of the most energy-efficient transport modes [1], accounting for approx. 8% of global freight and motorized passenger movements but only 2% of transport energy use, being the transport mode with highest percentage of electric penetration. Thus, the continuous decarbonization of power production will allow zero-emission rail transport in the medium term. This is especially relevant for urban environments, where fuel-based transport modes impact the most on people's health. For these reasons, urban rail transit (URT) plays a key role in a context of a significant rise of urban population, particularly in emerging economies, which increases pollution, congestion, and city-center traffic restrictions.

URT is ideally suited for high passenger throughput, and although investment is especially high per kilometer, costs per throughput capacity are lower than for urban road infrastructure [2]. Shifting passengers from private cars to public transport, particularly in large cities, is key to reducing net energy use and emissions to be able to meet the mobility challenges within the sustainable development goals (SDG) [3].

Nearly 200 cities worldwide have metro systems (URT with the highest capacity), whose length exceed 32,000 km, whereas around 400 cities have light rail systems (URT with less investment requirements, less speed, and more modest capacity). Most recent (in the 2010s decade) URT developments have been, in the case of metro, which requires the highest investments, in Asia (34 of 46 new cities with metro). In the case of light rail, 28 new projects have been developed in

Europe and another 37 new projects roughly equally among Asia, North America, the Middle East, and North Africa.

URT, on the one hand, has multiple benefits such as mitigating CO_2 emissions and local air pollution, and wider social and economic benefits. One of these is reducing commuting time, therefore expanding the urban/suburban areas to those directly communicated by URT. This way, labor force can live at a higher distance from the city center in new urban developments, which are more cost-effective (less expensive).

On the other hand, one of the limitations of URT networks, the high infrastructure investment and its capillarity degree, could be partially addressed through multi-modal communication, using URT for the main journey combined with another communication mode, typically walking for distances up to 1 km, cycling, bus, or private car from a park-and-ride facility. However, there is still room for improvement as living more than 1 km away from an URT station requires either a frequent bus service network or private car ownership and confronting parking costs.

The objective of this paper is to characterize the efficiency and sustainability of URT using a proposed methodology based on exploratory data analysis (EDA) [4] and data envelopment analysis (DEA) [5,6]. First, the available data is explored using EDA to identify the main factors influencing URT for, secondly, derive efficiency scores, both across different transport modes and different lines within a rail network.

The reminder of the paper is structured as follows: the current state of the research field is presented in Section 2 and our methodology for assessing the efficiency and sustainability of URT using EDA and DEA is explained in Section 3. Section 4 experimentally validates the approach using open data from Transport for London (TfL) URT, particularly its underground network data. Section 5 discusses how additional big-data sources can improve the efficiency and sustainability of URT. In Section 5, we draw the main conclusions, stressing the main contributions of this work.

2. State of the Art

The recent increase in quantity and quality of data in public transport systems has fueled the adoption of data-driven solutions, mainly based on EDA or artificial intelligence (AI)/machine learning, to make public transport systems more intelligent, green, and safe. However, research in this area is, compared to applications in roads and private services/vehicles, more challenging due to the scarcity of available data, and the difficulties in testing research hypotheses in the real world. Thus, it is relatively common to have projects such as TEMA Big-Data Platform [7], which monitored 28,000 fuel vehicles with on-board GPS in Modena and Firenze (Italy) for one month to obtain 4.5 million trips and parking events, whereas most public transport analysis are based on significantly lower data records.

In an earlier study [8] a method and a software has been developed to estimate an URT passenger origin–destination trip matrix using an automatic data collection system. The method was experimentally assessed with automatic fare collection (AFC) origin-only data from Chicago Transport Authority (CTA), inferring the destination to replace the manual and costly origin–destination surveys. TfL, although it has a different fare collection scheme, an origin–destination system vs. the origin-only of CTA, collaborated in this research. In [9] smart-card bus data (metro was not included), travel surveys, and passengers' addresses have been used to measure commuting efficiency in Beijing in 2008–2010 as a function of commuting time, and residence/work location. In [10] public transport users' behaviors have been explored, whereas [11] analyzes individual mobility choices in carpooling. In [12] multi-modal transportation systems are presented as a way to increase efficiency through economies of scale, claiming that a multi-modal system combining a fast and efficient URT with other mobility options can provide more potential gains than optimizing single modal transport systems. These early works are EDA or data mining single cases studies using only a few data sources.

As the available data has been increasing during the last decade, particularly thanks to the integration of sensors in intelligent transportation systems (ITS) [13], especially in roads and connected vehicles, the number of projects started growing exponentially. A recent review paper [14] presents almost a hundred EDA, data mining, AI, and machine-learning applications, challenges, and limitations, particularly for management, traffic safety, public transportation, and urban mobility. However, when it comes to public transport only tackles route planning, aviation, on-demand bus, and shared mobility, but no references to URT. Another reference paper [15] covers big-data projects and technologies in transportation and mobility, highlighting the scarcity of references in maritime and rail transport systems, with only a few works on predictive maintenance, risk management, and railway accidents.

URT, especially underground, faces important capacity limitations, especially in city centers at peak times. This has been the focus of [16] for forecasting passenger flows using Artificial Neural Networks (ANN) on a single metro line in Naples with a simulated dataset. Short-term forecasting on urban metros has also been studied along with other methods, such as Kalman filter in [17] and ARIMA (autoregressive integrated moving average) models in [18]. Li et al. [19] proposed a Multi-Scale Radial Basis Function (MSRBF) for forecasting short-term metro passenger flows on special occasions, such as sporting events and concerts. In this case, passenger flow is very irregular, and predictions are more difficult to obtain. Ling et al. [20] used smart-card data for predicting passenger flows in the subway of Shenzhen (China); they analyzed four predictive models: a historical average model, ANN, regression model, and a gradient-boosted regression tree model. Liu et al. [21] proposed a deep learning method for short-term forecasting of metro inbound/outbound passenger flows, while Wang et al. [22] proposed a Novel Markov-Grey model for solving the same problem.

A novel model of Multi-scale Mixture Feedback Wavelet Neural Network (MMFWNN) has been proposed in [23] to predict the short-term entrance flow of Shanghai subway stations, distinguishing passengers into commuter (more predictable) and non-commuter (more dependent on the weather). In [24] the factors affecting Seoul Metro boarding have been analyzed using regression analyses against the station environment (density, employment, commercial/office area), external connectivity (through metro and roads) and intermodal (bus and metro). This, and previous models, can predict highly accurately the short-term entrance flow, as it corresponds with regular patterns. However, the lack of historical data limits the behavior in special situations/events.

Relevant research works on other transportation modes are [25], where four classification algorithms have been used to model the relationship in London between weather and short cycling journeys using docked bikes. In addition, [26], proposing the application of deep learning methods to a Bus Rapid Transit (BRT) system (Xiamen, China) to forecast the hourly flow, adopting a three-stage architecture. This paper also analyses the literature, identifying four different approaches: (1) traditional classical algorithms; (2) regressive models; (3) machine-learning-based models, including ANNs; (4) hybrid models. In [27] a novel Context Neural Network framework has been proposed for the prediction of road traffic flow showing better long-term predictions than previous well-established models. All studied cases, however, refer to short-term or long-term time periods, without considering the spatial dimension.

The relevant number of references of the application of DEA to different public transport modes [28], contrasts with the scarcity of research on the efficiency of rail networks, [29], especially URT. These references in rail transport systems generally compare different public transport agencies at regional or national levels. In [30] 17 European URT networks have been evaluated using a two-stage methodology focusing on the relationship between the operational performance and their socioeconomic contexts. In [31] urban public transport systems of 652 Chinese cities have been analyzed, highlighting the high efficiency of URT. In [32] DEA has been used to assess the efficiency of 31 railway companies across multiple countries. In [33] the efficiency of 20 representative URT systems, among them London, Hong Kong, and New York, have been analyzed concluding that the higher the number of stations, the higher the efficiency. This conclusion is also supported by a recent study [34] on Chinese URT.

In [35] DEA has been used to assess the performance of the bus lines of a single transport authority in a suburban area in California Central Coast. In [36] the efficiency of Seoul Arterial Bus Route has been analyzed using DEA considering a wide variety of factors, including total rides, service satisfaction, and CO_2 emissions. This latter work was expanded in [37] with a network DEA model, also validated with bus companies in Seoul. In addition, finally, in [38], DEA has been used to compare different transport options and investments on a single route.

The selection of input and output variables in DEA is regarded as an important step that is normally conducted before the DEA model is implemented. Available techniques are, on the one hand, based on expert intervention, using heuristic decision-making, and expert judgement (e.g., using Delphi), and, on the other hand, fully automatic approaches [39] which in turn maximize efficiencies and lose discrimination power without a full understanding of the domain. There is a lack of data-based methodologies and use cases that avoid bias of experts and at the same time provide useful, repeatable, and interpretable results. The proposed methodology in this paper, using EDA for a thorough selection of a limited number of variables, addresses this need by combining both approaches.

A review of the related literature on efficiency analysis in urban public transport [28] shows a quite homogenous selection of input and output variables, guided by experts, with a fairly narrow perspective. Thus, state-of-the-art variables are (in parenthesis the percentage of the papers in the literature that reported each variable):

- Input (physical measure): number of vehicles (61%), number of employees (40%), fuel consumption (36%), worked hours (11%), drivers (3%), non-driving employees (2%), seat capacity (total seats of the fleet) (2%), and number of depots (1%).
- Input (CAPEX): Price of capital (21%), and investment (4%).
- Input (OPEX): Price of labor (39%), price of fuel (29%), OPEX (12%), material costs (11%), fuel costs (9%), operating cost of vehicles (11%), maintenance costs (6%), and operating labor expenses (4%).
- Output (Service supply): overall traveled kilometers by all vehicles (53%), seats offered multiplied by overall traveled kilometers (26%), vehicles multiplied by hours of operation (4%), and revenue-vehicle kilometer (3%).
- Output (Service consumption): passengers multiplied by traveled kilometers (24%), number of passengers (17%), and number of trips (6%).
- Output (revenue): operating revenues (11%).

Furthermore, in the literature there are additional variables, neither considered inputs nor outputs, but sometimes considered external variables, which characterize public transport systems. Representative examples of these variables are (listed together with their presence, in percentage, in the analyzed related literature):

- Quality and characteristics of service: length of the network (24%), average commercial speed (21%), average fleet age (15%), service frequency (7%), and number of stops (6%).
- Socio/demographic: population density (16%), location (13%), car ownership (7%), population served (7%), and area served (5%).
- Managerial: public company (15%), contract type of the operator (11%), size of the company (3%).
- Subsidies: subsidies from public funds (10%), subsidies to operating expenses (8%), local subsidy (2%).
- Externalities: number of accidents (7%), and emissions (4%).

With regards to URT, the variables used in the related literature are:

- In [30] the network length, the number of stations and cars are the inputs (CAPEX), whereas the number of employees is considered the only input (OPEX), due to the scarcity of materials and energy consumption information, two relevant inputs (OPEX). Additional variables considered to be inputs are ratios between these variables (e.g., the network length divided by the number of cars), historical data, as well as socioeconomic variables, such as area, population density of the core city, average household size, unemployment rate, GDP (Gross Domestic Product) per capita, and diesel pump price. In [30] two models are computed: (i) efficiency, using the number of cars-kilometers produced as output, and (ii) effectiveness, considering the number of transported passengers. The large number of variables and the limited number of analyzed URT networks (17) ends up with most of the evaluated systems considered highly efficient (here most URT networks excel in some, disjoint parameters, increasing its efficiency). The impact (elasticity) of the variables has also been considered, but the work fails in selecting the most representative ones.

- In [32] six inputs have been considered, the annual cost of operation as input (OPEX), and the network length, and the number of employees, traction vehicles, passenger cars, and cargo cars as inputs (CAPEX). Additionally, five outputs have been defined, revenues earned, transported passengers, transported passengers per kilometer, transported cargo tons, and transported cargo tons per kilometer.
- In [33] the number of employees and the labor costs are the selected inputs (OPEX), whereas the number of cars in operation and non-labor costs are used as inputs (both OPEX and CAPEX). The selected outputs are car-kilometers and transported passengers. Historical data has also been considered. Furthermore, additional variables have been used in the Tobit models phase, after DEA, such as population density, the number of stations, distance between stations, geographic location, and the type of URT (light/rapid or heavy).

So far, the use of input and output variables in DEA URT models relies on a wide range of state-of-the-art variables from the related literature, generally with limited selection and statistical analysis. Moreover, the access to these variables incurs relevant collection costs, such as accessing to unstructured reports, limiting the viability of comparing additional URTs.

This paper overcomes these latter limitations through:

- Selecting a limited number of representative variables through EDA, both state-of-the-art and new variables, increasing the discrimination power of DEA by bringing forward the statistical and visual analysis, prior to the variable selection (previous works [40] only suggested EDA after DEA, to understand the impact of variables on the models, so using EDA as first stage is one of the key contributions of this work).
- Automating data collection from public sources (e.g., open-data and online services), thus supporting the direct comparison across different URT systems.
- Comparing, for the first time, to the best of our knowledge, a single URT system at the line level, and also against other transport models from the traveler perspective, focusing on the efficiency and sustainability, and skipping the wide range of sociodemographic variables that require two-step modeling, as for [30,33].

Thus, the combination of EDA and DEA will be able to monitor, understand, and improve URT management.

3. Methods and Materials

Despite the relevance of URT for the development of sustainable cities, there is a lack of research on the efficiency and sustainability of URT systems and their management. The increasing availability of data, both personal (e.g., GPS location) and Internet-of-Things (IoT) big data, is expected to play a key role in the development of tailor-made mobility solutions, also known as Mobility-as-a-Service (MaaS) [41], based on convenience, sustainability, and resource efficiency to meet passengers' individual needs.

This paper introduces a methodology for assessing the efficiency and sustainability of an URT network based on large-scale data analytics consisting of four stages: (1st) EDA using state-of-the-art indicators; (2nd) EDA using new proposed indicators that deepen the analysis; (3rd) DEA using several original transport models, and (4th) rank transport modes and URT network elements according to efficiency measures, analyzing the results. Figure 1 summarizes graphically the methodology.

As URT systems, particularly in large cities, are a combination of complex interrelations, the proposed methodology aims at better capturing the most relevant efficiency and sustainability indicators to optimize transport infrastructures, from planning to real-time operation. Furthermore, as an additional outcome of this methodology, open-data repositories could be enriched with new data sources such as occupancy rates, queueing times, URT network elements capacities (e.g., stations) as well as CO_2 footprints.

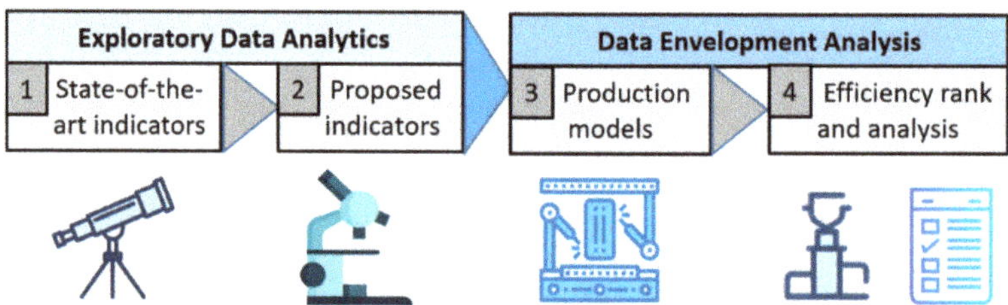

Figure 1. Overview of the proposed methodology.

3.1. Exploratory Data Analysis (EDA) of URT Data

The first stage of the proposed methodology uses EDA for deriving state-of-the-art quantitative indicators [30]: network length, number of stations, the number of trains, the number of frequencies, the number of employees, the number of operated kilometers, and the number of passengers. This data is usually publicly available at transport operator level, useful for comparing operator's efficiency, but it is more difficult to find at line level, limiting the analysis of the efficiency of rail network elements. However, thanks to big-data technologies (e.g., logging API requests/responses, queueing transport events, and web scraping) these indicators can be potentially estimated using models at a more fine-grained level. In the absence of data from operators (according to [28] only 9% of the research papers in this area has access to official data, generally open data) relying on big data is a much more scalable and cost-effective solution than ad hoc surveys. This approach will contribute to deepening the analysis of transport operators, thus increasing the limited number of research papers with city coverage (only 6% in [28]).

EDA, also known as Visual Analytics, is a heuristic search technique for finding significant relationships between variables in large datasets. Its simplicity and efficiency are key to derive insights from big data, in fact, it is usually the first technique when approaching data, particularly unstructured. According to Tufféry [42] EDA usually consists of six steps (see Figure 2) namely: (i) Distinguish/Identify Attributes; (ii) Univariate Data Analysis to characterize the data of the dataset; (iii) Detect Interactions Among Attributes performing bivariate and multivariate analysis; (iv) Detect and minimize impact of Missing and Aberrant Values; (v) Detect Outliers (further analysis or errors), and finally (vi) Feature Engineering, where features are transformed or combined to generate new features.

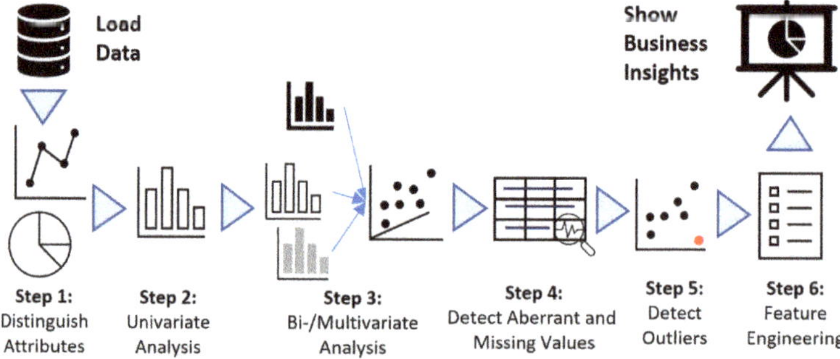

Figure 2. Exploratory data analysis (EDA) steps.

There is a large number of tools for performing EDA (50 of them are analyzed in [43]) with different functionalities to assist both with the identification of hidden patterns and correlations among attributes, but also with the formulation of hypotheses from the data and their validation. EDA can also be performed using R, python (used in our research work, programming ELTs—Extract, Load, and Transforms—followed by Datawrapper visualization) or any other programming language oriented to data preparation and exploration. Additionally, due to the geographical dimension of transport it is relevant that the tool includes Geographical Information Systems (GIS) support and a strong set of visualization capabilities.

3.2. Efficiency and Sustainability Key Performance Indicators (KPIs) for URT

The output of EDA is the estimation of state-of-the-art Key Performance Indicators (KPIs), as well as defining new ones based on large-scale data. For instance, new KPIs that can be defined are the number of trains per line that could be estimated based on the travel time and the rail frequencies. Moreover, another KPI, the number of passengers per line, can be estimated from the number of trains and the entry/exit numbers at the stations of a line. Finally, URT CO_2 footprint can be estimated from the annual supply (in GWh) and the breakdown by source of the consumed electricity and their CO_2 respective footprints.

Additional candidate KPIs that can be modeled after big-data sources are:

- Occupancy ratio: considering 100% occupancy ratio equals to all seated places plus 4 standing people per m^2.
- Trains Per Hour (TPH): the number of trains that enter and leave a given station per hour, monitoring real-time traffic conditions.
- Travel time: from origin to destination, involving an URT journey.
- Excess Journey Time: is the additional time on top of scheduled time.

The definition, measurement, and analysis of the evolution of KPIs is key to improve the efficiency, security, convenience, and sustainability of existing URT. In fact, the public availability of these KPIs might support that personalized preferences for route selection can be expanded to, for instance, the occupancy rate, for ensuring the availability of seating space, CO_2 footprint, or risk of an excess journey time higher than 10 min. Currently the preferences for route selection are quite rigid, the faster route, or manifest preference for a transport mode, although eventually passengers are considering additional factors, as seen when analyzing, anonymously, their routes using Wi-Fi data.

3.3. Data Envelopment Analysis (DEA) for Assessing Efficiency and Sustainability of Public Transport

DEA is a non-parametric method to measure the performance of entities, called Decision-Making Units (DMUs). A DMU can be a factory, a bank branch, a hospital and, as in our paper, a transport mode, an URT line, or an URT station. The initial DEA models consider Constant Return to Scale (CRS or CCR for Charnes, Cooper, and Rodhes), which ignores the fact that different DMUs could be operating at different scales. In our scenario, it would not make any distinction between two URT lines, one with 6 stations and another with 60 stations. To overcome the drawback the Variable Returns to Scale (VRS or BCC for Banker, Charnes, and Cooper) mode [44] was introduced, ensuring that DMUs are only benchmarked against DMUs of similar size. Figure 3 presents an example of four DMUs and both CRS and VRS efficiency frontiers. DMU 1 is the only one in CRS efficiency frontier (the only efficient in CRS), maximizing the output/input ratio, whereas DMUs 1, 2 and 3 are in VRS efficiency frontier (the three are efficient in VRS, DMU 2 in low input values and DMU 3 in high input values). Further to VRS, a wide range of DEA models have been designed for measuring efficiency and capacity specializing the original models into different types of problems.

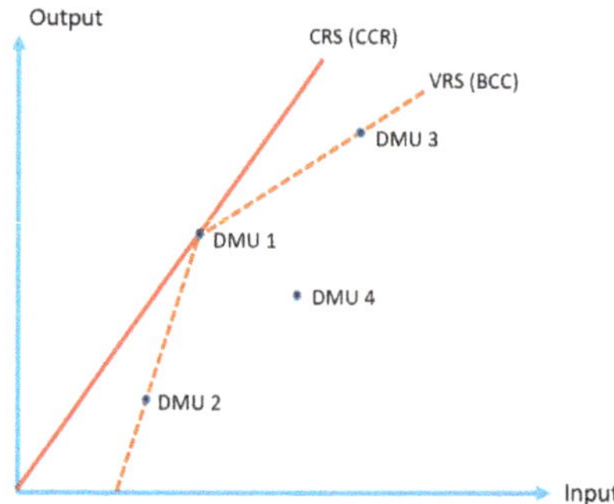

Figure 3. DEA CRS and VRS efficiency frontiers and four DMUs.

DEA models can be classified in either input-oriented or output-oriented models. Figure 4 shows an inefficient DMU (DMU 4 or C) to exemplify both approaches. Input-oriented efficiency is BA/CA. Output-oriented efficiency is CD/ED. With input-oriented DEA, a DMU computes the potential savings of inputs in case of operating efficiently (in Figure 4 reducing the inputs from C to B while providing the same output). In contrast, with output-oriented DEA, a DMU measures its potential output increase given its inputs do not vary (in Figure 4 increasing the outputs from C to E while using the same amount of input, D. If C were in the frontier, so C = B = E, the efficiency would be 1.

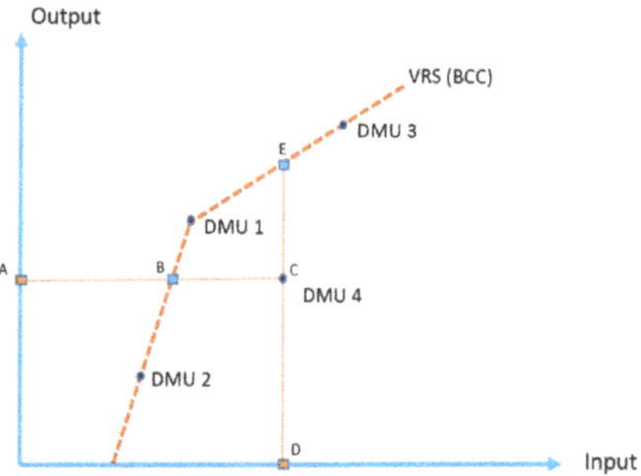

Figure 4. DEA VRS efficiency frontier and DMU 4 efficiencies.

The bad/undesirable outputs, in our case CO_2 emissions, have been treated as inputs reversing traditional DEA models [45,46]. This technique is based on the fact that undesirable outputs can be treated as inputs when there is a combination of undesirable and desirable outputs. The objective is to minimize the undesirable output, so considering it as input the function looks for its minimization.

A DEA model is a particular selection of inputs and outputs to analyze the efficiency of DMUs. In previous DEA assessments of transit lines, labor, capital, and energy have been used as inputs and vehicle-kms and passenger-kms have been used as outputs. In the absence of actual costs of labor, fuel/energy, and other operational expenses for individual transport lines, it is reasonable to assume that the cost of operating a line is related to its travel time, round-trip distance, and the number of stations/bus stops [35]. Additional, when alternative transport options are being considered, the cost is usually the single input whereas travel time savings, patronage (people for each transport mode), and car trips removed are outputs, as shown in [38], a study that implemented a constant returns of scale–output-oriented (CRS–O) model.

Figure 5 presents our candidate DEA models for: (a) assessing different transport modes from the traveler's viewpoint (for route planning), and (b) analyzing URT lines from the operator/local authority perspective. The analyzed DMUs are the available transport modes (e.g., URT, bus, car, taxi, walking, and cycling) for the first model, and the available URT lines, usually in the range of 1 to 24 lines (e.g.,

New York has the highest number of metro lines, 24, followed by Beijing 23, Seoul 23, Shanghai 17, Paris 16, Moscow 14, and Tokyo 13). CRSs are considered for both models, in the transport modes model because route planning is generally used for one traveler (or a small group) and DMUs operate in the same scale, whereas URT lines, for a given URT network, are usually directly comparable.

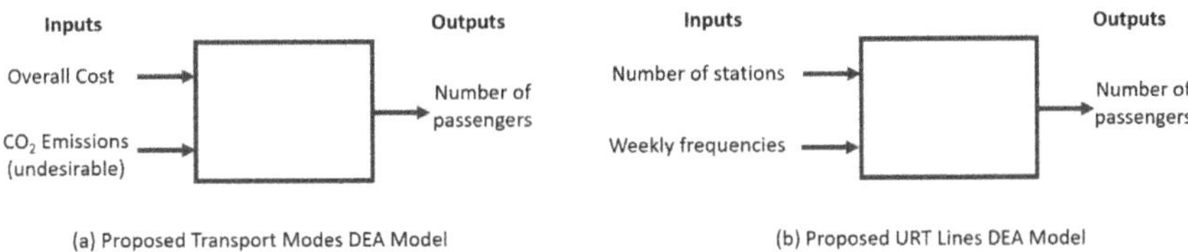

Figure 5. (**a**) Transport modes efficiency and (**b**) URT lines efficiency DEA models.

The selection of inputs and outputs is especially relevant in this scenario due to the large number of available variables and the modest sample size. Following the cardinality constraints introduced in [39], the recommended number of variables for these two CRS models is 3 (in case of considering VRS it would be 2). The selected variables depend eventually on EDA on the available data; however, a tentative output is the number of passengers, and the models can be considered input-oriented, designed for minimizing inputs when moving a given number of people.

In the first model CO_2 emissions, an undesirable output, have been treated as input, as already mentioned, while considering the overall cost the only true input [38]. Here, as the model is from the passenger viewpoint, the overall cost is the transport fare (or the direct costs incurred) plus the monetary value of the passenger time. As most of the mobility is associated with commuting to work, the passenger time value can be estimated at the cost of unskilled working time, although this can be configured on a per-passenger basis for personalized route planning. The selection of these two inputs, which combined with the output reaches the recommended number of variables (3), is original, selected after using EDA on the available data, which contrasts with state-of-the-art indicators for route planning such as travel time and fare cost.

With regards to the URT lines model, two tentative inputs, subject to change due to EDA conclusions on the available data for a given URT network, are considered: (1) the number of stations per line as estimate of the capital costs (CAPEX); and (2) weekly frequencies as operating costs (OPEX).

The related literature in public transport generally uses the actual investment as CAPEX; however, when considering URT lines it is neither directly disaggregated per lines, nor comparable across the time (e.g., 20th century vs. 21st century URT lines). The number of stations per line has been selected as input due to the wide availability of this KPI, although generally indirectly, derived from the longest URT route obtained from online route planning/maps services and applications. The line length, although it is a more popular metric and it is also widely available, has not been selected after EDA on the available data (data sourced from [47]) as long lines usually have generally lower investment due to a higher ratio of above-ground to underground construction, especially in suburban areas where distance between stations tend to be higher. In fact, in [47], a reference paper in CAPEX in Urban Rail only considers costs per kilometer, a state-of-the-art KPI, which shows a higher variability than the cost per station (e.g., in 16 European URT projects, after discarding 3 outliers, the cost per kilometer ranges from 26.7 to 88.3 M USD$, whereas the cost per station ranges, for the same projects, from 39.4 M to 83.1 USD$), with lower standard deviation. Additionally, stations have a share of 25–30% of the infrastructure costs, which favors the selection of the number of stations versus line length as CAPEX.

Regarding OPEX inputs, the related literature in public transport generally uses the price of labor and the price of fuel. However, they are not particularly useful for comparing different lines within the same URT system, as they are set at the operator level. Labor and energy consumption can vary per line, although this level of detailed data is generally not available. Nevertheless, a variable directly related to OPEX that is generally available per line is the number of weekly frequencies. EDA on route planning data shows different patterns for weekdays and for weekends, so the week is the selected period. This input, in combination with the number of stations (the other input), are the selected variables for this DEA model after EDA on publicly available data from URT systems.

The shortlisted inputs (e.g., number of stations and weekly frequencies) have a relevant positive correlation with most of the state-of-the-art inputs, such as the line length, labor force, and number of URT cars, as shown from EDA on [47] and validated in Section 4 (e.g., using LU lines key parameters), thus making it a highly representative selection, with higher discriminatory power and simplicity thanks to minimizing redundancy. Furthermore, the shortlisted inputs are directly obtained from route planning services and applications (e.g., Apple Maps, Bing Maps, Google Maps, and services such as Rome2rio.com that, as of today, includes worldwide 176,885 rail lines from 4151 operators), significantly easier than collecting data from other sources, some of them not available publicly. Finally, in case of availability of data, our candidate inputs for a more representative model would be car capacities, consider line branches, and breakdown passengers into time bands (a.m./p.m. peak versus off-peak). The selection of these additional candidate inputs, which add relevant information about URT efficiency, is one of the outcomes of the previous step, defining new indicators from EDA.

These DEA models have been computed using the solver software that comes with the reference DEA book by Cooper [6]. To illustrate DEA concepts this subsection concludes with an example of DEA analysis, computing the efficiency of London Underground (LU) lines, the use case to validate the proposed models, using for clarity purposes a simplified URT lines DEA model, with a single input, the number of stations, and a single output, the number of passengers. Table 1 summarizes the input and output data, as well as the results provided by the solver. As there is a single input/output the resolution is direct.

The DMU Victoria maximizes the production function (weekly passengers per station), 363,000, so it scores 1. Compared to the first DMU of the list, Bakerloo, with 98,000 passengers per station, 26.9% of 363,000, thus scoring 0.269. This is a CRS model, similar to the two proposed models, so the production function is the same for all DMUs, not varying at scale (as for VRS). Since there is a fixed number of stations, the key parameter is the number of passengers that maximizes the efficiency for each line, so the model has been computed as output-oriented. In fact, the highest ratio, 363,000 passengers per station, has been used to compute the projection of passengers, presented in Table 1, as well as the difference between the projection and the actual line passengers. Thus, for Bakerloo, ranking 6th in Efficiency, the projection is 9.08 million passengers, +272% over the actual number of passengers, 2.44 million passengers. Alternatively, models can be computed following an input-oriented approach, thus minimizing the required number of stations to achieve the maximum ratio. Thus, for Bakerloo line, it would need to carry 2.44 million passengers with 6.7 stations (2.44/0.353), which is 73.1% less stations (1 minus its efficiency score, 0.269).

Table 1. Results of simplified URT lines DEA model of LU lines.

LU Line (DMU)	Num. of Stations Longest Route (I)	Weekly Passengers (millions) (O)	Ratio Passeng./Station (thousands)	Efficiency Score (Solver Output)	Rank	Passengers Projection (millions)	Difference (%)
Bakerloo	25	2.44	98	0.269	6th	9.08	272%
Central	49	6.22	127	0.349	5th	17.8	186%
District	60	5.17	86	0.237	7th	21.8	322%
H&C and Circle	46	2.99	65	0.179	9th	16.71	459%
Jubilee	27	5.99	222	0.61	2nd	9.81	64%
Metropolitan	34	2	59	0.162	10th	12.35	517%
Northern	50	7.03	141	0.387	4th	18.17	158%
Piccadilly	53	4.4	83	0.229	8th	19.26	337%
Victoria	16	5.81	363	1	1st	5.81	0
Waterloo City	2	0.33	166	0.456	3rd	0.73	119%

Figure 6 represents graphically the 10 DMUs (URT LU lines) using their coordinates (number of passengers as y axis and number of stations as x axis). The production function, CRS, achieves its maximum value for Victoria, thus scoring 1 in efficiency. Please note that the CRS function starts at the origin (0,0). The remaining DMUs score below 1, depending on its ratio passengers/station compared to the optimal. The least efficient is Metropolitan, graphically it can be seen that it has the minimum slope to the origin. The figure also helps to understand how to measure inefficiency. Using Bakerloo as a sample, on the one hand, for input-oriented, the CRS optimal function requires 73,1% less stations (6.7 stations) for moving 2.44 million passengers. On the other hand, for output-oriented, CRS optimal function can move 9.08 million passengers, +272%, with 25 stations.

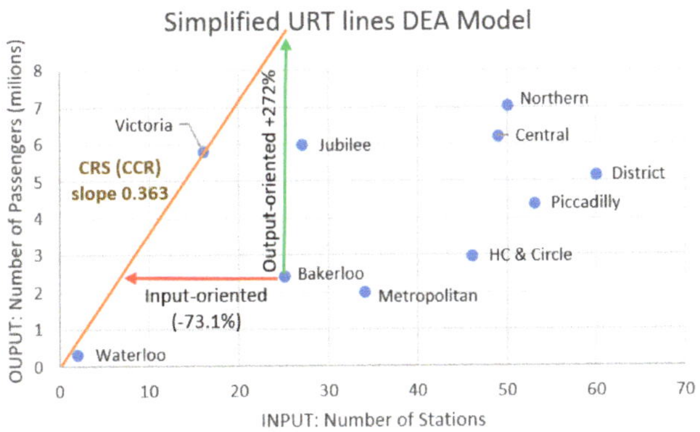

Figure 6. Simplified URT lines DEA model computed with 10 LU DMUs, CRS function, and efficiency measures for Bakerloo DMU.

3.4. Ranking DEA Models URT Lines According to Efficiency Indicators

The fourth and later stage of our methodology is to rank both transport modes and URT lines using the results of the DEA models. The efficiency of the transport models, from the traveler's viewpoint, can be used for personalized route planning, suggesting different transport modes depending on the time band, the travel distance and the user preferences (e.g., their own estimate of its value of time, and the usage of new mobility solutions such as private electric scooter, or bike/moto/car-sharing).

With regards to URT lines, ranking them according to their efficiency scores instead of less sustainable metrics, such as the number of car-kilometers or the increase in the number of passengers, contributes to align the public transport operation with sustainability goals. In fact, the most efficient URT lines will be those with a reduced number of stations and weekly frequencies that are able to transport more passengers. This model/rank can be complemented with the personalized route planning, as the frequency between URT services could be modified (increased/decreased) up to a point where URT is still the preferred transport choice.

3.5. Big Data and Sustainable URT

Public transport services, particularly URT systems, due to the economies of scale, are among the most efficient activities. However, they confront huge initial capital investments, and variables such as the number of stations, length, speed, are determined by this capital investment. Therefore, it is key to characterize their efficiency and sustainability, key to monitor its management.

Big data can gather, store, and process large amounts of heterogeneous, large-scale data to assist regulators, cities, transport operators, and travelers to improve the efficiency, regulation enforcement, and sustainability of their mobility solutions. So far route planning (e.g., Masivo model [48]) and public transport timetable optimization [49] are based on simulation models which can greatly benefit from the incorporation of big-data analysis into their models. Additional big-data applications are personalized route planning and smart taxation (based in the polluters-pay principle) such as dynamic tolling depending on the specific CO_2 footprint of cars and their usage (kilometers) in city centers, where air quality has one of the highest impacts on people's health.

4. Case Study: Efficiency and Sustainability of London Underground (LU)

This section presents the validation of the proposed methodology by analyzing the efficiency and sustainability of a reference URT network, the LU, selected because of the complexity of its network (3 million daily journeys, served by 540 trains across 10 lines covering 402 Km and 263 stations. Figure 7 presents the core of the LU network), and its open-data NUMBAT database (see Appendix A), one of the few publicly available and successful [50] datasets on URT.

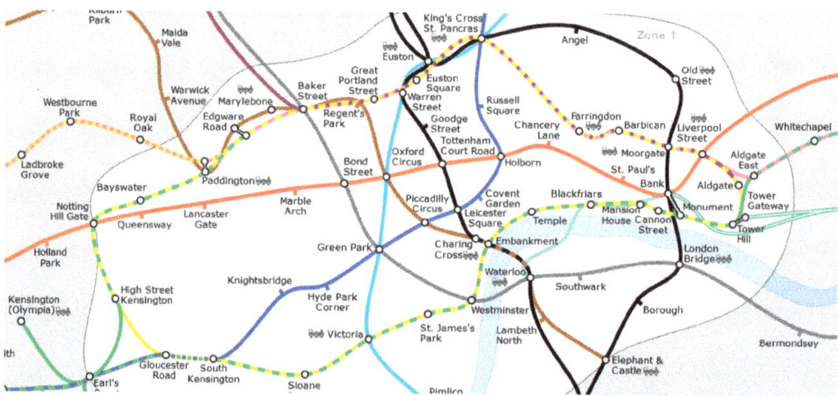

Figure 7. Map of LU lines (colored using the official palette) in Central London.

NUMBAT provides entry/exit/interchange passenger count for 263 stations and the number of trains per station every quarter hour. Additionally, it provides a 263 × 263 origin station–destination station matrix, covering all journeys and the annualized number of passengers for each line. However, NUMBA data is based on real data, but it is not real data. It is the output of a synthetic model used to research LU usage and travel patterns. Moreover, it assumes a perfect train schedule being operated and that all passengers board on the first train arriving at the station. This synthetic model is based on sampling real data from smartcards and gateline entry/exit totals for each station. Data is provided in quarter hours, grouped also by time bands (Early 3–7, AM Peak 7–10, Midday 10–16, PM peak 16–19, Evening 19–22, Late 22–3). Finally, data has been provided in a differentiated way for Fridays, Saturdays, Sundays and for the average of the remaining days (from Monday to Thursday).

As NUMBAT is quite limited (e.g., there is no information about schedules and LU lines, neither descriptive, nor the stations that belong to a line nor the capacity of the trains), we have extended this database with four major data incorporations: (i) train schedules; (ii) a table that relates lines with all their stations; (iii) a table that relates lines with their capacity (seated plus standing at 4 passengers per m^2), with data collected from TfL website (TfL open data does not include this data); and (iv) include GPS location for all the stations, obtained from Open StreetMap [51]. See Appendix A for further details. Figure 8 presents some key descriptive metrics of LU which are not originally available in its open-data repository.

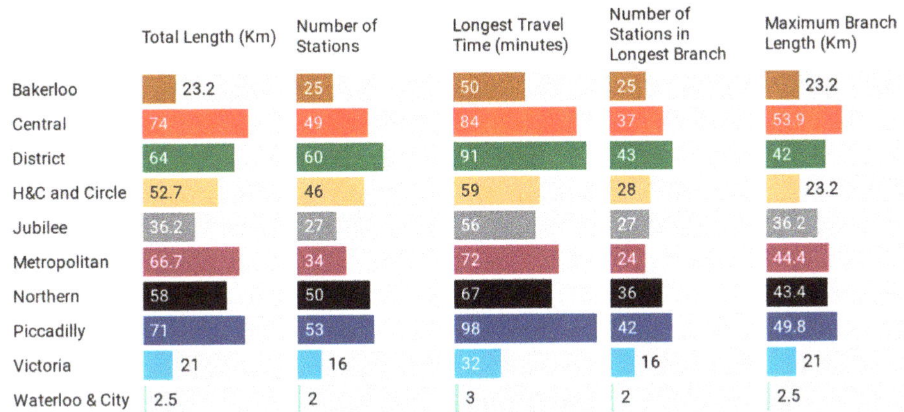

Figure 8. LU Key Descriptive Indicators.

4.1. Assessing the Efficiency and Sustainability of LU Using EDA

The first step of EDA is to distinguish attributes. Table 2 gathers LU key attributes: 3-letter LU line code (in the same order as Figure 8); the longest travel time in the line, it is the average scheduled time of the longest service, usually from the first until the last station of the line; and the length, in kilometers and stations, of the longest route. Additionally, the table contains the scheduled weekly LU frequencies at the station with the highest number of frequencies (usually stations at the middle part of the line), and the weekly passengers per line. A passenger counts as one passenger for each of the lines traveled. On average, a LU passenger uses 1.6 lines per journey (42.4 Weekly passengers in lines and 26 million weekly LU journeys).

The next parameters in Table 2 are metrics/KPI derived from the previous data. Figure 9 presents the scatter plot graphs of the number of passengers versus the number of stations (left), two variables that correlate positively with $R^2 = 0.55$ (the higher the number of stations, the more travelers it captures). Figure 9 also shows the number of passengers versus the line length (right), with $R^2 = 0.33$ (a long LU line might be reaching areas with less population density, so this correlation is weaker than the previous one). Additional parameters are the average number of passengers per service and station (included as it contributes to explain the variability with $R^2 > 0.5$, discarding the line length). Finally, Speed, in terms of km per hour and minutes per station is presented to illustrate key metrics of LU operation. Based on these analyses, two parameters, the number of stations of the longest route and the weekly frequencies, have been selected to be used in the second phase of the proposed methodology, efficiency scoring using DEA.

So far, the analyzed metrics are average numbers, not considering a relevant source of variability, the day of the week and especially the time band. Figure 10 presents the number of passengers per line and day of the week. The dataset provides an average number from Monday to Thursday. Fridays, except for the Metropolitan and Waterloo & City lines, is the busiest day, whereas Sundays is the day with the lowest number of passengers.

Table 2. LU lines key parameters.

LU Line	Longest Travel Time (minutes)	Longest Length (km)	Num. of Stations Longest Route	Weekly LU Frequencies (scheduled)	Weekly Passengers	Avg. Pass. per Service	Avg. Pass. per Service and Station	Speed (km/h)	Speed (min. per Station)
BAK	50	23.2	25	4836	2,444,910	506	21	28	2
CEN	84	53.9	37	6202	6,218,138	1003	27	39	2.3
DIS	91	42	43	4828	5,166,660	1070	25	28	2.1
HAM	59	23.2	28	3184	2,988,540	939	34	24	2.1
JUB	56	36.2	27	6624	5,985,450	904	33	39	2.1
MET	72	44.4	24	4718	2,003,527	425	18	37	3
NOR	67	43.4	36	10,610	7,028,737	662	18	39	1.9
PIC	98	49.8	42	5710	4,404,640	771	18	30	2.3
VIC	32	21	16	7480	5,813,439	777	49	39	2
WAC	3	2.5	2	3402	331,156	97	49	50	1.5

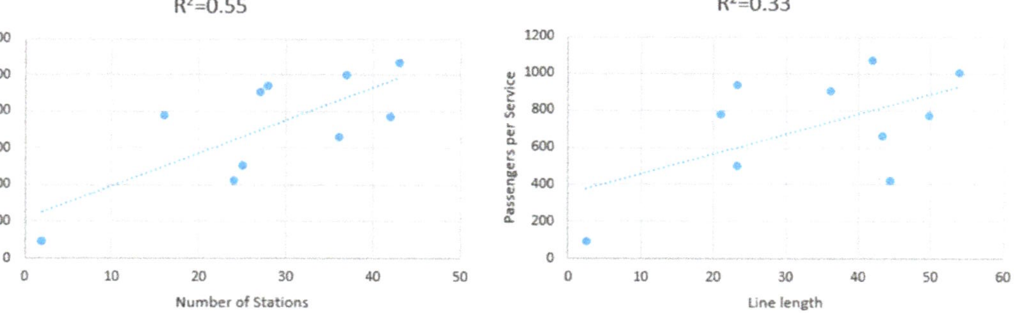

Figure 9. Linear regression of passengers per service versus line stations (**left**) and line length (**right**).

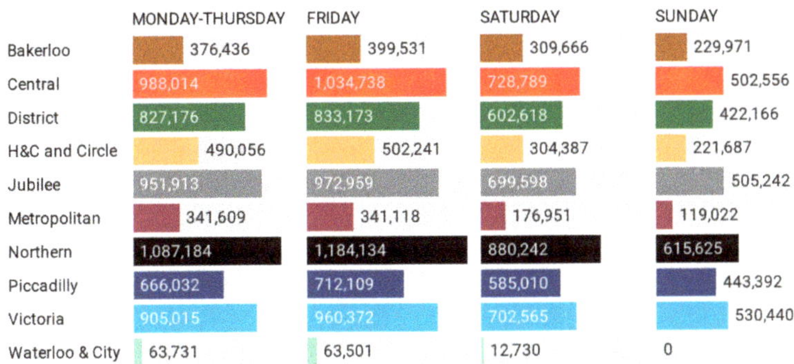

Figure 10. LU Daily passengers per Line.

Figure 11 presents the distribution of passengers per day of week and time bands (Early 3–7, AM Peak 7–10, Midday 10–16, PM peak 16–19, Evening 19–22, Late 22–3). AM and PM peak hours (3 h each) concentrate most of the use from Monday to Friday, whereas Midday (6 h) is the preferred time band for weekend passengers. WAC (Waterloo & City) only operates from Monday to Saturday and has the highest use during Monday to Friday peak hours. Late traffic is higher on Saturday and also Friday, which motivates the different traffic pattern of Friday versus Monday to Thursday, and the higher number of passengers of Friday than Monday to Thursday (except MET and WAC lines, see Figure 10).

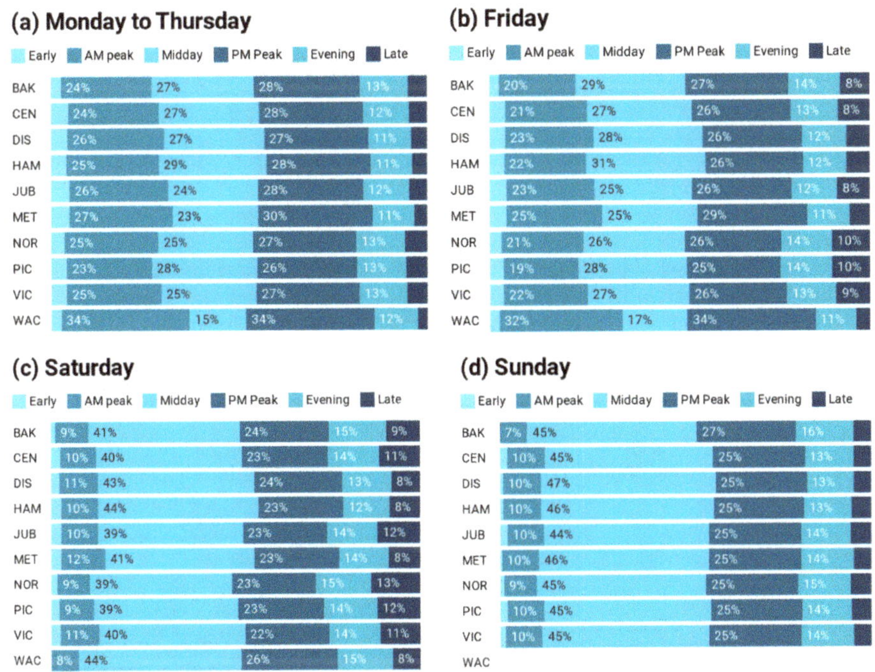

Figure 11. LU passengers per day of week, line, and time band.

A new metric, occupancy rate (usually not reported by URT operators), has been computed dividing the number of passengers by the capacity of the line by time band. To compute this KPI the underground capacity has been considered (seated spaces plus 4 standing passengers per m^2, see Appendix A). Figure 12 presents the occupancy rate, sometimes higher than 1 (e.g., Central and District lines). This means that a train, when going from the beginning to the end of the line, can move more passengers than its theoretical capacity. This is possible because these lines, Central and District, have branches and multiple exchanges with other lines, so each seat/standing space can be occupied by more than one passenger per service. A model that estimates the maximum capacity of a line based on an origin–destination trip matrix has been already suggested [52]. However, in our work we will capture these differences in the DEA efficiency model, without providing specific weights to the behavior of line travelers. However, the availability of actual origin–destination data (not the model-based NUMBAT dataset) would increase the interest of this research.

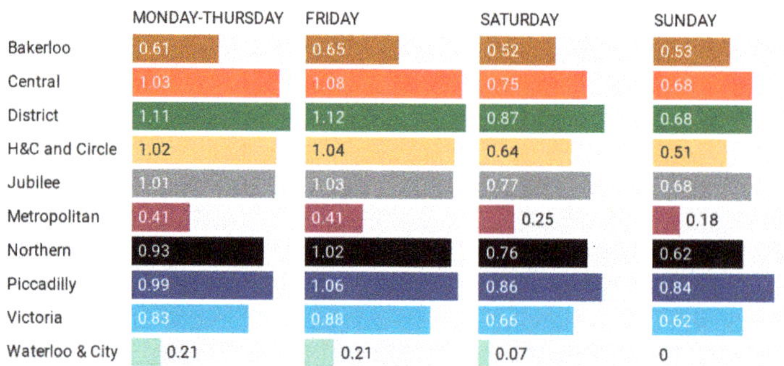

Figure 12. LU occupancy rate by line.

Figure 13 presents the occupancy rate by line, day of the week, and time band. On the one hand, the highest occupancy rates are in PM peak band (4–7 p.m.) from Monday to Thursday, particularly in Central, District and H&C and Circle lines, with rates over 2. As mentioned, on average a LU travel involves 1.6 lines, and these three lines cross Central London, so they might be capturing a relevant number of travels from/to an exchange to another line. In fact, the most crowded line, H&C at PM peak time, has lower traffic at Early time band (before 7 a.m.), which means that is a line close to weekday main destinations (Central London). On the other hand, the occupancy rate of Metropolitan and WAC is the lowest.

Next step is to explore the occupancy rate between two contiguous stations, to characterize the real occupancy rate experienced by travelers. The number of station links is the number of stations minus one for each line, thus 352 station links. The most relevant information analyzing occupancy rates are those extreme values, the lowest and highest, particularly the latter. Figure 14 shows the most crowded station links at the quarter hours with the highest occupancy rates during AM peak (left), 8:30–8:45 a.m., and PM peak (right), 5:30–5:45 p.m. These numbers have been derived from our dataset, combining passengers, line schedules, and line capacities. However, these are estimates as the real flow of passengers and train delays are not publicly available. As the objective of this paper is to characterize the efficiency and sustainability of LU, EDA finishes with the analysis of occupancy rates of stations links, relevant for assessing that the LU carriages theoretical capacity (with 4 standing people per m^2) can be considered its maximum capacity.

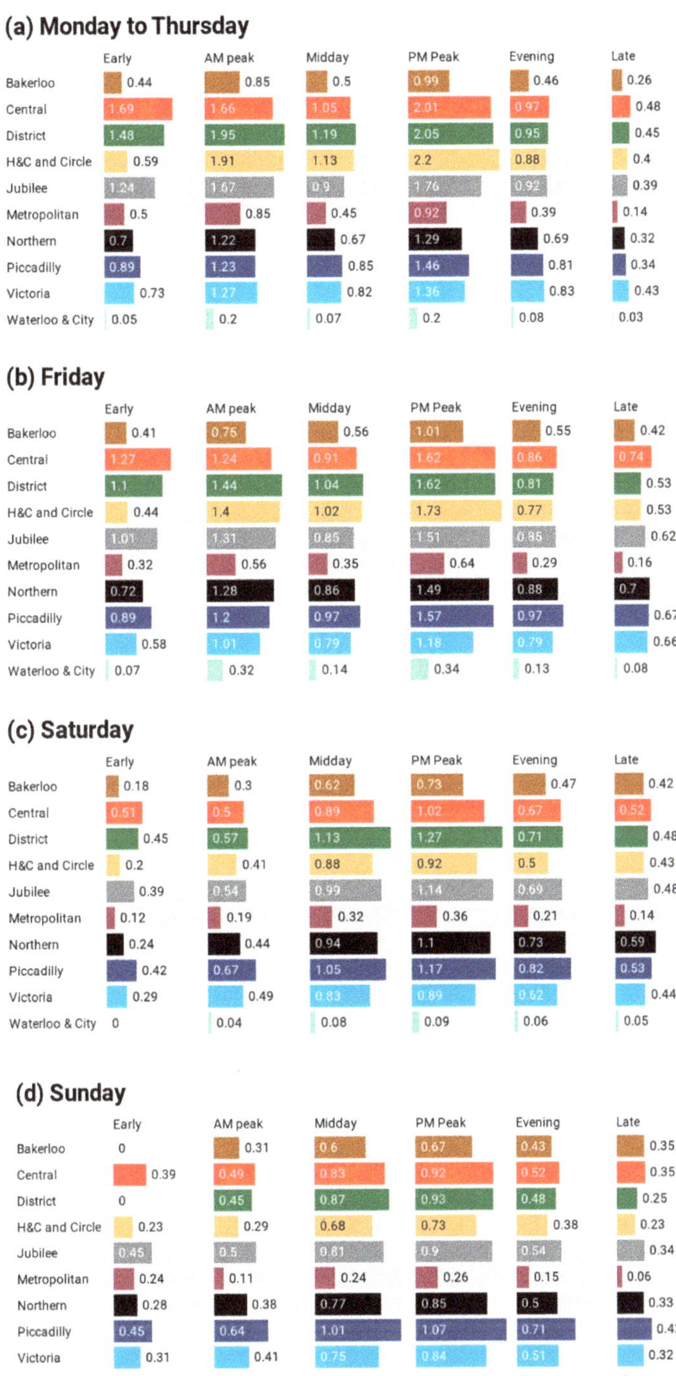

Figure 13. LU occupancy rate by line and time band.

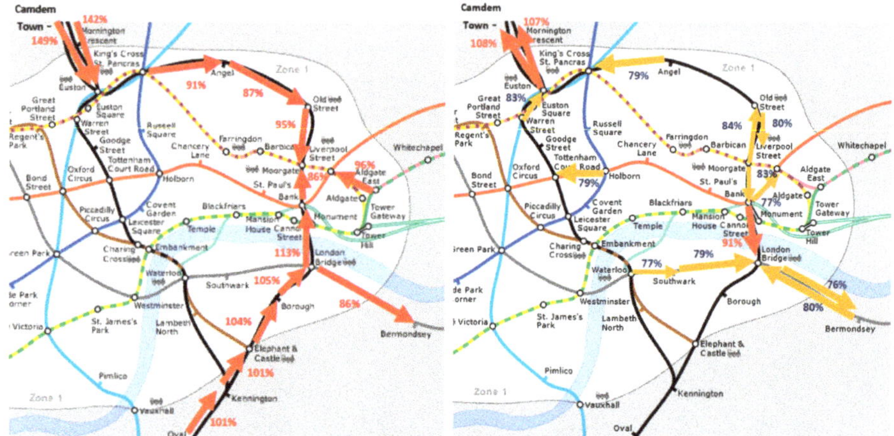

Figure 14. Map of Central London showing the highest occupancy rate of LU stations links at 8:30–8:45 a.m. (**left**) and 5:30–5:45 p.m. (**right**). Those links with 85% or higher occupancy rates are in red.

4.2. LU Additional KPIs

TfL considers additional LU KPIs in its reports [53], focused on service provision, reliability, and journey times, such as the percentage of scheduled kilometers operated (95.8% of the 88.7 million kilometers scheduled), and the excess journey time, and the average delay or (4.6 min, 11% of the average journey time which is 41.6 min). The average delay is formally defined as excess journey time, the additional time on top of scheduled time for access/egress/interchange, platform wait time and on train (the latest figure is 4.6 min for LU for 2018/2019 (table 12.5 in [53]). TfL has reduced the excess journey time since 2008/09, from 6.6 min to 4.6 min by increasing the frequency of the services around 20% higher.

Finally, from the attributable CO_2-equivalent emissions of operating LU (372,000 tons) and 12 billion annually passenger-km [53], a footprint of 31 g of CO_2-equivalent has been estimated by us. Previously, TfL released, outside of the open-data repository, its CO_2 footprints with out-of-date higher estimates [54]. Additional non-official estimates exist [55,56], although also out-of-date. Although the number of operated kilometers raised a 20% over the last 10 years, the CO_2 footprint has decreased far more than 20% (LU operates with power and UK National Grid has been reducing more than 20% its CO_2 footprint during this decade). Thus, LU is more sustainable than a decade ago, and more sustainable than buses (97% fuel-based), which have and 90 g CO_2 footprint per passenger per km (480 million vehicle-km, 4.45 billion passenger-km, and an average CO_2 emission of 822 g/km per vehicle, accounting for around 400,000 CO_2 tons).

4.3. Assessing the Efficiency and Sustainability of London Transport Modes Using DEA

This subsection presents the efficiency of the proposed DEA models, first transport modes, and second URT lines.

Figure 15 presents four routes to evaluate five transport modes (LU, bus, car/taxi, walking, and cycling), and potential combinations of these five transport modes, in Central London, from the shortest to the longest: (A) Bank–Covent Garden, (B) King's Cross St. Pancras–Waterloo, (C) Paddington–Liverpool Street, and (D) Notting Hill Gate–Liverpool Street. These are quite popular routes, connecting national rail stations, and commercial, leisure, and residential areas. However, apart from D, they are not directly connected via LU. Here the optimal route (minimizing travel time) for each transport mode, has been suggested by online services for multi-modal route planning (e.g., Rome2Rio, selected for reporting LU and bus distances and fares).

Figure 15. Routes under consideration for analyzing transport modes in Central London.

Table 3 presents the key parameters of the five analyzed transport modes for the Route A, and a sixth mode, the combination LU+bus. To be able to run DEA no missing values (or 0) are allowed, so it has been assigned a transport cost for cycling (0.20 GBP, the daily cost of an annual London cycle hiring subscription), and for walking (0.10 GBP per 2.4 km, an estimate of the cost of shoe wear). The estimated value of time is 12.00 GBP per hour, an estimate of unskilled pay rate in London, to consider the time factor. LU fare in Central London (Zone 1) is 2.40 GBP, and TfL Bus fare is 1.50 GBP. Costs are provided in the local currency. Moreover, for cycling and walking the additional physical activity has been also considered, estimating 1 g of CO_2-equivalent emission per additional Kcal of energy. This number varies with the diet and weight of the traveler, although it is usually in the range 0.5–2 g CO_2-equiv. per Kcal [57]. The additional cost of walking for a 70 Kg person at 5 km/h in a flat route has been estimated in 150 Kcal/h, and cycling at 15 km/h results in an additional consumption of 360 Kcal/h (these values are average of online calculators). Bus CO_2 emissions are 90 g per passenger per km and LU footprint 31 g per passenger per km. Private car/taxi estimates are 120 g per km, the maximum for driving within the Ultra-Low Emission Zone (ULEZ) of Central London. The number of passengers has been set to 1. These and other values are being used only for illustrative purposes, they can be adapted for personalized route planning and personalized efficiency analysis. Nevertheless, to the best of our knowledge they could be valid estimates.

Computed DEA efficiencies for Route A are 100% efficiency for cycling and walking, particularly for its lowest CO_2 footprint, followed by the combination LU+walking (there is no direct LU link for Route A). Although bus+walking has the second lowest overall cost, its emissions are more than double the most efficient and it scores 64% efficiency. Car/taxi is the least efficient. DEA shows that the limiting factor for improving the efficiency of bus+walking and car/taxi is CO_2 footprint, which can be seen graphically in Figure 16. Shifting from fossil fuel to electric transport can reduce emissions by 75% (according to CO_2 footprint of electricity mix in the UK). Thus, bus+walking would reach the efficiency line whereas car/taxi would increase its efficiency significantly.

Table 3. Route A. Results in descending order of efficiency of the transport modes DEA model.

Transport Mode	Distance (km)	Transport Cost (GBP)	Time (min)	CO_2-Equ. Emissions (I) (g)	Overall Cost (I) (GBP)	Passengers (O)	Efficiency (%)	Rank
Cycling	3.4	0.20	14	84	3.00	1	100%	1st
Walking	2.7	0.10	32	80	6.50	1	100%	1st
LU+Walking	1.7 + 1.3	2.40	20	92	6.40	1	88%	3rd
Bus+Walking	2.3 + 0.4	1.50	16	218	4.70	1	64%	4th
Car/Taxi	2.9	8.00	10	348	10.00	1	30%	5th

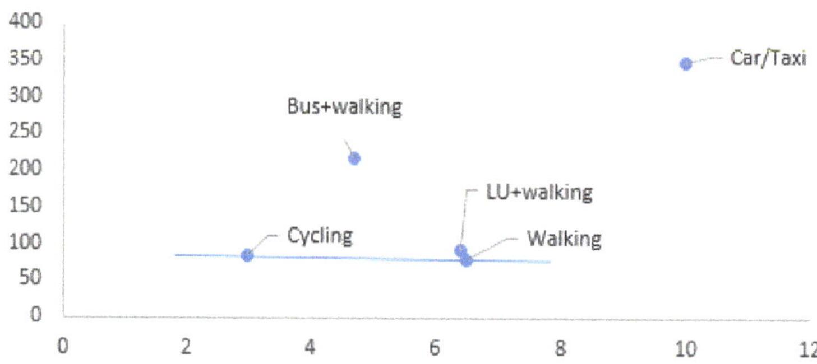

Figure 16. Scatter plot of the two inputs (output is always 1) of the transport modes DEA for Route A.

Table 4 presents the key parameters for Route B, in descending order of efficiency, from cycling, 100%, down to car/taxi, 21%. However, if 4 passengers go by car/taxi, CO_2 footprint is the same (for clarity purposes we will consider the same), and the overall cost rises from 16 to 30 GBP, whereas for the other transport modes both CO_2 footprint and costs are four times higher than the cost of one passenger. In this scenario, see Table 5, car/taxi jumps to the third efficiency position, rivaling with LU.

Table 4. Route B. Results in descending order of efficiency of the transport modes DEA model.

Transport Mode	Distance (km)	Transport Cost (GBP)	Time (min)	CO_2-Equ. Emissions (I) (g)	Overall Cost (I) (GBP)	Passengers (O)	Efficiency (%)	Rank
Cycling	4.5	0.20	18	108	3.80	1	100%	1st
Walking	4.5	0.20	46	120	9.40	1	90%	2nd
LU	4.3	2.40	23	133	6.80	1	81%	3rd
LU+Bus	1.6 + 2.0	2.40 + 1.50	18	230	7.50	1	51%	4th
Bus	3.4	1.50	35	306	8.50	1	48%	5th
Car/Taxi	4.4	14.00	20	528	18.0	1	21%	6th

Table 5. Route B with 4 passengers. Results of efficiency of the transport modes DEA model.

Transport Mode	Distance (km)	Transport Cost (GBP)	Time (min)	CO_2 Emissions (I) (g)	Overall Cost (I) (GBP)	Passengers (O)	Efficiency (%)	Rank
Cycling	4.5	0.20	18	432	15.20	4	100%	1st
Walking	4.5	0.20	46	480	37.60	4	90%	2nd
Car/Taxi	4.4	14.00	20	528	30.00	4	82%	3rd
LU	4.3	9.60	23	532	27.20	4	81%	4th
LU+Bus	1.6 + 2.0	9.60 + 6.00	18	920	30.00	4	51%	5th
Bus	3.4	6.0	35	1224	34.00	4	45%	6th

Table 6 presents the key parameters for Route C, in descending order of efficiency, from cycling, 100%, closely followed by LU, 94%. In this scenario the limiting factor is the cost. Considering 18 GBP per hour as time value then the fastest transport modes increase their efficiencies (LU rises to 100%, Car/Taxi to 40%), whereas slower transport modes reduce their efficiencies (Bus goes down to 52%).

Table 7 presents the key parameters for Route D, in descending order of efficiency, where LU and cycling are both 100% efficient. LU has the lowest overall cost (Cycling has 27% higher cost, 7.60 versus 6.00) and the second lowest CO_2 footprint (254 g., 14% higher than Cycling, the option with the lowest CO_2 emissions with 222 g.). In this scenario both CO_2 footprints and costs are the limiting factors. Thus, electrification of vehicles will have a limited impact if the overall cost remains unaltered.

The main cost reduction would come from reducing even more travel times in buses and car/taxi. This might be feasible reducing traffic in Central London, for instance imposing higher restrictions to polluting vehicles in ULEZ.

Table 6. Route C. Results in descending order of efficiency of the transport modes DEA model.

Transport Mode	Distance (km)	Transport Cost (GBP)	Time (min)	CO_2-Equ. Emissions (I) (g)	Overall Cost (I) (GBP)	Passengers (O)	Efficiency (%)	Rank
Cycling	8.3	0.20	33	198	6.80	1	100%	1st
LU	7.4	2.40	24	229	7.20	1	94%	2nd
Walking	7.2	0.30	86	215	17.50	1	90%	3rd
Bus	8.3	1.50	55	747	12.50	1	54%	4th
Car/Taxi	8.3	17.00	23	996	21.6	1	31%	5th

Table 7. Route D. Results in descending order of efficiency of the transport modes DEA model.

Transport Mode	Distance (km)	Transport Cost (GBP)	Time (min)	CO_2-Equ. Emissions (I) (g)	Overall Cost (I) (GBP)	Passengers (O)	Efficiency (%)	Rank
LU	8.2	2.40	18	254	6.00	1	100%	1st
Cycling	9.4	0.20	37	222	7.60	1	100%	1st
Walking	8.6	0.40	109	273	22.20	1	81%	3rd
Bus	8.9	1.50	55	801	12.50	1	48%	4th
Car/Taxi	8.6	19.00	26	1032	24.20	1	25%	5th

Figure 17 presents, considering the latter Route D, an analysis of the sensitivity of the value of time, the main factor impacting the transport mode efficiency. The range considered, from 0 GBP to 36 GBP per hour, shows that the fastest transport modes, Car/taxi and LU, gain efficiency as the value of time increases, slower for the Car/Taxi due to the higher transport cost of this mode. It is remarkable that walking, the slowest transport mode, keeps its efficiency due to its low CO_2 footprint. Thus, a shift from fuel to electric vehicles, reducing the CO_2 footprint by 75%, according to the energy generation mix in the UK, has been considered in Figure 18, also for Route D, together with a varying value of time. Now electric Bus is always 100% efficient, due to its low transport costs and low emissions, very similar to those of cycling, whereas LU, faster than LU and cycling but with a more expensive fare, is also efficient for passengers who value their time from 9 GBP/h on. In a scenario with electric cars/taxis this transport mode (Car/Taxi) is more efficient than walking from 1 GBP/h of value of time. As DEA is a relative (non-absolute) efficiency measure, improvements in some DMUs might impact the efficiency of other DMUs.

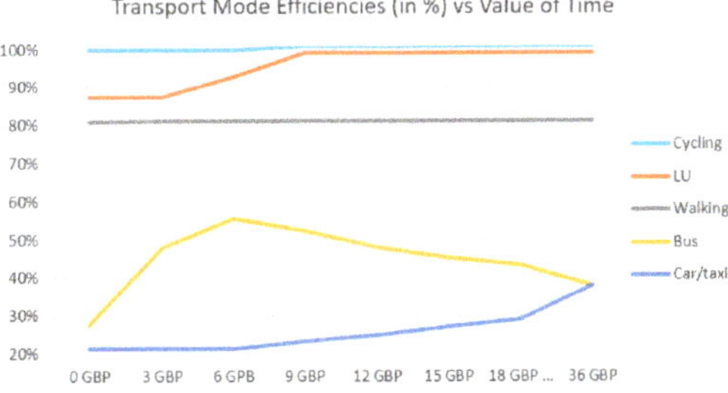

Figure 17. Route D: Impact of the value of time in the efficiency of the transport modes efficiency.

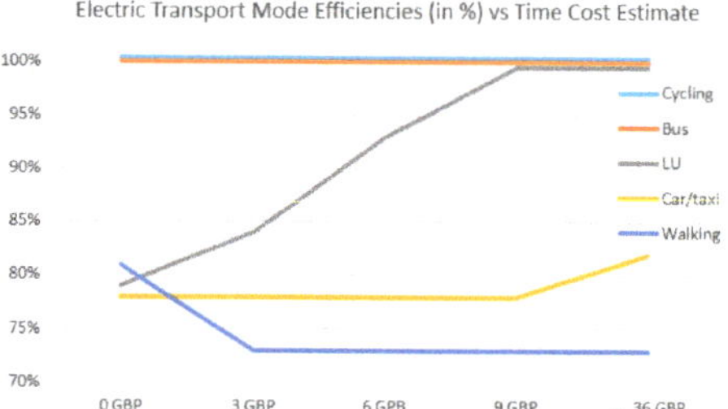

Figure 18. Route D: impact of the value of time and electricity in the transport modes efficiency.

4.4. Assessing the Efficiency and Sustainability of LU Lines Using DEA

This subsection presents the DEA efficiencies of the URT lines model, a CRS model computed with the same DEA software solver as in the previous subsection.

Table 8 presents the key parameters of the ten analyzed LU lines, the two input parameters, the number of stations of the longest route and weekly frequencies, and the output, weekly passengers. Then the efficiency, ranging from 44% for the Metropolitan line to four 100% efficient lines (Central, District, Jubilee, and Victoria line). In addition, finally, four KPIs considered in EDA to characterize and compare LU lines. Although DEA is a non-parametric technique, so efficiency is not a linear combination of the inputs, it looks as if the best performers are those lines with the highest average passengers per service, the highest passengers per service and station, and the highest speed. WAC efficiency (46%, 9th) is limited by the weekly frequencies, with just 426 weekly frequencies (87% lower than the current number), retaining the number of passengers, it would be 100% efficient. The rest of the inefficient lines, those scoring below 100%, are limited both by the number of stations and the weekly frequencies. Table 9 shows the optimal projections of the inputs of the LU lines.

An efficient line would maximize the number of passengers with the lowest number of stations (proxy variable of the capital expenses, CAPEX), and the lowest weekly frequencies (proxy variable of the operating expenses, OPEX), an analysis in tune with previous works [30]. However, to increase the efficiency, closing stations is not an option. URT management can only influence operating expenses, reducing/increasing the weekly frequencies. Thus, the efficiency of a LU line will increase if reducing a given percentage the number of weekly frequencies (e.g., 10%) the number of passengers reduces significantly less than the reduction of the frequencies. Further analysis of real transport data, actual number of passengers and actual schedule of LU trains, will help to understand the relationship between frequencies and the number of passengers of a line, particularly in such a complex network as LU, with multiple exchanges and different lines sharing the same rail section/station links.

Table 10 presents an alternative DEA model with two additional input variables, the longest travel time, and the longest length in km. The new ranking that comes out of this extended model only interchanges positions 8th and 9th, as the new variables are highly correlated with the previous input variables. Thus, now WAC ranks 8th and BAK ranks 9th as the new model favors the short length and travel time of WAC, although BAK also increases its efficiency.

Finally, Tables 11–13 present the efficiency of the proposed URT lines DEA model (the original with 2 input variables) using the data, frequencies, and passengers, for AM, Midday, and PM peak time bands, from Monday to Thursday, respectively. Efficiency results for the mentioned time bands (bands of 3, 6, and 3 h, respectively) are in tune with overall line efficiencies presented in Table 8. However, some differences arise, such as WAC is the 7th in efficiency during peak times, but the 10th during Midday. WAC connects a national rail station and transport hub, Waterloo Station, with Bank tube station, in the heart of the financial area in the City of London. Therefore, its traffic pattern shows more activity during AM and PM peak hours. Moreover, VIC is the only line 100% efficient in the three time bands. Finally, except for WAC, LU lines score similarly across the analyzed time bands.

Table 8. Results of the URT lines DEA model of LU lines in descending order of efficiency.

LU Line	Num. of Stations Longest Route (I)	Weekly LU Frequencies (Scheduled) (I)	Weekly Passengers (O)	Efficiency (%)	Rank	Avg. Pass. per Service	Avg. Pass./Service and Station	Speed (km/h)	Speed (min. per Station)
CEN	37	6202	6,218,138	100%	1st	1,003	27	39	2.3
DIS	43	4828	5,166,660	100%	1st	1,070	25	28	2.1
JUB	27	6624	5,985,450	100%	1st	904	33	39	2.1
VIC	16	7480	5,813,439	100%	1st	777	49	39	2
HAM	28	3184	2,988,540	88%	5th	939	34	24	2.1
NOR	36	10,610	7,028,737	77%	6th	662	18	39	1.9
PIC	42	5710	4,404,640	76%	7th	771	18	30	2.3
BAK	25	4836	2,444,910	53%	8th	506	21	28	2
WAC	2	3402	331,156	46%	9th	97	49	50	1.5
MET	24	4718	2,003,527	44%	10th	425	18	37	3

Table 9. Optimal LU input projections of the URT lines DEA model.

LU Line	Num. of Stations Longest Route (I)	Weekly LU Frequencies (Scheduled) (I)	Weekly Passengers (O)	Efficiency (%)	Rank	Optimal Num. of Stations	Diff. (%)	Optimal Weekly LU Freq.	Diff. (%)
CEN	37	6202	6,218,138	100%	1st	37	0%	6202	0%
DIS	43	4828	5,166,660	100%	1st	43	0%	4828	0%
JUB	27	6624	5,985,450	100%	1st	27	0%	6624	0%
VIC	16	7480	5,813,439	100%	1st	16	0%	7480	0%
HAM	28	3184	2,988,540	88%	5th	24.6	−12%	2799	−12%
NOR	36	10,610	7,028,737	77%	6th	27.8	−23%	8182	−22%
PIC	42	5710	4,404,640	76%	7th	28.8	−24%	4325	−24%
BAK	25	4836	2,444,910	53%	8th	13.2	−47%	2544	−47%
WAC	2	3402	331,156	46%	9th	0.9	−54%	426	−87%
MET	24	4718	2,003,527	44%	10th	10.6	−56%	2095	−56%

Table 10. Results of the URT lines DEA model of LU with two additional input parameters.

LU Line	Num. of Stations Longest Route (I)	Weekly LU Frequencies (Scheduled) (I)	Longest Travel Time (Minutes) (I)	Longest Length (km) (I)	Weekly Passengers (O)	Efficiency (%)	Rank
CEN	37	6202	84	53.9	6,218,138	100%	1st
DIS	43	4828	91	42	5,166,660	100%	1st
JUB	27	6624	56	36.2	5,985,450	100%	1st
VIC	16	7480	32	32	5,813,439	100%	1st
HAM	28	3184	59	23.2	2,988,540	94%	5th
NOR	36	10,610	67	43.4	7,028,737	79%	6th
PIC	42	5710	98	49.8	4,404,640	76%	7th
WAC	2	3402	3	2.5	331,156	60%	8th
BAK	25	4836	50	23.2	2,444,910	58%	9th
MET	24	4718	72	44.4	2,003,527	44%	10th

Table 11. URT lines DEA model scores of LU Monday to Thursday (MTT) AM peak traffic.

LU Line	Num. of Stations Longest Route (I)	MTT LU Frequencies (Scheduled) (I)	MTT Daily Passengers (O)	Efficiency (%)	Rank
JUB	27	176	250,037	100%	1st
VIC	16	210	227,188	100%	1st
DIS	43	128	212,294	100%	1st
HAM	28	74	120,077	98%	4th
CEN	37	170	239,754	94%	5th
NOR	36	262	273,060	77%	6th
WAC	2	128	21,875	77%	7th
PIC	42	144	150,140	65%	8th
MET	24	130	93,937	49%	9th
BAK	25	126	90,913	49%	10th

Table 12. URT lines DEA model scores of LU Monday to Thursday (MTT) Midday traffic.

LU Line	Num. of Stations Longest Route (I)	MTT LU Frequencies (Scheduled) (I)	MTT Daily Passengers (O)	Efficiency (O) (%)	Rank
CEN	37	294	262,604	100%	1st
VIC	16	328	229,735	100%	1st
DIS	43	218	220,130	100%	1st
HAM	28	146	140,918	96%	4th
JUB	27	298	227,774	95%	5th
PIC	42	258	187,219	76%	6th
NOR	36	482	274,669	74%	7th
BAK	25	240	102,527	51%	8th
MET	24	202	77,589	44%	9th
WAC	2	150	9537	33%	10th

Table 13. URT lines DEA model scores of LU Monday to Thursday (MTT) PM peak traffic.

LU Line	Num. of Stations Longest Route (I)	MTT LU Frequencies (Scheduled) (I)	MTT Daily Passengers (O)	Efficiency (%)	Rank
CEN	37	160	273,109	100%	1st
JUB	27	176	263,784	100%	1st
VIC	16	212	245,455	100%	1st
HAM	28	72	134,825	100%	1st
DIS	43	130	227,226	96%	5th
NOR	36	270	295,882	78%	6th
WAC	2	126	21,460	70%	7th
PIC	42	142	175,828	70%	8th
BAK	25	126	105,896	52%	9th
MET	24	132	103,654	50%	10th

5. Conclusions

This paper has analyzed the efficiency and sustainability of URT using EDA and DEA. The main contributions of this work are: (1) propose and compute new indicators for EDA of URT sustainability and efficiency (e.g., occupancy rate by URT line, station links, and time band, and CO_2 footprint per journey); (2) design and propose a methodology for DEA performance assessment based on the selection of input and output variables using EDA on publicly available data; (3) develop two original DEA production models, the first one for characterizing the sustainability of different transport modes, and the second one for measuring the efficiency of URT lines; (4) validating the methodology with open data from TfL and online services; and (5) ranking URT against other transport modes and analyzing DEA efficiency scores of URT lines.

The main conclusions of the paper are: (1) EDA plays a key role analyzing URT efficiency and sustainability indicators, as well and defining new indicators; (2) DEA variable selection can be done in a semi-automated and repeatable way relying on EDA; and (3) DEA is a simple and straightforward non-parametric technique to score multiple transport modes and URT lines efficiency to monitor, understand, and improve its management, even focusing on time bands and URT line sections for the latter scenario.

To sum up, the introduced big-data-based methodology supports the advance of efficiency and sustainability in public transport, particularly in URT, through disseminating data, KPIs, and assessments based on them. Thus, both operators and travelers alike are encouraged to improve their decision-making, from transport network management to route planning, to meet the Sustainable Development Goal target of having a more sustainable transport by 2030.

Author Contributions: Conceptualization, G.L.T. and L.H.; Investigation, G.L.T. and L.H.; Writing—original draft, G.L.T.; Writing—review & editing, G.L.T. All authors have read and agreed to the published version of the manuscript.

Funding: This research was funded by the Ministry of Economy, Industry and Competitiveness of Spain, Project TIN2016-75845-P (AEI/FEDER/EU) and SNEO-20161147 (CDTI) and by Xunta de Galicia and FEDER funds of the EU (Centro de Investigación de Galicia accreditation 2019–2022, ref. ED431G2019/01, and Consolidation Programme of Competitive Reference Groups, ref. ED431C 2017/04).

Acknowledgments: The authors would like to thank CITIC and Project ref. ED431G2019/01 for supporting the Postdoc visit of Guillermo L. Taboada to Manchester Metropolitan University to develop part of the collaborative work presented in this paper.

Conflicts of Interest: The authors declare no conflict of interest. The funders had no role in the design of the study; in the collection, analyses, or interpretation of data; in the writing of the manuscript, or in the decision to publish the results.

Appendix A

The sources of data used in the study are openly available online on TfL open data site: https://data.tfl.gov.uk (accessed on 25 June 2020).

The main source of information is Urban Rail Passengers Count and Travel Flow Dataset (codenamed project NUMBAT), https://crowding.data.tfl.gov.uk, which is based on data, from smartcards (Oyster and contactless bank/NFC cards), gatelines, and automatic passenger counters and services from timetables, combined through a model to assign journeys to routes using generalized journey time. Data has been obtained during the autumn of each year (at the time of writing, autumn 2018 is the last one) and provides an average for weekdays from Monday to Thursday, and additionally data for Friday, for Saturday and for Sunday, each of them independently. Includes only aggregated data (~100 MB data annually). Additional information from TfL origin–destination dataset is described here: https://data.london.gov.uk/dataset/tfl-rolling-origin-and-destination-survey.

This dataset, published as open data using an open TfL license, provides:

- Journeys per day by all URT modes in London, LU, London Overground (LO), DLR, Crossrail EZL, and London Trams.

-    – – Values for each 15-minute period of the day.
-    – – Number of passengers' entries/exits at all stations.
-    – – Number of interchanging passengers at all stations.

This data has been enriched including GPS location for all the stations, obtained from Open StreetMap [51], (ii) creating a table that relates lines with all their stations (TfL open data does not include this table), and (iii) a table that relates lines with their capacity (seated plus standing at 4 passengers per m^2), with data collected from TfL website (once again, TfL open data does not include this table). Table A1 presents LU lines and its associated train capacities.

Table A1. LU train capacity per line as of 2018.

Line Name	Capacity
Bakerloo	851
Central	1047
District	1045
H&C and Circle	1045
Jubilee	964
Metropolitan	1176
Northern	752
Piccadilly	798
Victoria	986
Waterloo & City	506

References

1.   International Association of Public Transport UITP. Energy Efficiency: Contribution of Urban Rail Systems. Available online: https://www.uitp.org/energy-efficiency-contribution-urban-rail-systems (accessed on 25 June 2020).
2.   International Energy Agency. "Tracking Transport". Available online: https://www.iea.org/reports/tracking-transport-2019/rail (accessed on 25 June 2020).
3.   Sustainable Mobility for All (Sum4all). Global Mobility Report 2017. Available online: https://sustainabledevelopment.un.org/content/documents/2643Global_Mobility_Report_2017.pdf (accessed on 25 June 2020).
4.   Tukey, J.W. *Exploratory Data Analysis*; Addison Wesley: Reading, MA, USA, 1977.
5.   Charnes, A.; Cooper, W.W.; Rhodes, E. Measuring the efficiency of decision making units. *Eur. J. Oper. Res.* **1978**, *2*, 429–444. [CrossRef]
6.   Cooper, W.W.; Seiford, L.M.; Tone, K. *Data Envelopment Analysis*; Springer: New York, NY, USA, 1999.
7.   Gennaro, M.D.; Paffumi, E.; Martini, G. Big Data for Supporting Low-Carbon Road Transport Policies in Europe: Applications, Challenges and Opportunities. *Big Data Res.* **2016**, *6*, 11–25. [CrossRef]
8.   Zhao, J.; Rahbee, A.; Wilson, N.H.M. Estimating a Rail Passenger Trip Origin-destination Matrix Using Automatic Data Collection Systems. *Comput. Aided Civ. Infrastruct. Eng.* **2007**, *22*, 376–387. [CrossRef]
9.   Zhou, J.; Murphy, E.; Long, Y. Commuting efficiency in the Beijing metropolitan area: An exploration combining smartcard and travel survey data. *J. Transp. Geogr.* **2014**, *41*, 175–183. [CrossRef]
10.   Tao, S.; Corcoran, J.; Mateo-Babiano, I.; Rohde, D. Exploring bus rapid transit passenger travel behaviour using big data. *Appl. Geogr.* **2014**, *53*, 90–104. [CrossRef]
11.   Galland, S.; Knapen, L.; Yasar, A.; Gaud, N.; Janssens, D.; Lamotte, O.; Koukam, A.; Wets, G. Multi-agent simulation of individual mobility behavior in carpooling. *Transp. Res. Part C Emerg. Technol.* **2014**, *45*, 83–98. [CrossRef]
12.   Zhang, M.; Wiegmans, B.; Tavasszy, L. Optimization of multimodal networks including environmental costs: A model and findings for transport policy. *Comput. Ind.* **2013**, *64*, 136–145. [CrossRef]
13.   Guerrero-Ibáñez, J.; Zeadally, S.; Contreras-Castillo, J. Sensor Technologies for Intelligent Transportation Systems. *Sensors* **2018**, *18*, 1212. [CrossRef]
14.   Abduljabbar, R.; Dia, H.; Liyanage, S.; Bagloee, S.A. Applications of artificial intelligence in transport: An overview. *Sustainability* **2019**, *11*, 189. [CrossRef]

15. Torre-Bastida, A.I.; Del Ser, J.; Laña, I.; Ilardia, M.; Bilbao, M.N.; Campos-Cordobés, S. Big data for transportation and mobility: Recent advances, trends and challenges. *IET Intell. Transp. Syst.* **2018**, *12*, 742–755. [CrossRef]

16. Gallo, M.; De Luca, G.; D'Acierno, L.; Botte, M. Artificial Neural Networks for Forecasting Passenger Flows on Metro Lines. *Sensors* **2019**, *19*, 3424. [CrossRef] [PubMed]

17. Jiao, P.; Li, R.; Sun, T.; Hou, Z.; Ibrahim, A. Three revised kalman filtering models for short-term rail transit passenger flow prediction. *Math. Probl. Eng.* **2016**, 9717582. [CrossRef]

18. Cai, C.; Yao, E.; Wang, M.; Zhang, Y. Prediction of urban railway station's entrance and exit passenger flow based on multiply ARIMA model. *J. Beijing Jiaotong Univ.* **2014**, *38*, 135–140.

19. Li, Y.; Wang, X.; Sun, S.; Ma, X.; Lu, G. Forecasting short-term subway passenger flow under special events scenarios using multiscale radial basis function networks. *Transp. Res. Part C Emerg. Technol.* **2017**, *77*, 306–328. [CrossRef]

20. Ling, X.; Huang, Z.; Wang, C.; Zhang, F.; Wang, P. Predicting subway passenger flows under different traffic conditions. *PLoS ONE* **2018**, *13*, e0202707. [CrossRef] [PubMed]

21. Liu, Y.; Liu, Z.; Jia, R. DeepPF: A deep learning based architecture for metro passenger flow prediction. *Transp. Res. Part C Emerg. Technol.* **2019**, *101*, 18–34. [CrossRef]

22. Wang, Y.; Ma, J.; Zhang, J. Metro Passenger Flow Forecast with a Novel Markov-Grey Model. *Period. Polytech. Transp. Eng.* **2019**, *48*, 70–75. [CrossRef]

23. Zhang, B.; Li, S.; Huang, L.; Yang, Y. An improved feedback wavelet neural network for short-term passenger entrance flow prediction in Shanghai subway system. In *Lecture Notes in Computer Science (LNCS), Proceedings of the International Conference on Neural Information Processing (ICONIP), Guangzhou, China, 14–18 November 2017*; Springer: Berlin/Heidelberg, Germany, 2017; Volume 10638, pp. 35–45.

24. Sohn, K.; Shim, H. Factors generating boardings at Metro stations in the Seoul metropolitan area. *Cities* **2010**, *27*, 358–368. [CrossRef]

25. Chin, J.; Callaghan, V.; Lam, I. Understanding and personalising smart city services using machine learning, The Internet-of-Things and Big Data. In Proceedings of the IEEE 26th International Symposium on Industrial Electronics (ISIE), Edinburgh, UK, 18–21 June 2017; pp. 2050–2055.

26. Liu, L.; Chen, R.-C. A novel passenger flow prediction model using deep learning methods. *Transp. Res. Part C Emerg. Technol.* **2017**, *84*, 74–91. [CrossRef]

27. Bartlett, Z.; Han, L.; Nguyen, T.T.; Johnson, P. A Novel Online Dynamic Temporal Context Neural Network Framework for the Prediction of Road Traffic Flow. *IEEE Access* **2019**, *7*, 153533–153541. [CrossRef]

28. Daraio, C.; Diana, M.; Di Costa, F.; Leporelli, C.; Matteucci, G.; Nastasi, A. Efficiency and effectiveness in the urban public transport sector: A critical review with directions for future research. *Eur. J. Oper. Res.* **2016**, *248*, 1–20. [CrossRef]

29. Wanke, P.; Azad, A.K. Efficiency in Asian railways: A comparison between data envelopment analysis approaches. *Transp. Plan. Technol.* **2018**, *41*, 573–599. [CrossRef]

30. Lobo, A.; Couto, A. Technical Efficiency of European Metro Systems: The Effects of Operational Management and Socioeconomic Environment. *Netw. Spat. Econ.* **2016**, *16*, 723–742. [CrossRef]

31. Han, J.; Hayashi, Y. A Data Envelopment Analysis for Evaluating the Performance of China's Urban Public Transport Systems. *Int. J. Urban Sci.* **2008**, *12*, 173–183. [CrossRef]

32. Kutlar, A.; Kabasakal, A.; Sarikaya, M. Determination of the efficiency of the world railway companies by method of DEA and comparison of their efficiency by Tobit analysis. *Qual. Quant.* **2013**, *47*, 3575–3602. [CrossRef]

33. Tsai, C.H.P.; Mulley, C.; Merkert, R. Measuring the cost efficiency of urban rail systems: An international comparison using DEA and Tobit models. *J. Transp. Econ. Policy* **2015**, *49*, 17–34.

34. Zhang, H.; You, J. An Empirical Study of Transport Efficiency of Urban Rail Transit Based on Data Envelopment Analysis and Tobit Model. *J. Tongji Univ.* **2019**, *46*, 1306–1311.

35. Lao, Y.; Liu, L. Performance evaluation of bus lines with data envelopment analysis and geographic information systems. *Comput. Environ. Urban* **2009**, *33*, 247–255. [CrossRef]

36. Hahn, J.-S.; Kim, H.-R.; Kho, S.-Y. Analysis of the efficiency of Seoul arterial bus routes and its determinant factors. *KSCE J. Civ. Eng.* **2011**, *15*, 1115–1123. [CrossRef]

37. Hahn, J.-S.; Kim, D.-K.; Kim, H.-C.; Lee, C. Efficiency analysis on bus companies in Seoul city using a network DEA model. *KSCE J. Civ. Eng.* **2013**, *17*, 1480–1488. [CrossRef]

38. Caulfield, B.; Bailey, D.; Mullarkey, S. Using Data Envelopment Analysis as a public transport project appraisal tool. *Transp. Policy* **2013**, *29*, 74–85. [CrossRef]
39. Peyrache, A.; Rose, C.; Sicilia, G. Variable selection in Data Envelopment Analysis. *Eur. J. Oper. Res.* **2020**, *282*, 644–659. [CrossRef]
40. Charnes, A.; Cooper, W.W.; Lewin, A.Y.; Seiford, L.M. The DEA Process, Usages, and Interpretations. In *Data Envelopment Analysis: Theory, Methodology, and Applications*; Springer: Dordrecht, The Netherlands, 1994.
41. Goodall, W.; Fishman, T.D.; Bornstein, J.; Bonthron, B. The rise of Mobility as a Service. *Deloitte Rev.* **2017**, *20*, 112–129.
42. Tufféry, S. *Data Mining and Statistics for Decision Making*; Wiley: Chichester, UK, 2011; Volume 2.
43. Ghosh, A.; Nashaat, M.; Miller, J.; Quader, S.; Marston, C. A comprehensive review of tools for exploratory analysis of tabular industrial datasets. *Vis. Inform.* **2018**, *2*, 235–253. [CrossRef]
44. Banker, R.D.; Charnes, A.; Cooper, W.W. Some models for estimating technical and scale inefficiencies in data envelopment analysis. *Manag. Sci.* **1984**, *30*, 1078–1092. [CrossRef]
45. Fare, R.; Grosskopf, S. Modelling undesirable factors in efficiency evaluation: Comment. *Eur. J. Oper. Res.* **2004**, *157*, 242–245. [CrossRef]
46. Seiford, L.M.; Zhu, J. Modeling undesirable factors in efficiency evaluation. *Eur. J. Oper. Res.* **2002**, *142*, 16–20. [CrossRef]
47. Flyvbjerg, B.; Bruzelius, N.; Wee, B.V. Comparison of Capital Costs per Route-Kilometre in Urban Rail. *Eur. J. Trans. Infrastruct. Res.* **2008**, *8*, 17–30.
48. Ruiz-Rosero, J.; Ramirez-Gonzalez, G.; Khanna, R. Masivo: Parallel Simulation Model Based on OpenCL for Massive Public Transportation Systems' Routes. *Electronics* **2019**, *8*, 1501. [CrossRef]
49. Hartmann Tolić, I.; Nyarko, E.K.; Ceder, A.A. Optimization of Public Transport Services to Minimize Passengers' Waiting Times and Maximize Vehicles' Occupancy Ratios. *Electronics* **2020**, *9*, 360. [CrossRef]
50. Stone, M.; Aravopoulou, E. Improving journeys by opening data: The case of Transport for London (TfL). *Bottom Line* **2018**, *31*, 2–15. [CrossRef]
51. Open Street Map. List of London Underground Station. Available online: https://wiki.openstreetmap.org/wiki/List_of_London_Underground_stations (accessed on 25 June 2020).
52. Shuai, H.; Haiying, L. A urban rail transport network carrying capacity calculation method based on the logit model. In Proceedings of the International Conference on Transportation, Mechanical, and Electrical Engineering (TMEE), Changchun, China, 16–18 December 2011; pp. 191–194.
53. TfL Travel in London Edition 12. Available online: http://content.tfl.gov.uk/travel-in-london-report-12.pdf (accessed on 25 June 2020).
54. TfL FOI Request Detail. Available online: https://tfl.gov.uk/corporate/transparency/freedom-of-information/foi-request-detail?referenceId=FOI-0880-1819 (accessed on 25 June 2020).
55. Carbon Independent Calculator Based on AEA Report. Available online: https://www.carbonindependent.org/20.html (accessed on 25 June 2020).
56. EC. AEA Report. Handbook of External Costs of Transport. Available online: https://ec.europa.eu/transport/sites/transport/files/handbook_on_external_costs_of_transport_2014_0.pdf (accessed on 25 June 2020).
57. Tom, M.S.; Fischbeck, P.S.; Hendrickson, C.T. Energy Use, Blue Water Footprint, and Greenhouse Gas Emissions for Current Food Consumption Patterns and Dietary Recommendations in the US. *Environ. Syst. Decis.* **2016**, *36*, 92–103. [CrossRef]

Article

A Data Mining and Analysis Platform for Investment Recommendations

Elena Hernández-Nieves [1,*], Javier Parra-Domínguez [1], Pablo Chamoso [1], Sara Rodríguez-González [1] and Juan M. Corchado [1,2,3,4]

1   BISITE Research Group, University of Salamanca, Edificio Multiusos I+D+i, 37007 Salamanca, Spain; javierparra@usal.es (J.P.-D.); chamoso@usal.es (P.C.); srg@usal.es (S.R.-G.); corchado@usal.es (J.M.C.)
2   Air Institute, IoT Digital Innovation Hub, Carbajosa de la Sagrada, 37188 Salamanca, Spain
3   Department of Electronics, Information and Communication, Faculty of Engineering, Osaka Institute of Technology, Osaka 535-8585, Japan
4   Pusat Komputeran dan Informatik, Universiti Malaysia Kelantan, Bachok 16300, Kelantan, Malaysia
*   Correspondence: elenahn@usal.es

Abstract: This article describes the development of a recommender system to obtain buy/sell signals from the results of technical analyses and of forecasts performed for companies operating in the Spanish continuous market. It has a modular design to facilitate the scalability of the model and the improvement of functionalities. The modules are: analysis and data mining, the forecasting system, the technical analysis module, the recommender system, and the visualization platform. The specification of each module is presented, as well as the dependencies and communication between them. Moreover, the proposal includes a visualization platform for high-level interaction between the user and the recommender system. This platform presents the conclusions that were abstracted from the resulting values.

Keywords: artificial intelligence; Big Data analytics; forecasting systems; recommender system; Fintech

Citation: Hernández-Nieves, E.; Parra-Domínguez, J.; Chamoso, P.; Rodríguez-González, S.; Corchado, J.M. A Data Mining and Analysis Platform for Investment Recommendations. *Electronics* **2021**, *10*, 859. https://doi.org/10.3390/electronics10070859

Academic Editor: Amir Mosavi

Received: 16 March 2021
Accepted: 31 March 2021
Published: 4 April 2021

Publisher's Note: MDPI stays neutral with regard to jurisdictional claims in published maps and institutional affiliations.

1. Introduction

Data analysis is a process of inspecting, cleaning, transforming, sorting, and modelling data for the purpose of finding useful information, reaching conclusions, and making appropriate decisions. In statistics, data analysis is divided into descriptive analytics, exploratory analytics, and predictive analytics.

Predictive analytics is defined as the branch of analytics that is used to make predictions regarding future events facing, for example, an organization. To do so, it will use various methods, such as data mining, text mining, artificial intelligence, statistics, or data modelling, among others. In addition, predictive analytics manages information technologies, analysis methods, and business process modelling with the purpose of anticipating future events that may happen to the organization in question.

In this research the focus is on predictive analytics with a specific approach to stock market analysis. It is assumed that a stock market prediction is considered successful if it achieves the best results using the minimum data input and the least complex stock market model [1]. Within the field of Artificial Intelligence, the emergence of Machine Learning and the increasing computing performance have allowed developing new services on the basis of traditional financial products, providing financial-economic instruments that provide higher versatility and greater speed [2]. As Jigar Patel et al. point out in [3], forecasting a stock's value is difficult because of the uncertainty of prediction due to the large number of potential determinants. The authors suggest a method that includes both fundamental and technical analysis combined with Machine Learning algorithms, which is an approach to prediction that tries to improve its efficiency.

In contrast to other research that focuses on a single model for investment recommendation such as Artificial Neural Networks (ANN) or optimised decision trees, in this research, a series of algorithms (Random Forest Regressor, Gradient Boosting Regressor, SVM-LinearSVR, MLP Regressor and kNNNeighbors Regressor) are applied. In addition, technical analysis is used, combining Momentum Indicators and Moving Averages.The proposed recommendation system will remove subjectivity from the process after evaluating and validating the algorithms and will provide the user with the algorithm with the best accuracy. However, the main advantage of the investigated system consists in the possibility of consulting the whole process that the system has carried out (analysis, prediction and investment recommendations).

The research conducted in this study has led to the development of a platform that integrates different modules. The modular approach favours not only the overall research, but it is also good for achieving scalability, flexibility, and usability. The modules that make up the system are:

1. Analysis and data mining. The initial objective was to draw up a document that breaks down the functioning of the Spanish continuous market. The goal of this analysis was to determine the needs to be met by the prediction and recommendation model. Therefore, the market analysis has served as a starting point for the development of the platform. Regarding data extraction, given that the operation of the prediction and recommendation platform is based on a dataset containing the historical data of the companies in the continuous market, it has been necessary to create a system that is in charge of extracting the data in real time. Thus, this implies the need to find a reliable data source that contains the information that is required by the platform. It is possible to either make use of an Application Programming Interface (API) to allow for data retrieval, or to develop a system based on Web Scraping for the extraction and formatting of data. The analysis of the data greatly facilitates the subsequent development of a forecasting and recommendation system and the calculation of the technical analysis factors.

2. The forecasting system. The objective of this system is to predict the closing value of a share in the Spanish continuous market from its opening value on the same day. This minimizes the error of the prediction model as much as possible, which will be presumably based on Machine Learning regression algorithms. The forecasting system will be developed on the basis of the extracted historical data of the shares of the Spanish continuous market companies.

3. The technical analysis. On the basis of the premise that the forecasting system relies on a series of historical market opening values to predict its closing values, the addition of a system that is based on the calculation of technical analysis factors (widely used in economics, specifically in the field of investment) is proposed, in order to combine Artificial Intelligence with the human calculation of technical factors. This brings a distinctive value to the prediction system, which is based on the combination of a series of techniques to determine the recommendation that is to be made to the user.

4. The recommendation system. It is proposed to create a recommender system, which, based on the values that result from the aforementioned objectives, is capable of recommending the decision to buy or sell a share in the Spanish continuous market to the user. Therefore, the recommender system is based on calculating the outputs of the rest of the modules and combining them in order to abstract a decision that benefits the user. Thus, the recommendation system is the most crucial and delicate phase of all the modules that make up the platform.

5. The visualization platform. In addition, we propose the creation of a platform that allows for the visualization of the information, recommendations, and predictions for each company in the Spanish continuous market that the end user wishes to consult. The visualization platform graphs the previously made calculations and predictions, so that the end user can consult how the platform operates.

The article is structured, as follows: Section 2 reviews the existing solutions for forecasting stock ratings, Section 3 considers the proposed system, including the data mining and analysis modules, the prediction system, the technical analysis, and the recommendation system. Section 4 outlines the results of the whole research process. Section 5 covers the discussion and the obtained results, as well as future research.

2. State of the Art

Throughout this section, the main contributions made in the field of stock prediction will be reviewed. The review begins with the study by Atsalakis, G.S. et al. in [1] who focused their study on stock forecasting through soft computing techniques. After classifying and processing the sample and applying the type of technique to the fuzzy set, the authors concluded that ANNs (Artificial Neural Networks) and neuro fuzzy models were valid for predicting stock market values. It should be noted that, despite being an exhaustive analysis, their research may be outdated as it was published in the period from 1992 to 2006. Another research that establishes ANNs as the best performing machine learning technique for stock market prediction is the research by Soni, S., in [4]. In the research it is compiled various studies applying machine learning and artificial intelligence techniques.

Beyond the research that proposes ANN as a method for stock prediction, during the review of the state of the art it was observed research that highlighted the need for historical stock market data after reviewing various machine learning techniques for stock prediction [5]. In [3] it was also highlighted that predicting stock market values is challenging due to the lack of certainty, perhaps in relation to the conclusions already drawn in the article discussed above. The authors attribute the lack of certainty to the unpredictability of a changing environment and provided a mixed approach that uses both machine learning algorithms and fundamental and technical analysis.

The research discussed above was the first mixed approach that was identified during the review of the state of the art. Once in this line and seeing that it was perhaps the starting point for the research proposed here, the research of [6] was found. On this occasion, the use of various techniques was focused on integrating collaborative and content-based filtering techniques, where the optimal investment recommendation was given by the investor's preferences, trends, macroeconomic factors, etc. To conclude the review, the research conducted in [7] where a mixed approach is also presented, is analysed. The research concerns the use of a decision tree of technical indicators optimised by GA-SVM. The result is a recommender system capable of detecting stock price fluctuations and suggesting a decision to the investor.

Table 1 summarises the contributions considered in this research.

Table 1. State of the art of AI applied to Stock investment recommendations.

References	Approaches
[1]	Stock forecasting through soft computing techniques. The authors concluded that ANNs and neuro fuzzy models were valid for predicting stock market values
[4]	compiled various studies applying machine learning and artificial intelligence techniques
[5]	Highlighted the need for historical stock market data after reviewing various machine learning techniques for stock prediction.
[3]	The author provided a mixed approach that uses both machine learning algorithms and fundamental and technical analysis.
[6]	The authors proposal mixed collaborative and content-based filtering techniques.
[7]	The research concerns the use of a decision tree of technical indicators optimised by GA-SVM.

3. Proposed Model

Once the state of the art has been reviewed, throughout this section the proposal is presented, more specifically the software architecture that results in the forecasting system. Specifically, the following modules are described and analyzed: data extraction package, data analysis package, forecasting system module, technical analysis module, recommender system package and the visualization platform.are described.

3.1. Software Architecture

This subsection presents the design specification of all the packages that form the software system and how they communicate and interrelate with each other. Figure 1 shows the dependencies between the packages that make up the proposed model, thus showing the different modules that make up the system and, consequently, the interrelationships and dependencies between them.

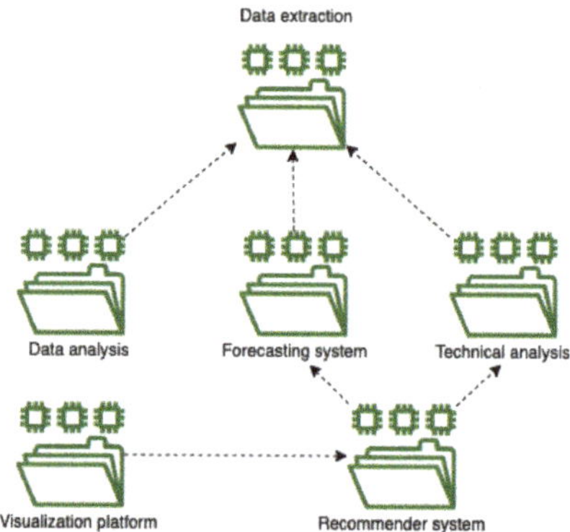

Figure 1. Software system package diagram.

Figure 2 shows the relationships between the use cases and the relationship between the different actors (user and system) and the resulting system. This is intended to provide a clearer and more exemplified understanding of the design specification of each and every one of the modules.

3.1.1. Data Extraction Package

The data extraction module works as follows: first, a user request is received through an API (Application Programming Interface) endpoint or through a request to a method in the package developed in Python for versions 3.x. Secondly, if the system has received that request, it will include in the header the name of the company and the date range (if the historical data has been requested) or it will only extract the name of the company (if the historical data has not been requested).

Figure 3, shows the Python package that has been created for data extraction from Investing.com (after prior authorisation from the company on 28 January 2019). It supports different versions and has been loaded into PyPI (Python Package Indexer). The second significant aspect is continuous integration. The package is monitored, also, unit tests and code coverage are checked (this functionality determines the number of lines of code, identifying unusable lines of code) through Travis CI. In addition, the developed Python package supports more banking products such as funds or ETFs (Exchange Traded Funds), enabling the future implementation of additional functionalities in the platform.

Once the HTML DOM Tree structure has been analyzed to determine which elements of the HTML are to be retrieved and how they can be identified, the development of the Web Scraper begins. Two main steps have been taken:

1. Web request: The HTML of the web have been retrieved and also requests of GET or POST type have been made; principally, urllib314 and requests15.
2. HTML parsing: consisted in recovering and formatting the data from the previously retrieved HTML. To parse and obtain the information from the HTML, two Python utilities have been employed, beautifulsoup416 and lxml17.

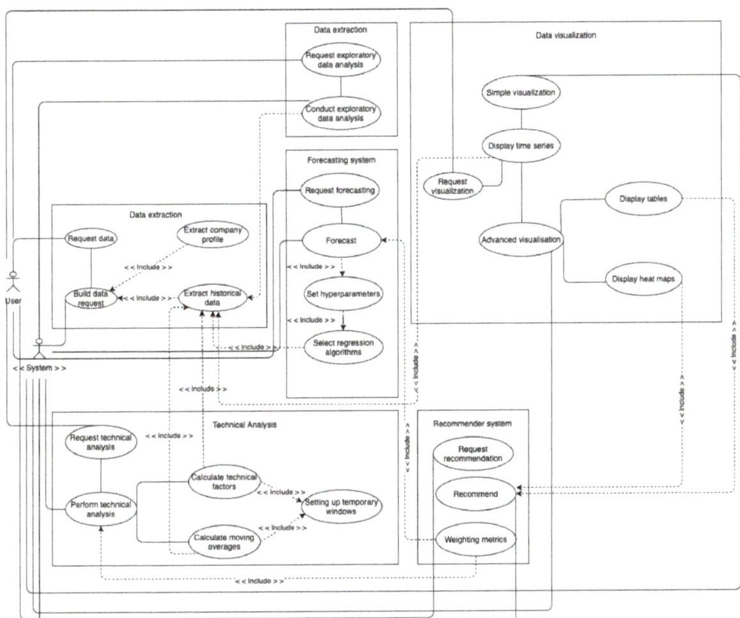

Figure 2. Related Use Case Packages.

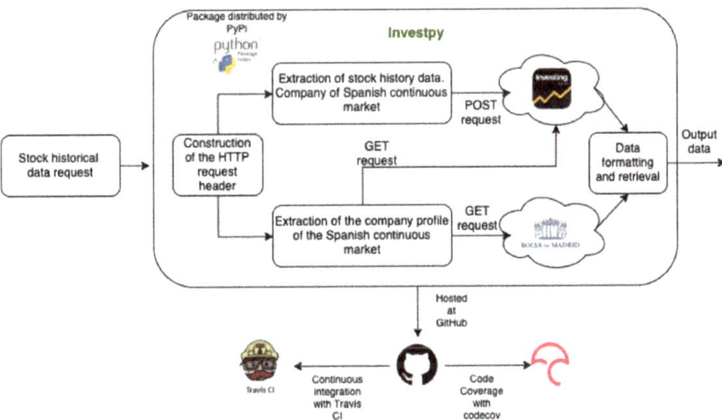

Figure 3. Data extraction Package Design Specification.

Figure 4 shows the graphic representation of the combination of possible Python package times for each of the different phases that are involved in Web Scraping. The combinations are shown in best to worst scaling, as follows: request-lxml, request-bs4, urllib3-lxml and urllib3-bs4. Therefore, to send the request to Investing.com and extract the HTML, either GET or POST type requests are optimal, while lxml is optimal for historical data extraction and parsing.

Finally, the resulting scripts give form to an extensible and open Python package, called investpy [8], intended for data extraction from investing.com. The package facilitates the extraction of data from various financial products, such as: stocks, funds, government bonds, ETFs, certificates, commodities, etc.

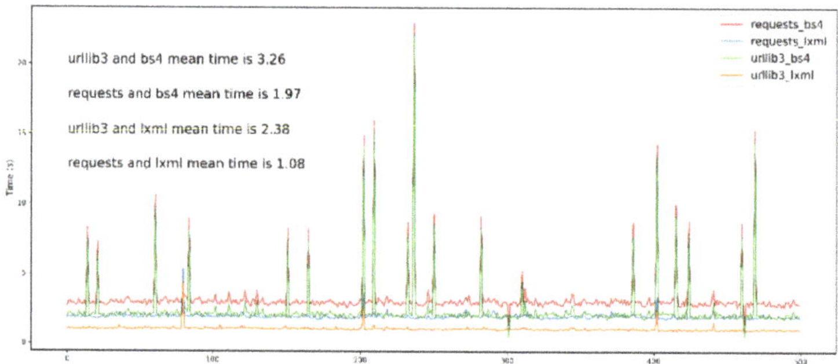

Figure 4. Best web scraping combination.

3.1.2. Data Analysis Package

Once the historical data for a stock has been extracted, the analysis of the data can be undertaken. All of the packages depend directly or indirectly on the data extraction package, as shown in Figure 1.

Exploratory data analysis is the set of graphical and descriptive tools used for the discovery of data behavior patterns and the establishment of hypotheses with as little structure as possible. Throughout this subsection, the design for the study of the structure of the data and the relationship between them is shown. A representation of how this module operates can be seen in Figure 5.

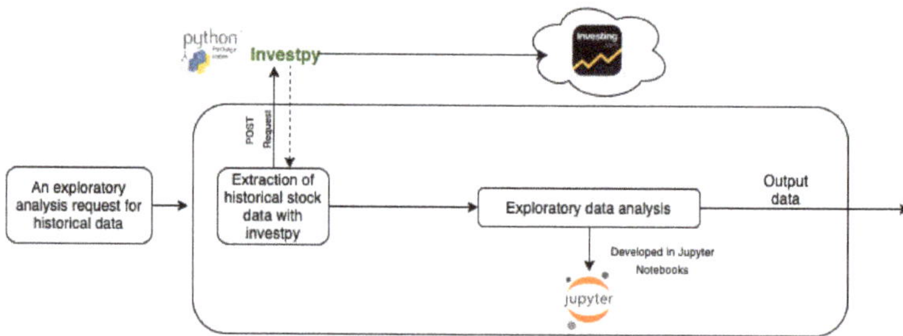

Figure 5. Data Analysis Package Design Specification.

3.1.3. Forecasting System

After obtaining the historical data from the last five years of a Spanish continuous market company share through the previously created Python package [8], the Prediction System's design specification is made, as shown in Figure 6.

To predict the future behavior of a stock, Machine Learning regression algorithms [4,9–11] are applied. The objective is to determine the closing price of the stock market, for this the set of opening values has been defined as the input variables and the set of closing values as the output variables, i.e. the closing values are the objective variable of the algorithm. Given the nature of the problem, regression algorithms must be applied. This is because when working with continuous data, regression algorithms can indicate the pattern in a given set of data. Consequently, these algorithms are applied in cases where the relationship between a scalar dependent variable or objective variable Y and one or more input variables X is to be modelled. The following section describes the algorithms that were used by the system to predict the last (unknown) closing value based on historical market data, from the last (known) opening value:

1. Random Forest Regressor: these algorithms are an automated learning method for classification, regression, and other tasks. A random forest is a meta-stimulus that fits

a series of classification decision trees into various sub-samples of the data set and uses the means to improve productive accuracy and control over fit.

2. Gradient Boosting Regressor: it is an automated learning technique that builds the model in a scenic way, just like methods that are based on reinforcement. It generalizes models allowing for the optimization of an arbitrary and differentiable loss function.

3. SVM-LinearSVR: learning models that analyze data for classification and regression analysis. An SVM training algorithm builds a model that assigns new examples to one or another category, which makes it a non-probabilistic binary linear classifier. In SVR we try to adjust the error within a certain threshold. In this case, it is similar to SVR with the kernel = linear parameter.

4. MLP Regressor: a kind of artificial feedback neural network. MLP uses a supervised learning technique, called backpropagation, for the construction of the network. In addition, its multiple layers and non-linear activation distinguish MLP from a linear perceptron. It also allows for distinguishing data that are not linearly separable.

5. KNNeighbors Regressor: non-parametric method used for classification and regression

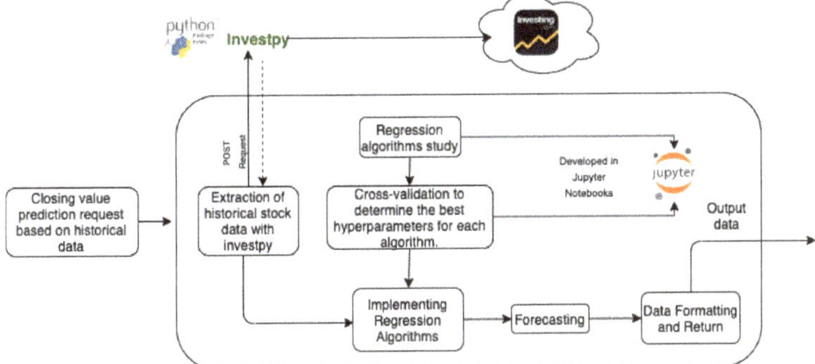

Figure 6. Forecasting System Package Design Specification.

3.1.4. Technical Analysis

Based on the Spanish continuous market companies' historical stock data, a technical analysis of the market is carried out, in this case combining Momentum Indicators and Moving Averages. This is done for several previously defined time windows for each of the different factors to be calculated based on the standard of the size of the time windows; Figure 7 describes its design specification.

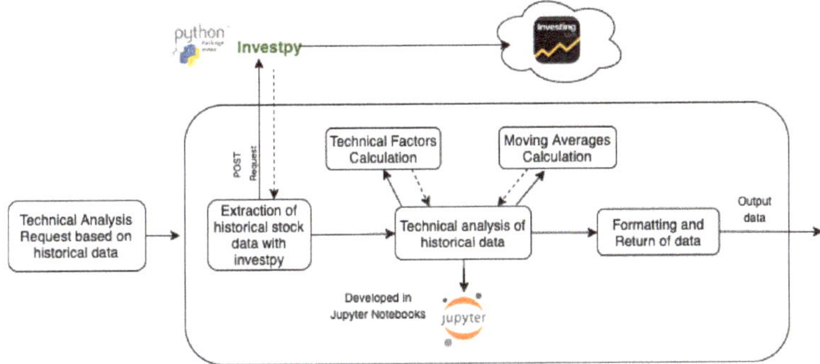

Figure 7. Technical Analysis Package Design Specification.

To calculate the factors for the technical analysis, the TA-Lib library has been used through the wrapper written in Python with the same name. Pandas' utilities have been used to calculate the moving averages. Technical Analysis is an analysis that is used to weigh and evaluate investments. It identifies opportunities to acquire or sell stocks based on market trends. Unlike fundamental analysis, which attempts to determine the exact price of a stock, technical analysis focuses on the detection of trends or patterns in market behavior for the identification of signals to buy or sell assets, along with various graphical representations that help to evaluate the safety or the risk of a stock [12]. This type of analysis can be used in any financial product as long as historical data are available. It is required to include both share prices and volume. Technical analysis is very often employed when a short-term analysis is required, thus, it can help to adequately address the problem presented in this research, where the closing value of a share in a day is predicted. The following indicators are considered in the analysis [13]:

1. Relative Strength Index (RSI): it is a Momentum Indicator (these indicators reflect the difference between the current closing price and the closing price of the previous N days), which measures the impact of frequent changes in the price of a stock, identifying the signs of overbuying or overselling. The representation of the RSI is shown on an oscillator, which is, a line whose value oscillates between two extremes, which, in this case, is between 0 and 100.

$$\text{RSI}_{\text{step one}} = 100 - \left[\frac{100}{1 + \frac{\text{Average gain}}{\text{Average loss}}} \right] \tag{1}$$

2. Stochastic Oscillator (STOCH): it is a Momentum Indicator that compares the closing price of a stock on a given day with the range of closing values of that stock over a certain period of time, defined by the time window. It also allows to adjust the sensitivity of the oscillator either by adjusting the time window or by calculating the moving average of the STOCH result. Like RSI, it identifies the signals of over-bought or oversold stock within a range of 0 to 100 possible values.

$$\%K = 100 - \left(\frac{C - L14}{H14 - L14} \right) \times 100 \tag{2}$$

where C is the most recent closing price, L14 is the lowest price traded of the 14 previous trading sessions, H14 is the highest price traded during the same 14-day period, and %K is the current value of the stochastic indicator.

3. Ultimate Oscillator (ULTOSC): it is a Momentum Indicator used to measure the evolution of a stock over a series of time frames using a weighted average of three different windows or time frameworks. Therefore, it acquires a lower volatility and identifies fewer buy-sell signals than other oscillators that only depend on a single time frame. When the lines generated by ULTOSC diverge from the closing values of a stock, buy and sell signals are identified for it.

$$UO = \left[\frac{(A_7 \times 4) + (A_{14} \times 2) + A_{28}}{4 + 2 + 1} \right] \times 100 \tag{3}$$

where UO is the Ultimate Oscillator and A is the average. The average calculation follows the next formulas.

$$A_7 = \left[\frac{\sum_{p=1}^{7} BP}{\sum_{p=1}^{7} TR} \right] \tag{4}$$

$$A_{14} = \left[\frac{\sum_{p=1}^{14} BP}{\sum_{p=1}^{14} TR} \right] \tag{5}$$

$$A_{28} = \left[\frac{\sum_{p=1}^{28} BP}{\sum_{p=1}^{28} TR} \right] \tag{6}$$

where BP is the Buying Pressure and PC is the Prior Close

$$BP = \text{Close} - \text{Min(Low, PC)} \tag{7}$$

where TR is the Ture Range

$$TR = \text{Max(High, Prior Close)} - \text{Min(Low, Prior Close)} \tag{8}$$

where TR is the Ture Range

4. Williams %R (WILLR): also known as the Williams Percent Range, is a Momentum Indicator that fluctuates between -100 and 0 and measures and identifies levels of stock overbuying or overselling. WILLR is very similar to the STOCH in its use, as it is used for the same purpose. This indicator compares the closing value of a stock with the range between the maximum and minimum values within a given time frame.

$$\text{Williams\%K} = \frac{\text{Highest High} - \text{Close}}{\text{Highest High} - \text{Lowest Low}} \tag{9}$$

where the Highest High is the highest price in the look-back period, typically 14 days, Close is the most recent closing price, and Lowest Low is the lowest price in the look-back period, typically 14 days.

Moving averages are also used in Technical Analysis, as it also represents the Momentum or value change in a timeframe N. Hence, moving averages help to understand the market trend and, like Momentum Indicators, allow to identify buy and sell signals from the historical data of a stock in a previously mentioned timeframe N. In this research, we have applied the simple moving average (SMA) and the exponential moving average (EMA) for timeframes of 5, 10, 20, 50, 100, and 200 days, so there will be indicators in different periods.

1. Simple Moving Average (SMA): it is an arithmetic moving average. It is calculated by adding the recent closing values of an action for a window of size N and dividing that sum by the size of the window. Thus, when the size of the timeframe N is low, it responds quickly to changes in the value of the stock; if the size of the window N is high, it responds more slowly.

$$\text{SMA} = \frac{A_1 + A_2 + ... + A_n}{n} \tag{10}$$

where A_n the price of an asset at period n and n is the number of total periods.

2. Exponential Moving Average (EMA): also called Exponentially Weighted Moving Average, since it weights recent observations, i.e., closing prices of a stock closer to the current one. It can be said that EMAs respond better than SMAs to recent changes in a share's price.

$$\text{EMA}_{Today} = \left(Value_{Today} \times \left(\frac{Smoothing}{1 + Days} \right) \right) + \left(EMA_{Yesterday} \times \left(\frac{Smoothing}{1 + Days} \right) \right) \tag{11}$$

where EMA is the exponential moving average. The smoothing factor is calculated, as follows:

$$Smoothing = \frac{2}{n + 1} \tag{12}$$

where n represents the number of periods the EMA uses.

Because both the algorithmic predictions and the results of the technical factor and moving average calculations result in the next closing value of a stock, the recommendation is based on identifying buy and sell signals based on the comparison of the predicted value with the value that the stock has at the current time.

3.1.5. Recommender System

Based on the results that are obtained from the forecasting and technical analysis systems, the Recommendation System design specification proceeds, in which the obtained results are weighted to identify buy/sell signals in order to be able to make a recommendation. Figure 8 shows the functionality of the Recommender System and, consequently, the process of signal extraction and the dependencies/relationships between it and the rest of the modules on which it depends.

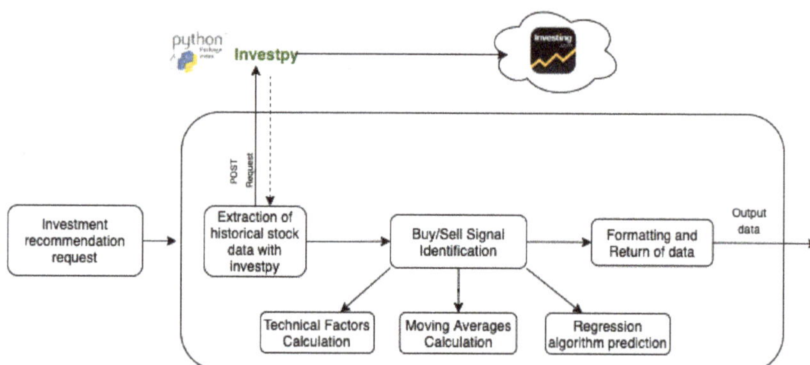

Figure 8. Recommender System Package Design Specification.

The package design proposes the creation of a neutral system, which, based on the analysis of buy/sell signals, determines the action to be taken for/with a stock. This is intended to eliminate the burden of subjectivity.

In addition to the calculation of moving averages and technical analysis ratios, an analysis using regression algorithms is also included, as can be seen in Figure 8. Regression algorithms are used when a prediction is to be made on a continuous dataset. This is the case with the historical time series data of a stock. The output of the algorithm is a quantity that can be measured in a flexible way, depending on the inputs that are passed to the algorithm. Sorting algorithms would be limited to a set of labels.

Linear regression can be defined as an approach to modelling the relationship between a dependent scalar variable y, and one or more explanatory variables, named x. Mathematically, it is expressed in the form that is presented in Equation (13).

$$y_i = \beta_0 + \beta_1 x_1 \tag{13}$$

where the variable to be predicted y_i is distinguished, as well as the constant β_0, the slope β_1, and the input variable x_i. In the current scenario, given that a stock's historical data set is available, the explanatory variable x gives the market opening values, and the target variable y gives the market closing values. Thus, the model input is x and the expected output y, where y is the dependent variable and x the independent variable, so that the market opening value conditions the market closing value.

The machine learning algorithms that are used by the system to predict the last (unknown) closing value, based on historical market data, from the last (known) opening value are:

1. Random Forest Regressor
2. Gradient Boosting Regressor
3. SVM-LinearSVR
4. MLP Regressor
5. KNNeighbors Regressor

These algorithms are applied using sklearn library (Python library that compiles machine learning algorithms) by means of cross-validation (a technique that is used to evaluate the results of a statistical analysis and ensure that they are independent of the partition between train and test data). In this way, the best hyperparameters of the algorithms can be determined and, thus, the best combination can be identified.

The results have generated a series of heat maps have been generated. The accuracy of the algorithm is represented by the hue. Lighter shading corresponds to a worse result, so darker areas indicate that the resulting hyperparameter combination is better. However, sometimes it is not known which is the best combination because the shades are very similar. The results of these combinations for each action are stored in a JSON file that will be used later by the platform when applying the models of the action prediction system. In this way, the result of applying the cross-validation of the hyperparameters to all of the stocks in the Spanish continuous market is a data file with the best hyperparameters and they are shown in the respective heat maps, to justify the decision. This allows for a more accurate decision to be made, as the user can compare the effectiveness of some hyperparameters against others.

Figures 9–13 present the heat maps resulting from the cross-validation of BBVA's historical data from the last five years. These figures represent the accuracy of the model with each combination of hyperparameters for the algorithms: MLPRegressor, SVMLinearSVR, RandomForestRegressor, GradientBoostingRegressor, and KNNNeighborsRegressor.

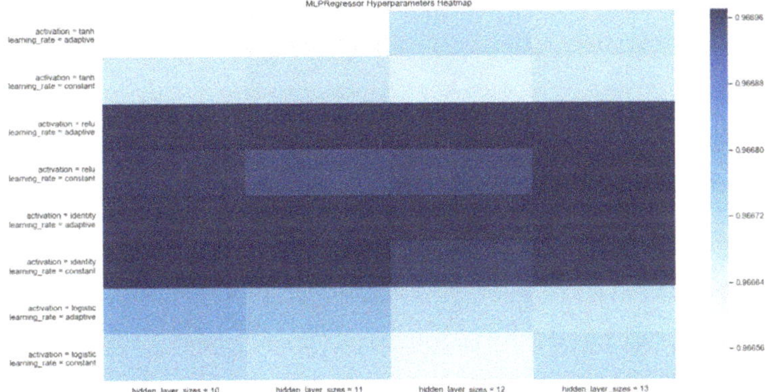

Figure 9. MLPRegressor Hyperparameter Heat Map.

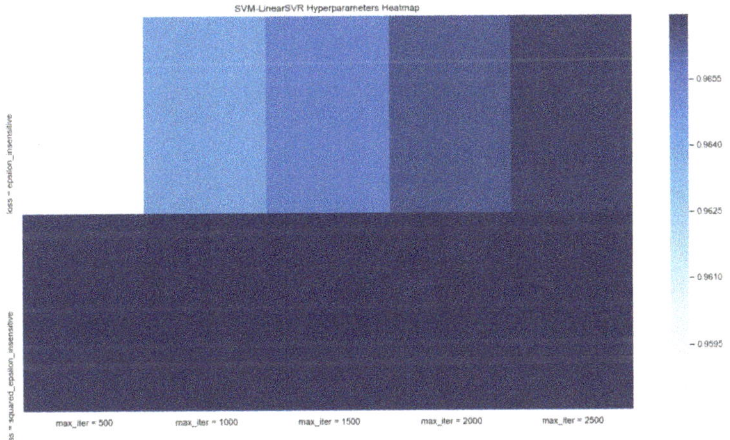

Figure 10. SVM-LinearSVR Hyperparameter Heat Map.

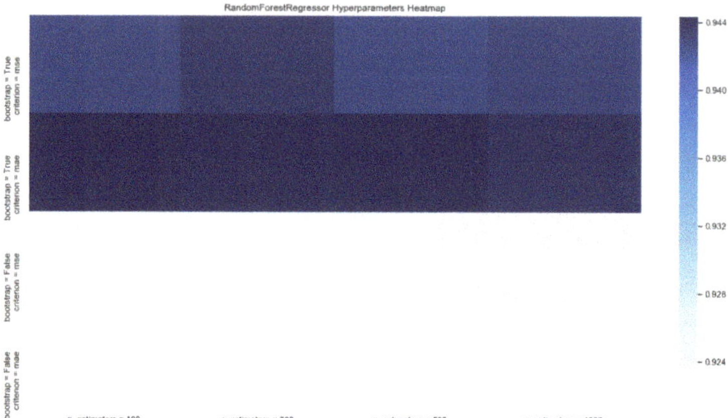

Figure 11. RandomForestRegressor Hyperparameter Heat Map.

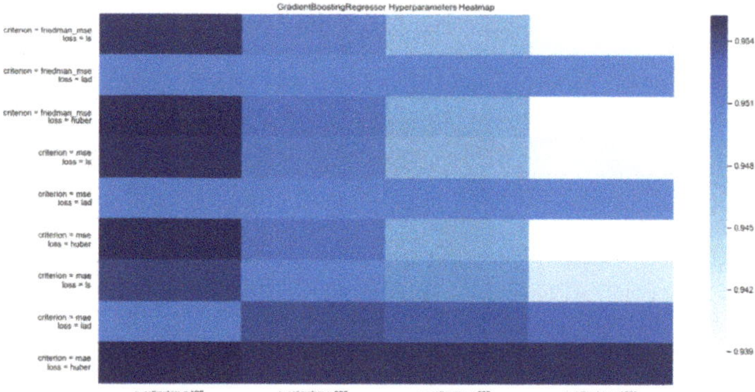

Figure 12. GradientBoostingRegressor Hyperparameter Heat Map.

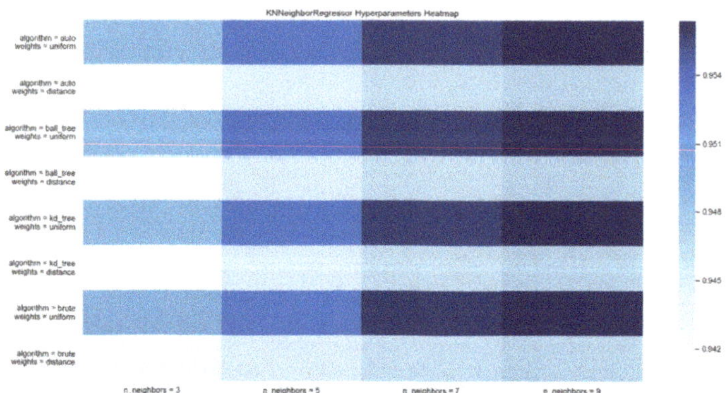

Figure 13. KNNeighborsRegressor Hyperparameter Heat Map.

Once the evaluation process and the cross-validation of the algorithms have been completed, a graph showing the best algorithm has been drawn up. The top-25 equities of the Spanish Continuous Stock Market have been chosen. Figure 14 shows that the algorithms that best fit the proposed problem are SVM-LinearSVR and MLP Regressor.

Thus, these algorithms are the ones that have the highest accuracy after being trained and tested with an 80/20 split of the dataset.

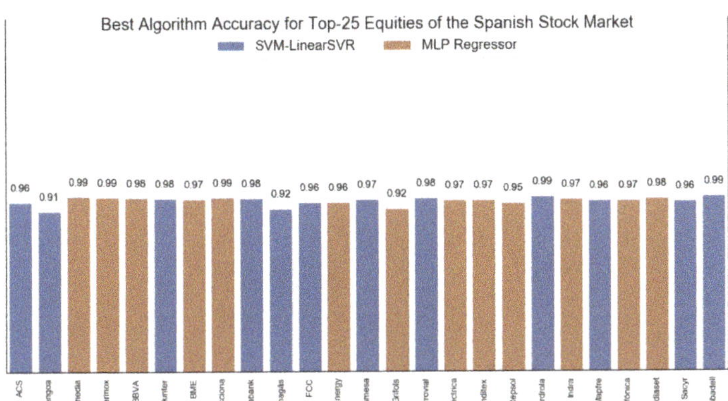

Figure 14. Best algortihm accuaracy for top-25 equities of the Spanish Continuous Stock Market.

4. Platform Visualization

Finally, after detailing each of the software modules created, the description of the visualization platform follows as a deduction of the integration of the most relevant aspects of the rest of the modules, so that the end user can interact with the platform.

Therefore, the visualization platform takes up the conclusions of the research carried out in the rest of the modules, so that only two options are given at user level for visualization, either an overview as a result of the exploratory analysis of the data, or the result of the underlying recommender system.

In this way, the different phases or tools that are used for the development of the platform architecture are detailed, based on the results of the study of the rest of the modules.

It is worth mentioning that the development of a visualization platform only aims to bring the results closer to the user, without being the central part of the proposed system.

The result is a platform that provides a user interface for both data visualization, analysis, prediction, and investment recommendation. It has been determined that the platform will be developed using Django, as shown in Figure 15. Django is a Python framework for creating web services, in this case it has been used to communicate the back-end with the front-end. The web application created combines the use of Python for data management and communication, and HTML, CSS, and JavaScript for the visualization of both the platform and the data.

The design pattern used, called MVC (Model-View-Controller), focuses on the division of the web project according to the functionalities of each of its parts. However, Django does not use the MVC pattern, but rather the MVT (Model-View-Template), which is an abstraction of the MVC model. It is worth mentioning that Django works with templates, not with views, being oriented to the development of web applications, as explained in [14], where the author not only teaches aspects of using of Django, but it also lists the different design patterns that can be followed in order to structure to follow the web application to be developed.

The platform's objective is not only to be usable and intuitive, but also to enable any user, whether an expert or not in the stock market, to abstract their own conclusions from the data and evaluate the information analyzed by the system. The created platform completely depends on the Python package developed for data extraction: investpy. The web platform initially shows a screen where the overview option is given on one side and the overview and recommendation option on the other (Figure 16). The overview functionality covers the extraction and basic visualization of the data. The system retrieves the company profile and the historical data for the last five years of the stock. On the basis of those data, it produces a series of representations:

1. Time series: offers a graphic representation of the retrieved historical data, where the X and Y axes represent the value of the stock in euros, and the date on which the stock reached that value, respectively.
2. Candlestick chart: this representation shows the opening and closing values for each date and the difference between the maximum and minimum values for the same date.
3. Data table: represents the available values. They are called OHLC (Open-High-Low-Close).

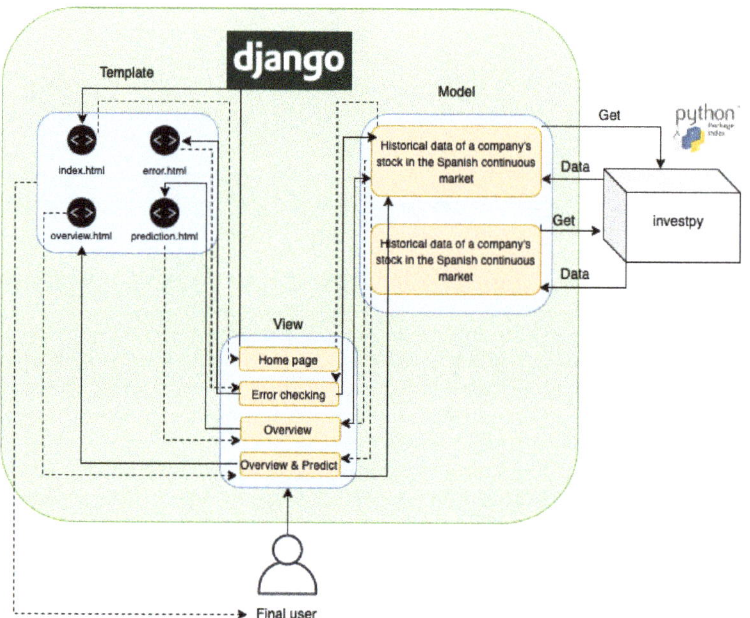

Figure 15. Django Design Architecture.

Figure 16. Main view of the web platform.

The Overview & Recommendation functionality is the same as the user input check, in that it also extracts the company profile and historical data. However, this functionality also includes technical factors and moving averages with the consequent buy/sell recommendation. The generated graphs are visualized on the platform, among them are graphs that compare the different algorithms that the system has applied to make the prediction. This enables the user to identify those that have had a better precision. The platform presents the conclusions abstracted from the resulting values. It shows the buy/sell recommendation that is based on those values. The process of prediction and recommendation made by the system is transparent to the user.

The novelties that are presented by the module are the graphs generated, in which a comparison between the different algorithms applied by the system to make the prediction can be observed, thus being able to contrast which one has had the best accuracy (Figure 17).

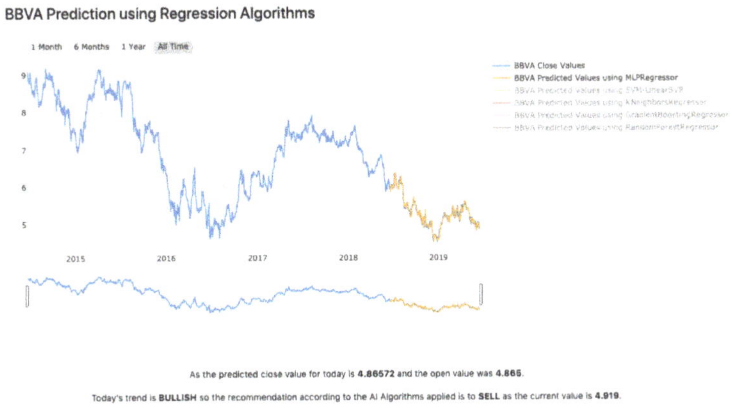

Figure 17. Prediction using regression algorithms.

Additionally, there is an option that allows the user to observe which algorithms have been applied, what they consist of, and which hyperparameters have been used based on the results in the form of a heat map of the cross validation carried out by the system.

Once the justification of the regression algorithms used in the platform by the system has been shown, the results of the different algorithms applied are displayed, where the "best" algorithm (the one with better precision than the rest) is the one that shows its results by default (Figure 18). Even so, the platform gives the option of displaying different time windows and visualising the results of all the algorithms. Finally, the platform displays a paragraph, in which it indicates the conclusions drawn from the study of the values resulting from the prediction and, therefore, shows the buy/sell recommendation based on these values.

Therefore, the platform displays the recommendations based on the results of the prediction, which it will combine with the results of the financial technical analysis, which includes the calculation of moving averages and technical factors. Finally, the system calculates the technical factors, called Momentum Indicators, which indicate the market trend based on calculations taking different time windows (Figure 19).

In this way, the system not only makes the recommendation, but it also supports this recommendation and each of the predictions and calculations that give rise to it, with the data used throughout the process. Therefore, the prediction and recommendation process carried out by the system is transparent to the user at a technical level, so that the user is aware of what has happened in each of the stages of the process, being able to trust that the prediction has not been altered for the benefit of third parties, for example.

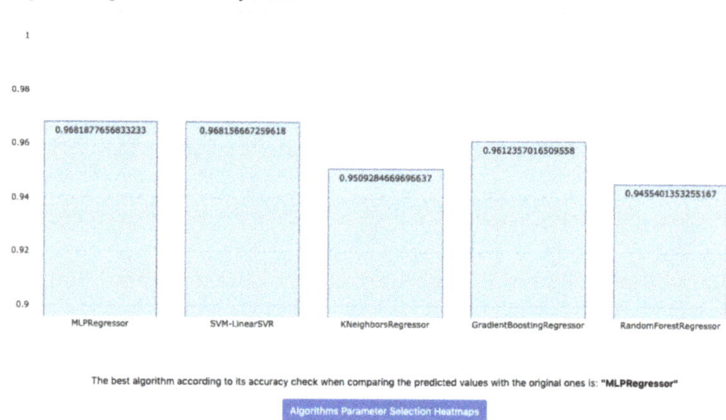

Figure 18. Regression Algorithms Accuracy Check.

• Technical Factors

Tech. Factor	Result	Signal
RSI - 14 days	43.63047	SELL
STOCH - 9,6 days	38.29643	SELL
ULTOSC - 7,14,28 days	53.80424	BUY
WILLR - 14 days	-55.2381	SELL

Figure 19. Buy/sell recommendations.

5. Discussion and Results

The conducted research provides an initial approach to data analysis and the combined use of Machine Learning algorithms and techniques, with traditional market analysis. Their use enables the proposed platform to arrive at conclusions regarding future market behavior. Thus, it can be concluded that, when Machine Learning algorithms are trained with a sufficiently large amount of data, it is possible to successfully predict the closing value on the basis of the current opening value of the market. Thus, after identifying buy and sell signals, it has been possible to create a system that recommends the user to buy, hold, or sell a stock at a certain time of day, according to the prediction obtained by the regression algorithms.

Although the recommender system operates well and meets the initial objectives of this study, system extensions will be considered in future research. The breadth of the platform in terms of functionalities was the most significant complication that arose during the research, therefore it was decided to approach it with a modular architecture. Thanks to the modular, highly scalable design it is possible to provide the system with more functionalities; the combination of Natural Language Processing (NLP) techniques could be used in an opinion mining process, the recommender system will be able to abstract the future market trend based on the sentiment analysis. In addition, the use of NLP techniques is also proposed for the classification of companies into sectors based on their company profiles, thus being able to group companies into sectors based on the description that each company in the Spanish continuous market proposes. Therefore, additional functionality must be added to the Python package that was created for the extraction of the Investing data, called investpy. The enhancement will consist of retrieving all the data provided freely by Investing.com. Additionally, a study of the algorithms applied to other markets should be carried out, as the proposed system is oriented towards a very specific market; the Spanish continuous market. It will be necessary to carry out a study to determine the best algorithms for the stock markets of each of the countries to be incorporated. It is considered to be viable given that all historical stock data previously go through a GridSearchCV, which consists in cross validating the optimal hyperparameters to be used by an algorithm from a specific dataset. In addition, further research is considered on event identifications that can be used to better choose the operation performed (buy/sell) and the social characteristics of the different communities [15,16].

Author Contributions: Conceptualization, E.H.-N., J.P.-D. and P.C.; Funding acquisition, E.H.-N.; Investigation, E.H.-N., J.P.-D. and P.C.; Methodology, S.R.G. and J.M.C.; Software, P.C.; Supervision, S.R.-G. and J.M.C.; Writing—original draft, E.H.-N. All authors have read and agreed to the published version of the manuscript.

Funding: This research was partially Supported by the project "Computación cuántica, virtualización de red, edge computing y registro distribuido para la inteligencia artificial del futuro", Reference: CCTT3/20/SA/0001, financed by Institute for Business Competitiveness of Castilla y León, and the European Regional Development Fund (FEDER). The research of Elena Hernández-Nieves is funded by Ministry of Education of the Junta de Castilla y León and the European Social Fund grant number EDU/556/2019.

Conflicts of Interest: The authors declare no conflict of interest.

References

1. Atsalakis, G.S.; Valavanis, K.P. Surveying stock market forecasting techniques–Part II: Soft computing methods. *Expert Syst. Appl.* **2009**, *36*, 5932–5941. [CrossRef]
2. Nicoletti, B.; Nicoletti, W.; Weis. *Future of FinTech*; Palgrave Studies in Financial Services Technology: Rome, Italy, 2017.
3. Patel, J.; Shah, S.; Thakkar, P.; Kotecha, K. Predicting stock and stock price index movement using trend deterministic data preparation and machine learning techniques. *Expert Syst. Appl.* **2015**, *42*, 259–268. [CrossRef]
4. Soni, S. Applications of ANNs in stock market prediction: A survey. *Int. J. Comput. Sci. Eng. Technol.* **2011**, *2*, 71–83.
5. Yoo, P.D.; Kim, M.H.; Jan, T. Machine learning techniques and use of event information for stock market prediction: A survey and evaluation. In Proceedings of the International Conference on Computational Intelligence for Modelling, Control and Automation and International Conference on Intelligent Agents, Web Technologies and Internet Commerce (CIMCA-IAWTIC'06), Vienna, Austria, 28–30 November 2005; Volume 2, pp. 835–841.
6. Taghavi, M.; Bakhtiyari, K.; Scavino, E. Agent-based computational investing recommender system. In Proceedings of the 7th ACM Conference on Recommender Systems, Hong Kong, China, 12–16 October 2013; pp. 455–458.
7. Nair, B.B.; Mohandas, V. An intelligent recommender system for stock trading. *Intell. Decis. Technol.* **2015**, *9*, 243–269. [CrossRef]
8. Del Canto, A. Investpy—Financial Data Extraction from Investing.com with Python. Available online: https://github.com/alvarobartt/investpy (accessed on 4 April 2020).
9. Arora, N. Financial analysis: Stock market prediction using deep learning algorithms. In Proceedings of the International Conference on Sustainable Computing in Science, Technology and Management (SUSCOM), Amity University Rajasthan, Jaipur, India, 26–28 February 2019.
10. Khaidem, L.; Saha, S.; Dey, S.R. Predicting the direction of stock market prices using random forest. *arXiv* **2016**, arXiv:1605.00003.
11. Pimprikar, R.; Ramachadran, S.; Senthilkumar, K. Use of machine learning algorithms and twitter sentiment analysis for stock market prediction. *Int. J. Pure. Appl. Math.* **2017**, *115*, 521–526.
12. Edwards, R.D.; Magee, J.; Bassetti, W.C. *Technical Analysis of Stock Trends*; Routledge, Taylor & Francis Group: New York, NY, USA, 2018.
13. Dash, R.; Dash, P.K. A hybrid stock trading framework integrating technical analysis with machine learning techniques. *J. Financ. Data Sci.* **2016**, *2*, 42–57. [CrossRef]
14. Ravindran, A. *Django Design Patterns and Best Practices*; Packt Publishing Ltd.: Birmingham, UK, 2015.
15. Di Girolamo, R.; Esposito, C.; Moscato, V.; Sperlí, G. Evolutionary game theoretical on-line event detection over tweet streams. *Knowl. Based Syst.* **2021**, *211*, 106563. [CrossRef]
16. Mercorio, F.; Mezzanzanica, M.; Moscato, V.; Picariello, A.; Sperli, G. DICO: A graph-db framework for community detection on big scholarly data. *IEEE Trans. Emerg. Top. Comput.* **2019**. [CrossRef]

 *electronics*

Article

A Mathematical Study of Barcelona Metro Network

Irene Mariñas-Collado [1], Elisa Frutos Bernal [2], Maria Teresa Santos Martin [3], Angel Martín del Rey [4,*], Roberto Casado Vara [5] and Ana Belen Gil-González [5]

[1] Department of Statistics and Operations Research and Mathematics Didactics, University of Oviedo, 33007 Oviedo, Spain; marinasirene@uniovi.es
[2] Department of Statistics, University of Salamanca, 37007 Salamanca, Spain; efb@usal.es
[3] Department of Statistics, Institute of Fundamental Physics and Mathematics, University of Salamanca, 37007 Salamanca, Spain; maysam@usal.es
[4] Department of Applied Mathematics, Institute of Fundamental Physics and Mathematics, University of Salamanca, 37007 Salamanca, Spain
[5] BISITE Research Group, University of Salamanca, 37007 Salamanca, Spain; rober@usal.es (R.C.V.); abg@usal.es (A.B.G.-G.)
* Correspondence: delrey@usal.es

Abstract: The knowledge of the topological structure and the automatic fare collection systems in urban public transport produce many data that need to be adequately analyzed, processed and presented. These data provide a powerful tool to improve the quality of transport services and plan ahead. This paper aims at studying, from a mathematical and statistical point of view, the Barcelona metro network; specifically: (1) the structural and robustness characteristics of the transportation network are computed and analyzed considering the complex network analysis; and (2) the common characteristics of the different subway stations of Barcelona, based on the passenger hourly entries, are identified through hierarchical clustering analysis. These results will be of great help in planning and restructuring transport to cope with the new social conditions, after the pandemic.

Keywords: complex network analysis; centrality measures; network robustness; ridership patterns; clustering analysis; passenger flow; Barcelona underground

Citation: Mariñas-Collado, I.; Frutos Bernal, E.; Santos Martín, M.T.; Martín del Rey, A.; Casado Vara, R.; Gil-González, A.B. A Mathematical Study of Barcelona Metro Network. *Electronics* **2021**, *10*, 557. https://doi.org/10.3390/electronics10050557

Academic Editor: Myung-Sup Kim

Received: 29 January 2021
Accepted: 22 February 2021
Published: 27 February 2021

Publisher's Note: MDPI stays neutral with regard to jurisdictional claims in published maps and institutional affiliations.

1. Introduction

Sustainable urban mobility is one of the most distinct characteristics of Smart Cities. Specifically, intelligent public urban transport planning plays an important role in the design of the future cities and in the sustainable development of the environment (in this sense, it has become one of the most powerful tools in the fight against air pollution in cities); moreover, it is well known that efficient mass transit systems have a highly beneficial impact on economic development and social integration. Particularly, the subway is the best choice in big cities since it exhibits many advantages including reducing traffic congestion, saving energy and non-renewable resources, reducing the number of traffic accidents and therefore deaths, large capacity, time reliability, etc. [1].

Hundreds of millions of passengers commute in public transport daily in large cities, hence failures in the network can cause major problems to commuters and business activities with significant economic and social losses. In addition, the COVID-19 pandemic has changed the security measures on the transport network in order to maintain the sanitary requirements. Proper social distancing between passengers is hard to ensure in public transport if it is not well planned (taking into account the different characteristics of the different stations and lines). To avoid overcrowded stations and trains, it is crucial to know transit trip patterns. This will also allow better network planning, demand forecasting and, ultimately, a more effective use of the available resources in general.

Two main goals are addressed in this work: (1) study the structural and robustness characteristics of Barcelona subway network; and (2) identify ridership patterns at its stations. In the first case, the basic techniques of Complex Network Analysis are used (centrality measures, structural indices, robustness coefficients, etc.), whereas, in the second case, a hierarchical cluster analysis is performed to group stations according to their boarding patterns. Barcelona's metro is Spain's second largest city subway system: there are a total of 13 lines and 151 stations in the network. Its length is 119 km, and during 2018 more than 400 million people used it.

In recent years, the complex network approach has been used to analyze the subway rail networks of several cities around the world. Since 2002, when Latora and Marchiori studied the topological properties of the Boston subway [2], many other works have appeared. Lu and Shi found that the public transportation network in China had scale-free and small world characteristics [3]. Zhang et al. studied the topological characteristics of some subway networks around the world and investigated network failures to discuss the vulnerability of these subway networks [4]. Liu and Song [5] studied the topology of Guangzhou subway network using L-space method, and the value and distribution of the network's degree, clustering coefficient and average shortest path length were computed and analyzed. Cats [6] conducted a longitudinal analysis of the topological evolution of a multimodal rail network by investigating the dynamics of its topology for the case of Stockholm during 1950–2025.

The robustness of subway networks has also been discussed by many other researches. For example, Derrible and Kennedy studied the complexity and robustness of 33 metro networks [7]. Using network science and graph theory, ten theoretical and four numerical robustness metrics and their performance in quantifying the robustness of metro networks under random failures or targeted attacks were investigated by Wang et al. [8]. Zhang et al. [9] investigated the connectivity, robustness and reliability of the Shanghai subway network of China. Forero-Ortiz et al. [10] gave insights for stakeholders and policymakers to enhance urban flood risk management, as a reasonable approach to tackle this issue for Metro systems worldwide. De Bona et al. [11] proposed a novel methodology called Reduced Model as a simple method of network reduction that preserves the network skeleton (backbone structure) by properly removing 2-degree nodes of weighted and unweighted network representations. In [12], a new perspective for understanding vulnerability of metro networks is shown with the aims of improving operation reliability and stability of the network, designing emergency strategies to protect the network, etc.

In this work, the topological characteristics of the metro network are investigated considering the complex network approach. Specifically a brief analysis of the Barcelona subway network is provided from the computation of the most important centrality measures: (i) degree centrality C_D; (ii) average degree $E[D]$; (iii) degree distribution $p(k)$; (iv) average path length L; (v) closeness centrality C_{CL}; and (vi) betweenness centrality C_B. In addition, to assess the robustness of the subway network, eight theoretical robustness metrics are investigated: (i) normalized robustness indicator $\bar{r}^T$; (ii) effective graph conductance C_G; (iii) average efficiency $E[\frac{1}{H}]$; (iv) clustering coefficient C_{CG}; (v) normalized algebraic connectivity $\bar{\mu}_{N-1}$; (vi) average degree $E[D]$, (vii) normalized natural connectivity $\bar{\lambda}$; and (viii) degree diversity κ.

Most public transit networks use automated fare collection (AFC) systems. The interest in this kind of technology is because it is perceived as a secure method of user validation and fare payment. Moreover, it improves the quality of the data, gives transit a more modern look and provides new opportunities for innovative and flexible fare structuring [13]. While the main purpose of AFC systems is to collect revenue, they also produce very large quantities of very detailed data of on-board transactions. These data are very useful to transit planners, from the day-to-day operation of the transit system to the strategic long-term planning of the network [14].

AFC systems are classified into two types according to the fare charge mode of transit: flat-rate fare systems and distance-based fare systems. In flat-rate fare systems, only entry swipes are registered, while, in distance-based fare systems, entry and exit swipes are registered. Barcelona metro uses a flat-rate fare system, therefore only metro boarding is available in this study. This has the inconvenience of not knowing where the passenger's journey ends, e.g., the trip's purpose. The destination of the trip helps understand peak hours. For instance, most of the work and education trips start in the morning peak from home and return back to home in the evening peak. While not within the scope of this paper, the destination estimation of public transport is one of the major concerns for the implementation of smart card data and there exist several approaches (see, e.g., [15–18]).

Every day, depending on the size of the network, millions of transactions are registered by the AFC systems, which can be used to analyze human mobility. It has been determined that human trajectories and trips generated with human mobility show a high degree of temporal and spatial regularity [19]. Passenger flow of the urban subway varies according to time and space, including working days, holidays, seasons, residential areas, business centers, workplaces and other factors such as weather, as well as other forms of transportation that connect to the subway network. In this regard, several methods have been developed in the literature for this type of analysis, most using clustering approaches [20].

Two viewpoints can be considered when a cluster analysis using smart card data is performed. The first one clusters stations based on the temporal-spatial distribution characteristics of subway ridership. The second one identifies groups of passengers that have similar boarding times aggregated into weekly profiles [21].

From the first point of view, Chen et al. [22] studied the diurnal pattern of subway ridership in New York City using the k-means algorithm. Wang et al. [23] analyzed eight metro stations in the central area of Hong Kong using the hierarchical cluster analysis. The k-means algorithm was also employed by Kim et al. [24] to identify the daily travel patterns at subway stations of Seoul Capital Area. Ding et al. [25] applied gradient boosting decision trees to investigate the non-linear effects of built environment variables on station boarding in the Washington metropolitan area. Langlois et al. [26] proposed a longitudinal representation of user's multi-week activity and identified 11 travel patterns from London's public transport network.

The study and analysis of different characteristics of subway networks have been tackled by means of other different paradigms. For example, risk analysis has been addressed in some recent works (see, e.g., [10,27–29]), the GIS-based technologies improves the analysis performed using mathematical methods [30], modern statistical and mathematical techniques can be also applied [31–34], the study of bus–metro transfers is considered in [35,36], etc. Moreover, techniques based on the Artificial Intelligence paradigm have also been used to study different aspects of subway networks (see, e.g., [37–39]).

The rest of the paper is organized as follows. Section 2 describes the data used in the study. Section 3 is devoted to presenting the methodology used for the analysis of travel patterns. Finally, the results obtained and the discussion are presented in Section 4 and the conclusions in Section 5.

2. Structural and Transit Data of Barcelona Subway Network

2.1. Study Area

Barcelona is considered a significant success in urban development across Europe. As the second largest city of Spain, it has been growing and transforming itself to be a knowledge-intensive city and, more importantly, a pioneer in being a smart city [40]. In addition, it has been one of the Spanish cities with the most confirmed cases of coronavirus. This is why it is an excellent case to explore.

Barcelona has an area of 102 km^2 and a resident population of more than 1.62 million. The city has a diverse public transport system composed of metro, urban and intercity buses, commuter trains, tramway, funicular cable tramway and taxis.

The Barcelona Metro is a metropolitan railway network that gives service to Barcelona and the municipalities of its metropolitan area: Badalona, Cornellà de Llobregat, L'Hospitalet de Llobregat, Montcada i Reixac, El Prat de Llobregat, Sant Adrià de Besòs , Sant Boi de Llobregat and Santa Coloma de Gramanet. It comprises 13 lines with a length of 119 km (see Figure 1):

- L1: Hospital de Bellvitge–Fondo
- L2: Paral-lel–Badalona Centre
- L3: Zona Universitària–Trinitat Nova
- L4: La Pau–Trinitat Nova
- L5: Cornellà Centre–Vall d'Hebron
- L6: Plaça Catalunya–Reina Elisenda
- L7: Plaça Catalunya–Avinguda Tibidabo
- L8: Plaça Espanya -Molí Nou Ciutat Cooperativa
- L9 Nord: La Sagrera–Can Zam
- L9 Sud: Aeroport T1–Zona Universitària
- L10 Nord: La Sagrera–Gorg
- L10 Sud: Foc–Collblanc
- L11: Trinitat Nova–Can Cuiàs

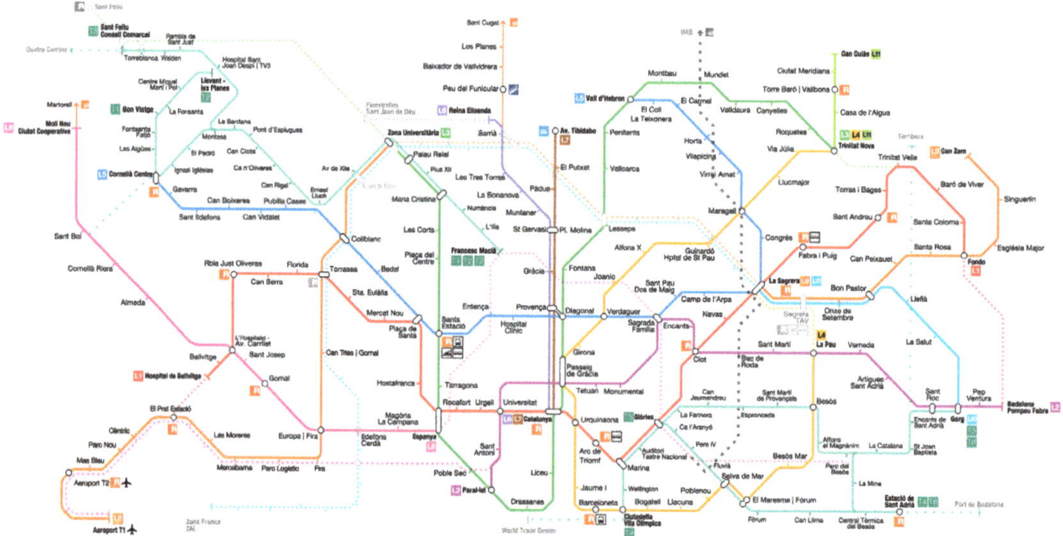

Figure 1. The 2019 Barcelona subway (Available online: https://www.metrobarcelona.es/mapas.html (accessed on 15 February 2021)).

2.2. Transit Data

The data used in this research correspond to the ridership (number of entries) in each station from 5 March 2018 to 11 March 2018. The reason this week was selected is because it is a week without public holidays or summer or winter holidays, and, therefore, it can reflect the general station ridership characteristics under normal circumstances. There was no extreme weather associated with that week either (e.g., heavy storms or very hot temperatures).

A statistical analysis of daily transit data was performed to analyze hourly inbound ridership of the 151 stations of Barcelona subway. The Barcelona metro operates from Sunday to Thursday from 5:00 to 24:00. On Fridays, the metro schedule is extended until 2:00, while on Saturdays it offers continuous service for 24 h. Thus, there are 140 variables for each station.

There are some aspects that need to be taken into account when addressing the analysis. First, it is important to notice there are two time-related patterns: the inbound ridership patterns on weekdays and at weekends. While they are both highly correlated on their own, the correlation between the ridership on weekdays and on the weekend is relatively low (see Figure 2). Second, from the analysis of the inbound ridership, it can be deduced that the highest peak hour during weekday mornings is between 7:00 and 8:00. During the evening rush hour, the highest peak hours are between 14:00 and 15:00 and between 18:00 and 19:00. Meanwhile, the rush hours during the weekend are from 13:00 to 14:00 and from 18:00 to 19:00 (see Figure 3). Figure 4, where the total number of entries at each hour is added up for all the days in the selected week for 35 randomly selected stations, illustrates how the different rush hours change depending on the station, and that both the time and the number of validations that represent a peak for a station vary. In addition, the total number of passengers significantly differs from one station to another. For instance, taking the daily ridership of 5 March, Diagonal station has a total of 54,636 passengers, while, at Casa de l'agua, there were only 207 boardings that day. These are the stations with the maximum and minimum total number of boardings and illustrate the huge difference there can be. Finally, as shown in Figure 5, the distribution of passenger flow decreases significantly on Saturdays and Sundays, which is why it was decided to focus on the data from Monday to Friday.

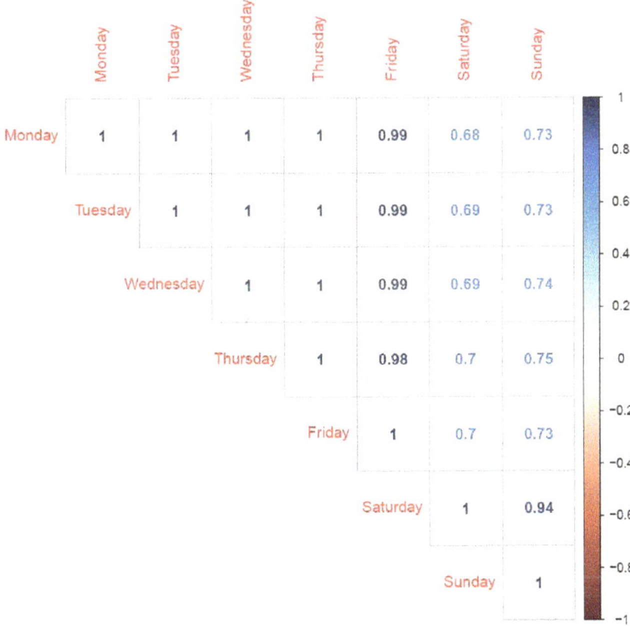

Figure 2. Pearson's correlation coefficients of daily ridership.

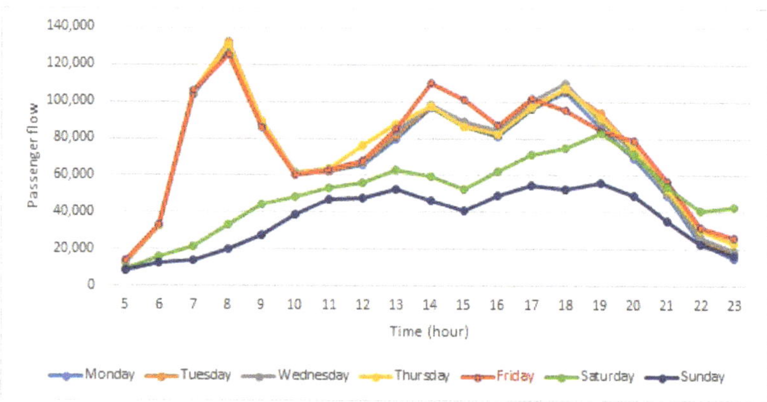

Figure 3. Time-varying diagram of passenger flow (total counts of boarding).

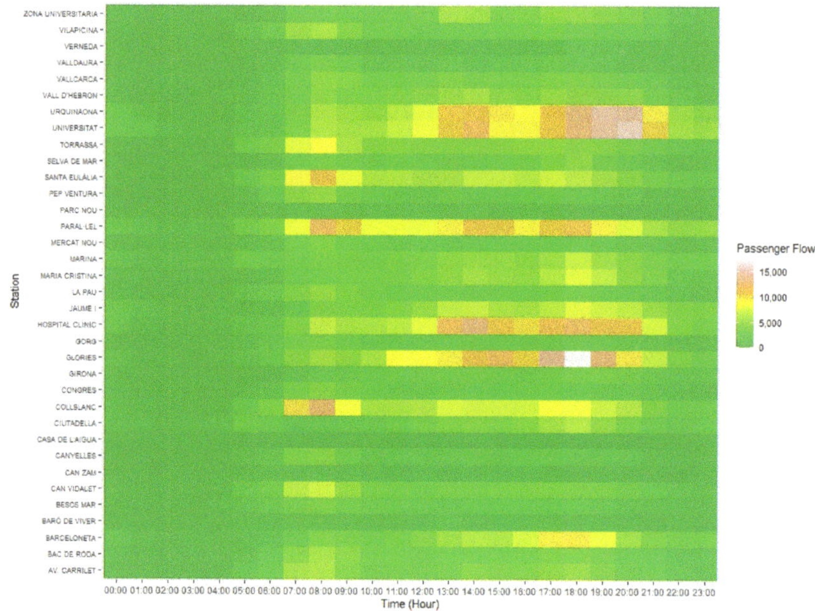

Figure 4. Heatmap with the total number of validations per hour for 35 randomly selected stations.

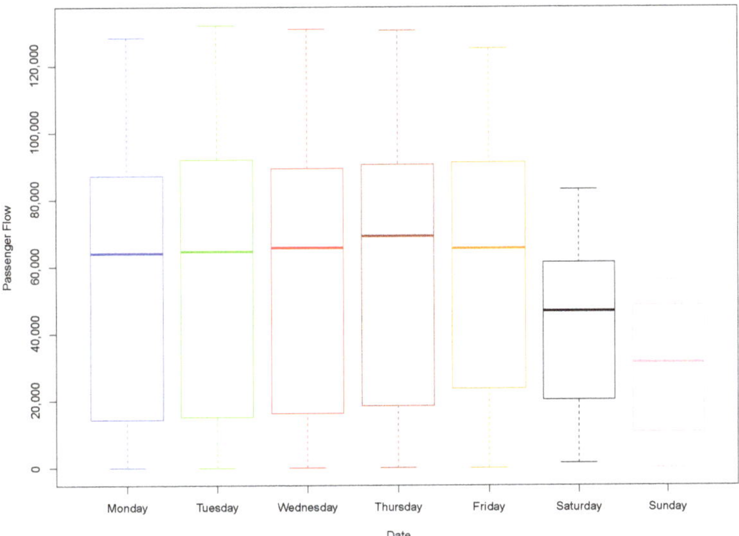

Figure 5. Passenger flow boxplots.

3. Methodology

3.1. Complex Network Analysis

In this study, the L-space representation of the network is considered. Hence, the stations of the subway network are represented by nodes of a graph and the tracks connecting two stations are represented by edges of the graph. Therefore, the subway network is represented by a undirected graph $G = (V, E)$ where $V = \{v_1, v_2, \ldots, v_N\}$ is the set of nodes, and $E = \{e_{ij} = (v_i, v_j), v_i, v_j \in V\}$ is the set of edges, where $|E| = M$.

The adjacency matrix of G, $A_G = (a_{ij})_{1 \leq i,j, \leq N}$, is a $N \times N$ symmetric matrix such that the coefficient a_{ij} takes the value 1 or 0 depending on whether or not there is a link between nodes v_i and v_j. The degree of a node v_i is the number of adjacent nodes to v_i and can be computed as follows: $d_i = \sum_{j=1}^{N} a_{ij}$.

The Laplacian matrix $Q_G = \Delta - A_G$ is an $N \times N$ matrix, where $\Delta = \text{diag}(d_1, \ldots, d_N)$ is the $N \times N$ diagonal degree matrix. The eigenvalues of Q_G play a very important role in robustness analysis; they are non-negative and can be ordered as $0 = \mu_N \leq \mu_{N-1} \leq \cdots \leq \mu_1$.

3.1.1. Centrality Measures

The analysis of a complex network is performed through the computation and analysis of several structural coefficients of the network topology. Specifically, the most important are the following [41]: degree centrality, average degree, degree distribution, average path length, closeness centrality and betweenness centrality.

The degree centrality of v_i is the average number of incident edges to v_i:

$$C_D(v_i) = \frac{d_i}{N},$$

(1)

and the normalized average degree of the network G is given by:

$$\overline{E}[D] = \frac{\sum\limits_{i=1}^{N} d_i}{N(N-1)}.$$

(2)

Moreover, the degree distribution of the network, $P(k)$, is the probability distribution of degrees over the whole network.

The shortest path length or distance between two nodes $v_i, v_j \in V$ is denoted by $d(v_i, v_j)$ and is defined as the minimum number of links necessary to go from node v_i to node v_j. The average path length of the network is defined as the average distance between two nodes:

$$L = \frac{2}{N(N-1)} \sum_{1 \le i < j \le N} d(v_i, v_j). \tag{3}$$

The diameter D of G is the greatest distance between any pair of nodes:

$$D = \max\{d(v_i, v_j), v_i, v_j \in V\}. \tag{4}$$

The closeness centrality of the node v_i measures the mean distance from v_i to the rest of the nodes of the network:

$$C_{CL}(v_i) = \frac{1}{\sum_{i \ne j} d(v_i, v_j)}. \tag{5}$$

The greater is the value of closeness centrality, the smaller is the length of the shortest paths to all other nodes.

Finally, the *betweenness centrality* of the node $v_i \in V$ measures the number of shortest paths between two nodes that run through node v_i. Mathematically it is defined as follows:

$$C_B(v_i) = \frac{2}{(N-1)(N-2)} \sum_{r \ne s \ne i} \frac{\ell_{rs}(v_i)}{\ell_{rs}}, \tag{6}$$

where ℓ_{rs} is the total number of shortest paths from v_r to v_s and $\ell_{rs}(v_i)$ is the the number of shortest paths between v_r and v_s that pass through v_i. In networks, the greater is the number of paths that pass through a node, the greater is the importance of this node and more central it is.

3.1.2. Theoretical Robustness Metrics

Robustness can be defined as the network's ability to survive random failures or deliberate attacks consisting of the elimination of nodes and/or edges [42]. In this sense, several robustness measures have been proposed to quantitatively determine this characteristic. The most important ones are described in what follows:

The *normalized robustness indicator* $\bar{r}^T$ measures the ratio between the number of alternative paths in the network topology and the total number of stations [8]:

$$\bar{r}^T = \frac{\ln(M - N + 2)}{\ln\left(\frac{N(N-1)}{2} - N + 2\right)}. \tag{7}$$

Note that $\bar{r}^T$ is higher in the case there are alternative routes to reach a destination and it is smaller in large systems.

The *effective graph resistance* R_G estimates the robustness of a network from the number of parallel paths (i.e., redundancy) and the length of each path between each pair of nodes. The effective graph resistance is calculated in terms of the eigenvalues of the Laplacian matrix as follows:

$$R_G = N \sum_{i=1}^{N-1} \frac{1}{\mu_i}. \tag{8}$$

In this work, the normalized version of the the effective graph resistance, called *effective graph conductance* [43], is used:

$$C_G = \frac{N-1}{R_G}. \tag{9}$$

Note that $0 \leq C_G \leq 1$ and a larger C_G indicates a higher level of robustness.
The average efficiency $E[\frac{1}{H}]$ is defined as follows [44]:

$$E\left[\frac{1}{H}\right] = \frac{2}{N(N-1)} \sum_{i,j=1,i\neq j}^{N} \frac{1}{d(v_i, v_j)}. \tag{10}$$

Note that the greater is the value of the average efficiency, the greater is the robustness of the network (recall that the global efficiency of the complete network is 1).

The *clustering coefficient* is used to assess how the neighbors of a node are connected with one another [41]. For node v_i, it is mathematically defined as follows:

$$C_C(v_i) = \frac{2E_i}{d_i(d_i - 1)}, \tag{11}$$

where E_i is the number of edges linked to the neighbors of node v_i. The clustering coefficient shows the fault tolerance characteristic: in a subway network, when one station is out of function, the traffic will not be affected if the neighboring stations are connected. Thus, a larger value of C_C implies a better tolerance to fault in a local scale. The *average clustering coefficient* is the average of all the individual clustering coefficients:

$$C_{CG} = \frac{1}{N} \sum_{i=1}^{N} CC(v_i). \tag{12}$$

The *algebraic connectivity* μ_{N-1} is the second smallest eigenvalue of the Laplacian matrix A_G. It has been shown that the larger μ_{N-1} is, the higher the robustness of a network is [43]. The *normalized algebraic connectivity* is obtained dividing by the total number of nodes: $\bar{\mu}_{N-1} = \frac{\mu_{N-1}}{N}$.

The *normalized natural connectivity* $\bar{\lambda}$ is defined as:

$$\bar{\lambda} = \frac{\ln\left[\frac{1}{N} \sum_{i=1}^{N} e^{\lambda_i}\right]}{N - \ln N}, \tag{13}$$

where λ_i is the ith eigenvalue of the adjacency matrix A_G. It measures the redundancy in terms of alternative paths and is considered as a measure of structural robustness [45].

Finally, the *degree diversity* κ is defined as:

$$\kappa = \frac{\sum_{i=1}^{N} d_i^2}{\sum_{i=1}^{N} d_i}. \tag{14}$$

The greater κ is, the more nodes must be removed from the network to disintegrate it [46]. In this work, we take the inverse of the degree diversity $\bar{\kappa} = \frac{1}{\kappa}$ in order to scale the value in the interval $[0, 1]$.

3.2. Normalization and Dimensionality Reduction

Given the large differences in the number of passengers from station to station, the entries are normalized. The normalization consists in using the ratio of hourly passengers to the total number of passengers that day at each station, instead of the total amount of passengers per hour [24].

On the other hand, the number of variables used to classify the stations is large and they are also highly correlated; therefore, it was decided to perform a Principal Component Analysis (PCA). PCA is a technique for reducing the dimensionality of large datasets, increasing interpretability and minimizing information loss [47]. PCA is defined as an orthogonal linear transformation which transforms the data into a new system of coordinates such that the first coordinate (called the first principal component) represents the largest variance, the second coordinate the second greatest, etc. PCA can be thought of as fitting an n-dimensional ellipsoid to the data, where each axis of it represents a principal component. If an axis of the ellipse is small, then the variance along that axis is also small. To find the axes of the ellipse, first the mean of each variable from the dataset must be subtracted to center the data around the origin. Then, the covariance matrix of the data is computed. The covariance between two data is calculated as:

$$\sigma_{jk} = \frac{1}{n-1} \sum_{i=1}^{n} (x_{ij} - \bar{x}_j)(x_{ik} - \bar{x}_k) \tag{15}$$

The principal components are calculated from the eigen-vectors and eigenvalues of this matrix. The eigenvectors represent the directions, whereas the eigenvalues are the numbers representing how much variance there is in the data in each particular direction. The eigenvector with the highest eigenvalue is taken as the first principal component. More details can be found in the work of Dunteman [48].

3.3. Clustering Analysis

Cluster analysis is an exploratory technique which is used to classify objects into groups, known as clusters, in such way that observations belonging to a cluster are more similar to each other than observations assigned to different clusters. Nevertheless, clustering is rather a subjective statistical analysis and there are several possible algorithms that may be used. The decision of which technique to apply should be made depending on the kind of data or the type of problem to be solved. The k-means algorithm is known to be computationally fast and has the ability to handle large datasets. However, one needs to know the number of clusters in advance, it is sensitive to outliers and different initial centroids produce different results [49]. Hierarchical clustering is one of the most popular clustering techniques. Although it may be computationally slower when the dataset size increases and clusters depend on the distance metric used, the authors consider that the result of a hierarchical clustering is a structure that is more informative and interpretable than the unstructured set of flat clusters returned by k-means. Hence, it is easier to determine the optimal number of clusters by looking at the dendrogram of a hierarchical clustering than trying to predict this optimal number in advance in case of k-means. For these reasons, the agglomerative hierarchical clustering technique is used [50]. The basic algorithm consists of the following steps:

1. Initially, each observation is considered as a single-element cluster.
2. An iterative process is then initiated in which the two clusters that are the most similar are combined into a new bigger cluster. This is done by computing the dissimilarities between every pair of observations. This procedure is iterated until all points are members of one single big cluster.
3. Finally, one needs to determine where to cut the hierarchical tree into clusters. This creates a partition of the data.

The distance between clusters can be calculated using different methods [51,52]. In this study, the Ward method was used, which has been very widely used since its first description by Ward Jr [53], it and has outperformed other methods in several comparison studies [54,55]. The Ward method is the only one among the agglomerative clustering methods that is based on a classical sum-of-squares criterion, producing groups that minimize within-group dispersion at each binary [56]. In the Ward method, the distance between two clusters, *A* and *B*, is how much the sum of squares will increase once they are merged:

$$\Delta(A, B) \;=\; \sum_{i \in A \cup B} \| \vec{x}_i - \vec{m}_{A \cup B} \|^2 - \sum_{i \in A} \| \vec{x}_i - \vec{m}_A \|^2 - \sum_{i \in B} \| \vec{x}_i - \vec{m}_B \|^2$$

$$= \;\; \frac{n_A n_B}{n_A + n_B} \| \vec{m}_A - \vec{m}_B \|^2, \tag{16}$$

where $\vec{m}_j$ is the center of cluster j and n_j is the number of points in it. Δ is called the merging cost of combining the clusters A and B. In this method, in each step, the variability within clusters is minimized.

In addition, the agglomerative coefficient (AC), measuring the clustering structure of the dataset, is calculated [57]. For each observation i, let $m(i)$ represent its dissimilarity to the first cluster it is merged with, divided by the dissimilarity of the merger in the final step of the algorithm. The AC is the average of all $1 - m(i)$. Generally speaking, the AC describes the strength of the clustering structure that has been obtained by group average linkage. However, the AC tends to become larger when n increases, so it should not be used to compare datasets of very different sizes. The coefficient takes values from 0 to 1, and it is actually the mean of the normalized lengths at which the clusters are formed. A coefficient close to 1 points to a pretty reasonable cluster structure in the data.

4. Mathematical and Statistical Analysis

4.1. Structural Network Analysis

As previously mentioned, the topology of Barcelona subway network is established using the L-space method, where each station stands for a node of the graph and the edges are defined by means of the direct connections by rail ways between the stations. The number of nodes is $N = 151$ and the number of edges is $M = 177$ and therefore the density of the subway network is $d \approx 0.0157$. In Figure 6, the graph corresponding to Barcelona subway network using Mathematica is shown (note that the exact placing and positioning of the stations is not taken into account).

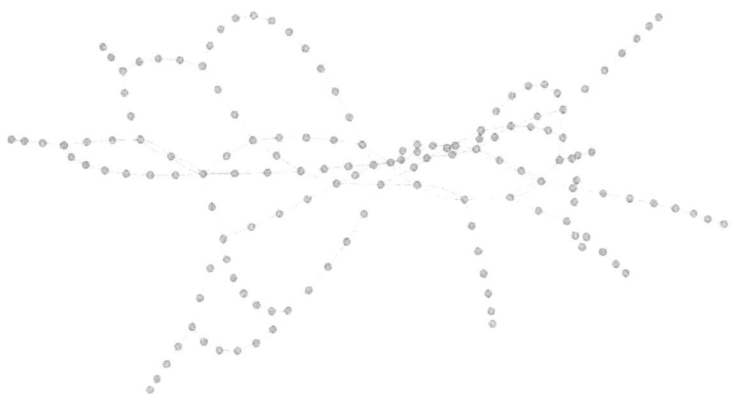

Figure 6. The graph representing the Barcelona subway network using Mathematica.

4.1.1. Basic Structural Characteristics

In this subsection, the most usual coefficients and centrality measures, introduced in Section 3.1.1, are computed and associated to the Barcelona subway network.

As shown in Table 1 the five stations with the highest degree are "Passeig de Gràcia" with degree 6 and "Diagonal", "Espanya", "Catalunya" and "La Sagrera" with degree 5.

Note that the first four stations belong to Line 3; in addition, three of the top five are on Line 1.

Table 1. The five stations with the highest degree.

Station	Subway Lines	Degree
Passeig de Gràcia	2, 3, 4	6
Diagonal	3, 5	5
Espanya	1, 3	5
Catalunya	1, 3	5
La Sagrera	1, 5, 9N, 10N,	5

The average degree of the network is $E[D] \approx 2.2649$ and the degree distribution $p(k)$ is shown in Figure 7, while the cumulative degree distribution is illustrated in Figure 8. A simple calculus shows that the fitting function of the cumulative degree distribution is $h(x) = 4.0834e^{-1.4796x}$.

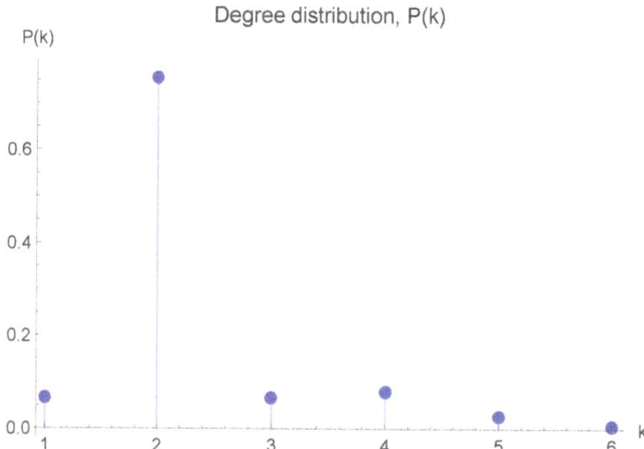

Figure 7. Degree distribution of Barcelona subway network.

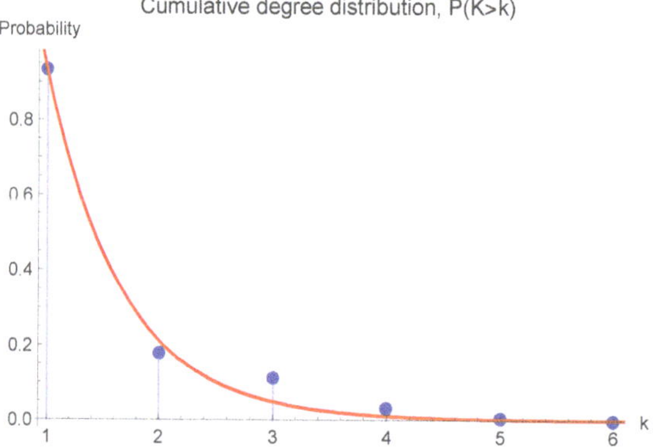

Figure 8. Cumulative degree distribution of Barcelona subway network.

The maximum travel distance of the network is no more than 31 stops (diameter), while the average shortest path is 11.0032 stops.

Table 2 shows the results obtained from the computation of the closeness centrality. The station with the highest closeness centrality is "Diagonal" with $C_{CL} \approx 0.1424$, and the next four stations ("Verdaguer", "Hospital Clínic", "Passeig de Gràcia" and "Provença") have similar closeness centrality. In this case, the most centrality subway line is Line 5 and, to a lesser extent, Line 3.

Table 2. The five stations with the highest closeness centrality.

Station	Subway Lines	Closeness Centrality
Diagonal	3, 5	0.1424
Verdaguer	4, 5	0.1372
Hospital Clìnic	5	0.1362
Passeig de Gràcia	2, 3, 4	0.1358
Provença	3, 5, 6, 7	0.1323

Finally, the results obtained when the betweenness centrality was computed are displayed in Table 3. It is important to note that all the stations with the highest coefficient belong to Line 5.

Table 3. The five stations with the highest betweenness centrality.

Station	Subway Lines	Betweenness Centrality
Diagonal	3, 5	0.4298
Verdaguer	4, 5	0.3333
Sants Estaciò	3, 5	0.2610
Hospital Clìnic	5	0.2593
Entença	5	0.2553

From these results, it can be seen that some specific stations play a central role in the structural definition of the network. For example, "Diagonal" and "Verdaguer" are very important structural pieces of the subway network since they have the highest values of closeness and betweenness centralities. In addition the most central lines are Lines 5, 3 and 1.

4.1.2. Network Robustness

Failures of subway networks can have enormous impact on our society, so the analysis of the robustness is very important when studying subway networks. The robustness of networks reflects the extent to which the networks can solve possible (intentional or unintentional) failures by offering alternative routes that overcome the attacked edges or nodes.

In this section, eight robustness metrics (introduced in Section 3.1.2 are computed for the Barcelona subway network and compared with those obtained for the Madrid subway network.

In Table 4, the stations with the highest clustering centrality are illustrated. The most central are "Catalunya" ($C_C = 0.2$), "Universitat" and "Urquinaona" with $C_C \approx 0.1666$ and "Passeig de Gràcia" with $C_C \approx 0.1333$. As a consequence, they have better tolerance to fault in a local scale. The first three stations belong to Line 1, and Lines 2–4 have a couple of stations on this list. Moreover, the mean clustering coefficient is 0.0044, which is significantly lower than that of other metro networks such as London ($C_C = 0.0409$), Tokyo ($C_C = 0.0285$) or Paris ($C_C = 0.0163$) [58].

Table 5 shows the values of the eight robustness metrics computed using Equations (7)–(14) for the Barcelona subway network and the Madrid subway network [59].

According to the reduced robustness indicator $\bar{r}^T$, the Barcelona metro network is slightly more robust than the Madrid metro network, probably because there are more alternative paths between any pair of nodes.

Table 4. Stations with non-zero clustering centrality.

Station	Subway Lines	Clustering Centrality
Catalunya	1, 3	0.2
Universitat	1, 2	0.1666
Urquinaona	1, 4	0.1666
Passeig de Gràcia	2, 3, 4	0.1333

According to the effective graph conductance C_G, the Barcelona subway network also has a slightly higher value than that of Madrid. Note that the effective graph conductance takes into account not only the number of alternative paths but also the length of each alternative path, hence effective graph conductance favors networks with the smallest length of the shortest paths.

In general, according to all the metrics except the clustering coefficient C_{CG} and the normalized degree diversity $\overline{\kappa}$, Barcelona has a higher robustness level than Madrid.

Table 5. Robustness metrics in Barcelona and Madrid subway networks.

Coefficients	Barcelona Subway	Madrid Subway
Nodes, N	151	243
Edges, M	177	280
Normalized robust indicator, $\overline{r}^T$	0.35747	0.35635
Efective graph conductance, C_G	0.00221	0.00086
Average efficiency, $E[\frac{1}{H}]$	0.13524	0.10533
Average clustering coefficient, C_{CG}	0.00441	0.00774
Normalized algebraic connectivity, $\overline{\mu}_{N-1}$	0.00006	0.00001
Normalized average degree, $\overline{E}[D]$	0.01562	0.00952
Normalized natural connectivity, $\overline{\lambda}$	0.00770	0.00441
Normalized degree diversity, $\overline{\kappa}$	0.35975	0.37135

4.2. Data Analysis Results

Principal component analysis was performed to study the data from the working days (Monday to Friday) of the selected week. The first three principal components are able to explain 66.32% of the variability in the data (PC1 = 46.56%, PC2 = 12.88% and PC3 = 6.89%). Figure 9 shows the total variability explained by each principal component.

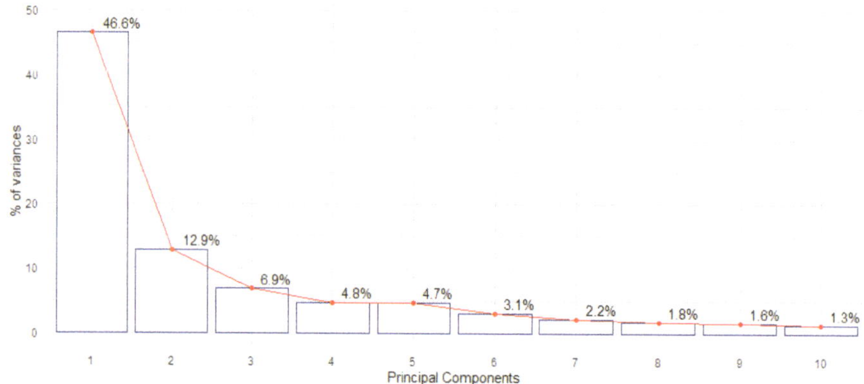

Figure 9. Total variance explained by each principal component (weekdays).

In Figure 10, the top plot shows the contributions of 18 variables to the first three components. The six variables which most contribute to each component are chosen. In the bottom plot, the correlations of these 18 variables to each component are shown. The contribution is represented both by the color scale and the circle size, while, for the correlation, the direction of the correlation is represented by color and the circle size represents the strength of the relationship. The variables which contribute the most to the first component are those corresponding to 7 a.m., and they are strongly negatively correlated with it. Regarding the second component, the variables which contribute the most are the ones corresponding to 11 a.m. and noon. Finally, the variables contributing to the third component are the ones from 1 and 11 p.m. The second and third components have a positive correlation with the variables that contribute the most to them.

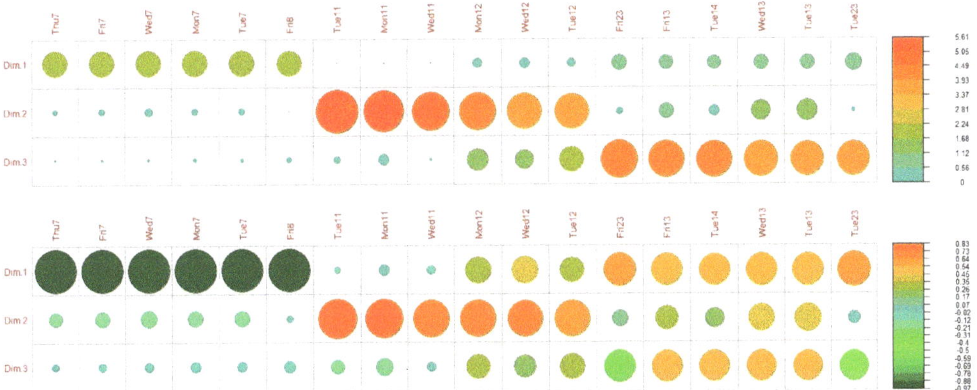

Figure 10. Contributions (**top**) and correlation (**bottom**) for the first three components.

A hierarchical cluster analysis was performed over the coordinates from the first three principal components. The resulting AC is 0.9811, which indicates a pretty reasonable cluster structure in the data. The dendrogram in Figure 11 shows that two clustering solutions are possible. The four-cluster solution is chosen as it provides a more detailed segmentation of the stations.

Statistical properties of the four clusters are summarized in Table 6. The diameters represent the maximum within cluster distances. The average and median distances are the within cluster average and median distances. Separation is the minimum distance of a point in the cluster to a point of another cluster and average to other is the average distance of a point in the cluster to the points of other clusters.

Table 6. Statistical properties of the four clusters (weekdays).

Cluster	1	2	3	4
Size	49	41	35	4
Diameter	17.49	9.39	9.32	8.46
Average distance	7.55	4.15	3.58	5.42
Median distance	7.30	4.05	3.45	5.21
Separation	1.64	1.08	1.08	7.70
Average to other	12.76	9.32	11.79	19.93

In Table 7, the stations belonging to each cluster are listed. For a better understanding of the clusters, the different stations of each cluster are located in the Barcelona map, making use of a Voronoi diagram (based on Euclidean distance) to partition the city map. In Figure 12, each Voronoi cell representing a station is colored by cluster. It may be noted that stations from the same cluster are not necessarily close in space, but their behavior pattern is similar. This may be due to, e.g., the business activities taking place in the area or being residential neighborhoods.

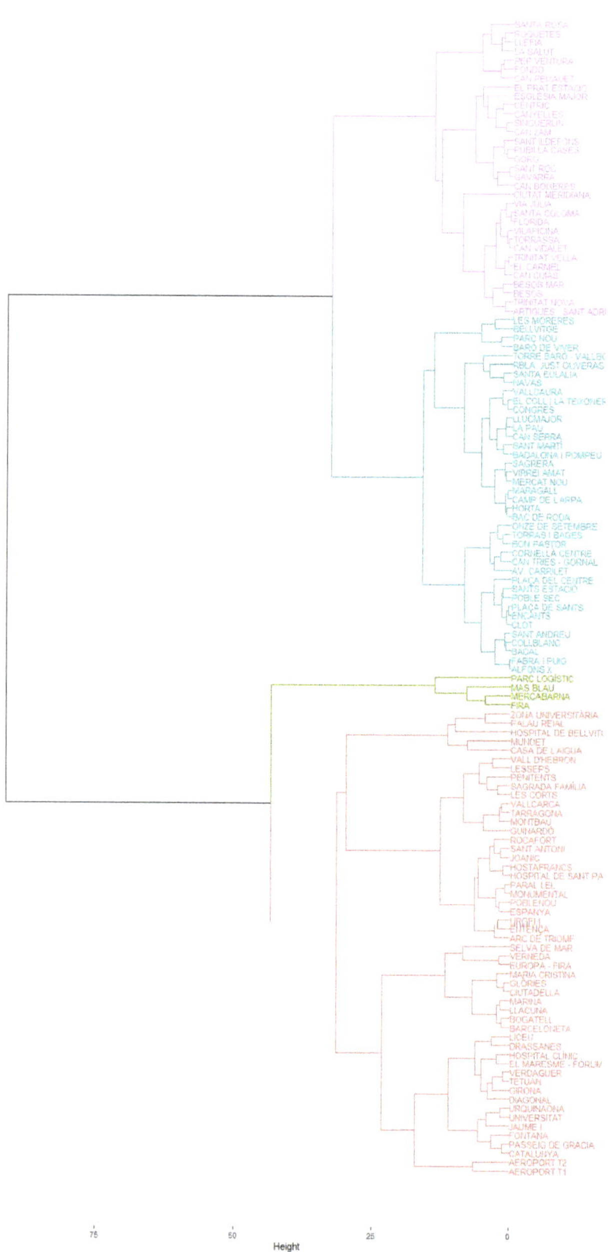

Figure 11. Clusters: Hierarchical clustering.

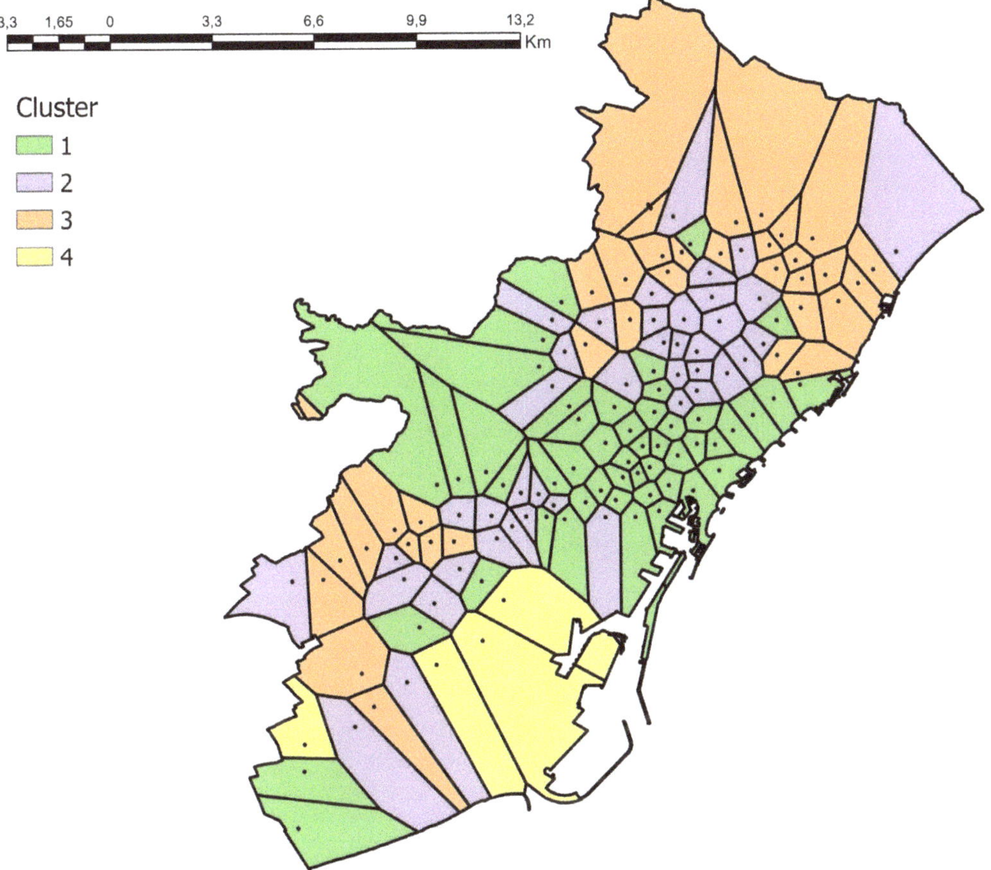

Figure 12. Map of the different stations colored by cluster (weekdays).

In the case of Cluster 1, most of the stations are located in the district of L'Eixample, Ciutat Vella and Sant Martí, where some of the most popular beaches of Barcelona are located, as well as important monuments such as Casa Milà, popularly known as *La Pedrera*, the Cathedral, Park Güell and Casa Batlló. Moreover, this cluster includes the zoo, the Maritime Museum of Barcelona and the museum *Poble Espanyol*. The hospital stations Vall d'Hebron, Hospital Clínic, Sant Pau and Hospital de Bellvitge are included in this cluster too, as well as those belonging to university campuses, such as Mundet, Palau Reial, Universitat and Zona Universitària. There are also two stations from the airport and some stations from the districts Les Corts, Sants, Montjuic and Gracia, all of them located in the city center. In Figure 13, passenger flow per hour is shown for some of the stations in Cluster 1. All of them have peak hours at 8 a.m., 2 p.m. and 7 p.m.

Table 7. Results of station classification.

Cluster	Stations	Number
Cluster 1	Aeroport T1, Aeroport T2, Arc de Triomf, Barceloneta, Bogatell, Casa de l'aigua, Catalunya, Ciutadella, Diagonal, Drassanes, El Maresme-Fórum, Entença, Espanya, Europa-Fira, Fontana, Girona, Glories, Guinardó, Hospital Cliníc, Hospital de Bellvitge, Hospital de Sant Pau, Hostafrancs, Jaume I, Joanic, Les Corts, Lesseps, Liceu, LLacuna, Maria Cristina, Marina, Monumental, Mundet, Palau Reial, Parallel, Passeig de Gràcia, Penitents, Poble-nou, Rocafort, Sagrada Familia, Sant Antoni, Selva de Mar, Tetuan, Universitat, Urgell, Urquinaona, Vall d'Hebron, Verdaguer, Verneda, Zona Universitària	49
Cluster 2	Alfons X, Av. Carrilet, Bac de roda, Badal, Badalona - Pompeu Fabra, Baró de viver, Bellvitge, Bon pastor, Camp de l'arpa, Can tries - Gornal, Clot, Collblanc, Congrés, Cornellà Centre, El Coll - La Teixonera, Encants, Fabra I Puig, Horta, La Pau, Les Moreres, Llucmajor, Maragall, Mercat Nou, Montbau, Navas, Onze De Setembre, Parc Nou, Plaça De Sants, Plaça Del Centre, Poble Sec, Rbla. Just Oliveras, Sagrera, Sant Andreu, Sant Martí, Santa Eulàlia, Sants Estació, Tarragona, Torras I Bages, Torre Baró - Vallbona, Vallcarca, Virrei Amat	41
Cluster 3	Artigues - Sant Adrià, Besòs, Besòs Mar, Can Boixeres, Can Cuiós, Can Peixauet, Can Serra, Can Vidalet, Can Zam, Canyelles, Cèntric, Ciutat Meridiana, El Carmel, El Prat Estació, Església Major, Florida,Fondo, Gavarra, Gorg, La Salut, Llefià, Pep Ventura, Pubilla Cases, Roquetes, Sant Ildefons, Sant Roc, Santa Coloma, Santa Rosa, Singuerlín, Torrassa, Trinitat Nova, Trinitat Vella, Valldaura, Via Júlia, Vilapicina	35
Cluster 4	Fira, Mas Blau, Mercabarna, Parc Logístic	4

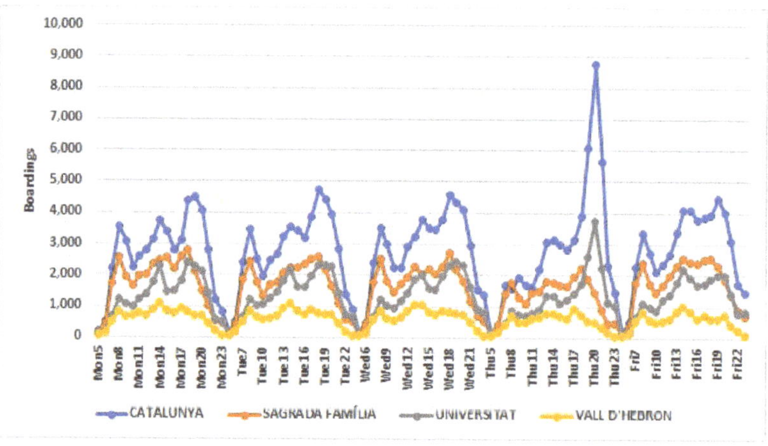

Figure 13. Pattern of boardings in stations of Cluster 1 (weekdays).

The stations in Cluster 2 are mainly around the central area of Barcelona, with some in the north and some in the south. These are traditional, residential, well-communicated neighborhoods, with many markets and shops. The stations in the north are from the districts Sant Andreu, Horta-Ginardó and Nou Barris. The stations in the south belong to L'Eixample and are the furthest from the city center together with the stations from Sant Andreu, one of the entrances to Barcelona with a large cultural and sports offer. The hours with the largest number of passengers in this cluster are 7 a.m., 8 a.m. and 6 p.m. The pattern of boarding per hour for some stations in this cluster is shown in Figure 14.

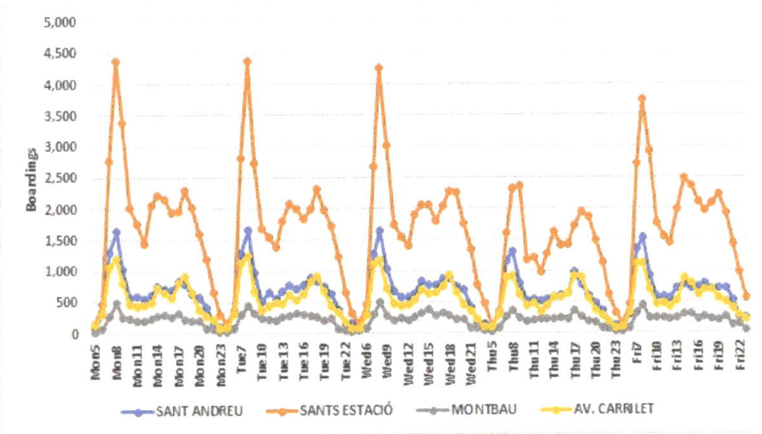

Figure 14. Pattern of boarding in stations of Cluster 2 (weekdays).

Cluster 3 contains mostly stations located outside of the city. There are two stations in El Prat de Llobregat and eight bordering the north side of L'Hospitalet de Llobregat. The rest are gathered in the north urban periphery of the city, linking to different small municipalities or towns, such as Badalona, Santa Coloma de Gramenet or Sant Adrià de Besòs. These belong to what is known as the metropolitan area of Barcelona, which is a geographical area that goes beyond the administrative area. Given the growth of the city of Barcelona, some of these municipalities are now essentially suburbs of Barcelona. Badalona is, however, the third largest city in Catalonia. Moreover, there are also stations in Ciutat Meridiana, which is the poorest neighborhood of the city. In Figure 15, the peak hours of the stations of this cluster can be seen. The hours with the highest number of boardings are 8 a.m., 2 p.m. and 6 p.m.

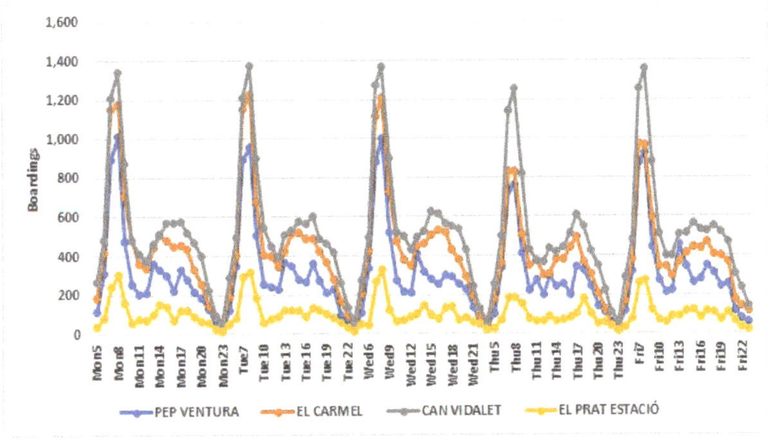

Figure 15. Pattern of boarding in stations of Cluster 3 (weekdays).

The stations that form Cluster 4 have the particular characteristics of the area they give access to: Fira is the entry to one of the largest and most modern fairgrounds of Europe; Mas Blau corresponds with the industrial park closest to Barcelona's airport; Mercabarna is considered the most important central market in Europe, as it is a reference center in the Mediterranean Sea for the distribution of fresh products at the international level; and Parc Logístic serves the logistics park of the city's Free economic zone. Overall, 2 p.m., 5 p.m.

and 6 p.m. have the highest number of boardings. The peak hours of these stations are shown in Figure 16.

All the analysis here presented were performed with RStudio Team [60].

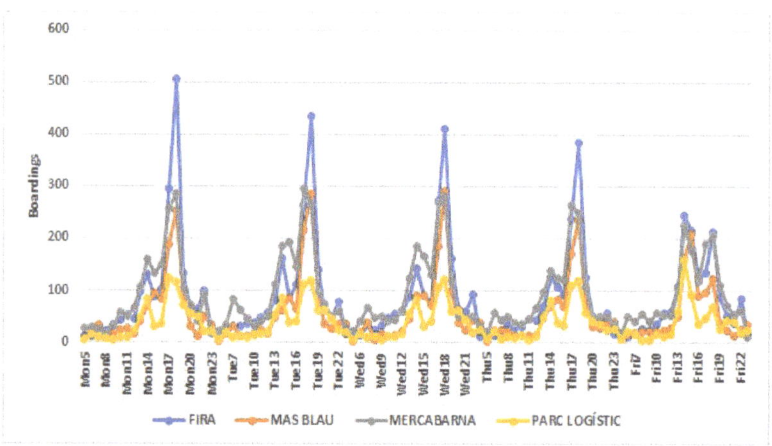

Figure 16. Pattern of boarding stations of Cluster 4 (weekdays).

5. Conclusions

In Barcelona, as in any major urban area, many people use the public transport network, which is why it is necessary to have as much information as possible to forecast and plan the subway trip.

Moreover, in the bibliography studied, there are no previous studies that analyze not only the structural and robustness characteristics but also travel patterns of the Barcelona metro network.

In this study, a detailed analysis of Barcelona subway network was done using Complex Network Analysis. To achieve this goal, the most important centrality measures and coefficients were computed. In this sense, the important role of stations such as "Diagonal" and "Verdaguer" to control the flow of passengers was shown. It was also shown that the stations "Catalunya", "Universitat", "Urquinaona" and "Passeig de Gràcia" have high fault tolerance in a local scale. Moreover, L5 and L3 are the most central subway lines.

In addition, the robustness of the Barcelona subway network was investigated by analyzing several robustness metrics and compared with the robustness of the Madrid subway network. The results indicate that the Barcelona subway network is slightly more robust than the Madrid subway network according to most of the robustness metrics. A previous study [8] analyzed Barcelona subway robustness using ten theoretical robustness metrics, but only taking into account terminals and transfer stations. The results in the former study cannot be compared with ours since in our study all Barcelona subway stations are used.

The data collected at the entry of the metro stations in Barcelona provide a vast quantity of data with very valuable information about the ridership patterns in them. The set of real data was provided by the Barcelona Metropolitan Network, providing information on the number of entries per hour in each of the 151 stations. There are no data related to the passenger's journey or personal data (age, sex, fare, etc.).

The statistical techniques used in this study allowed observing the following: in the first place, there are differences in behavior between working days, which are highly correlated with each other, and over the weekend, with which the correlation decreases. The hours with the highest number of passengers correspond mainly to the hours of entry and exit of work and school hours. However, these rush hours are not the same at all stations, nor are the number of passengers each have, reaching a difference of more than

54,000 daily entries between some stations. It is because of this reason that the data were normalized, using the proportion of passengers per hour with respect to the total number of entries in that particular day at each particular station.

The principal component analysis performed reduced the dimensionality of the dataset. The first three principal components explain most of the variability in the data. Moreover, it was observed which hours have a higher effect in each of them.

The cluster analysis carried out revealed, for working days, the existence of four groups with similar characteristics. The first conglomerate gathers the stations of the downtown area, the most touristic and monumental. In the second cluster, the stations that surround the center of Barcelona are grouped. They are, mainly, traditional and residential neighborhoods. The periphery stations, which link the center with the nearest municipalities, are those found in the third cluster. In the fourth cluster, the stations of the fairgrounds, large markets and logistics parks appear. Within each cluster, one can see the same pattern of behavior that reflects the similarities of the stations that form it, as can be seen at peak times, which differ between clusters.

The patterns observed reflect the daily activities of the urban area of Barcelona, which are related to the spatial structuring of the city and its characteristics, and are highly correlated with general daily routines.

The results of this work provide relevant information for the "Transports Metropolitan of Barcelona" company for public transport planning. These studies allow us to discover patterns of behavior needed to make decisions to improve the metro service. Nowadays, in the new post-pandemic normality, it is imperative to travel safely so as to stop the coronavirus spreading. It is important to avoid rush hours travels; people may choose to get on and off at subway stations with fewer travelers and do part of their journey by foot. Moreover, it is the task of public transport companies to increase the number of subway cars at a certain time if it gets too crowded, improve the infrastructure of stations with high passenger flow and reduce the time in-between metro services, among other security measures. For instance, the station "Sant Andreu", from Cluster 2, has the highest number of passengers between 7:00 and 8:00 a.m., and, therefore, it is one of the stations where increasing the number of subway cars or the frequency of the service would be imperative. On the other hand, the station "Fira", from Cluster 4, has peak hours at 14:00, 17:00 and 18:00 (p.m.), although with a much smaller number of passengers than "Sant Andreu", and, thus, depending on the capacity of the station, the measures may not be as crucial as in the first one.

Future work involves relating these results to population, climate and economic variables that reflect other social circumstances that may influence the characteristics of the metro network stations. Moreover, annual data shall be analyzed to detect seasonality in behavior patterns. Further lines of investigations will also include a structural and robustness analysis of the network, using complex network analysis to determine critical nodes using different centrality measures. In addition, a detailed analysis of the structural characteristics of this subway network considering other different topological representations such as reduced L-space, P-space, C-space, etc. must be tackled. In addition, a theoretical framework must be proposed in which the notion of "subway line" is used as the basis to define new structural and robustness coefficients. Furthermore, additional transport lines (light rail network, bus network, etc.), can be considered in the analysis to obtain more realistic results. It would also be interesting to analyze the data post-COVID-19 and compare how the use of the public transport has changed, once the data become available.

Author Contributions: Conceptualization, E.F.B., M.T.S.M. and A.M.d.R.; methodology, E.F.B., I.M.-C. and R.C.V.; software, I.M.-C.; writing—original draft preparation, I.M.-C., E.F.B., A.M.d.R. and A.B.G.-G.; and writing—review and editing, M.T.S.M., A.M.d.R., R.C.V. All authors have read and agreed to the published version of the manuscript.

Funding: This research was funded by Ministerio de Ciencia, Innovación y Universidades (MCIU, Spain), Agencia Estatal de Investigación (AEI, Spain), and Fondo Europeo de Desarrollo Regional

(FEDER, UE) under project NOTREDAMME and by Scientific Research Grant of the "Fundación Memoria D. Samuel Solórzano Barruso", University of Salamanca.

Data Availability Statement: Not Applicable.

Acknowledgments: The authors extend their gratitude to the Transport Metropolitans of Barcelona.

Conflicts of Interest: The authors declare no conflict of interest. The funders had no role in the design of the study; in the collection, analyses, or interpretation of data; in the writing of the manuscript, or in the decision to publish the results.

Abbreviations

The following abbreviations are used in this manuscript:

AFC	Automated Fare Collection
AC	Agglomerative Coefficient
PCA	Principal Component Analysis

References

1. Pternea, M.; Kepaptsoglou, K.; Karlaftis, M.G. Sustainable urban transit network design. *Transp. Res. Part A Policy Pract.* **2015**, *77*, 276–291. [CrossRef]
2. Latora, V.; M, M. Is the Boston subway a small-world network? *Phys. A* **2002**, *314*, 109–113. [CrossRef]
3. Lu, H.P.; Shi, Y. Complexity of public transport network. *Tsinghua Sci. Technol.* **2007**, *12*, 204–213. [CrossRef]
4. Zhang, J.H.; Zhao, M.W.; Liu, H.K.; Xu, X.M. Networked characteristics of the urban rail transit networks. *Phys. A* **2013**, *392*, 1538–1546. [CrossRef]
5. Liu, Z.; Song, R. Reliability analysis of Guangzhou subway with complex network theory. *J. Transp. Syst. Eng. Inf. Technol.* **2010**, *10*, 194–200.
6. Cats, O. Topological evolution of a metropolitan rail transport network: The case of Stockholm. *J. Transp. Geogr.* **2017**, *62*, 172–183. [CrossRef]
7. Derrible, S.; Kennedy, C. The complexity and robustness of metro networks. *Phys. A* **2010**, *389*, 3678–3691. [CrossRef]
8. Wang, X.; Koc, Y.; Derrible, S.; Ahmad, S.N.; Kooij, R.E. Multi-criteria robustness analysis of metro networks. *Phys. A* **2017**, *474*, 19–31. [CrossRef]
9. Zhang, J.H.; Xu, X.M.; Hong, L.; Wang, S.; Fei, Q. Networked analysis of the Shanghai subway network, in China. *Phys. A* **2011**, *390*, 4562–4570. [CrossRef]
10. Forero-Ortiz, E.; Martinez-Gomariz, E.; Canas Porcuna, M.; Locatelli, L.; Russo, B. Flood Risk Assessment in an Underground Railway System under the Impact of Climate Change-A Case Study of the Barcelona Metro. *Sustainability* **2020**, *12*, 5291.10.3390/su12135291. [CrossRef]
11. De Bona, A.; de Oliveira Rosa, M.; Ono Fonseca, K.; Lüders, R. A reduced model for complex network analysis of public transportation systems. *Phys. A Stat. Mech. Its Appl.* **2021**, *567*, 125715. [CrossRef]
12. Wang, Y.; Tian, C. Measure Vulnerability of Metro Network under Cascading Failure. *IEEE Access* **2021**, *9*, 683–692. [CrossRef]
13. Dempsey, P.S. *Privacy Issues with the Use of Smart Cards*; The National Academies Press: Washington, DC, USA, 2007.
14. Pelletier, M.P.; Trépanier, M.; Morency, C. Smart card data use in public transit: A literature review. *Transp. Res. Part C Emerg. Technol.* **2011**, *19*, 557–568. [CrossRef]
15. Li, T.; Sun, D.; Jing, P.; Yang, K. Smart card data mining of public transport destination: A literature review. *Information* **2018**, *9*, 18. [CrossRef]
16. Alsger, A.; Tavassoli, A.; Mesbah, M.; Ferreira, L.; Hickman, M. Public transport trip purpose inference using smart card fare data. *Transp. Res. Part C Emerg. Technol.* **2018**, *87*, 123–137. [CrossRef]
17. Alexander, L.; Jiang, S.; Murga, M.; González, M.C. Origin-destination trips by purpose and time of day inferred from mobile phone data. *Transp. Res. Part C Emerg. Technol.* **2015**, *58*, 240–250. [CrossRef]
18. Jun, C.; Dongyuan, Y. Estimating smart card commuters origin-destination distribution based on APTS data. *J. Transp. Syst. Eng. Inf. Technol.* **2013**, *13*, 47–53. [CrossRef]
19. Gonzalez, M.C.; Hidalgo, C.A.; Barabasi, A.L. Understanding individual human mobility patterns. *Nature* **2008**, *453*, 779–782. [CrossRef]
20. Briand, A.S.; Côme, E.; Trépanier, M.; Oukhellou, L. Analyzing year-to-year changes in public transport passenger behaviour using smart card data. *Transp. Res. Part C Emerg. Technol.* **2017**, *79*, 274–289. [CrossRef]
21. El Mahrsi, M.K.; Come, E.; Oukhellou, L.; Verleysen, M. Clustering Smart Card Data for Urban Mobility Analysis. *IEEE Trans. Intell. Transp. Syst.* **2017**, *18*, 712–728. [CrossRef]
22. Chen, C.; Chen, J.; Barry, J. Diurnal pattern of transit ridership: A case study of the New York City subway system. *J. Transp. Geogr.* **2009**, *17*, 176–186. [CrossRef]

23. Wang, W.; Lo, S.; Liu, S. Aggregated metro trip patterns in urban areas of Hong Kong: Evidence from automatic fare collection records. *J. Urban Plan. Dev.* **2015**, *141*, 05014018. [CrossRef]
24. Kim, M.K.; Kim, S.P.; Heo, J.; Sohn, H.G. Ridership patterns at subway stations of Seoul capital area and characteristics of station influence area. *KSCE J. Civ. Eng.* **2017**, *21*, 964–975. [CrossRef]
25. Ding, C.; Cao, X.; Liu, C. How does the station-area built environment influence Metrorail ridership? Using gradient boosting decision trees to identify non-linear thresholds. *J. Transp. Geogr.* **2019**, *77*, 70–78. [CrossRef]
26. Langlois, G.G.; Koutsopoulos, H.N.; Zhao, J. Inferring patterns in the multi-week activity sequences of public transport users. *Transp. Res. Part C Emerg. Technol.* **2016**, *64*, 1–16. [CrossRef]
27. Lu, Y.; Zhang, Y. Toward a Stakeholder Perspective on Safety Risk Factors of Metro Construction: A Social Network Analysis. *Complexity* **2020**, *2020*, 8884304. [CrossRef]
28. Zhou, C.; Kong, T.; Jiang, S.; Chen, S.; Zhou, Y.; Ding, L. Quantifying the evolution of settlement risk for surrounding environments in underground construction via complex network analysis. *Tunn. Undergr. Space Technol.* **2020**, *103*, 103490. [CrossRef]
29. Niu, K.; Fang, W.; Song, Q.; Guo, B.; Du, Y.; Chen, Y. An Evaluation Method for Emergency Procedures in Automatic Metro Based on Complexity. *IEEE Trans. Intell. Transp. Syst.* **2021**, *22*, 370–383. [CrossRef]
30. Chen, S.; Zhuang, D. Evolution and evaluation of the Guangzhou metro network topology based on an integration of complex network analysis and GIS. *Sustainability* **2020**, *12*, 538. [CrossRef]
31. Bernal, E.; del Rey, A.; Villardón, P. Analysis of madrid metro network: From structural to HJ-biplot perspective. *Appl. Sci.* **2020**, *10*, 5689. [CrossRef]
32. Moreno-Pulido, S.; Pavón-Domínguez, P.; Burgos-Pintos, P. Temporal evolution of multifractality in the Madrid Metro subway network. *Chaos Solitons Fractals* **2021**, *142*, 110370. [CrossRef]
33. Meng, Y.; Tian, X.; Li, Z.; Zhou, W.; Zhou, Z.; Zhong, M. Exploring node importance evolution of weighted complex networks in urban rail transit. *Phys. A: Stat. Mech. Its Appl.* **2020**, *558*, 124925. [CrossRef]
34. Yu, W.; Ye, X.; Chen, J.; Yan, X.; Wang, T. Evaluation indexes and correlation analysis of origination-destination travel time of Nanjing metro based on complex network method. *Sustainability* **2020**, *12*, 1113. [CrossRef]
35. Wang, J.; Ren, J.; Fu, X. Research on Bus and Metro Transfer from Perspective of Hypernetwork- A Case Study of Xi'an, China (December 2020). *IEEE Access* **2020**, *8*, 227048–227063. [CrossRef]
36. Wang, W.; Wang, Y.; Correia, G.; Chen, Y. A Network-Based Model of Passenger Transfer Flow between Bus and Metro: An Application to the Public Transport System of Beijing. *J. Adv. Transp.* **2020**, *2020*, 6659931. [CrossRef]
37. Han, Y.; Wang, S.; Ren, Y.; Wang, C.; Gao, P.; Chen, G. Predicting Station-Level Short-Term Passenger Flow in a Citywide Metro Network Using Spatiotemporal Graph Convolutional Neural Networks. *ISPRS Int. J. Geo-Inf.* **2019**, *8*, 243. [CrossRef]
38. Huh, J.; Seo, Y. Understanding Edge Computing: Engineering Evolution With Artificial Intelligence. *IEEE Access* **2019**, *7*, 164229–164245. [CrossRef]
39. Li, W.; Luo, Q.; Cai, Q. A Smart Path Recommendation Method for Metro Systems with Passenger Preferences. *IEEE Access* **2020**, *8*, 20646–20657. [CrossRef]
40. Bakıcı, T.; Almirall, E.; Wareham, J. A smart city initiative: The case of Barcelona. *J. Knowl. Econ.* **2013**, *4*, 135–148. [CrossRef]
41. Kolaczyk, E.D. *Statistical Analysis of Network Data*; Springer Science+Business Media: Berlin/Heidelberg, Germany, 2009.
42. Barabási, A.L. *Network Science*; Cambridge University Press: Cambridge, UK, 2016.
43. Van Mieghem, P. *Graph Spectra for Complex Networks*; Cambridge University Press: Cambridge, UK, 2011.
44. Newman, M.E.J. *Networks: An Introduction*; Oxford University Press: Oxford, UK, 2010.
45. Wu, J.; Barahona, M.; Tan, Y.J.; Deng, H.Z. Spectral measure of structural robustness in complex networks. *IEEE Trans. Syst. Man Cybern. Part A* **2011**, *41*, 1244–1252. [CrossRef]
46. Li, C.; Wang, H.; de Haan, W.; J, S.C.; Van Mieghem, P. The correlation of metrics in complex networks with applications in functional brain networks. *J. Stat. Mech. Theory Exp.* **2011**, *2011*, P11018. [CrossRef]
47. Jolliffe, I.T.; Cadima, J. Principal component analysis: A review and recent developments. *Philos. Trans. R. Soc. A Math. Phys. Eng. Sci.* **2016**, *374*, 20150202. [CrossRef] [PubMed]
48. Dunteman, G.H. *Principal Components Analysis*; Number 69 in Quantitative Applications in the Social Sciences; Sage Publications: Thousand Oaks, CA, USA, 1989.
49. Govender, P.; Sivakumar, V. Application of k-means and hierarchical clustering techniques for analysis of air pollution: A review (1980–2019). *Atmos. Pollut. Res.* **2020**, *11*, 40–56. [CrossRef]
50. Johnson, S.C. Hierarchical clustering schemes. *Psychometrika* **1967**, *32*, 241–254. [CrossRef]
51. Bouguettaya, A.; Yu, Q.; Liu, X.; Zhou, X.; Song, A. Efficient agglomerative hierarchical clustering. *Expert Syst. Appl.* **2015**, *42*, 2785–2797. [CrossRef]
52. Day, W.H.; Edelsbrunner, H. Efficient algorithms for agglomerative hierarchical clustering methods. *J. Classif.* **1984**, *1*, 7–24. [CrossRef]
53. Ward, J.H, Jr. Hierarchical grouping to optimize an objective function. *J. Am. Stat. Assoc.* **1963**, *58*, 236–244. [CrossRef]
54. Mojena, R. Hierarchical grouping methods and stopping rules: An evaluation. *Comput. J.* **1977**, *20*, 359–363. [CrossRef]
55. Blashfield, R.K. Mixture model tests of cluster analysis: Accuracy of four agglomerative hierarchical methods. *Psychol. Bull.* **1976**, *83*, 377. [CrossRef]

56. Murtagh, F.; Legendre, P. Ward's hierarchical agglomerative clustering method: Which algorithms implement Ward's criterion? *J. Classif.* **2014**, *31*, 274–295. [CrossRef]
57. Kaufman, L.; Rousseeuw, P.J. *Finding Groups in Data: An Introduction to Cluster Analysis*; John Wiley & Sons: Hoboken, NJ, USA, 2009; Volume 344.
58. Wu, X.; Tse, C.; Dong, H.; Ho, I.; Lau, F. A Network Analysis of World's Metro Systems. In Proceedings of the 2016 International Symposium on Nonlinear Theory and Its Applications (NOLTA2016), Yugawara, Japan, 27–30 November 2016; The Institute of Electronics, Information and Communication Engineers: Tokyo, Japan, 2016; pp. 606–609.
59. Frutos Bernal, E.; Martín del Rey, A. Study of the Structural and Robustness Characteristics of Madrid Metro Network. *Sustainability* **2019**, *11*, 3486. [CrossRef]
60. RStudio Team. *RStudio: Integrated Development Environment for R*; RStudio, PBC: Boston, MA, USA, 2020.

 *electronics*

MDPI

Article

Two Advanced Models of the Function of MRT Public Transportation in Taipei

You-Shyang Chen [1,*], Chien-Ku Lin [2,3], Su-Fen Chen [4,*] and Shang-Hung Chen [1]

[1] Department of Information Management, Hwa Hsia University of Technology, New Taipei City 235, Taiwan; trtc042@gmail.com
[2] Department of Business Management, Hsiuping University of Science and Technology, Taichung City 412, Taiwan; cklin@hust.edu.tw
[3] Department of Computer Science and Information Management, Hungkuang University, Taichung City 433304, Taiwan
[4] National Museum of Marine Science & Technology, Keelung City 202010, Taiwan
[*] Correspondence: ys_chen@go.hwh.edu.tw (Y.-S.C.); 10201a@gapps.nou.edu.tw (S.-F.C.)

Abstract: Tour traffic prediction is very important in determining the capacity of public transportation and planning new transportation devices, allowing them to be built in accordance with people's basic needs. From a review of a limited number of studies, the common methods for forecasting tour traffic demand appear to be regression analysis, econometric modeling, time-series modeling, artificial neural networks, and gray theory. In this study, a two-step procedure is used to build a predictive model for public transport. In the first step of this study, regression analysis is used to find the correlations between two or more variables and their associated directions and strength, and the regression function is used to predict future changes. In the second step, the regression analysis and artificial neural network methods are assessed and the results are compared. The artificial neural network is more accurate in prediction than regression analysis. The study results can provide useful references for transportation organizations in the development of business operation strategies for managing sustainable smart cities.

Keywords: passenger traffic; artificial neural network; regression analysis

Citation: Chen, Y.-S.; Lin, C.-K.; Chen, S.-F.; Chen, S.-H. Two Advanced Models of the Function of MRT Public Transportation in Taipei. *Electronics* **2021**, *10*, 1048. https://doi.org/10.3390/electronics10091048

Academic Editors: Juan M. Corchado, Josep L. Larriba-Pey, Pablo Chamoso and Fernando De la Prieta

Received: 2 April 2021
Accepted: 26 April 2021
Published: 29 April 2021

1. Introduction

Taipei has the highest population density and traffic capacity in Taiwan. The construction of the Mass Rapid Transit (MRT) [1] system has relieved long-standing traffic problems in Taipei's urban area. In addition to its safety and convenience, the public MRT may also control the growth of private vehicles, reduce carbon emissions, and save energy, allowing for the creation of a low-carbon and green energy-based city. The preliminary road network of the public MRT system in the urban area of Taipei was approved by the Executive Yuan in 1986. The Taipei City Government established the Department of Rapid Transit Systems in 1987 and launched the construction of a preliminary road network or extension, following a revision. Taipei Rapid Transit Corporation was incorporated in 1994. The Muzha Line, the first driverless medium-capacity rapid transit line in Taiwan, was opened in March 1996, turning over a new leaf for public transportation in Taiwan. In March 1997, the first high-passenger-capacity system, the Tamsui Line, was opened, with a service scope that extended from Taipei City to New Taipei City. Following the continuous opening of road networks, 21 administrative regions across Taipei (with 12) and New Taipei City (with 9) all came to be included in the MRT routes after the opening of the Songshan Line in November 2014. According to the statistics of the department of the account of the Ministry of Transportation and Communications (MOTC), for traffic indicators from January to June 2015, the daily passenger capacity was 1.943 million on average, indicating that the MRT system is frequently used now and poses an interesting/important positive issue relating to traffic volume forecasting. However, autonomous vehicles (or self-driving cars) pose another challenging public transport-related issue, and they have been the subject of much research in recent years. According to the study of Wiseman [2], although it is noteworthy that the overall consequences of autonomous vehicles are still unknown, their impacts on other means of transportation and on the transportation infrastructure remain unclear. In particular, the point at which autonomous vehicles penetrated the transportation market needs to be determined, and more information about them needs to be available, as this information will enable more profound studies about the future. More seriously, Wiseman [3] indicated that public transportation systems will soon be obsoleted due to the benefits of autonomous vehicles. While this serious topic is valuable for sustainability studies, it is not the focus of this study. This study is concerned with traffic forecasting.

It is problematic that the rising number of passengers affects the passenger traffic of each MRT station, resulting in the spread or concentration of some passengers in certain transit stations. Therefore, passenger traffic forecasting and analysis are required. In addition to providing a train control and transportation plan, the data on MRT passenger traffic, with the e-ticket or passenger traffic information about the surrounding High-Speed Rail (HSR) [4], Taiwan Railway, and city bus, may also be used as the basis of the public transportation system plan and operational management to enhance the efficiency and service quality of all public transportation tools and derive the maximum benefits from passenger traffic forecasting. Since the volume of passenger traffic has a significant impact on the transportation industry, past passenger traffic trends may be used as the basis for future operational plans and decisions in order to derive maximum benefit. The forecasting of MRT passenger traffic is correlated with the positive or negative economic development of the greater Taipei area in the future, while a precise forecast depends on a full understanding of the environment of urban areas, factors affecting demand, and the proper forecasting tool selection. The primary purpose is to collect past research data and sort out the best variables affecting the passenger traffic demand forecast from a review of the literature. The aims of this study are as follows: (1) to analyze important variables that affect the selection and forecasting of passenger traffic and develop a forecast model using regression analysis [5]; (2) to select important variables affecting MRT passenger traffic using an artificial neural network [6] and create a forecast model; and (3) to compare the results of the regression analysis and artificial neural network research, evaluate the forecast performance of the two projection methods, and then select the best forecasting model.

In this study, a two-step procedure is used to build a predictive model for public transport. In the first step of this study, regression analysis is used to find the correlations between two or more variables and their associated directions and strength, and the regression function is used to predict future changes. In the second step, the regression analysis and artificial neural network methods are assessed, and the results are compared.

2. Literature Review

This section introduces the passenger traffic forecast and application, selection of variables, artificial neural network, regression analysis, and mean absolute percentage error (MAPE) evaluation indicators.

2.1. Passenger Traffic Forecasts and Application

Passenger traffic forecasts are an important basis for developing a traffic capacity and transportation plan for a transportation organization, as well as the foundation for constructing and modifying various forms of transportation equipment. The transportation demand and qualitative and quantitative features determine the transportation supply plan. Some common approaches to forecasting passenger traffic demand include regression analysis, econometric modeling, time-series modeling, artificial neural networks, and gray theory. This study adopts an artificial neural network to create the model. The artificial neural network is more flexible in terms of its parameter design without limitations relating to statistical assumptions, it is highly capable of learning, and it is available to precisely forecast passenger traffic under the conditions of a sufficient number of training samples and appropriate parameter design. The other research model used is regression analysis. The approach adopts mathematical statistics to create independent and dependent variables based on massive observation data in order to understand whether two or more variables are correlated and their correlative direction and strength before establishing a regression function to predict future movement.

The literature related to passenger traffic discussed in this study includes the following: Mazanec [7] suggested that the artificial neural network model is better than discriminate analysis. Furthermore, the artificial neural network can also classify a group of input variables (such as the social, economic, demographic, or behavioral attributes) of train passengers, including output data. Law and Au [8] predicted the passenger traffic of the air route between Hong Kong and Japan, using the GDP, foreign exchange, population, marketing expenses, service price, and hotel rate as input variables. The authors randomly picked 20 trained datasets among the data from the period of 1967–1997 and another 10 datasets for comparison. In the meantime, the MAPE, normalized correlation coefficient, and acceptable output percentage were used as the standards to compare the actual value with the forecast value derived from an artificial neural network, exponential smoothing, the moving average, and regression analysis. As a result, the MAPE value of an artificial neural network is only 10%, meaning that it has the best forecast result.

Kulendran and Witt [9] used the leading indicator transfer function (TF) to predict the needs for travel from England to six other countries (including the U.S., Germany, the Netherlands, Spain, Portugal, and Greece) in the period of 1993–1995. The authors first selected the six most common economic indicators for travel testing—relative price, exchange rate, relative exchange, domestic real income, and GDP—to find the correlation between the need to travel to these six countries and all variables. Next, they found the formula, using data from 1978 to 1992 as the input variable and the passenger number as the output variable to forecast the demand in the period of 1993–1995. Meanwhile, the MAPE was used as the criteria to compare the results of the autoregressive integrated moving average (ARIMA) and error correlation models (ECMs). The results show that ARIMA is more precise than TF when the quarter unit is one, four, and eight, while TF is more precise under other circumstances. On the other hand, the ECMs are more precise than TF in terms of long-term conditions (by a unit of four quarters and eight quarters). Grosche et al. [10] used two gravity models to predict the passenger traffic between two cities. The input variables of the first gravity model were population, consumption ability, GDP, distance, and average travel time. The second one was the extended gravity model, which added the number of competitive airports and average distances of competitive airports as input variables. They were verified by the data on Germany and 29 other European countries, and the results show that both gravity models were able to precisely predict the passenger traffic volume.

2.2. Variable Selection

The input variables (independent variables) adopted in this study are summarized in Table 1 based on a literature review and an analysis of the demand for public transportation. X1–X20 are independent variables based on the total passenger number of the Taipei MRT, i.e., the output variables. Among them, each datum shall be based on the "year".

Table 1. Variable selection.

Variable	Name of Variable	Variable Code
Y1	MRT passenger traffic	Passengers
X1	Gross National Product	GNP
X2	Gross Domestic Product	GDP
X3	Economic growth rate	Economic
X4	Number and density of registered financial businesses	InFinanden
X5	Number of major tourists	Visitors
X6	Number of listed companies	Companies
X7	Employment population	Employed
X8	Consumer Price Index	CPI
X9	Population density	InPopden
X10	Personal income	PCI
X11	Import Price Index	IPI
X12	Total cargo input amount	CargoInput
X13	MRT train trips	Trains
X14	Number of administrative regions	Administrative
X15	MRT train kilometers	TrainsKM
X16	MRT passenger traffic income	Income
X17	Number of MRT stations	Stations
X18	MRT operation kilometers	OperationsKM
X19	MRT passenger kilometers	PassengersKM
X20	Round trip transfer discount volume	Transfers

The input variables of the study are shown in Table 1, and their definitions are as follows:

1. Y1: The training samples are from the passenger traffic data of the Muzha Line from 1996 to 2015, and the artificial neural network and regression analysis are adopted to forecast the Taipei MRT passenger traffic.
2. X1: The "nominal market value" of "final products and services" produced by "all people in a specific region" within a "certain period". A prosperous economy will lead the passenger traffic.
3. X2: The total market value of all the "services" and "final products" produced by a country within a certain period. In general, a higher number means higher productivity in the country, which means it is more prosperous and helpful in relation to passenger traffic.
4. X3: The most important economic indicator to determine the macroeconomic situation, which may reflect the change in the scale of the total economic output, and the most physical symbol of economic prosperity, which may affect the capacity of public transportation tools.
5. X4: The higher the density of a financial industry, the more prosperous the business development, which may increase the demand for MRT.
6. X5: The tourism business is an important service industry that contributes much to the national income, employment, and foreign exchange earnings. Meanwhile, the number of tourists is also helpful in evaluating the increase in passenger traffic.
7. X6: The good operation of listed companies with sufficient working capital may promote business activities and increase the demand for transportation.
8. X7: A local employee population may reflect the level of local economic prosperity. The people of a country with greater employment opportunities may have a certain consumption ability and will affect the public transportation volume.
9. X8: The price variation indicator, calculated based on the prices of products and services relating to living in a residence, is one of the major indicators for measuring inflation. The increase in price will affect the public transportation volume.
10. X9: The total population density of Taipei and New Taipei City, obtained using the land area of Taipei and New Taipei City divided by the population, is adopted as the variable. The population will affect the consumption ability of the public.

The consumption ability will automatically increase with an increasing population. Therefore, the population may affect the demand for transportation in a certain area.

11. X10: Personal income affects the consumption ability of the public. The higher the income, the greater the consumption ability. Therefore, personal income may affect the transportation demand.

12. X11: This factor measures the change in prices of imported and exported products. The higher the consumption amount of the public, the stronger the consumption ability of the nationals in the country, which may affect the passenger traffic.

13. X12: The consumption of the public in the country could be observed from the total amount imported to Taiwan. A higher amount means a stronger consumption ability of the people in the country, which may affect the passenger traffic.

14. X13: The density of the total MRT train trips and time spent waiting for mass transportation are important factors affecting the willingness of commuters to take public transportation. The shorter the waiting time, the higher the willingness of commuters to take public transportation.

15. X14: The accessibility of the transportation network to administrative regions may likely affect the willingness of the public to take public transportation. If different mass transportation tools can be effectively integrated, commuters will be more willing to take public transportation.

16. X15: The total travel mileage of all MRT trains within a specific time and period. The higher the passenger traffic is, the more train trips and travel mileage of trains there will be.

17. X16: Operation income status is critical for the sustainable operation of a transportation organization. If the organization has operation losses, its transportation supply efficiency must be affected.

18. X17: The actual number of stops in all the routes of the MRT operation and the expansion of the transportation network will increase the accessibility of the expected destinations. If different mass transportation tools can be effectively integrated, commuters will be more willing to take public transportation.

19. X18: Shorter distances of transportation routes will increase the accessibility of the planned destinations. If different mass transportation tools can be effectively integrated, commuters will be more willing to take public transportation.

20. X19: The total number of passengers multiplied by the mileage traveled. The mileage for the calculation of passenger kilometers will be based on the mileage for the freight calculation.

21. X20: The changes in MRT, bus transportation, and transfer volume are used to analyze the effects of decreases in transfers. The time spent waiting for public transportation tools is an important factor affecting the willingness of commuters to take public transportation. The shorter the waiting time, the higher the willingness of the commuters to take public transportation.

2.3. Artificial Neural Network

An artificial neural network [11] is a parallel calculation model, which is similar to the human neural structure. It is an information-processing technology inspired by brain and nervous system research, and it is also usually referred to as a parallel distributed processing model or a connectionist model. An artificial neural network uses a lot of simply connected artificial neurons [12] to simulate the ability of the biological neural network. An artificial neuron is a simple simulation of a biological neuron, which acquires information from the external environment or other artificial neurons. It performs very simple calculations and then exports the results to the external environment or other artificial neurons for further action. The basic structure of a general neural network is divided into the neuron, layer, and network parts. The layer consists of basic neurons, and the network is constituted by layers. The neuron is also called the processing unit and the basic unit of the artificial neural network. The operation model of an artificial neural network is mainly divided into the training phase and the recall phase: the training phase refers to learning from training to adjust the weight of the network so that the network can become stable. The recall phase involves determining the output value induced by the network to test whether the output is close to the target output value. The artificial neural network adopted by this study is the back-propagation network [13] in a supervised learning network [14], which is applicable to classification, forecasting, system control, noise filtering, sample identification, and data compression. The input layer units are different in each step. The number of processing units in the hidden layer is determined by the number of input and output layers, and the final output result will be either one or two. The back-propagation network is the most representative and popular model among the currently available artificial neural networks. The basic theory of artificial neural networks involves minimizing the error function, using the gradient steepest descent method (GSDM) [15] to achieve the learning purpose.

2.4. Regression Analysis Method

Regression analysis [16] is a method involving the analysis of data in statistics. It is mainly used to analyze whether there is a specific relationship between one or more independent variables and dependent variables. Regression analysis is a model for establishing the relationship between a response variable Y and independent variables X. The purpose is to understand whether two or more variables are related and their correlative direction and strength, as well as to establish a mathematical model that allows for the observation and prediction of specific variables. Since the purpose of prediction regression is not to clarify but rather to establish the best formula, the primary consideration in variable selection is whether there is a maximum practical value, as opposed to the theoretical appropriateness. The theory primarily explains the value of the regression model in practical applications and its mechanism for solving problems in prediction regression. It is expected to achieve the maximum practical value with the lowest cost. The first job of explanation regression is to carefully review the features and relationships among all variables, that is, to examine the correlation among the variables.

2.5. Mean Absolute Percentage Error

MAPE [17–19] refers to the average absolute percentage error, which is the evaluation indicator for whether a prediction model is good or bad. Since MAPE is a relative value that is not affected by the measurement value and estimated value, it can observe the difference between the estimated and evaluated values objectively. The estimation effect is better if the MAPE value is closer to 0. The standards to evaluate the precision of a forecast are shown in Table 2.

Table 2. Standards for the precision of MAPE evaluation.

MAPE Value	Standard
MAPE < 10%	Excellent prediction power (the closer to 0, the better)
10% < MAPE < 20%	Good prediction power
20% < MAPE < 50%	Reasonable prediction power
50% < MAPE	Poor prediction power

3. Materials and Methods

This section introduces the research structure and research design and steps.

3.1. Research Structure

The flow of this study is shown in Figure 1, below, and the relevant steps are as follows:

1. The literature and selected variables that may affect the MRT passenger traffic are collected via a literature review.
2. The artificial neural network and regression analysis are adopted as research methods for training and establishing a prediction model of the MRT passenger traffic: (1) the

years of training are selected; (2) the input variables of the MRT passenger traffic are deleted one by one, and the best variable is found; (3) the possible input variables that are likely to affect the MRT passenger traffic in the training are added, and the best variable is found; and (4) the MRT passenger traffic forecast model is established via an artificial neural network and regression analysis.

3. The MRT passenger traffic forecast model is established.
4. The advantages and disadvantages of the input variables selected are compared and analyzed, considering the results of the artificial neural network and regression analysis.
5. The research conclusions and suggestions based on the research results are provided.

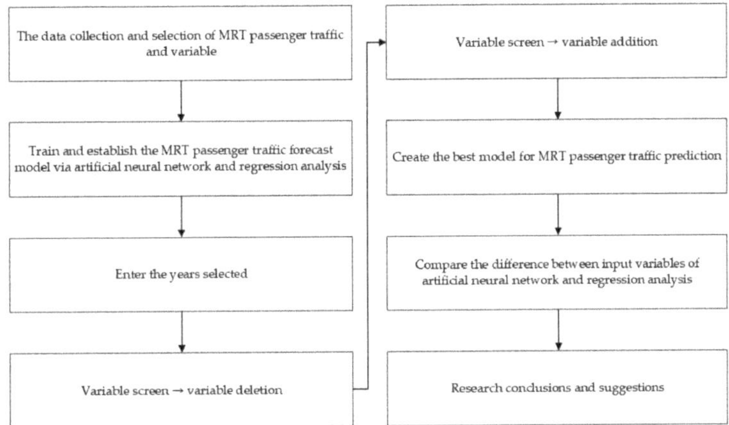

Figure 1. Research structure.

3.2. Research Design and Steps

We first collected the relevant literature and selected variables that could potentially affect the MRT passenger traffic via a literature review, as shown in Table 1, and then selected the years of training, as shown in Table 3.

The main structure of this study is established by the idea of an artificial neural network structure, including the input layer, hidden layer, and output layer. X1 to X10 are variables that may affect the Taipei MRT passenger traffic, while Y1 is the Taipei MRT passenger traffic. We first tested independent variables X1 to X10 and then carried out the deletion test, before adding independent variables X11 to X20 one by one. The purpose of this step was to screen the input variables with a lower correlation to the Taipei MRT passenger traffic to avoid interfering with the prediction results. The regression analysis is primarily used to discuss the causal result relationship among the variables and conduct the prediction via a line chart [20,21], a scatter chart [22], correlation analysis [23–25], the enter method [26,27], and stepwise regression [28–30].

Table 3. Annual information of the variables in relation to passenger traffic.

Variable	X1	X2	X3	X4	...	X17	X18	X19	X20	Y1
Unit/Year	Million	Million	%	Km² (Train)	...	Station	Km	Persons/Km	Persons (Thousand)	Total Persons
1996	8,146,092	8,036,590	6.18	500	...	12	10.5	57,226,810	102	11,174,359
1997	8,806,852	8,717,241	6.11	509	...	32	32.4	243,676,517	3282	31,081,395
1998	9,449,692	9,381,141	4.21	501	...	39	40.3	512,282,678	12,229	60,737,782
1999	9,906,113	9,815,595	6.72	495	...	56	56.4	1,031,342,472	21,203	126,952,122
2000	10,490,818	10,351,260	6.42	491	...	62	65.1	2,042,303,171	38,138	268,716,740
2001	10,350,233	10,158,209	-1.26	448	...	62	65.1	2,223,486,596	44,368	289,642,714
2002	10,923,385	10,680,883	5.57	437	...	62	65.1	2,469,037,312	53,093	324,433,557
2003	11,294,739	10,965,866	4.12	433	...	62	65.1	2,440,488,934	72,399	316,189,128
2004	12,021,744	11,649,645	6.51	428	...	63	67	2,680,355,529	125,350	350,141,956
2005	12,383,120	12,092,254	5.42	422	...	63	67	2,742,372,258	127,424	360,729,803
2006	12,952,502	12,640,803	5.62	415	...	69	74.4	3,002,988,957	130,916	384,003,220
2007	13,739,828	13,407,062	6.52	417	...	69	74.4	3,298,870,463	140,044	416,229,685
2008	13,465,596	13,150,950	0.70	418	...	70	75.8	3,513,969,060	152,643	450,024,415
2009	13,375,650	12,961,656	-1.57	428	...	82	90.6	3,720,991,244	153,679	462,472,351
2010	14,548,852	14,119,213	10.63	425	...	93	100.8	4,123,189,518	162,098	505,466,450
2011	14,700,572	14,312,200	3.80	425	...	94	101.9	4,607,801,411	170,963	566,404,489
2012	15,141,108	14,686,917	2.06	427	...	102	112.8	4,973,666,983	177,918	594,864,715
2013	15,646,211	15,221,201	2.23	428	...	109	121.3	5,234,160,050	182,193	634,961,083
2014	16,621,378	16,081,798	3.74	428	...	116	129.2	5,589,414,250	180,133	679,506,401
2015	17,485,375	16,881,614	3.78	429	...	117	131.1	5,880,980,257	173,877	717,511,809

3.2.1. Line Chart and Scatter Chart

The line chart demonstrates that the growth trend of the Taipei MRT passenger traffic (Figure 2) is relatively similar to the variable X1 GNP (Figure 3), and its effect is therefore assumed to be more important. From the scatter chart (Figure 4), if the variable X1 GNP and the Taipei MRT passenger traffic have a linear distribution, they should be more correlated, and the effect of variable X1 GNP on the Taipei MRT passenger traffic will be obvious.

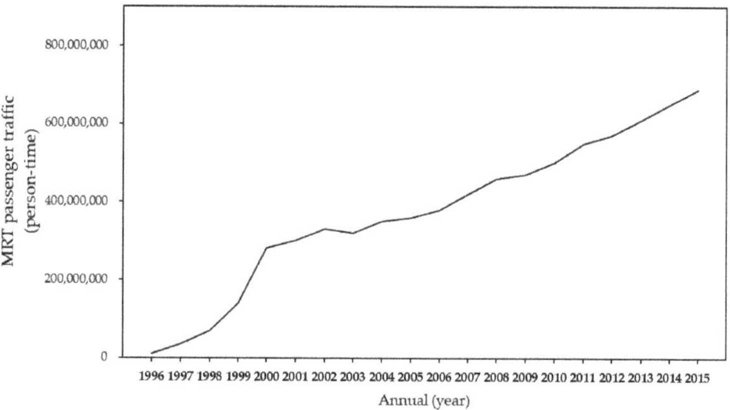

Figure 2. The line chart of the total Taipei MRT passenger traffic.

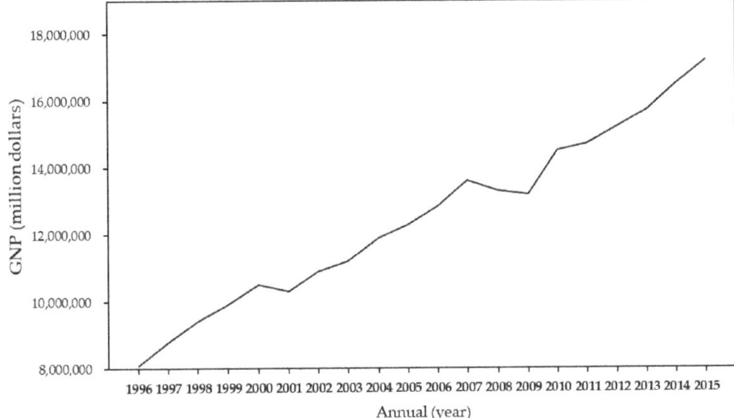

Figure 3. The line chart of the annual GNP.

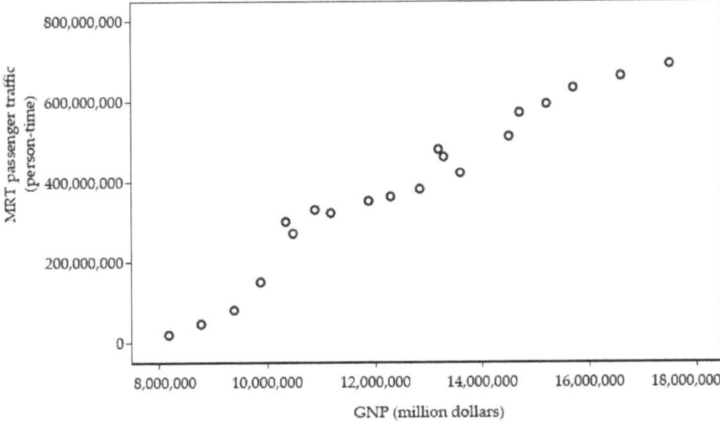

Figure 4. The scatter chart of the MRT passenger traffic and GNP.

3.2.2. Correlation Analysis

In correlation analysis, if two variables are significantly correlated, this only means that the strength and direction between the variables are significant. When the coefficient is significant, it only explains that the two variables are correlated at a certain level, including the strength and direction, instead of indicating the existence of a causal relationship. They may both be the cause and result at the same time, or a causal relationship may actually exist.

3.2.3. Enter Method

The enter method means that all the prediction variables must be entered into the regression formula, regardless of the significance of the individual variable.

3.2.4. Stepwise Regression

When variables are entered into the regression equation, the backward selection method is used to eliminate unimportant variables.

4. Empirical Analysis

This study uses regression analysis and class neural network methods to select and predict the variables affecting passenger traffic using the Taipei Metro passenger traffic trends from March 1996 to December 2015. Finally, we compare the results of the neural network and regression analysis prediction methods, evaluate and analyze the prediction performance of the two methods, and then select the best prediction model.

4.1. Regression Analysis

We adopted statistical software to analyze 20 parametric variables from March 1996 to December 2015 to explore the causal relationship between the variables as well as to make predictions through line graphs, scatter plots, correlation analysis, forced entry, and stepwise regression analysis to explore the causal relationship between the variables.

4.1.1. Line Chart

If the line graphs of the variables are found to be more similar to the growth trend line graphs of Taipei Metro's passenger volume, it is speculated that the effect should be more important (Figure 5).

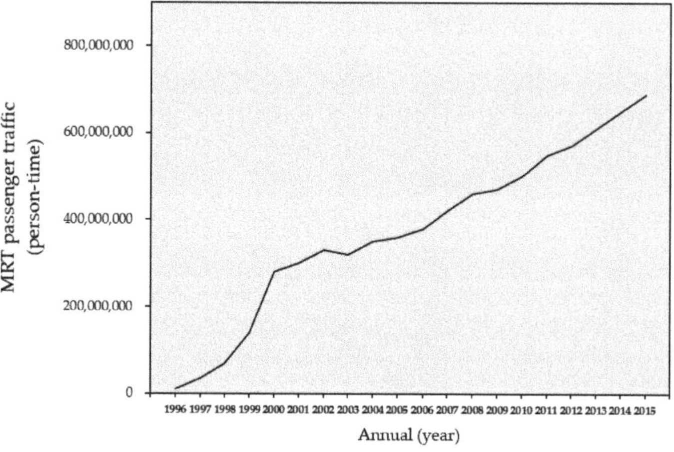

Figure 5. Line chart of the MRT passenger traffic.

4.1.2. Scatter Chart

From the scatter plot, we can see that the correlation between the variables and Taipei Metro's passenger volume is greater if they are clearly distributed in a straight line (Figure 6).

4.1.3. Correlation Analysis

The correlation coefficient primarily refers to the degree of correlation between the variables, while it does not verify the influence of the "independent variable" on the "dependent variable". Therefore, the obtained correlation coefficient (R value) can only indicate that the two variables are positively correlated, negatively correlated, or independent. It cannot be interpreted as the effect of the independent variable on the dependent variable. In the interpretation of the correlation coefficient, positive and negative indicate the direction of the correlation, not the degree of the correlation. If the degree of the correlation is between the R values, i.e., between plus or minus 0.3 (i.e., between 0.3 and −0.3), it is considered a low correlation; it is considered a moderate correlation if the value is between plus or minus 0.3 to 0.6 (that is, between 0.3 and 0.6, or between −0.3 and −0.6); and if it is between plus or minus 0.6 to 0.9 (i.e., 0.6 to 0.9, or −0.6 to −0.9), it is considered a high correlation. If the R value is plus or minus 1, this indicates a complete correlation.

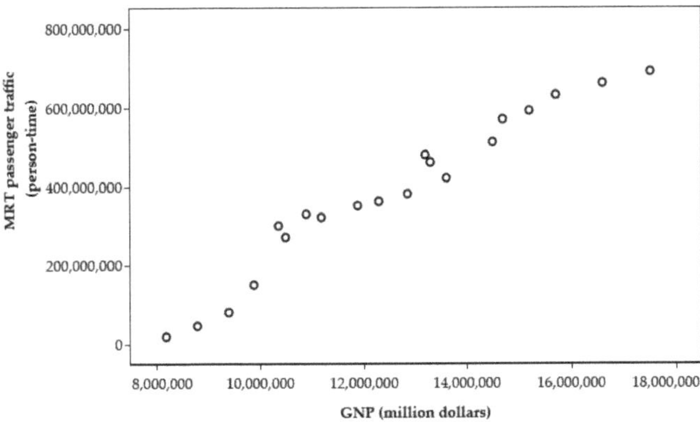

Figure 6. Scatter chart of GNP and MRT passenger traffic.

4.1.4. Entry Method

All the predictive variables are incorporated into the regression equation at the same time, and the least squares method and statistical software are used to calculate the complex regression model. The analysis of the explanatory variables is as follows. Table 4, showing the entry variables inputted/removed, provides a list of variables that are subjected to regression analysis, including a total of 16 independent variables.

Table 4. Forced entry method variables inputted/removed.

Model	Variable Inputted	Variable Removed	Method
1	X3, X4, X5, X6, X7, X8, X9, X10, X11, X12, X13, X14, X15, X16, X17, X20		Entry

The variance analysis result of the forced entry method shows the overall verification of the model significance to verify the significance of the overall regression model. The quadratic sum of the regression model is 830,852,340,051,751,420, the total sum of squares is 830,902,477,461,923,070, the F value = 3107.157, and the p value = 0.000 < 0.05, which reaches significance. Furthermore, F is used to verify the regression model, F (α, K, N-K-1) = F(0.05, 16, 3) $\doteqdot$ 8.7, and the F value of this regression equation = 3107.157 > critical value F value of 8.7, thus rejecting the null hypothesis and indicating that the explanatory power of the overall regression model reaches a significant level.

In Table 5, the coefficients of the forced entry method are the verification results of the multicollinearity of the output independent variables using statistical software. This model has a total of 16 independent variables and 17 eigenvalues, and the minimum tolerance is 0.000. If the tolerance is lower than 0.2, it is determined that there is collinearity between this independent variable and other independent variables. The highest VIF is 4167.205. In general, if the VIF is greater than 4, this means that the independent variable is collinear with other independent variables.

The variables excluded by the forced entry method demonstrate that there are four variables, including the total GNP, GDP, MRT mileage, and MRT extended passenger mileage, that did not enter the regression model. To reduce the collinearity problem of the multiple regression model, we adopted the stepwise method. The advantage of this method is that the variables are selected one by one according to the influence of the independent variable on the dependent variable, thus eliminating the collinearity problem.

Table 5. Coefficients of the entry method.

Model		T	Significance	Collinearity Statistical Material	
				Permissible Deviation	VIF
1	(Constant)	−2.374	0.098		
	X1	0.685	0.543	0.094	10.601
	X3	2.787	0.069	0.004	263.957
	X4	1.500	0.231	0.002	457.889
	X5	1.429	0.248	0.001	1047.998
	X6	3.584	0.037	0.000	4167.205
	X8	−1.309	0.282	0.001	835.575
	X9	−3.509	0.039	0.001	1728.353
	X10	−1.115	0.346	0.007	143.916
	X11	2.919	0.062	0.005	206.473
	X12	1.151	0.333	0.007	142.958
	X13	−3.006	0.057	0.003	293.111
	X14	−1.913	0.152	0.003	313.741
	X15	0.030	0.978	0.002	617.120
	X16	11.990	0.001	0.003	374.019
	X17	−0.098	0.928	0.002	656.926
	X20	−2.325	0.103	0.003	348.165

4.1.5. Stepwise Method

To reduce the collinearity problem of multiple regression models, we adopted the stepwise method and performed calculations using statistical software. The following Table 6 was obtained.

Table 6. Stepwise regression variables inputted/removed.

Model	Variable Inputted	Method
1	X16	Step by step (criteria: F-to-enter probability $\leq$ 0.050, F-to-remove probability $\geq$ 0.100)
2	X20	Step by step (criteria: F-to-enter probability $\leq$ 0.050, F-to-remove probability $\geq$ 0.100)

Table 6 shows the stepwise regression variables inputted/removed and a list of stepwise method variables. The criteria for selection are: (F-to-enter probability $\leq$ 0.050, F-to-remove probability $\geq$ 0.100). In total, two variables are selected for the regression equation in two steps (models). In Model 1, MRT passenger revenue is selected. In Model 2, two-way transfer preferential volume is added. Therefore, the two variables of the MRT passenger revenue and two-way interchange preferential volume are selected for the model.

The analysis result of stepwise regression variance using Model 1 of the MRT passenger revenue variables shows that the F value of the overall regression model is 15,598.145, $p = 0.000 < 0.05$, reaching significance and indicating a significant correlation between the independent variable (the MRT passenger revenue) and the dependent variable (the Taipei MRT passenger traffic). Taking Model 2 as an example, the F value of the regression model of Model 2 is 12,741.962, $p = 0.000 < 0.05$, which reaches significance and indicates a significant correlation between the independent variable (the preferential volume of two-way interchange) and the dependent variable (the Taipei MRT passenger traffic). The two variables can effectively predict the passenger traffic of the Taipei MRT.

Table 7 demonstrates a positive relationship between the MRT passenger revenue and the preferential volume of the two-way passenger transfer—two independent variables in the table of stepwise regression coefficients, both of which were consistent with the prediction and significant.

Table 7. Stepwise regression coefficients.

	Model	Unstandardized Coefficients		Standardized Coefficients	T	Significance
		B	Standard Error	Beta		
1	(Constant)	−12,082,965.667	3,520,486		−3.432	0.003
	X16	0.047	0	0.999	124.893	0
2	(Constant)	−9,989,402.575	2,818,390		−3.544	0.002
	X16	0.044	0.001	0.942	53.482	0
	X20	193.099	54.85	0.062	3.52	0.003

Table 8 demonstrates the stepwise mode analysis results of the variable table excluded by stepwise regression, with two variables remaining: the MRT passenger revenue and two-way interchange preferential volume. From Table 8, it can also be understood that the variables may enter the regression model because the correlation value of the dependent variables is the highest in Model 1, where the preferential volume of two-way interchange is 0.649. In Model 2, the correlation value of the total amount of goods imported into Taiwan is −0.310, which is the highest.

Table 8. Variables excluded by the stepwise method.

Model		Beta	T	Significance	Partial Correlation
1	X1	0.031	0.799	0.436	0.190
	X2	0.024	0.641	0.530	0.154
	X3	0.007	0.780	0.446	0.186
	X4	−0.031	−2.705	0.015	−0.549
	X5	−0.020	−1.234	0.234	−0.287
	X6	0.100	2.949	0.009	0.582
	X7	0.085	2.668	0.016	0.543
	X8	0.009	0.371	0.715	0.090
	X9	0.080	1.411	0.176	0.324
	X10	0.002	0.071	0.944	0.017
	X11	0.025	1.981	0.064	0.433
	X12	0.016	0.782	0.445	0.186
	X13	−0.008	−0.204	0.841	−0.049
	X14	0.012	0.481	0.637	0.116
	X15	−0.039	−0.831	0.418	−0.197
	X17	−0.058	−1.788	0.092	−0.398
	X18	−0.063	−1.873	0.078	−0.414
	X19	0.220	1.473	0.159	0.336
	X20	0.062	3.520	0.003	0.649
2	X1	−0.026	−0.750	0.464	−0.184
	X2	−0.030	−0.900	0.382	−0.219
	X3	0.004	0.562	0.582	0.139
	X4	−0.015	−1.232	0.236	−0.294
	X5	−0.007	−0.519	0.611	−0.129
	X6	0.050	1.239	0.233	0.296
	X7	0.021	0.471	0.644	0.117
	X8	−0.017	−0.804	0.433	−0.197
	X9	−0.076	−1.158	0.264	−0.278
	X10	−0.018	−0.708	0.489	−0.174
	X11	−0.007	−0.447	0.661	−0.111
	X12	−0.026	−1.306	0.210	−0.310
	X13	−0.032	−1.015	0.325	−0.246
	X14	−0.011	−0.519	0.611	−0.129
	X15	−0.024	−0.637	0.533	−0.157
	X17	−0.027	−0.945	0.359	−0.230
	X18	−0.033	−1.093	0.290	−0.264
	X19	0.090	0.698	0.495	0.172

We used statistical software to analyze the inputting of 20 parameter variables from 1996 to 2013 and outputting of the predictive value of the passenger traffic of the Taipei MRT (Table 9). The predicted value was calculated according to the predicted value table of the progressive regression analysis using the MAPE formula. The result obtained using the regression analysis prediction method was MAPE = 6.47%.

Table 9. Predictive value of regression analysis.

Year of Observation	Actual Number of MRT Passengers	Predictive Value of Regression Analysis	Residual Value
1996	11,174,359	3,305,117	7,869,242
1997	31,081,395	36,987,970	−5,906,575
1998	60,737,782	68,054,760	−7,316,978
1999	126,952,122	129,264,711	−2,312,589
2000	268,716,740	270,052,789	−1,336,049
2001	289,642,714	287,079,395	2,563,319
2002	324,433,557	319,060,611	5,372,946
2003	316,189,128	312,874,254	3,314,874
2004	350,141,956	351,632,592	−1,490,636
2005	360,729,803	359,697,272	1,032,531
2006	384,003,220	385,978,346	−1,975,126
2007	416,229,685	421,166,927	−4,937,242
2008	450,024,415	449,687,066	337,349
2009	462,472,351	457,533,201	4,939,150
2010	505,466,450	495,603,126	9,863,324
2011	566,404,489	560,822,898	5,581,591
2012	594,864,715	607,214,292	−12,349,577
2013	634,961,083	638,210,638	−3,249,555

4.2. Artificial Neural Network

Using the training data from 1996 to 2013 as the artificial neural network training materials and those from 2014 to 2015 as the sample data, we were able to use a total of 20 parameters as the input variables of the artificial neural network; the output variables were the passenger traffic prediction values of the Taipei MRT. Statistical software was used to analyze 20 parametric variables from 1996 to 2015, before the input variables were converted into an artificial neural network to investigate the causal relationship between the variables and make predictions based on a line chart, scatter chart, correlation analysis, forced entry method, and stepwise method. Therefore, the artificial neural network does not temporarily delete any variables and inputs all of them into the artificial neural network. Then, the optimal combination of input variables was analyzed.

4.2.1. Year of Selection

Basic information on the year of selection for variables X1 to X20 and the output variable Y1 is shown in Table 3, where Y1 stands for the Taipei MRT passenger traffic. Using the backward propagation artificial neural network algorithm model, the average absolute percentile error was used to analyze 7 years/2 years of training, 6 years/2 years of training, and 5 years/2 years of training from 1996 to 2015, and the training results are shown in Table 10. This study uses the training model of 5 years/2 years of training prediction for empirical research.

Table 10. Year selection of the artificial neural network.

Evaluation Indicators	7 Years/2 Years of Training (1996–2002/2003–2004)	6 Years/2 Years of Training (2003–2008/2009–2010)	5 Years/2 Years of Training (2009–2013/2014–2015)
MAPE	8.26%	4.35%	2.45%

4.2.2. Deleted Variables

First, we used the 5-year and 2-year training modes (2009 to 2013 and 2014 to 2015) and then deleted the X1 to X10 variables in order, selecting the variables with the highest MAPE value for deletion. The purpose of this was to screen out variables that have little influence on the passenger traffic of the Taipei MRT to avoid interference with the predicted results. After the deletion was complete, we then started the training process to determine the best variables and improve the accuracy of the prediction results.

4.2.3. The Output Results of the Artificial Neural Network

Using the training data from 1996 to 2013 as the artificial neural network training materials and the data from 2014 to 2015 as the sample data, we used a total of 20 parameters as the input variables of the artificial neural network; the output variables were the passenger traffic prediction values of the Taipei MRT (Table 11).

Table 11. Predictive values of the artificial neural network.

Year of Observation	Actual Number of MRT Passengers	Predictive Values of the Artificial Neural Network	Residual Value
1996	11,174,359	15,320,813	4,146,454
1997	31,081,395	30,226,981	−854,414
1998	60,737,782	60,188,236	−549,546
1999	126,952,122	126,305,437	−646,685
2000	268,716,740	267,922,813	−793,927
2001	289,642,714	289,012,656	−630,058
2002	324,433,557	324,181,085	−252,472
2003	316,189,128	327,612,737	11,423,609
2004	350,141,956	349,893,508	−248,448
2005	360,729,803	360,534,282	−195,521
2006	384,003,220	383,858,990	−144,230
2007	416,229,685	415,952,914	−276,771
2008	450,024,415	285,870,926	−164,153,489
2009	462,472,351	462,125,697	−346,654
2010	505,466,450	505,270,408	−196,042
2011	566,404,489	565,977,673	−426,816
2012	594,864,715	616,442,418	21,577,703
2013	634,961,083	630,207,775	−4,753,308

According to the prediction values, shown in Table 10, of the artificial neural network, calculated by the MAPE formula, the result obtained by the neural network prediction method is MAPE = 4.82%. To compare the predicted values of regression analysis with those of the artificial neural network, the results showed that the MAPE value of the regression analysis was 6.47%. The MAPE value of the artificial neural network was 4.82%, so the artificial neural network has the best prediction outcomes.

4.3. Comparison of Regression Analysis and Artificial Neural Networks

4.3.1. Deletion of the Forced Entry Method of Regression Analysis to Exclude Four Variables

Using the 5-year training for 2 years (2009–2013/2014–2015) model, we deleted the four eliminated variables of the regression analysis, which were the GNP, GDP, MRT mileage, and MRT extended passenger mileage. Using a total of 16 parameters as the input variables for the regression analysis and artificial neural network comparison, it was found that the MAPE value predicted by regression analysis was 0.94%, and the MAPE value of the artificial neural network was 0.54%. Both methods showed the best prediction results.

4.3.2. Deletion of Four Variables with High MAPE Values in Artificial Neural Networks

Using the model of 5-year training for 2 years (2009–2013/2014–2015), we deleted the four variables with the highest MAPE values in the artificial neural network, which included the economic growth rate, personal income, total amount of goods imported into Taiwan, and number of MRT stations. In total, 16 parameters were used as the input variables for the regression analysis and artificial neural network comparison, and it was found that the MAPE value predicted by regression analysis was 0.62%. The MAPE value of the artificial neural network was 0.31%. Both methods showed the best prediction results.

4.3.3. Collinearity Verification

In order to verify whether the independent variable of the MRT passenger revenue is collinear with the dependent variable of the MRT passenger traffic, statistical software was used to analyze the inputting of 19 parameter variables from 1996 to 2013 and outputting of the passenger traffic prediction value of the Taipei MRT. Again, the training data from 1996 to 2013 were used as the artificial neural network training materials, and the data from 2014 to 2015 were used as the sample data. In total, 19 parameters were used as the input variables of the artificial neural network, while the output variable was the passenger traffic prediction value of the Taipei MRT. Twenty variables and 19 parameter variables were used to compare the predictive values of the regression analysis and artificial neural network after removing the independent variable, the MRT revenue. The results showed that the MAPE residual value of the regression analysis was 1.28%, and the MAPE residual value of the artificial neural network was 0.19%. We observed no significant difference between the two, which proves that there was no collinearity problem after deleting the independent variable, the MRT passenger revenue.

4.3.4. Monthly Passenger Traffic Experiment

In this study, we collected data on passenger traffic (month) and related variables (month); however, because some data were recorded at different times, 167 regression analyses were conducted from January 2000 to December 2015 (excluding the incomplete data on Cyclone Nari in September 2001). The neural network used the mode of 2000–2013 training 2014–2015 and the following seven parameter variables: the MRT operation mileage, MRT station number, MRT train number, MRT extended vehicle mileage, MRT extended passenger mileage, MRT passenger revenue, and two-way transit preferential volume. The output variable is the passenger traffic prediction value of the Taipei MRT. The results showed that the MAPE value of the regression analysis was 1.45%, and the MAPE value of the artificial neural network was 0.42%. Both methods showed the best prediction results.

4.4. Summary of the Empirical Results

The results of this study are shown in Table 12, which summarizes the empirical results of the regression analysis and artificial neural network.

Table 12. Summary of the empirical results of the regression analysis and artificial neural network.

Experiment Module	Regression Analysis MAPE Value	Artificial Neural Network MAPE Value	Results
20 variables	6.47%	4.82%	The artificial neural network is better.
The four variables excluded in the regression analysis were deleted: the GNP, GDP, MRT mileage, and MRT extended passenger mileage.	0.94%	0.54%	Both had the best prediction results.
The four variables with the highest MAPE values were deleted: the economic growth rate, personal income, total amount of goods imported into Taiwan, and number of MRT stations.	0.62%	0.31%	Both had the best prediction results.
19-variable collinearity verification (deletion of the MRT passenger revenue).	5.19%	4.63%	There is no collinearity problem.
Monthly passenger traffic experiment.	1.45%	0.42%	Both had the best prediction results.

5. Analysis of Empirical Results

The results and research contributions of this paper in relation to passenger traffic prediction provide a valuable reference for both academics and practitioners. They are summarized in this section.

5.1. Analysis of Results

In this study, we used an artificial neural network and regression analysis to construct the Taipei MRT passenger traffic prediction model. Seven findings of this study are worth summarizing:

1. A total of 20 parameters from 1996 to 2013 were used as the input variables of the regression analysis, and the output value was the passenger traffic prediction value of the Taipei MRT. The stepwise regression of the regression analysis, predicting the MAPE value of the Taipei MRT passenger traffic, is 6.47%, which demonstrates an excellent predictive performance.

2. Using the training data from 1996 to 2015 as artificial neural network training materials and the data from 2013 to 2015 as the sample data, a total of 20 parameters were used as the input variables of the artificial neural network; the output value was the passenger traffic prediction value of the Taipei MRT. Using an artificial neural network to predict the MAPE value of the Taipei MRT passenger traffic, the result was 4.82%, which demonstrates an excellent predictive performance.

3. We used the 5-year training for 2 years (2009–2013/2014–2015) model and deleted the already eliminated four variables of regression analysis, which were the GNP, GDP, MRT mileage, and MRT extended passenger mileage. In total, 16 parameters were used as the input variables for regression analysis and artificial neural network comparison. The MAPE value predicted by regression analysis was found to be 0.94%, and the MAPE value of the artificial neural network was 0.54%. Both methods had the best prediction results.

4. We used the model of 5-year training for 2 years (2009–2013/2014–2015) and deleted the four variables with the highest MAPE values in the artificial neural network, including the economic growth rate, personal income, total amount of goods imported into Taiwan, and number of MRT stations. In total, 16 parameters were used as the input variables for regression analysis and artificial neural network comparison. The

MAPE value predicted by regression analysis was 0.62%, and the MAPE value of artificial neural networks was 0.31%. Both methods had the best prediction results.

5. In order to verify whether the independent variable MRT passenger revenue is collinear with the dependent variable MRT passenger traffic, we applied statistical software to analyze the inputting of 19 parameter variables from 1996 to 2013 and outputting of the passenger traffic prediction value of the Taipei MRT. Again, we used the training data from 1996 to 2015 as the artificial neural network training materials and the data from 2013 to 2015 as the sample data. We used a total of 19 parameters as the input variables of the artificial neural network and the passenger traffic prediction value of the Taipei MRT as the output value in order to compare the predictive values of the regression analysis and artificial neural network. The result showed that the MAPE value of the regression analysis was 5.19%, and the MAPE value of the artificial neural network was 4.63%. Twenty variables and 19 parameter variables were used to compare the predictive values of the regression analysis and artificial neural network after removing the independent variable, the MRT revenue. The results showed that the MAPE residual value of regression analysis was 1.28%, and the MAPE residual value of the artificial neural network was 0.19%. No significant difference was observed between them, which proves that there is no collinearity problem after deleting the independent variable, the MRT passenger revenue.

6. We collected data on the passenger traffic (month) and related variables (month); however, because some data were recorded at different times, 167 regression analyses were conducted from January 2000 to December 2015 (excluding the incomplete data on Cyclone Nari in September 2001). The neural network used the mode of 2000–2013 training 2014–2015 and the following seven parameter variables: the MRT operation mileage, MRT station number, MRT train number, MRT extended vehicle mileage, MRT extended passenger mileage, MRT passenger revenue, and two-way transit preferential volume. The output value is the passenger traffic prediction value of the Taipei MRT. The results show that the MAPE value of regression analysis was 1.45%, and the MAPE value of the artificial neural network was 0.42%. Both methods had the best prediction results.

7. This study investigated the passenger traffic prediction of the Taipei MRT and analyzed and constructed the prediction model based on two prediction methods: regression analysis and artificial neural networks. The results are presented to allow transportation organizations to maximize their profits by making plans and decisions relating to future operations based on past passenger traffic trends.

5.2. Research Contributions

To improve the utilization rate of the metro area's transportation system and reduce environmental pollution by combining it with the public transportation system of the MRT station, the following research contributions are provided:

1. Choosing the right forecasting tools for business planning and decision making.

The amount of passenger traffic significantly impacts the transportation industry, and accurate prediction depends on a full understanding of the metropolitan environment and analysis of the factors that influence demand. The commonly used methods for forecasting passenger traffic demand include statistical regression analysis, econometric modeling, time-series modeling, neural-network modeling, gray theory, etc.; selecting the appropriate forecasting tools can allow for the maximization of profits by aiding in planning and decision making relating to future operations based on past passenger traffic trends.

2. Construction and planning of transport systems.

Passenger traffic prediction is a very important factor in the construction and planning of transportation systems and serves as the essential basis for putting forward requirements for the construction and expansion of transportation equipment. Therefore, the results of this study can be used as a reference for transportation organizations, such as operation management, manpower allocation, shift distance, transportation demand, etc.

3. Curbing the growth of private vehicles.

The excessively increased use of private vehicles will cause such problems as air pollution, noise, road congestion, traffic accidents, etc., and the social costs shall be shared by the public. Therefore, strengthening the control on the increase of private vehicles is necessary, and passenger traffic prediction cooperates perfectly with public transportation transfer system construction, encouraging people to transfer from private vehicles to public transport systems and then improving ridership on public transportation systems.

4. Public transportation leads to urban development.

With the public transport system as the backbone of urban development, passenger traffic prediction can establish a different planning method and procedure from traditional urban development, implement the priority concept of public transportation, look toward the future, and actively promote subsequent MRT-related construction based on the existing construction. In addition to being safe and convenient, public transportation systems can also inhibit the growth of private transportation, reduce carbon while promoting energy saving, and build low-carbon and green energy cities.

5. Improvement of the efficiency of all public transport vehicles.

In this study, the data on MRT passenger traffic prediction not only provide traffic dispatching and transport strategic planning data but also combine them with the electronic ticket or passenger traffic data of public transport vehicles, such as the high-speed railways, Taiwan's railways, urban buses, and UBikes around the MRT stations, to conduct the overall passenger traffic analysis. This is used as the basis for the planning and operation management of the public transport system to improve the efficiency and service quality of each public transport vehicle and thus maximize the benefit of passenger traffic prediction.

6. Conclusions

This study investigated the passenger traffic prediction of the Taipei MRT and analyzed and constructed the prediction model based on two prediction methods: regression analysis and artificial neural networks. The results are presented as a reference for transportation organizations to allow them to maximize profits by making plans and decisions relating to future operations based on past passenger traffic trends.

During this research, we discussed with senior executives of MRT companies and found that accurate passenger traffic prediction can effectively reduce costs. The management implications of this study are as follows:

1. The forecasting of passenger traffic has a great influence on the transportation industry, and transportation organizations can make plans and decisions for future operations based on past passenger traffic trends. Therefore, it is very important for transportation organizations to have a clear and objective forecasting method.
2. Based on past research data and a literature review, we determined the variable that affects the forecasting of passenger traffic demand the most. The forecasting of the MRT passenger traffic affects the future economic development of Taipei City and New Taipei City, so it is necessary to select the appropriate passenger traffic forecasting tool.
3. The data on passenger traffic forecasting in this study not only allow for the planning of traffic dispatching, operation management, manpower allocation, shift distance, transportation demand, etc., but also serve as an essential basis for transportation organizations, enabling them to put forward requirements for the construction and expansion of transportation equipment.

Despite the limitations of this study, we believe that the findings and management implications of our study are intriguing enough to invite future research on the topic of passenger traffic forecasting, as well as future research on other traffic-related topics.

Author Contributions: Conceptualization, Y.-S.C. and S.-H.C.; methodology, Y.-S.C.; software, S.-H.C.; validation, S.-H.C.; writing—original draft preparation, S.-H.C.; writing—review and editing, Y.-S.C., S.-F.C. and C.-K.L. All authors have read and agreed to the published version of the manuscript.

Funding: This research was partially supported by the Ministry of Science and Technology of Taiwan, grant number MOST 109-2221-E-146-003.

Conflicts of Interest: The authors declare no conflict of interest.

References

1. Verbavatz, V.; Barthelemy, M. Access to mass rapid transit in OECD urban areas. *Sci. Data* **2020**, *7*, 1–6. [CrossRef]
2. Wiseman, Y. Autonomous vehicles. In *Encyclopedia of Information Science and Technology*, 5th ed.; IGI Global: Hershey, PA, USA, 2021; pp. 1–11.
3. Wiseman, Y. Driverless cars will make passenger rail obsolete [Opinion]. *IEEE Technol. Soc. Mag.* **2019**, *38*, 22–27. [CrossRef]
4. Yang, Z.; Li, C.; Jiao, J.; Liu, W.; Zhang, F. On the joint impact of high-speed rail and megalopolis policy on regional economic growth in China. *Transp. Policy* **2020**, *99*, 20–30. [CrossRef]
5. Lio, W.; Liu, B. Uncertain maximum likelihood estimation with application to uncertain regression analysis. *Soft Comput.* **2020**, *24*, 9351–9360. [CrossRef]
6. Toraman, S.; Alakus, T.B.; Turkoglu, I. Convolutional capsnet: A novel artificial neural network approach to detect COVID-19 disease from X-ray images using capsule networks. *Chaos Solitons Fractals* **2020**, *140*, 110122. [CrossRef]
7. Mazanec, J.A. Classifying tourists into market segments: A neural network approach. *J. Travel Tour. Mark.* **1992**, *1*, 39–60. [CrossRef]
8. Law, R.; Au, N. A neural network model to forecast Japanese demand for travel to Hong Kong. *Tour. Manag.* **1999**, *20*, 89–97. [CrossRef]
9. Kulendran, N.; Witt, S.F. Leading indicator tourism forecasts. *Tour. Manag.* **2003**, *24*, 503–510. [CrossRef]
10. Grosche, T.; Rothlauf, F.; Heinzl, A. Gravity models for airline passenger volume estimation. *J. Air Transp. Manag.* **2007**, *13*, 175–183. [CrossRef]
11. Shams, S.R.; Jahani, A.; Moeinaddini, M.; Khorasani, N. Air carbon monoxide forecasting using an artificial neural network in comparison with multiple regression. *Model. Earth Syst. Environ.* **2020**, *6*, 1467–1475. [CrossRef]
12. Liu, S.; Mocanu, D.C.; Matavalam, A.R.R.; Pei, Y.; Pechenizkiy, M. Sparse evolutionary deep learning with over one million artificial neurons on commodity hardware. *Neural Comput. Appl.* **2020**, 1–16. [CrossRef]
13. Sun, W.; Huang, C. A carbon price prediction model based on secondary decomposition algorithm and optimized back propagation neural network. *J. Clean. Prod.* **2020**, *243*, 118671. [CrossRef]
14. Liu, R.; Mancuso, C.A.; Yannakopoulos, A.; Johnson, K.A.; Krishnan, A. Supervised learning is an accurate method for network-based gene classification. *Bioinformatics* **2020**, *36*, 3457–3465. [CrossRef] [PubMed]
15. Shirokanev, A.S.; Kirsh, D.V.; Kupriyanov, A.V. Research of an algorithm for crystal lattice parameter identification based on the gradient steepest descent method. *Comput. Opt.* **2017**, *41*, 453–460. [CrossRef]
16. Hemilä, H.; Chalker, E. Vitamin C may reduce the duration of mechanical ventilation in critically ill patients: A meta-regression analysis. *J. Intensive Care* **2020**, *8*, 15. [CrossRef]
17. Şahin, U.; Şahin, T. Forecasting the cumulative number of confirmed cases of COVID-19 in Italy, UK and USA using fractional nonlinear grey Bernoulli model. *Chaos Solitons Fractals* **2020**, *138*, 109948. [CrossRef]
18. Prado, F.; Minutolo, M.C.; Kristjanpoller, W. Forecasting based on an ensemble autoregressive moving average-adaptive neuro-fuzzy inference system–neural network-genetic algorithm framework. *Energy* **2020**, *197*, 117159. [CrossRef]
19. Longo, G.A.; Righetti, G.; Zilio, C.; Ortombina, L.; Zigliotto, M.; Brown, J.S. Application of an Artificial Neural Network (ANN) for predicting low-GWP refrigerant condensation heat transfer inside herringbone-type Brazed Plate Heat Exchangers (BPHE). *Int. J. Heat Mass Transf.* **2020**, *156*, 119824. [CrossRef]
20. Feng, Y.; Yang, T.; Niu, Y. Subpixel computer vision detection based on wavelet transform. *IEEE Access* **2020**, *8*, 88273–88281. [CrossRef]
21. Sohn, C.; Choi, H.; Kim, K.; Park, J.; Noh, J. Line Chart understanding with convolutional neural network. *Electronics* **2021**, *10*, 749. [CrossRef]
22. Shibzukhov, Z.M. Robust method for finding the center and the scatter matrix of the cluster. *J. Phys. Conf. Ser.* **2020**, *1479*, 012045. [CrossRef]
23. Peng, Q.; Wen, F.; Gong, X. Time-dependent intrinsic correlation analysis of crude oil and the US dollar based on CEEMDAN. *Int. J. Financ. Econ.* **2021**, *26*, 834–848. [CrossRef]
24. Shi, J.; Liu, B.; Qin, J.; Jiang, J.; Wu, X.; Tan, J. Experimental study of performance of repair mortar: Evaluation of in-situ tests and correlation analysis. *J. Build. Eng.* **2020**, *31*, 101325. [CrossRef]
25. Ibrahim, M.S.; Sidiropoulos, N.D. Reliable detection of unknown cell-edge users via canonical correlation analysis. *IEEE Trans. Wirel. Commun.* **2020**, *19*, 4170–4182. [CrossRef]
26. Di Tella, M.; Romeo, A.; Benfante, A.; Castelli, L. Mental health of healthcare workers during the COVID-19 pandemic in Italy. *J. Eval. Clin. Pract.* **2020**, *26*, 1583–1587. [CrossRef]

27. Moskowitz, S.; Dewaele, J.M. Is teacher happiness contagious? A study of the link between perceptions of language teacher happiness and student attitudes. *Innov. Lang. Learn. Teach.* **2021**, *15*, 117–130. [CrossRef]
28. Liu, B.; Zhao, Q.; Jin, Y.; Shen, J.; Li, C. Application of combined model of stepwise regression analysis and artificial neural network in data calibration of miniature air quality detector. *Sci. Rep.* **2021**, *11*, 1–12.
29. Hornby, T.G.; Henderson, C.E.; Holleran, C.L.; Lovell, L.; Roth, E.J.; Jang, J.H. Stepwise regression and latent profile analyses of locomotor outcomes poststroke. *Stroke* **2020**, *51*, 3074–3082. [CrossRef]
30. Żogała-Siudem, B.; Jaroszewicz, S. Fast stepwise regression based on multidimensional indexes. *Inf. Sci.* **2021**, *549*, 288–309. [CrossRef]

MDPI

St. Alban-Anlage 66

4052 Basel

Switzerland

Tel. +41 61 683 77 34

Fax +41 61 302 89 18

www.mdpi.com

Electronics Editorial Office

E-mail: electronics@mdpi.com

www.mdpi.com/journal/electronics

9 783036 539805